Introduction to Insurance Mat

Annamaria Olivieri · Ermanno Pitacco

Introduction to Insurance Mathematics

Technical and Financial Features of Risk Transfers

Annamaria Olivieri
Università di Parma
Dipartimento di Economia
via Kennedy 6
43125 Parma
Italy
annamaria.olivieri@unipr.it

Ermanno Pitacco
Università di Trieste
Dipartimento di Scienze
Economiche, Aziendali,
Matematiche e Statistiche
piazzale Europa 1
34127 Trieste
Italy
ermanno.pitacco@econ.units.it

ISBN 978-3-642-16028-8 e-ISBN 978-3-642-16029-5
DOI 10.1007/978-3-642-16029-5
Springer Heidelberg Dordrecht London New York

Library of Congress Control Number: 2010938965

Mathematics Subject Classification (2010): 91B30

Cover design: WMXDesign GmbH, Heidelberg

Printed on acid-free paper

Springer is part of Springer Science+Business Media (www.springer.com)

Preface

This book aims at introducing technical and financial aspects of the insurance business, with a special emphasis on the actuarial valuation of insurance products. While most of the presentation concerns life insurance, also non-life insurance is addressed, as well as pension plans.

The book has been planned assuming as target readers:

- advanced undergraduate and graduate students in Economics, Business and Finance;
- advanced undergraduate students in Mathematics and Statistics, possibly aiming at attending, after graduation, actuarial courses at a master level;
- professionals and technicians operating in insurance and pension areas, whose job may regard investments, risk analysis, financial reporting, and so on, hence implying communication with actuarial professionals and managers.

Given the assumed target, the use of complex mathematical tools has been avoided. In this sense, the book can be placed at some "midpoint" of the existing literature, part of which adopts more formal approaches to insurance problems, which implies the use of non-elementary mathematics and calculus, whereas another part addresses practical questions totally avoiding even basic mathematics (which, in our opinion, can conversely provide effective tools for presenting technical and financial features of the insurance business).

We assume that the reader has attended courses providing basic notions of Financial Mathematics (interest rates, compound interest, present values, accumulations, annuities, etc.) and Probability (probability distributions, conditional probabilities, expected value, variance, etc). As mentioned, Mathematics has been kept at a rather low level. Indeed, almost all topics are presented in a "discrete" framework, thus not requiring analytical tools like differentials, integrals, etc. Some Sections in which differential calculus has been used can be skipped, without significant losses in understanding the following material.

Some details concerning the chapters of the book can help in explaining the "rationale" underlying its structure and the choice of the materials therein included.

Chapter 1 first aims at presenting the concept of risk, focussing in particular on the (negative) consequences of some events which can concern a person, a family, a firm, and so on. Secondly, the Chapter describes the role of an insurance company, which takes individual risks, builds up a pool of risks, and bears the risk of losses caused by large numbers of events within the pool or unexpected severity of the claims.

In Chapter 2 various aspects of the risk pooling process are addressed. The effects of cross-subsidy (and, in particular, mutuality and solidarity) are illustrated. Then, referring to a simple portfolio structure, reinsurance arrangements, solvency and capital allocation are dealt with.

Hence, the first two Chapters provide the reader with an introduction to risk and insurance. Indeed, a risk-management oriented approach should underpin, in our opinion, the teaching of the insurance technique and finance. It is worth stressing that these two Chapters can fulfill the syllabus of a very short course (say, 20-25 hours) aiming to present the basics of risk identification, risk assessment, and risk management actions.

Chapters 3 to 7 focus on life insurance. Although many topics dealt with are rather traditional (life tables, discounting cash-flows, premiums and reserves for various insurance products), several issues of great current interest have been included; for example: mortality trends, best-estimate reserving, risk margins, profit assessment, linking life insurance benefits to the investment performance, unit-linked products, and so on.

Chapter 8 addresses problems related to the post-retirement income. In particular, defined contribution pension plans are addressed. The protection that an individual can obtain by underwriting appropriate benefits and financial guarantees, before and after retirement, is examined. Special emphasis is placed on life annuities as an element in post-retirement income arrangements. Risks emerging for the provider are described, with particular regard to the financial and longevity risks.

Finally, Chapter 9 deals with non-life insurance. First, an overview of the contents of non-life insurance products is provided. Then, premium calculation and related statistical bases are focused. Issues presented in Chapter 1 are progressed, in order to introduce the stochastic modeling of claim frequency, claim severity and aggregate claim amounts. An introduction to technical reserves and profit assessment concludes the Chapter.

Each chapter concludes with a section providing bibliographic references and suggestions for further reading. The list of references only includes textbooks and monographs, while disregarding papers in scientific journals, congress proceedings, research and technical reports, and so on. Our choice aims at limiting the number of citations, in line with the teaching orientation of this work.

We have successfully tested the logical structure and the contents of the book in various recent courses. In particular: a course of Insurance technique and finance for graduate students in Finance at the University of Parma; a course of Life insurance mathematics for undergraduate students in Statistics and undergraduate students in Mathematics at the University of Trieste; courses of Risk and Insurance, Life insurance technique, Non-life insurance technique and a distance-learning course of

Insurance technique for employees of a European insurance company, at the MIB School of Management in Trieste. Part of the material included in the book has been used also in CPD (Continuing Professional Development) courses of Life insurance technique for non-actuaries organized by the Italian actuarial professional body. Further, some specific topics have been delivered in short seminars and other teaching initiatives (for example: risk-management approach to insurance problems, stochastic mortality, linking life insurance benefits to the investment performance, etc).

Risks must be carefully identified, assessed and managed by all the agents (individuals, households, firms, public institutions, and so on). Risk transfer constitutes an effective tool for managing risks, and the importance of insurers in this transfer process is self-evident. Actually, the insurance business constitutes a growing market. Appropriate risk management solutions must be taken also by insurers, due to the risks they assume through their products.

If this book helps to better understand the technical and financial features of the insurance activity, the role of insurers as intermediaries in the risk pooling process and as financial intermediaries, and the basics of the risk management of an insurance business, then we have achieved our objective.

Annamaria Olivieri
Ermanno Pitacco

Trieste, July 2010

Contents

Chapter 1
Risks and insurance

1.1 Introduction

The main purposes of this Chapter are:

- to present the concept of individual "risk", focussing in particular on quantitative aspects of the (negative) financial consequences of some events which can concern a person, a family, a firm, and so on;
- to introduce the role of an insurance company (briefly, an insurer), which takes individual risks, building up a "pool", and then bears the risk of losses caused by an unexpected number of events in the pool, or by an unexpected severity of the financial consequences of such events.

While basic ideas concerning the risk transfer process and the related construction of a pool of risks are presented in this Chapter, more complex issues regarding the management of a pool will be discussed in Chap. 2.

1.2 "Risk": looking for definitions

1.2.1 Some preliminary ideas

A number of definitions have been proposed for the term "risk", some of which concern the common language, whereas others relate to the more specific business language, and the language of insurance business in particular.

A rather general definition can be provided in mathematical terms. In fact, a *risk* can be defined as a *random number*, X, whose actual *outcome* (or *realisation*) is unknown. Yet, a set of possible outcomes has to be specified, and *probabilities* over this set have to be assigned.

As regards the set of possible outcomes, consider the following examples.

A. Olivieri, E. Pitacco, *Introduction to Insurance Mathematics*,
DOI 10.1007/978-3-642-16029-5_1,

- Assume that X denotes the spot on the face which will appear by tossing a dice; clearly, the possible outcomes are the numbers $1,2,\dots,6$.
- X can represent, in financial terms, the damage consequent a fire in an industrial building. So, $X = 0$ denotes the absence of a damage, whilst $X = x_{\max}$ denotes the total loss of the building, whose value is $x_{\max}$; thus, the interval $[0, x_{\max}]$ is the set of possible outcomes of the risk X.
- If X represents the annual economic result of a firm, as at the end of the year, $X > 0$ denotes a profit whilst $X < 0$ denotes a loss. The maximum possible loss, x' ($x' < 0$), and the maximum possible profit, x'' ($x'' > 0$), should be estimated, so that the possible outcomes of X are given by the interval $[x', x'']$.

Note that in the first example above, the random number X refers to a "physical" result, whereas in the other examples X is a random amount describing economic consequences of some events. In what follows we will be involved just in the financial consequences of events, and hence the risk will be expressed in monetary terms.

We now move to a set of examples, that we call "cases", specifically concerning the fields of finance and insurance. In this Section we just aim at describing various types of risk, looking at the *sources* from which risks originate (namely, the financial scenario, some demographical aspects, and so on). In the following sections, we will turn again several times on these cases, in particular for assessing the impact of risks on significant results (profits, cash-flows, and so on). Thus, we will follow a stepwise process, starting from the discussion of various features of risks and aiming at the description of important risk transfer opportunities.

1.2.2 Transactions with random results

We consider a set of transactions, denoted by A, B, ... (for example: purchase of zero-coupon bonds, investment in equities, and so on), each transaction leading to a random *result* at a stated time. We denote with X_{A}, X_{B}, ..., the results produced by the various transactions. For instance, transaction A leads to the result X_{A}, whose possible outcomes are $x_{\mathrm{A},1}, x_{\mathrm{A},2}, \dots$.

The actual outcome of each random result depends on which *state of the world* will occur, out of a given set of mutually exclusive states, $S_1, S_2, \dots$. Each state of the world summarizes aspects of the economic-financial scenario, which can affect the results. Table 1.2.1 illustrates the link between possible outcomes and a (finite) set of states of the world.

In what follows, we will focus on two special cases, denoted as Case 1a and Case 1b. Although these cases do not involve insurance issues, they constitute a good starting point for discussion about risk assessment, as we will see in Sect. 1.4.7.

Case 1a - Zero-coupon bonds We refer to a zero-coupon bond, whose pay-off at maturity (say, in one year) depends on the state of the world at that time. When the bond is purchased, the state of the world at maturity is unknown, namely random. We assume, for simplicity, two possible states only, denoted by S_1, S_2. Further, we

Table 1.2.1 A set of transactions with random results

	S_1	S_2	...
X_A	$x_{A,1}$	$x_{A,2}$	...
X_B	$x_{B,1}$	$x_{B,2}$	...
X_C	$x_{C,1}$	$x_{C,2}$	...
...	...	...	...

assume that 50 and 150 are the corresponding pay-offs of the bond. In particular, we can assume that the outcome 50 implies a loss in relation to the amount invested in purchasing the bond, whereas 150 leads to a profit. We denote with X_A the random pay-off at maturity.

Another zero-coupon provides at maturity a pay-off, $X_B = 100$, which is independent of the state of the world.

The former zero-coupon bond is a *risky bond*, whereas the latter is a *risk-free bond*. Table 1.2.2 illustrates the relation between pay-offs and states of the world.

Table 1.2.2 Payoffs of two zero-coupon bonds

	S_1	S_2
X_A	50	150
X_B	100	100

Remark It is worth stressing the true meaning of the expression "risk-free" referred to the bond with pay-off X_B. It actually means that risk is regarded as negligible, in the sense that we are (almost) sure that the pay-off will be 100, whatever the scenario. For example, the *counter-party risk*, i.e. the risk of default of the bond issuer, is considered negligible and hence disregarded so far in our model. Indeed, all models should provide simplified, yet unbiased, representations of a highly complex reality.

□

Case 1b - Random yields The possible yields (per 100 monetary units) provided by four investments are represented in Table 1.2.3. For any given investment choice, each outcome is linked to a state of the world. Low yields (e.g. 0%, in the example) can be considered as "losses", if compared to an appropriate benchmark.
□

In Case 1a (Zero-coupon bonds) and Case 1b (Random yields), the presence of risk may lead either to a profit or a loss. Risks of this type are usually called *speculative risks*.

Table 1.2.3 Investments with random yields

	S_1	S_2	S_3
X_1	5.0	6.0	7.0
X_2	0.0	6.0	12.0
X_3	5.2	6.1	6.1
X_4	5.0	6.0	6.5

1.2.3 A very basic insurable risk

The case discussed in this Section can be considered the simplest example of a situation in which the presence of a risk can only cause a loss. Thanks to its simplicity, this case will be often referred to in the following sections, as the starting point for introducing assessment procedures, construction of "pools" of risks, and so on.

Case 2 - Possible loss with fixed amount An "agent" (a person, a company, an institution) may suffer a loss, because of an event (an *accident*) occurring within a stated period. For example, the event could consist in the total loss of a cargo moved by an aircraft. We denote with $\mathscr{E}$ the event which causes the financial loss, and with x the amount of the loss itself. Thus, we are assuming that, if the event occurs, the amount of the loss is certain. In the example above, no partial damage of the cargo is accounted for. We can formally represent the potential loss with the random number X, defined as follows:

$$X = \begin{cases} x & \text{if } \mathscr{E} \\ 0 & \text{if } \bar{\mathscr{E}} \end{cases} \tag{1.2.1}$$

This basic model can be used to represent also other situations of risk. For example, the death (say, within a one-year period) of a person, who sustains her family with an income, may have dramatic consequences on the availability of financial resources. Although the financial impact of the death could be assessed in terms of the present value of future expected incomes, a huge degree of uncertainty in defining the amount x obviously remains.

Another example is given by a permanent and total disablement, because of an accident or a body injury, which causes the working incapacity of an individual. A high degree of uncertainty in determining the amount x affects also this case.
□

1.2.4 Random number of events and random amounts

More realistic situations can be depicted by generalizing the risk described as Case 2 (Possible loss with fixed amount). The five following cases constitute generalizations of Case 2, as they include a larger set of random items, or a longer time horizon. In particular:

- the event causes a random loss, instead of a deterministic loss (Case 3a below);
- a random number of events, instead of one event at most, may occur within the stated period, each event implying a deterministic loss (Case 3b and 3c below);
- a longer time horizon is addressed (Case 3c below), so that the time-value of money cannot be disregarded;
- a random number of events may occur within the period, each event implying a random loss (Cases 3d and 3e below).

Figure 1.2.1 provides an illustration of the various generalizations, which are dealt with in this Section.

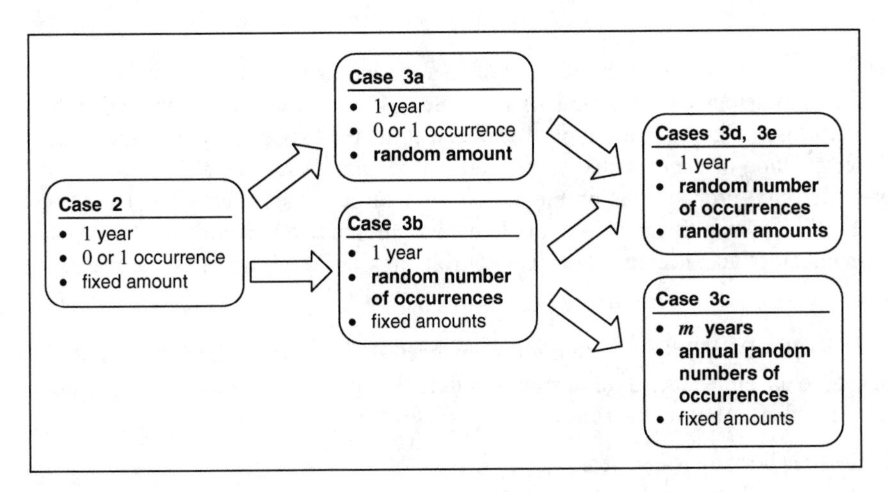

Fig. 1.2.1 From the basic risk to more general situations

Case 3a - Damage / loss of a cargo Unlike in Case 2 (Possible loss with fixed amount), we also allow for partial damage of a cargo during the transport. Thus, the damage is a random amount X. As regards its possible outcomes, we can choose either a discrete setting, namely

$$X : 0, x_1, \dots, x_{\max} \tag{1.2.2}$$

or a continuous setting, namely

$$0 \le X \le x_{\max} \tag{1.2.3}$$

Typically, the maximum amount $x_{\max}$ will be given by the value of the cargo. The outcome $X = 0$ denotes the absence of damage thanks to the absence of accident.

A formal representation of the discrete setting, as described by the outcomes listed in (1.2.2), can be as follows:

$$X = \begin{cases} x_1 & \text{if } \mathscr{E}_1 \\ x_2 & \text{if } \mathscr{E}_2 \\ \dots & \dots \\ x_m & \text{if } \mathscr{E}_m \\ 0 & \text{if } \bar{\mathscr{E}} \end{cases} \tag{1.2.4}$$

where $x_m = x_{\max}$, and the events $\mathscr{E}_1, \mathscr{E}_2, \dots, \mathscr{E}_m$ are scaled according to increasing severity of the consequences in terms of amount of the loss. The event $\mathscr{E}$, namely the occurrence of an accident whatever its severity may be, can be then represented as the union

$$\mathscr{E} = \mathscr{E}_1 \cup \mathscr{E}_2 \cup \dots \cup \mathscr{E}_m \tag{1.2.5}$$

whereas $\bar{\mathscr{E}}$ still represents the absence of an accident (as in Case 2).

For various purposes (as we will see in Sect. 1.3), it is useful to summarize the random loss by using some typical value, such as the expected value. Obviously, the calculation of the expected value (and other typical values as well) requires, in principle, that probabilities on the set of possible outcomes (given by (1.2.2), or (1.2.3)) have been assigned. As an alternative, in practice we can just assume an estimation of the expected value, derived from previous (and similar) experiences.
□

Case 3b - Disability benefits; one-year period An employer takes the risk of paying to the employees a lump-sum benefit in the case of permanent disability due to an accident. Assume the following hypotheses:

1. the time horizon is one year;
2. n employees are exposed to the risk of accident;
3. for each employee, the amount of the benefit is C.

Let K denote the random number of accidents within a given year. Hence, the total benefit paid by the employer is given by

$$X = CK \tag{1.2.6}$$

The possible outcomes of K are $0, 1, \dots, n$, so that the corresponding outcomes of X are $0, C, \dots, nC$.

Also in this case, the random payment can be summarized by using some typical value, in particular the expected value. For the expected value, $\mathbb{E}[X]$, of the total benefit, we clearly have

$$\mathbb{E}[X] = C\,\mathbb{E}[K] \tag{1.2.7}$$

If we replace hypothesis 3 with the following one

4. for the j-th employee, $j = 1, 2, \dots, n$, the amount of the benefit is $C^{(j)}$ (e.g. related to the employee's salary)

then, the total benefit paid by the employer does not depend on the number of accidents only, as it also depends on which employees enter the disability state. In formal terms, with reference to employee j we define the random amount $X^{(j)}$ as follows:

$$X^{(j)} = \begin{cases} C^{(j)} & \text{in the case of accident} \\ 0 & \text{otherwise} \end{cases} \tag{1.2.8}$$

Then, the total random payment of the employer is given by

$$X = \sum_{j=1}^{n} X^{(j)} \tag{1.2.9}$$

Note that, if we assume hypothesis 4, the expression of the total payment X is more complex than that given by (1.2.6), as various amounts of benefit are in general involved.

□

It is worth noting that, in Case 3b (Disability benefits; one-year period), the risk borne by the employer, which leads to the random payment X, is actually a set (or a "pool") of individual risks, each one represented by the possible disability of an employee and the related payment C (or $C^{(j)}$) by the employer. Interesting features of the risk pooling will be analyzed, in general terms, in the following sections starting from Sect. 1.6.1.

Case 3c - Disability benefits; multi-year period We generalize Case 3b (Disability benefits; one-year period) by assuming that the time horizon consists of m years, and in particular we are interested in setting $m > 1$ (say, $m = 5$ or $m = 10$). We still assume that n employees (at the beginning of the m-year period) are exposed to the disability risk. Moreover, we suppose that each employee who suffered an accident implying permanent disability in any given year is replaced, at the beginning of the following year, by another employee. Further new entrants are not allowed. Hence, n employees are exposed to risk at the beginning of each year. The individual lump-sum benefit paid, at the end of the year in which the accident occurs, is C, whatever the year may be (within the stated period).

We denote with $K_1, K_2, \ldots, K_m$ the random number of accidents occurring in the various years, so that

$$X_t = C K_t \tag{1.2.10}$$

is the random amount paid by the employer at time t, namely at the end of year t, for $t = 1, 2, \ldots, m$. Note that, if we defined the total random payment of the employer simply as follows

$$X = X_1 + X_2 + \cdots + X_m \tag{1.2.11}$$

we would disregard the time-value of the money (i.e. we would assume a zero interest rate). We will return on this aspect in Sect. 1.4.5.

□

Case 3d - A fire in a factory Referring to a given period (say, one year), we assume that a factory can be damaged, one or more times within the stated period, by fire. In each occurrence, the amount of the related damage is random. Note that, in this case, features of Case 3a (Damage / loss of a cargo), i.e. randomness of the loss, and Case 3b (Disability benefits; one-year period), i.e. random number of occurrences, are merged together.

In formal terms, we first define the random number N as the number of occurrences of fire within the stated period. Then, we denote with X_k the damage caused by the k-th fire. Hence, the total random damage X is defined as follows:

$$X = \begin{cases} 0 & \text{if } N = 0 \\ X_1 + \cdots + X_N & \text{if } N > 0 \end{cases} \tag{1.2.12}$$

For each k, we have $X_k > 0$, as the case of a zero damage is expressed by $N = 0$. Further, for each X_k, a minimum amount $x_{\min}$ and a maximum amount $x_{\max}$ should be stated. In particular, the maximum amount could be the value of the factory. However, it is unlikely that, in the case of multiple occurrence of fire, each event completely destroys the factory (which, in the meanwhile, should have been completely rebuilt). This aspect can be dealt with by properly assigning the probabilistic structure of the random numbers $N, X_1, \ldots, X_N$.

As regards the random number N, we can assume in principle that the possible outcomes are all the integer numbers $0, 1, 2, \ldots$. Conversely, we can assume a maximum (reasonable) outcome $n_{\max}$, so that the possible outcomes are $0, 1, 2, \ldots, n_{\max}$. Note that, in Case 3b (Disability benefits; one-year period), the maximum number of accidents is, of course, n.

In order to summarize the random quantities mentioned above by using, for example, the expected value, the probabilities related to the possible outcomes of these random quantities should be available. In practice, as said above, we can just assume estimations of the expected values, derived from previous (and similar) experiences. We denote with $\mathbb{E}[N]$ the expected value of the random number of occurrences (fire, in this example) in the given period, $\mathbb{E}[X_k]$ the expected value of the damage resulting from the k-th occurrence, and $\mathbb{E}[X]$ the expected value of the total damage.

If we assume appropriate hypotheses (which will be specified in Sect. 1.4.4), in particular if we assume that all the random amounts X_k have the same expected value, namely

$$\mathbb{E}[X_1] = \mathbb{E}[X_2] = \cdots = \mathbb{E}[X_{n_{\max}}] \tag{1.2.13}$$

then, we find that

$$\mathbb{E}[X] = \mathbb{E}[X_1]\,\mathbb{E}[N] \tag{1.2.14}$$

Damages to the factory (and in particular to buildings, machineries, equipments, and so on) constitute an example of *direct losses* caused by fire. Conversely, *indirect losses* arise as a consequence of direct losses. For example, damages to the machineries may cause an indirect loss by reducing the normal production level, and hence by reducing the profit usually generated by the factory. Clearly, for each possible occurrence of fire, also indirect losses should be taken into account.

□

Case 3e - Car driver's liability A car driver may cause damage to the property (for example, cars) of others, or injury to persons (pedestrians, or other car drivers). Then, a third-party liability arises. Referring to a given period (say, one year), we can define, as in Case 3d (A fire in a factory), the random number N as the number of damages or injuries caused by the driver within the period. We still denote with X_k the damage caused by the k-th occurrence. Hence, the total random damage, X, in the given period, is defined by formula (1.2.12).

Note however that, in this case, maximum amounts cannot be stated, as the damaged property (or the type and severity of the injury) is not predefined, whereas it is well defined in Case 3d (A fire in a factory).

□

It is worth stressing that in Cases 2 (Possible loss with fixed amount) and 3a (Damage / loss of a cargo) to 3e (Car driver's liability), the presence of a risk can only cause losses (or damages, liabilities, and so on). Hence, in these situations we refer to *pure risks*.

1.2.5 Risks inherent in the individual lifetime

Any individual, while managing her financial resources, should account for various risk sources. We now focus on those risks which are directly related to the randomness of the individual lifetime.

The lifetime of an (adult) individual can be split into two economic periods, namely the *working period* and the *retirement period*. During the working period, while getting an income from her working activity, the individual should accumulate resources in order to finance the post-retirement income. Thus, as regards the management of resources aiming to provide an income at old ages, the working period corresponds to the *accumulation phase*, whereas the retirement period corresponds to the *decumulation phase* (see Fig. 1.2.2).

In both the phases various risks affect the management of resources, among which financial risks, arising from randomness in the investment yield, should not be disregarded. Further, some needs (and then the impact of risks) are related to the presence of dependants. In what follows we focus on risks inherent in the individual lifetime, singling out the following aspects:

- accumulation of resources to be used during the retirement period;
- risk of early death, and specifically the risk of dying during the working period;
- income during the retirement period.

Case 4a - The need for resources at retirement An individual, during her working period, is aware that at retirement she will need an amount, say S, to be converted, at that time, into a sequence of periodic amounts, so that a regular income will be available from retirement onwards.

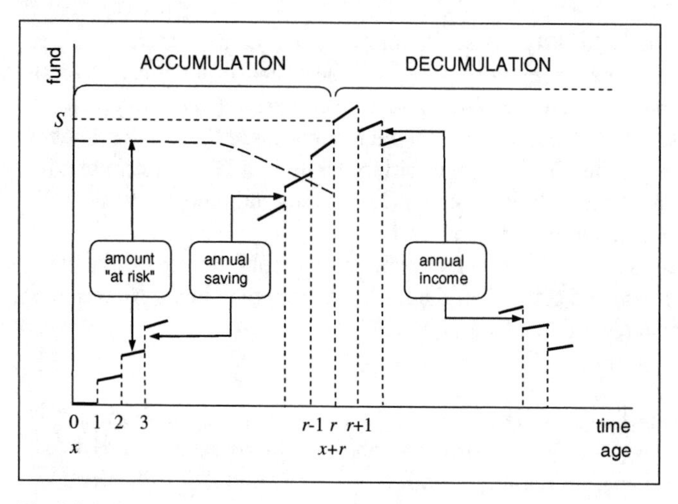

Fig. 1.2.2 Accumulation and decumulation periods

Assume the year as the time unit. Let r denote the retirement time. A *saving plan* is designed, in order to find a suitable sequence of annual savings $c_1, c_2 \dots, c_r$, which, during the working period, progressively constitute a fund. A financial institution manages the fund itself. Note that the amounts $c_1, c_2 \dots, c_r$, deposited at times $1, 2, \dots, r$ respectively, are represented by the jumps (during the accumulation period) in Fig. 1.2.2. The slope of each segment approximately represents the effect of interest credited to the fund. The resulting piece-wise profile shows the behavior of the fund throughout the accumulation period.

In formal terms, the value, S, of the fund at time r is given by:

$$S = \phi(c_1, c_2, \dots, c_r) \tag{1.2.15}$$

where the function ϕ depends on the interests credited to the fund. In particular, denoting with i an estimate of the (constant) annual interest rate credited to the accumulated fund, we have

$$S = c_1 (1+i)^{r-1} + c_2 (1+i)^{r-2} + \dots + c_r \tag{1.2.16}$$

Assume that the actual sequence of deposits exactly follows the saving plan. If the interest rate i is guaranteed by the financial institution, the accumulated value S is certain. Thus, the investment risk is borne by the financial institution, whilst the accumulation process is risk-free for the individual.

Conversely, if the financial institution does not provide the individual with any guarantee, the accumulation process could result in an amount lower than S (given by formula (1.2.16)), because of changes in interest rates, in the value of equities purchased, and so on. In particular note that, because of these possible changes, all the increases in the accumulation profile between two consecutive jumps (see Fig. 1.2.2) should be considered as random quantities. Then, also the final result of the accumulation process should be considered as a random amount. Hence, the risk is linked to the conditions of the accumulation process, and, in particular, is driven by the guarantees provided by the financial intermediary.

□

Case 4b - Early death of the individual Assume that the accumulation process described in Case 4a (The need for resources at retirement) is in progress. However, in the case of early death of the individual (namely, before the retirement time r), the accumulation process is interrupted (and the accumulated amount is lower than the target amount S). Of course, the death can cause a financial distress to the individual's family, in particular in the presence of one or more dependants. In practice, it is almost impossible to quantify in monetary terms the impact of the early death, in particular because of the unknown value of future incomes lost. Thus, the financial impact is represented by a random amount. Assume that no estate, other than the accumulated fund, is available to face family's future needs, and that the dashed line in Fig. 1.2.2 represents a tentative estimation of the random impact. Then, at any point in time, the amount resulting as the difference (if positive) between the estimated impact of the early death and the accumulated fund is an amount "at risk", because of the lack of resources.

Note that this case generalizes the basic risk, namely Case 2 (Possible loss with fixed amount), as a multi-year period of exposure to risk is allowed for.

□

Case 4c - Outliving the resources available at retirement Assume that a given amount S is available to an individual at her retirement, i.e. at time r (see Fig. 1.2.2), presumably as the result of an accumulation process (see Case 4a). Further, assume that now S is the initial amount of a fund, managed by a financial institution which guarantees a constant annual rate of interest i. In order to get her post-retirement income, the retiree withdraws from the fund at time t the amount b_t ($t = r+1, r+2, \dots$).

Let F_t denote the fund at time t, immediately after the payment of the annual amount b_t. Clearly:

$$F_t = F_{t-1}(1+i) - b_t \quad \text{for } t = r+1, r+2, \dots \tag{1.2.17}$$

with $F_r = S$. Thus, the annual variation in the fund is given by

$$F_t - F_{t-1} = F_{t-1}\, i - b_t \quad \text{for } t = r+1, r+2, \dots \tag{1.2.18}$$

Figure 1.2.3 illustrates the causes explaining the behavior of the fund throughout time, formally expressed by Eq. (1.2.18); note that, the (usual) case $b_t > F_{t-1}\, i$ is referred to.

The behavior of the fund obviously depends on the sequence of withdrawals $b_{r+1}, b_{r+2}, \dots$. In particular, if for all t the annual withdrawal is equal to the annual interest credited by the fund manager, that is

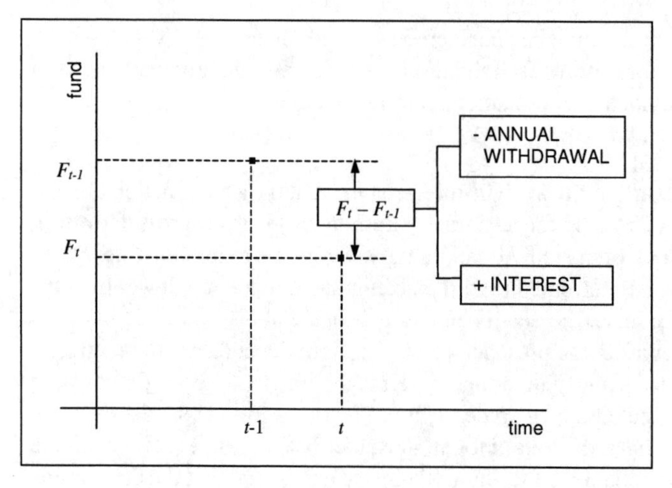

Fig. 1.2.3 Annual variation in the fund providing an annuity-certain

$$b_t = F_{t-1}\, i \tag{1.2.19}$$

then, from (1.2.18) we immediately find

$$F_t = S \tag{1.2.20}$$

for all t, and hence a constant withdrawal

$$b = S i \tag{1.2.21}$$

follows.

Conversely, if we assume a constant withdrawal greater than the annual interest (as probably needed to obtain a reasonable post-retirement income), namely

$$b > S i \tag{1.2.22}$$

the drawdown process will exhaust, sooner or later, the fund (of course, provided that the retiree is still alive). Indeed, from Eq. (1.2.18) we have

$$F_t < F_{t-1} \text{ for } t = r+1, r+2, \ldots \tag{1.2.23}$$

and we can find a time $t_{\max}$ such that

$$F_{t_{\max}} \geq 0 \text{ and } F_{t_{\max}+1} < 0 \tag{1.2.24}$$

Clearly, the exhaustion time $t_{\max}$ depends on the annual amount b (and the interest rate i as well), as it can be easily understood from Eq. (1.2.18).

The sequence of $t_{\max} - r$ constant annual withdrawals b (with $t_{\max}$ defined by conditions (1.2.24), and possibly completed by the exhausting withdrawal at time $t_{\max} + 1$) constitutes an *annuity-certain.*

Example 1.2.1. Assume $S = 1000$. Figure 1.2.4 illustrates the behavior of the fund when $i = 0.03$ and for different annual amounts b. Conversely, Fig. 1.2.5 shows the behavior of the fund for various interest rates i, assuming $b = 100$.

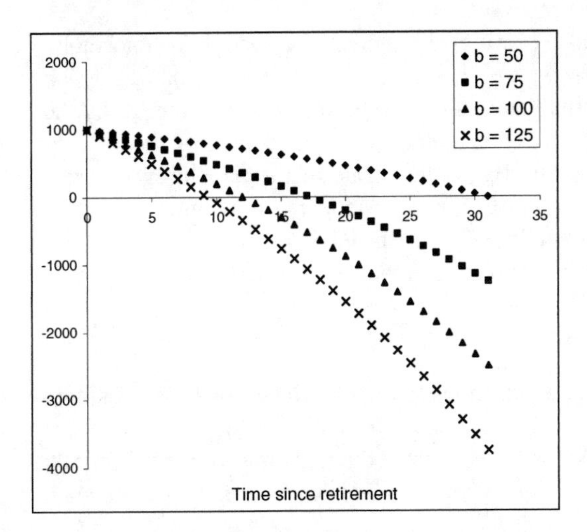

Fig. 1.2.4 The fund providing an annuity-certain ($i = 0.03$)

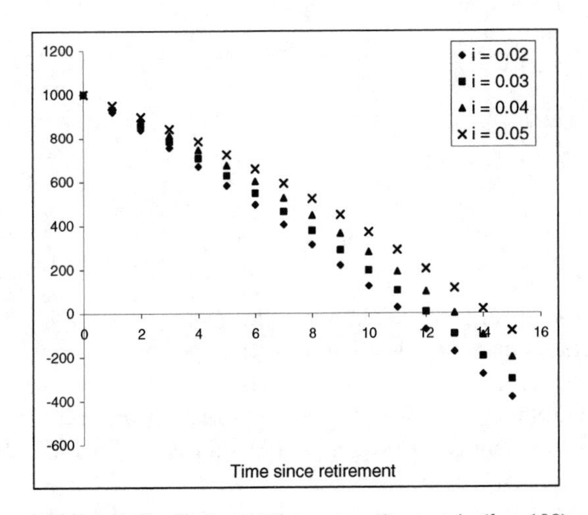

Fig. 1.2.5 The fund providing an annuity-certain ($b = 100$)

□

It is interesting to compare the exhaustion time $t_{\max}$ with the lifetime of the retiree. Her age at retirement is $x+r$ (see Fig. 1.2.2), for example $x+r=65$. Of course the lifetime is a random number. Denote with T_{x+r} the remaining random lifetime for a person age $x+r$. Let ω denote the maximum attainable age (or *limiting age*), say $\omega = 110$. Hence, T_{x+r} can take all values between 0 and $\omega - x - r$. If $T_{x+r} < t_{\max} - r$, then the balance of the fund at the time of death is available as a bequest. On the contrary, if $T_{x+r} > t_{\max} - r$ there are $\omega - x - t_{\max}$ years with no possibility of withdrawal (and hence no income).

In practice, the annual amount b (for a given interest rate i) could be chosen by comparing the related number of withdrawals $t_{\max} - r$ with some quantity which summarizes the remaining lifetime. For example, a typical value is provided by the expected remaining lifetime $\mathbb{E}[T_{x+r}]$. As an alternative, we can focus on the remaining lifetime with the highest probability, i.e. the mode of the remaining lifetime, $\text{Mode}[T_{x+r}]$. Note that, in order to find $\mathbb{E}[T_{x+r}]$ or $\text{Mode}[T_{x+r}]$, assumptions about the probability distribution of the lifetime T_{x+r} are needed.

For example, the value b may be chosen, such that

$$t_{\max} - r \approx \text{Mode}[T_{x+r}] \tag{1.2.25}$$

Thus, with a high probability the exhaustion time will coincide with the residual lifetime. Notwithstanding, events like $T_{x+r} > t_{\max} - r$, or $T_{x+r} < t_{\max} - r$, may occur and hence the retiree bears the risk originating from the randomness of her lifetime, and in particular the risk of outliving her resources. Conversely, the choice of b such that

$$t_{\max} = \omega - x \tag{1.2.26}$$

obviously removes the risk of remaining alive with no withdrawal possibility, but this choice would result in a very low amount b.

□

1.3 Managing risks

1.3.1 General aspects

Although "insurance" is the main scope of this book, it is worth stressing that transferring risks via insurance contracts constitutes just one possibility within a very wide range of actions which can be taken in order to manage risks.

Risk management is the name of the discipline which aims to analyze the risks borne by a firm, a bank, a public institution, and so on, and to suggest actions in order to face risks (the insurance transfer included).

The expression risk management is commonly referred to business entities; Cases 3a (Damage / loss of a cargo) to 3d (A fire in a factory) provide examples of risky situations involving such entities. Nevertheless, the ideas underlying the analysis of risks and the choice of the appropriate actions can, and in principle should, be applied also to individuals (or families), so that a *personal risk management* framework can also be defined. Cases 3e (Car driver's liability), 4a (The need for resources at retirement), 4b (Early death of an individual), and 4c (Outliving the resources available at retirement) constitute examples of situations to be dealt with in this context.

In this Section, some basic ideas about risk management are introduced. Obviously, type of risks and actions to be taken to manage the risks themselves depend, to a large extent, on the particular business involved (or family needs concerned). So, a bank bears some types of risks connected to its specific activity, whereas other risks affect an industry, or an aviation company, and so on. In our presentation, we only address some general issues, without focussing on technical details concerning the various fields of activity.

The implementation of risk management principles takes place via the *risk management process*, which basically consists of five phases, namely

- risk identification;
- risk assessment;
- analysis of possible actions;
- choice of (a combination of) actions;
- monitoring.

It should be noted that the risk management process is a "never-ending" process. In fact, the monitoring phase aims at checking the results of the actions, and possibly suggesting a revision of the four steps previously performed (see Fig. 1.3.1).

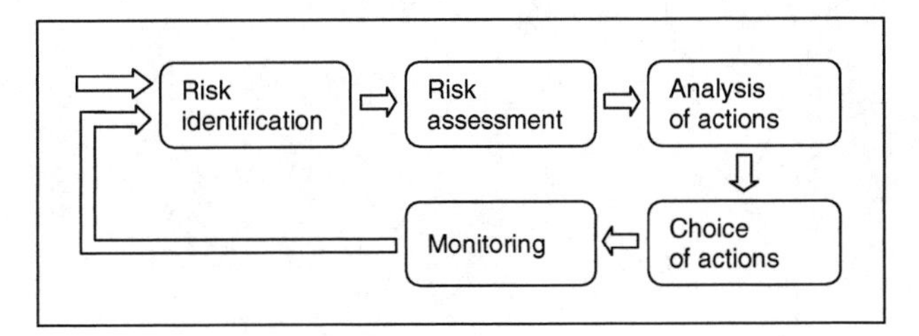

Fig. 1.3.1 The Risk Management process

In the following sections, the five phases of the risk management process are described.

1.3.2 Risk identification and risk assessment

The *risk identification* phase aims to single out the potential loss exposures of the business (or the family, or the individual). As already mentioned, types of risks concerned depend on the particular business under analysis. Nevertheless, some risks are borne by rather broad categories of activity. For example, any factory bears risks as fire, liability towards workers arising from possible accidents, liability towards the population arising from potential air and water pollution, and so on.

In the *risk assessment* phase, the potential number and severity of (direct and indirect) losses should be evaluated. Frequently, just the average number (or the frequency) and the average severity of losses are estimated. A more accurate (and complex) approach to risk assessment should involve the use of probabilistic models, in particular related to the random number of cases of loss, and to the random amount (namely, the severity) of each potential loss.

The risk assessment phase turns out to be itself a process, consisting of a sequence of steps. First the range of some variables should be stated. Then, appropriate probability distributions should be assumed. Finally, some typical values (expected value, variance, mode, and so on) should be focussed on, as these can help in comparing various situations and then taking decisions. Some examples of risk assessment will be presented in Sect. 1.4.

1.3.3 Risk management actions

Actions in risk management are not mutually exclusive; usually, a combination of two or more actions is chosen to face risks.

These actions can be classified as follows.

1. Loss control:
   a. loss prevention:
   b. loss reduction;
   c. risk avoidance.
2. Loss financing:
   a. retention;
   b. insurance;
   c. hedging;
   d. other contractual risk transfers.
3. Internal risk reduction:
   a. diversification;
   b. investment in information.

In order to illustrate various risk management actions, we can refer to **Case 3d** (A fire in a factory) presented in Sect. 1.2.4. *Loss control* actions (also called *risk control* actions) generally aim at reducing the expected total loss $\mathbb{E}[X]$. In particular, actions which tend to lower the expected number of occurrences, $\mathbb{E}[N]$, are known as *loss prevention* methods, whereas actions aiming to reduce the expected severity of each damage, $\mathbb{E}[X_k]$, $k = 1,2,\ldots,N$, are called *loss reduction* methods.

For example, appropriate electric equipments can contribute in reducing the expected number of fire occurrences (loss prevention), whereas fire protection measures (e.g. doors) can lower the risk of fire propagation and hence the expected amount of damages (loss reduction).

Loss control can also be realized by reducing the level of risky activities, in particular by shifting to less risky product lines. Clearly, the cost of this action is given by a reduction in the profits produced by the risky activities. The limit case is given by the total elimination of these activities: this action is usually called *risk avoiding*.

The expression *loss financing* (sometimes *risk financing*) denotes a wide range of methods which aim at obtaining financial resources to cover possible losses, anyhow unavoidable.

First, the business can choose the *retention* of the obligation to pay losses. Retention is often called *self-insurance*. Instead of retaining a risk, the business can transfer it to another business. The usual transfer consists in the *insurance* of the risk, and thus involves, as the counterpart, an insurance company. Nevertheless, other risk transfer arrangements can be conceived. More details on this topic are provided in Sect. 1.3.4.

Hedging is based on the use of financial derivatives, such as futures, forwards, swaps, options, and so on. These derivatives can be used to offset potential losses caused by changes in commodity prices, interest rates, currency exchange rates, and so on. For example, a factory which uses oil in the production process is exposed to losses due to unanticipated increases in the oil price. This risk can be hedged by entering into a forward contract, according to which the oil producer must provide the user with a specified quantity of oil on a specified date at a price stated in the contract.

Finally, we turn to actions aiming at *internal risk reduction*. *Diversification* typically relates to investment strategies and related risks, and consists in investing relatively small amounts of wealth in a number of different stocks, rather than putting all of the wealth into one stock. Diversification makes the investment results not totally depending on the economic results of just one company, and hence aims at the reduction of investment risks.

Investment in information is the second major form of internal risk reduction. Appropriate investments can improve the "quality" of estimates and forecasts. A reduced variability around expected values follows, so that more accurate actions of, for example, loss financing can be performed.

The analysis of alternative actions must be followed by the choice of a set of specific actions to be implemented. As already mentioned, risk management actions are not mutually exclusive, so that the strategy actually adopted is usually an appropriate mix of several actions. For example, loss prevention and loss reduction can be accompanied by an appropriate insurance transfer, which, in its turn, will be less expensive if an effective loss control can be proved.

1.3.4 Self-insurance versus insurance

The results achieved throughout the risk assessment phase should provide the risk manager with data supporting important decisions, and, in particular:

1. what (pure) risks can be retained and what risks should be transferred;
2. how to finance potential losses produced by the retained risks;
3. what kind of risk transfer should be chosen.

As regards point 1, basic guidelines for the decision can follow a *frequency-severity logic* as sketched in Fig. 1.3.2. Risks generating potential losses with low severity (i.e. losses which can be faced thanks to the financial capacity of the firm) can be retained. In particular, as regards point 2, if the frequency of occurrence is low, the losses do not constitute an important concern and thus can be financed either via internal resources, or via external funds, i.e. borrowing money. Internal resources consist of current cash-flows produced by ordinary activities, and shareholders' capital (namely, the assets exceeding the liabilities). High frequency of losses, on the contrary, suggests funding in advance via specific capital allocation.

Risks generating potential losses with a low frequency but a high severity (and then a high impact on the firm) should be transferred, in particular to an insurance company. Activities implying potential losses with high frequency and high severity should be avoided, because of the possible dramatic costs, likely leading to bankruptcy.

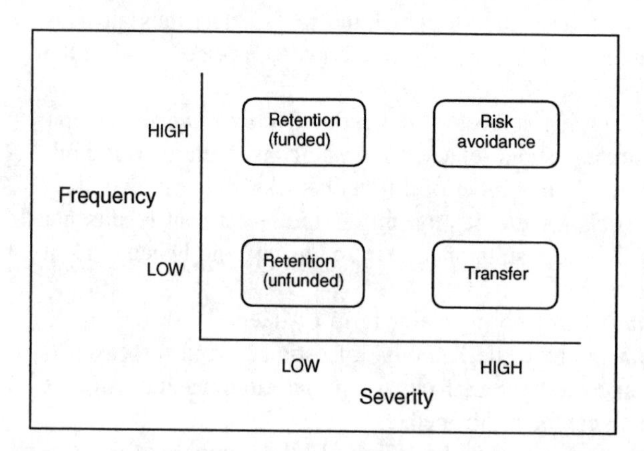

Fig. 1.3.2 How to manage risks according to their possible impact

As regards point 3, we note that the term "transfer" should be understood in a rather broad sense: first, it simply denotes "not a full retention" of the risk; secondly, various counterparts, i.e. agents taking (part of) the risk, can be involved to this purpose.

As far as the first aspect is concerned, risks can be partially transferred and, more precisely, only the heaviest part of a potential loss can be transferred whereas amounts which can be faced thanks to the financial capacity of the business can be retained.

In particular, the rationale of a risk transfer involving an insurer is the splitting of losses into two parts, one retained by the insured and the other paid by the insurer. In formal terms, and still referring to **Case 3d** (A fire in a factory), the random loss for the k-th occurrence is split as follows:

$$X_k = X_k^{[\text{ret}]} + X_k^{[\text{transf}]} \tag{1.3.1}$$

The random amount paid by the insurer, $X_k^{[\text{transf}]}$, is determined according to the policy conditions stated in the insurance contract, and can be usually represented as a function of the loss X_k, namely

$$X_k^{[\text{transf}]} = \psi(X_k) \tag{1.3.2}$$

An example of the function ψ is provided by a "proportional" retention, also called *fixed-percentage deductible*. In this case, we have

$$X_k^{[\text{ret}]} = \theta X_k \tag{1.3.3a}$$

$$X_k^{[\text{transf}]} = (1-\theta) X_k \tag{1.3.3b}$$

where θ is a given percentage. See Fig. 1.3.3.

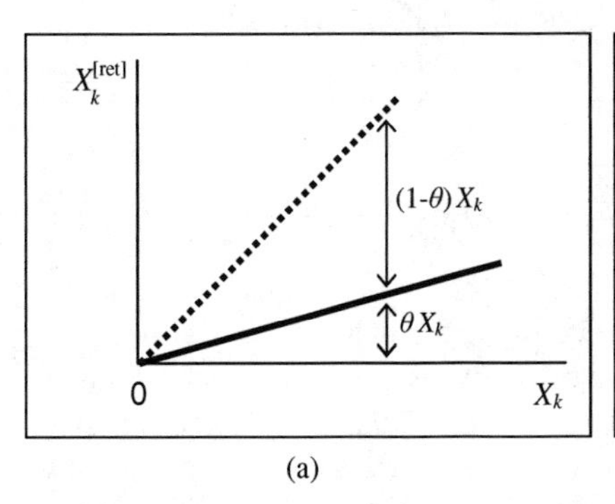

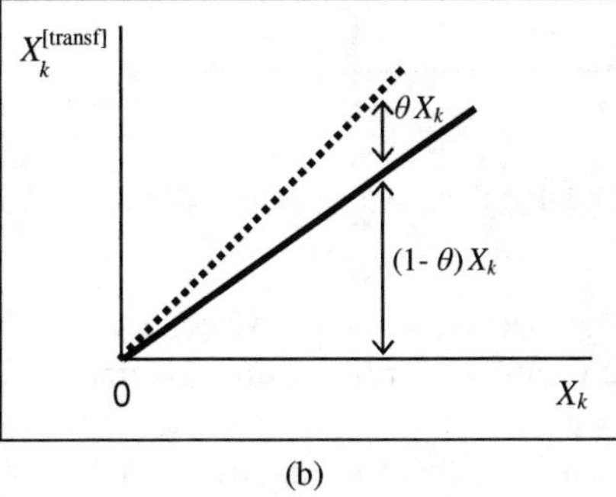

Fig. 1.3.3 An example of risk transfer: the fixed-percentage deductible

Another example of the function ψ is provided by the *fixed-amount deductible*, which is a condition included in a number of insurance contracts. When a fixed-amount deductible d works, any loss under the amount d is fully retained, whilst losses higher than d are transferred only for the amount exceeding d. Thus

$$X_k^{[\mathrm{ret}]} = \min\{X_k, d\} \tag{1.3.4a}$$
$$X_k^{[\mathrm{transf}]} = \max\{X_k - d, 0\} \tag{1.3.4b}$$

See Fig. 1.3.4.

The risk transfer based on a fixed-amount deductible can be more interesting, for various reasons. In particular, we note what follows:

- small losses do not originate insurer's payments, thus saving the related costs;
- for large losses, the insurer pays the whole amount net of the deductible (provided that no upper limit is stated).

Note that, on the contrary, if a fixed-percentage deductible is stated, the insurer intervenes also for small losses, whereas, in the case of large losses, an important part of the loss is suffered by the insured.

Other transfer arrangements, of great practical interest, will be described in Chap. 9.

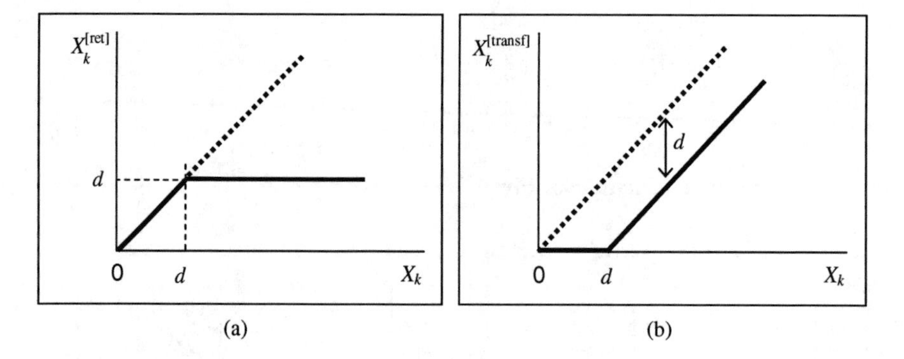

Fig. 1.3.4 An example of risk transfer: the fixed-amount deductible

As regards the second aspect, namely the counterparts in a risk transfer deal, we note what follows. The usual risk transfer involves, as the counterpart, an insurance company (or even more insurance companies). In the practice of risk management, a deep analysis of all the available insurance opportunities should be performed. Appropriate insurance covers should be chosen for each type of risk (fire, third-party liability, and so on) borne by the firm. The ultimate result is the construction of an *insurance programme*, possibly involving several insurance companies.

Despite the prominent importance of insurance arrangements, other transfer solutions are feasible. In particular, large organizations can transfer risks to financial markets by issuing specific bonds; also insurers can resort to this "alternative" solution for transferring risks taken from other agents.

A particular form of risk transfer relies on the so-called *captive insurers*. Some large corporations have established their own insurance companies, namely the "captives", to fulfill insurance requirements of various companies inside the group.

The captive insurer can be interpreted as a profit center within the group. It is worth noting that an insurance company cannot be regarded as a captive simply because completely owned by one or more companies within the group. Conversely, the discriminating feature is whether the majority of its insurance business comes from the companies of the group, rather than from the market or companies outside the group.

Another solution to risk transfer problems is provided by *pools* which share the same type of risks (without resorting to an insurance company). Examples can be found in professional associations which build up pools to manage specific types of risk, like those related to medical expenses. It should be noted that, as these pools do not imply the existence of an insurance company, their establishment is subject to constraints stated by the current legislation.

1.3.5 Monitoring and the Risk Management cycle

The choice of actions is the fourth phase of the risk management process, and, in its turn, this phase originates from the previous three phases (see Fig. 1.3.1). The results consequently obtained must be carefully monitored. The *monitoring* phase has two main objectives:

- checking the effectiveness of the undertaken actions;
- determining whether changes in the scenario suggest novel solutions.

Thus, as clearly appears from Fig. 1.3.1, monitoring is not the "final" phase of the risk management process: indeed, no final phase exists because, after monitoring, the risk management process starts again, with the re-identification of risks in a possibly changed scenario, and so on. Hence, the risk management process is actually a never-ending "cycle".

1.4 Quantifying risks: some models

1.4.1 Some preliminary ideas

As already mentioned in Sect. 1.3.2, risk assessment (or risk quantification) mainly aims at expressing in quantitative terms the impact of risks on significative target results (monetary outgoes, profits, and so on). To this regard, the following points should be stressed.

- The impact of each risk is, because of the nature of the risk itself, random.
- Although useful, the assessment of just the minimum and the maximum impact is clearly insufficient for operational purposes.

- Typical values, like measures of "location" and "dispersion", are much more useful in general, together with probabilities of events like "the result is worse than a given benchmark", or "the loss is higher than a stated (critical) threshold".
- Typical values, for example the expected value or the mode (typical measures of location) and the variance (a measure of dispersion), are particularly useful for the following purposes:
  - comparisons among various risky situations, and related decisions, for example choices between risk retention and risk transfer, e.g. via insurance (see Sect. 1.3.4);
  - in the case of retention, decisions about loss financing, either via specific capital allocation or via current assets and cash-flows (see Sect. 1.3.4);
  - from the insurer's point of view, "pricing" of risky situations via an appropriate tool-kit for premium calculation.

Typical values could be drawn from previous (and analogous) experiences. Similarly, probabilities like those mentioned above, could also be estimated from frequency data. Notwithstanding, the construction of a "complete" probabilistic model (underpinned by statistical experience), also including convenient hypotheses (for example, correlation versus independence among the various random inputs of the model itself), is the most appropriate approach to a sound risk assessment.

Of course, the complexity of a probabilistic model strictly depends on the specific risk and the result dealt with (as we will see in the following sections). It is worth noting that, anyhow, a model should constitute a simplified representation of the reality, and hence it should include all the elements which have an important role in the assessment process, conversely disregarding those elements which (at least according to the opinions of the modeler and the decision maker) do not sensibly affect the results.

1.4.2 A very basic model

We refer to **Case 2** (Possible loss with fixed amount), presented in Sect. 1.2.3. The random loss is expressed, in financial terms, by (1.2.1). Then, the construction of the probabilistic model simply requires to specify the probability of the event $\mathscr{E}$. Let

$$p = \mathbb{P}[\mathscr{E}] \tag{1.4.1}$$

denote this probability. The expected value of the potential loss X is then given by

$$\mathbb{E}[X] = x\,p \tag{1.4.2}$$

and the variance by

$$\mathbb{V}\mathrm{ar}[X] = x^2\,p\,(1-p) \tag{1.4.3}$$

The standard deviation, $\sigma[X]$, is defined as the square root of the variance; hence:

$$\sigma[X] = \sqrt{\mathbb{V}\mathrm{ar}[X]} = x\sqrt{p(1-p)} \tag{1.4.4}$$

1.4.3 Random number of events and random amounts

We first refer to **Case 3a** (Damage / loss of a cargo), described in Sect. 1.2.4. A *probability distribution* should be assigned to the random amount X, describing the severity of the loss. The distribution may be of a discrete or a continuous type. The latter case, rather common in the actuarial practice of non-life insurance, will be dealt with in Chap. 9. In the former case, a set of possible outcomes must be conveniently chosen. We denote with $x_0 = 0, x_1, \dots, x_{\max}$ the possible outcomes of the random amount X. We recall that the outcome $x_0 = 0$ means the absence of accident and hence the absence of damage. Then, the probability distribution is specified by assigning the following probabilities:

$$p_h = \mathbb{P}[X = x_h]; \quad h = 0, 1, \dots, m \tag{1.4.5}$$

where $x_m = x_{\max}$. The obvious constraint is $\sum_{h=0}^{m} p_h = 1$. The expected value is then given by

$$\mathbb{E}[X] = \sum_{h=0}^{m} x_h\, p_h \tag{1.4.6}$$

and the variance by

$$\mathbb{V}\mathrm{ar}[X] = \sum_{h=0}^{m} (x_h - \mathbb{E}[X])^2\, p_h \tag{1.4.7}$$

Of course, $\sigma[X] = \sqrt{\mathbb{V}\mathrm{ar}[X]}$.

Example 1.4.1. A possible accident causes a loss, whose amount depends on the severity of the accident itself. We assume that the outcomes of the random loss X are:

0, 100, 200, 300, 400, 500

with the following probabilities

0.99, 0.002, 0.004, 0.002, 0.001, 0.001

The outcome $X = 0$ denotes that the accident does not occur. Then:

$$\begin{aligned} \mathbb{E}[X] &= 2.5 \\ \mathbb{V}\mathrm{ar}[X] &= 763.76 \\ \sigma[X] &= 27.64 \end{aligned}$$

Assume now that an accident (whatever its severity may be) does occur. The outcomes of the random losses are now restricted to the positive values, namely

$$100,\ 200,\ 300,\ 400,\ 500$$

Of course, the sum of the relevant probabilities must be equal to one. Hence, keeping their "relative" values, we have:

$$0.2,\ 0.4,\ 0.2,\ 0.1,\ 0.1$$

From these probabilities we can calculate the expected value of the loss under the hypothesis that an accident occurs:

$$\bar{x} = 250$$

The expected value of the loss (whether an accident occurs or not) can be expressed as follows:

$$\mathbb{E}[X] = \bar{x} \times \mathbb{P}[\text{accident}] = 250 \times 0.01 = 2.5$$

□

What has emerged from Example 1.4.1 can be formalized as follows. The probability p of an accident (whatever its severity may be) can be expressed, according to the notation adopted in Sect. 1.2.4, as

$$p = \mathbb{P}[\mathscr{E}] = \mathbb{P}[\mathscr{E}_1 \cup \mathscr{E}_2 \cup \cdots \cup \mathscr{E}_m] \tag{1.4.8}$$

and is clearly given by

$$p = \sum_{h=1}^{m} p_h \tag{1.4.9}$$

whereas

$$p_0 = 1 - p \tag{1.4.10}$$

According to the theorem of conditional probabilities, we have, for $h = 1, 2, \ldots, m$:

$$\mathbb{P}[X = x_h] = \mathbb{P}[X = x_h \,|\, \mathscr{E}]\, \mathbb{P}[\mathscr{E}] \tag{1.4.11}$$

Then, the *probability distribution* of the amount of the loss, *conditional* on the occurrence of an accident, is the following one:

$$\mathbb{P}[X = x_h \,|\, \mathscr{E}] = \frac{\mathbb{P}[X = x_h]}{\mathbb{P}[\mathscr{E}]} = \frac{p_h}{p};\quad h = 1, 2, \ldots, m \tag{1.4.12}$$

We can define the *expected value* of the loss, *conditional* on the occurrence of an accident:

$$\bar{x} = \mathbb{E}[X \,|\, \mathscr{E}] = \frac{1}{p} \sum_{h=1}^{m} x_h\, p_h \tag{1.4.13}$$

Note that, as $x_0 = 0$, we have:

$$\mathbb{E}[X \mid \mathscr{E}]\,\mathbb{P}[\mathscr{E}] = \sum_{h=1}^{m} x_h\, p_h = \sum_{h=0}^{m} x_h\, p_h = \mathbb{E}[X] \tag{1.4.14}$$

Thus, the *unconditional expected value* can be expressed as follows:

$$\mathbb{E}[X] = \mathbb{E}[X \mid \mathscr{E}]\,\mathbb{P}[\mathscr{E}] = \bar{x}\, p \tag{1.4.15}$$

Remark The factorization of the expected value $\mathbb{E}[X]$ of the loss, as shown by formula (1.4.15), reflects the format in which statistical data are commonly available. Namely, the quantity $\bar{x}$ can be estimated relying on the observed mean damage per accident, whereas the probability p can be estimated on the basis of the frequency of accident.

We now move to **Case 3b** (Disability benefits; one-year period). Assuming that the same benefit C is paid, in the case of disability, to anyone of the n employees, then the risky situation is completely described by the random number, K, of accidents implying disability. As the possible outcomes of K are $0, 1, \dots, n$, a finite probability distribution should be assigned. In particular, if we assume that

1. for each employee the probability of accident is p;
2. the accidents are independent events;

then, the probability distribution of K is a binomial distribution with parameters n, p, shortly

$$K \sim \mathrm{Bin}(n, p) \tag{1.4.16}$$

Thus

$$\pi_k = \mathbb{P}[K = k] = \binom{n}{k} p^k (1-p)^{n-k}; \quad k = 0, 1, \dots, n \tag{1.4.17}$$

Note, however, that assumption 2 may be controversial, as events like accidents occurring, for example, inside a factory could be considered positively correlated.

According to the probability distribution (1.4.17), we have:

$$\mathbb{E}[K] = n\, p \tag{1.4.18}$$

$$\mathbb{V}\mathrm{ar}[K] = n\, p\, (1-p) \tag{1.4.19}$$

$$\sigma[K] = \sqrt{n\, p\, (1-p)} \tag{1.4.20}$$

As regards the total benefit, X, paid by the employer, we obviously have:

$$\mathbb{P}[X = kC] = \mathbb{P}[K = k] = \pi_k; \quad k = 0, 1, \dots, n \tag{1.4.21}$$

and then we find:

$$\mathbb{E}[X] = C\, \mathbb{E}[K] = C\, n\, p \tag{1.4.22}$$

$$\mathbb{V}\mathrm{ar}[X] = C^2\, \mathbb{V}\mathrm{ar}[K] = C^2\, n\, p\, (1-p) \tag{1.4.23}$$

$$\sigma[X] = C \sqrt{n\, p\, (1-p)} \tag{1.4.24}$$

From (1.4.17) and (1.4.21) we have, in particular:

$$\mathbb{P}[X = 0] = \mathbb{P}[K = 0] = (1-p)^n \tag{1.4.25}$$

Thus, the probability of no accident, and hence of zero payment, decreases as n increases.

It is worth stressing that formulae (1.4.18) and (1.4.22) for the expected values do not require the independence hypothesis.

As noted in Sect. 1.2.4, in Case 3b the risk borne by the employer is actually a pool of n individual risks. The size of the pool plays an important role in the riskiness of the pool itself, as we will see in Example 1.4.2. For a better understanding of the role of n, a relative measure of "risk" can be defined, namely the *coefficient of variation*:

$$\mathbb{CV}[X] = \frac{\sigma[X]}{\mathbb{E}[X]} \tag{1.4.26}$$

An extensive discussion about the meaning of the coefficient of variation and the relevant applications will follow in Sects. 1.5.2, 1.6.1, and 2.3.3.

Example 1.4.2. A benefit $C = 1\,000$ is paid in the case of permanent disability to anyone of the employees of a firm. The probability of an accident causing permanent disability is $p = 0.005$ for each employee. The accidents are assumed to be independent events. Table 1.4.1 shows various results in the cases $n = 10$, $n = 100$ and $n = 1\,000$ respectively. Note that $\mathbb{E}[X \mid X > 0]$ is the expected value of the random payment conditional on the occurrence of at least one accident; see the analogy with the expected value in (1.4.13).

The role of the pool size can be perceived by looking at various quantities as functions of n. In particular, we note what follows.

- When n is "small" ($n = 10$ or $n = 100$, in our example), the expected value $\mathbb{E}[K]$ does not correspond to any possible outcome of the random number K. Then, an interpretation can be as follows: on average, an accident every $\frac{1}{\mathbb{E}[K]}$ years (that is, every 20 years or every 2 years, respectively) will occur.
- The expected value and variance of both K and X increase linearly as n increases (as it results from Eqs. (1.4.18), (1.4.19), (1.4.22), and (1.4.23)), whereas the standard deviation increases proportionally to $\sqrt{n}$ (see Eqs. (1.4.20) and (1.4.24)). It follows that the relative riskiness, expressed by $\mathbb{CV}[X]$ as regards the total benefit, decreases as n increases.
- The probability of no accident is very high when n is small, while it is very low for large values of n. Note also the consequent variation of $\mathbb{E}[X \mid X > 0]$.

□

We now refer to the problems described as **Case 3d** (A fire in a factory) and **Case 3e** (Car driver's liability) which, as already noted, combine features of Cases 3a (Damage / loss of a cargo) and 3b (Disability benefits; one-year period). A number of modeling alternatives are available for these problems. While a detailed analysis of these issues will be presented in Chap. 9, here we just focus on a rather simple choice.

Table 1.4.1 Disability benefits (one-year period)

	$n = 10$	$n = 100$	$n = 1000$
$\mathbb{E}[K]$	0.05	0.5	5
$\mathbb{V}\text{ar}[K]$	0.04975	0.4975	4.975
$\sigma[K]$	0.22305	0.70534	2.23047
$\mathbb{E}[X]$	50	500	5000
$\mathbb{V}\text{ar}[X]$	49750	497500	4975000
$\sigma[X]$	223.047	705.34	2230.47
$\mathbb{CV}[X]$	4.4609	1.41068	0.44609
$\mathbb{P}[X=0] = \mathbb{P}[K=0]$	$0.995^{10} = 0.9511$	$0.995^{100} = 0.6058$	$0.995^{1000} = 0.0067$
$\mathbb{P}[X>0] = \mathbb{P}[K>0]$	$1-0.9511 = 0.0489$	$1-0.6058 = 0.3492$	$1-0.6058 = 0.9933$
$\mathbb{E}[X \mid X>0]$	$50\,\frac{1}{0.0489} = 1022.71$	$500\,\frac{1}{0.3942} = 1268.30$	$5000\,\frac{1}{0.9933} = 5033.49$

Assume the same probability distribution for all the random amounts, $X_1, X_2, \ldots, X_N$, describing the damages (or liabilities). As for Case 3a (Damage / loss of a cargo), the distribution may be of a discrete or a continuous type. We assume the discrete setting, and denote with $x_1, x_2, \ldots, x_m$ the possible outcomes for every random amount X_k, $k = 1, 2, \ldots, N$. Then, the common probability distribution (that is, the same distribution for $k = 1, 2, \ldots, N$) is specified by assigning the probabilities

$$f_h = \mathbb{P}[X_k = x_h]; \quad h = 1, 2, \ldots, m \tag{1.4.27}$$

The expected value $\bar{x} = \mathbb{E}[X_k]$, $k = 1, 2, \ldots, N$, is then given by:

$$\bar{x} = \mathbb{E}[X_k] = \sum_{h=1}^{m} x_h f_h \tag{1.4.28}$$

and the variance by:

$$\mathbb{V}\text{ar}[X_k] = \sum_{h=1}^{m} (x_h - \bar{x})^2 f_h \tag{1.4.29}$$

Note that the probabilities (1.4.27) correspond to the conditional probabilities (1.4.12) in Case 3a; similarly, the expected value (1.4.28) corresponds to the conditional expected value (1.4.13). The case of no accident and hence damage equal to 0 is now accounted for by the outcome $N = 0$ of the random number of accidents.

As regards the random number N, a discrete distribution should be obviously assigned. In particular, a finite distribution requires the choice of a reasonable maximum outcome $n_{\max}$. As an alternative, the Poisson distribution is frequently used, as we will see in Chap. 9. In the finite setting, the following probabilities must be assigned

$$\pi_h = \mathbb{P}[N = h]; \quad h = 0, 1, \ldots, n_{\max} \tag{1.4.30}$$

and then the expected value, $\bar{n}$, and the variance can be derived as follows:

$$\bar{n} = \mathbb{E}[N] = \sum_{h=0}^{n_{\max}} h\,\pi_h \tag{1.4.31}$$

$$\mathbb{V}\mathrm{ar}[N] = \sum_{h=0}^{n_{\max}} (h - \bar{n})^2 \pi_h \tag{1.4.32}$$

The probability distribution of the total loss X, defined by (1.2.12), and the related typical values are of great interest, as X represents the random cost referred to the stated period (say, one year). In Sect. 1.4.4 we will describe some assumptions commonly adopted in insurance technique, which in particular allow us to express the expected value of the total loss, $\mathbb{E}[X]$, by using formula (1.2.14).

Example 1.4.3. Assume that a factory can be damaged by fire, possibly more times within a year. As regards the random damages, X_k, $k = 1, 2, \dots$, assume $m = 5$ and the following possible outcomes:

$$x_1 = 100;\ x_2 = 200;\ x_3 = 300;\ x_4 = 400;\ x_5 = 500$$

with the related probabilities

$$f_1 = 0.2;\ f_2 = 0.4;\ f_3 = 0.2;\ f_4 = 0.1;\ f_5 = 0.1$$

We find, for $k = 1, 2, \dots$:

$$\bar{x} = \mathbb{E}[X_k] = 250$$

For the random number N, assume $n_{\max} = 4$, and the following probabilities:

$$\pi_0 = 0.9934;\ \pi_1 = 0.0040;\ \pi_2 = 0.0020;\ \pi_3 = 0.0004;\ \pi_4 = 0.0002$$

We obtain:

$$\bar{n} = \mathbb{E}[N] = 0.01$$

and (under the appropriate hypotheses):

$$\mathbb{E}[X] = \bar{n}\,\bar{x} = 2.5$$

We note that the expected value $\mathbb{E}[X]$ coincides with that found in Example 1.4.1. Notwithstanding, different interpretations should be given to the two results, because of different structures of the two problems.

□

1.4.4 Random sums: a critical assumption

The total random damage X, defined by (1.2.12), is a *random sum*, since the number N of terms in the summation as well as the individual values of the terms are random variables. The probability distribution, the expected value and the variance of X are of great practical interest, both in risk assessment in general and in pricing insurance products in particular. However, probabilistic assumptions about the random variables N and X_k, $k = 1, 2, \dots$ are needed in order to get to workable calculation procedures. We now describe a set of assumptions, which are commonly adopted for calculating, in particular, the expected value $\mathbb{E}[X]$.

Assume that:

1. the random variables X_k are independent of the random number N;
2. whatever the outcome n of N, the random variables $X_1, X_2, \dots, X_n$
   a. are mutually independent;
   b. are identically distributed (and hence with a common expected value, say $\mathbb{E}[X_1]$).

We have, in general:

$$\mathbb{E}[X] = \sum_{h=0}^{n_{\max}} \pi_h \mathbb{E}[X \mid N = h] = \sum_{h=1}^{n_{\max}} \pi_h \mathbb{E}[X \mid N = h] = \sum_{h=1}^{n_{\max}} \pi_h \left[\sum_{i=1}^{h} \mathbb{E}[X_i \mid N = h] \right] \tag{1.4.33}$$

Thanks to assumption 1, we have:

$$\mathbb{E}[X_i \mid N = h] = \mathbb{E}[X_i] \quad \text{for all } i \tag{1.4.34}$$

and thanks to assumption 2b we obtain

$$\mathbb{E}[X_i] = \mathbb{E}[X_1] \quad \text{for all } i \tag{1.4.35}$$

and finally:

$$\sum_{i=1}^{h} \mathbb{E}[X_i \mid N = h] = h\,\mathbb{E}[X_1] \tag{1.4.36}$$

Then, we obtain:

$$\mathbb{E}[X] = \sum_{h=1}^{n_{\max}} \pi_h\, h\, \mathbb{E}[X_1] = \mathbb{E}[X_1]\,\mathbb{E}[N] \tag{1.4.37}$$

Although frequently adopted in the insurance technique, the assumptions described above may be rather unrealistic. For example, the assumption of independence between the random variables X_k and the random number N may conflict with those situations in which a very high total number of damages is likely associated to a prevailing number of damages with small amounts.

1.4.5 Introducing time into valuations

While dealing with risks defined on a multi-year horizon, as in the Case 3c (Disability benefits; multi-year period) described in Sect. 1.2.4, the role of time, and in particular the time-value of the money, can have a dramatic importance. This is especially true when risks arising from randomness of the individual lifetime (i.e.

in the framework of personal risk management) are focussed; see for example the need for resources at retirement (Case 4a in Sect. 1.2.5), or the problem of outliving the resources available at retirement (Case 4c). However, these and similar problems will be dealt with in depth while describing life insurance products designed to cover the related risks, namely in Chap. 4. Now, to introduce the role of time in risk assessment, we only focus on a specific example.

We refer to the problem described as **Case 3c** (Disability benefits; multi-year period). As regards probabilities related to payment of disability benefits, we denote with p the probability of an employee suffering an accident during a one-year interval; we assume that this probability is constant over the whole m-year period.

Under the hypotheses we have assumed for Case 3b (Disability benefits; one-year period), the probability distribution of K_t, for $t = 1, 2, \ldots, m$, is binomial (see (1.4.16) and (1.4.17)); thus

$$\pi_k = \mathbb{P}[K_t = k] = \binom{n}{k} p^k (1-p)^{n-k}; \quad k = 0, 1, \ldots, n \tag{1.4.38}$$

and hence

$$\mathbb{E}[K_t] = n\,p \tag{1.4.39}$$

$$\mathbb{V}\mathrm{ar}[K_t] = n\,p\,(1-p) \tag{1.4.40}$$

As $X_t = C\,K_t$, we obviously have

$$\mathbb{E}[X_t] = C\,n\,p \tag{1.4.41}$$

We assume that the employer decides to fund her liability, related to the group of employees, by allocating at the beginning of the m-year period (i.e. at time 0) an amount of assets meeting the expected value of the disability benefits. Further, we assume that the assets provide the employer with an interest, at the annual interest rate i. We denote with A_t the share of assets, allocated at time 0, to fund the benefits payable at time t. Then, the following relation must hold:

$$A_t\,(1+i)^t = \mathbb{E}[X_t] \tag{1.4.42}$$

that is

$$A_t = C\,n\,p\,(1+i)^{-t} \tag{1.4.43}$$

Hence, the total amount of assets to allocate at time 0 is given by

$$A = \sum_{t=1}^{m} A_t = C\,n\,p \sum_{t=1}^{m} (1+i)^{-t} \tag{1.4.44}$$

The quantity A can also be read in an alternative manner. We define the random amount, Y, as follows:

$$Y = \sum_{t=1}^{m} X_t\,(1+i)^{-t} \tag{1.4.45}$$

Thus, Y is the *random present value* of the benefits. Then, we calculate the *expected present value* (shortly, the *actuarial value*) of the benefits:

$$\mathbb{E}[Y] = \sum_{t=1}^{m} \mathbb{E}[X_t]\,(1+i)^{-t} = Cnp\sum_{t=1}^{m}(1+i)^{-t} \tag{1.4.46}$$

Finally, from (1.4.44) we find that $A = \mathbb{E}[Y]$, that is, the amount of assets to allocate at time 0 is equal to the actuarial value of the benefits.

Remark The allocation of assets to fund the payment of disability benefits to the employees constitutes an example of risk retention, or self-insurance; see Sect. 1.3.4.

Example 1.4.4. Refer to the disability benefit arrangement described in Example 1.4.2. Assume $n = 100$, and a time horizon of $m = 5$ years. Table 1.4.2 shows the allocations A_t needed to fund (at the beginning of the period) the employer's liability, if the interest rate is $i = 0.02$, or $i = 0.03$ respectively.

Table 1.4.2 Disability benefits (m-year period)

year t	$\mathbb{E}[X_t]$	Allocation A_t	
		$i = 0.02$	$i = 0.03$
1	500	490.20	480.77
2	500	480.58	462.28
3	500	471.16	444.50
4	500	461.92	427.40
5	500	452.87	410.96
Total A		2356.73	2225.91

□

1.4.6 Comparing random yields

We now refer to **Case 1b** (Random yields). First, we note that the investment with yield X_4 can be disregarded because *dominated* by the investment with yield X_1 (see Table 1.2.3): indeed, in all the states of the world the outcome of X_1 is not worse than the corresponding outcome of X_4, and in at least one state (state S_3, in the example), the outcome of X_1 is better than the corresponding outcome of X_4. Thus, the choice can be restricted to the first three investments.

We assume, for simplicity, that the three states of the world have the same probability, i.e.

$$\mathbb{P}[S_1] = \mathbb{P}[S_2] = \mathbb{P}[S_3] = \frac{1}{3}$$

Table 1.4.3 Expected value and variance of random yields

	$\mathbb{E}[X_i]$	$\mathbb{V}\mathrm{ar}[X_i]$
X_1	6.0	0.667
X_2	6.0	24.000
X_3	5.8	0.180

Expected values and variance of the random yields X_1, X_2 and X_3 are given in Table 1.4.3, from which we can note what follows:

a. investments 1 and 2 can be considered to be *equivalent* in terms of expected value;
b. investment 1 is less risky than investment 2, as the former has a lower variance;
c. from (a) and (b), investment 2 turns out to be *dominated in mean-variance* by investment 1 (albeit it is not dominated in terms of the items of the pay-off matrix);
d. investment 3 is less profitable than investment 1 in terms of expected value, but, at the same time, it is less risky; hence, a *risk averse* investor could prefer investment 3 to investment 1, and then "pay" the lower riskiness by accepting a lower expected yield.

The analysis of Table 1.4.3, according to a mean-variance approach, leads to the following conclusion: while investment 2 can be excluded from further analysis, both investments 1 and 3 are candidates, the preference being driven by the risk aversion of the investor.

In more general terms, the set of "solutions", each of which consists in the choice of an investment, can be split into two subsets, namely the set of dominated solutions and the set of *mean-variance efficient* solutions (see Fig. 1.4.1). According to the mean-variance approach, the choice should be restricted to efficient solutions. Of course, the choice of a specific solution depends on the investor's risk aversion.

Expected value and variance can be summarized by choosing an appropriate function, which associates a real number to each investment choice. The value of the function should increase as the expected value increases, and decrease as the variance increases. For example, the following function can be adopted:

$$\mathrm{Q}[X_i] = \mathbb{E}[X_i] - \alpha\, \mathbb{V}\mathrm{ar}[X_i] \tag{1.4.47}$$

The (positive) parameter α quantifies the risk aversion. If $\alpha = 0$, there is no risk aversion, and the choice relies on the expected values only. The higher is α, the more importance is attributed to the riskiness expressed by the variance.

Another function which balances expectation and riskiness is the following one:

$$\mathrm{Q}[X_i] = \mathbb{E}[X_i] - \beta\, \sigma[X_i] \tag{1.4.48}$$

where β expresses the risk aversion.

Table 1.4.4 refers to the example discussed above (see Table 1.4.3). We see that, for a (relatively) high value of the parameter, i.e. for $\alpha = 1$, namely for a (relatively)

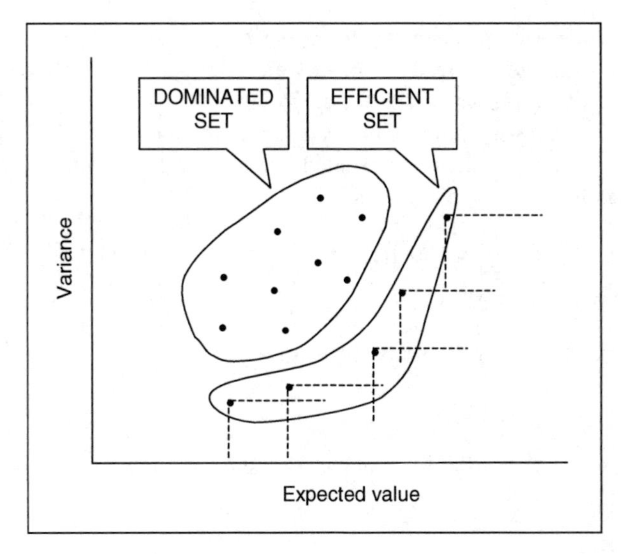

Fig. 1.4.1 Dominated versus efficient solutions according to the mean-variance criterion

high risk aversion, investment 3 is preferred to investment 1, despite its lower expected yield.

Table 1.4.4 Summarizing expected value and variance of the efficient solutions

	$Q[X_i] = \mathbb{E}[X_i] - \alpha\,\mathbb{V}\text{ar}[X_i]$			
α	0	0.01	0.1	1
X_1	6.000	5.993	5.933	5.333
X_3	5.800	5.798	5.782	5.620

1.4.7 Risk-adjusted valuations

We now attack the following problem: how can we evaluate the future cash-flows of a random financial transaction,

- allowing for risk,
- adopting a valuation criterion only based on
  - discount factors,
  - expected values.

This means that the presence of risks will be accounted for, in the evaluation model, via an appropriate choice of the discount factor and/or the ingredients in the expected value calculation. Hence, variance and other specific risk measures do not enter the model (unlike the examples in Sect. 1.4.6, in which the variance or the standard deviation are explicitly accounted for; see (1.4.47) and (1.4.48)).

For brevity, we only deal with a very simple financial transaction and, in particular, we refer to **Case 1a** (Zero-coupon bonds) presented in Sect. 1.2.2. We denote the probabilities of the states of the world as follows:

$$p = \mathbb{P}[S_1] \tag{1.4.49}$$

$$1 - p = \mathbb{P}[S_2] \tag{1.4.50}$$

These probabilities are usually called *natural* (or *realistic*, or *physical*) *probabilities*. In particular, we assume

$$p = 1 - p = \tfrac{1}{2} \tag{1.4.51}$$

For the pay-off of the risk-free bond (described in the second row of Table 1.2.2), we obviously have

$$\mathbb{E}_p[X_{\mathrm{B}}] = 100 \tag{1.4.52}$$

For the pay-off of the risky bond (see the first row of Table 1.2.2), we find the same expected value, i.e.

$$\mathbb{E}_p[X_{\mathrm{A}}] = 50\,p + 150\,(1 - p) = 100 \tag{1.4.53}$$

(note that the suffix p recalls that the expected value is calculated using the natural probabilities). Clearly, the expected values do not account for the different risk degrees. Conversely, the prices of the two bonds should reflect the absence / presence of risk.

We denote with P_{A} the price (at time 0) of the risky bond, and P_{B} the price of the risk-free bond. Further, we denote with r_{f} the risk-free rate, and set $r_{\mathrm{f}} = 0.03$. As regards the price of the risk-free bond, we assume that it is given by the present value of its pay-off. Thus, we have

$$P_{\mathrm{B}} = 100\,(1 + r_{\mathrm{f}})^{-1} = 97.09 \tag{1.4.54}$$

For the price of the risky bond, P_{A}, it is reasonable to assume

$$P_{\mathrm{A}} < P_{\mathrm{B}} \tag{1.4.55}$$

because of *risk aversion*. In particular, let $P_{\mathrm{A}} = 95$ be the price observed on the financial market. How can this price be formally "explained"? The three following approaches can be adopted.

1. Calculate (by using the natural probabilities) the expected value of the pay-off, and discount this value by using the *risk-adjusted discount rate* (briefly, the *risk discount rate*) ρ, $\rho > r_{\mathrm{f}}$. Thus, we find

$$P_{\mathrm{A}} = \mathbb{E}_p[X_{\mathrm{A}}]\,(1 + \rho)^{-1} \tag{1.4.56}$$

The quantity $\rho - r_\mathrm{f}$ is known as the *risk premium*. In our numerical example, we find that $P_\mathrm{A} = 95$ implies a risk rate $\rho = 0.05263$.

2. Calculate the expected value of the pay-off by using *risk-adjusted probabilities* $p', 1 - p'$ (instead of the natural probabilities):

$$\mathbb{E}_{p'}[X_\mathrm{A}] = 50\,p' + 150\,(1 - p') \tag{1.4.57}$$

The terms $50\,p', 150\,(1 - p')$ are called *risk-adjusted expected cash-flows*. As the presence of risk has been allowed for via adjusted probabilities, we adopt the risk-free rate for discounting. Hence

$$P_\mathrm{A} = \mathbb{E}_{p'}[X_\mathrm{A}]\,(1 + r_\mathrm{f})^{-1} \tag{1.4.58}$$

In our numerical example, the price $P_\mathrm{A} = 95$ implies $\mathbb{E}_{p'}[X_\mathrm{A}] = 97.85$; from Eq. (1.4.57), we then find the risk-adjusted probabilities $p' = 0.5215$, $1 - p' = 0.4785$. Note that adjusting for risk leads to a higher "weight" attributed to the worst result.

3. Allow for riskiness by "transforming" the amounts of the cash-flows of the risky bond. Denote with $u(X_\mathrm{A})$ the transformed random cash-flow, whose possible outcomes are $u(50)$ and $u(150)$. In particular, as transform u we can take a *utility function*, expressing our risk aversion. The expected value of $u(X_\mathrm{A})$ is then called the *expected utility* of X_A, and is denoted with $\mathbb{U}[X_\mathrm{A}]$. Thus, we have

$$\mathbb{U}[X_\mathrm{A}] = \mathbb{E}_p[u(X_\mathrm{A})] = u(50)\,p + u(150)\,(1 - p) \tag{1.4.59}$$

We define the *certainty equivalent* of the random result X_A as the amount A which, if received certainly, is regarded as equivalent to the random result. In formal terms:

$$u(A) = u(50)\,p + u(150)\,(1 - p) \tag{1.4.60}$$

Note that, because of the risk aversion (which should be expressed by the function u), we will find:

$$A < \mathbb{E}[X_\mathrm{A}] = 50\,p + 150\,(1 - p) \tag{1.4.61}$$

(see Fig. 1.4.2a; the graphs are just indicative). Finally, the price P_A is given by the present value of A, i.e.

$$P_\mathrm{A} = A\,(1 + r_\mathrm{f})^{-1} \tag{1.4.62}$$

For example, the quadratic function

$$u(x) = -0.000005507\,x^2 + 0.007493\,x \tag{1.4.63}$$

leads (with $p = \frac{1}{2}$) to $A = 97.85$, and then $P_\mathrm{A} = 95$. We note that, if we choose the risky bond, we will get at maturity either 50 or 150, instead of 100 provided by the risk-free bond. In terms of the utility function, we find:

$$u(50) = 0.360867$$
$$u(100) = 0.6942$$
$$u(150) = 1$$

Hence, because of risk aversion we attribute to the negative difference $50 - 100$ an absolute "value", $u(100) - u(50)$, greater than the value we attribute to the positive difference $150 - 100$ (again, see Fig. 1.4.2a). Under specific hypotheses, the risk aversion reflects on a concave curve (the graph of the utility function), which associates values to all monetary amounts (see Fig. 1.4.2b).

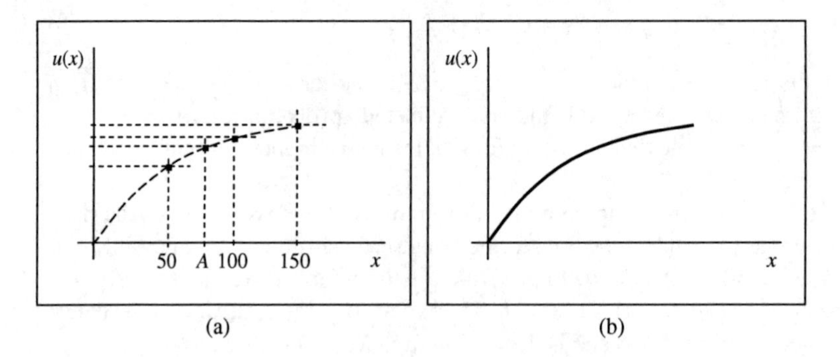

Fig. 1.4.2 Risk aversion and utility function

Figure 1.4.3 summarizes the approaches we have now described. Note that the term "values" denotes either the amounts of cash-flows, or some transform (for example, the utility) of the amounts.

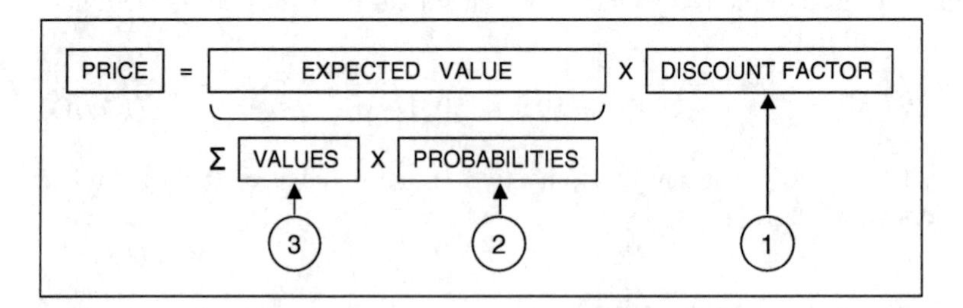

Fig. 1.4.3 Risk-adjusted pricing. Interpretations

Another approach to the evaluation of future cash-flows, allowing for risk, relies on the idea of balancing expectation and riskiness by using an appropriate function, as discussed in Sect. 1.4.6 (see, in particular, formulae (1.4.47) and (1.4.48)). Hence, this approach consists in adding to the expected value, calculated according to the natural probabilities, a (negative) term based on an appropriate measure of risk.

However, it should be stressed that such an approach (explicitly involving a risk measure) cannot be placed into the framework described at the beginning of this Section. Referring to the example above, the approach is as follows.

4. Calculate (by using the natural probabilities):

   - the expected value of the pay-off,
   - the value of a function chosen to express the randomness of the pay-off.

   Then, for the price of the risky bond we set:

$$P_{\mathrm{A}} = (\mathbb{E}_p[X_{\mathrm{A}}] - \gamma \Psi_p[X_{\mathrm{A}}])\,(1 + r_{\mathrm{f}})^{-1} \tag{1.4.64}$$

   where $\Psi_p[X_{\mathrm{A}}]$ quantifies the risk of the transaction. Examples (as seen in Sect. 1.4.6) are given by the variance $\mathbb{V}\mathrm{ar}[X_{\mathrm{A}}]$ and the standard deviation $\sigma[X_{\mathrm{A}}]$ (both calculated with the natural probabilities). The parameter γ expresses the risk aversion. Note that the risk-free rate has been used for discounting, as the adjustment for risk is already expressed by the term $-\gamma \Psi_p[X_{\mathrm{A}}]$.

The four approaches have been adopted in various application fields. Moreover, combining two or more approaches is a rather common practice in financial and actuarial calculations. Several examples will be provided in the following chapters.

1.5 Risk measures

1.5.1 Some preliminary ideas

While expressions such as "quantifying risks" and "risk assessment" have a broad meaning, denoting, for example, a whole phase of the risk management process (see Sects. 1.3.2 and 1.4), the expression *risk measures* has a rather specific meaning. Actually, it denotes a set of typical values which can be used in order to express the variability of a random quantity.

A number of risk measures belong to the field of probability theory and statistics (see Sect. 1.5.2), although in that context the expression "risk measures" is not commonly used. Other measures have been proposed in more recent times, and specifically oriented to risk management problems, and to capital allocation strategies (see Sect. 1.5.4).

In what follows, we refer to a random amount Z which represents some result originating from a transaction. In particular, Z can refer to a speculative risk (see Sect. 1.2.2); thus

$$\begin{aligned} Z < 0 &\quad \Leftrightarrow \quad \text{loss} \\ Z > 0 &\quad \Leftrightarrow \quad \text{profit} \end{aligned}$$

For the random amount Z, we assume a probability distribution (based on statistical experience, or hypotheses about the underlying causes of risk, and so on).

Further, we assume that the probability distribution can be described in terms of the probability density function (briefly, the pdf). Hence, denoting with $f(z)$ the pdf, we have in general terms

$$\mathbb{P}[a \leq Z \leq b] = \int_a^b f(z)\,\mathrm{d}z \tag{1.5.1}$$

The expected value, μ, of Z is given by the following expression:

$$\mu = \mathbb{E}[Z] = \int_{-\infty}^{+\infty} z f(z)\,\mathrm{d}z \tag{1.5.2}$$

Clearly, if the possible outcomes of Z constitute a limited interval, say $[z_{\min}, z_{\max}]$, the integration interval should be consequently modified.

1.5.2 Traditional risk measures

The *variance* (that we have already used in a discrete context; see, for example, formulae (1.4.7) and (1.4.32)) is defined as follows:

$$\mathbb{V}\mathrm{ar}[Z] = \mathbb{E}[(Z-\mu)^2] = \int_{-\infty}^{+\infty} (z-\mu)^2 f(z)\,\mathrm{d}z \tag{1.5.3}$$

As is well known, the square root of the variance, usually denoted with $\sigma[Z]$, is called the *standard deviation*:

$$\sigma[Z] = \sqrt{\mathbb{V}\mathrm{ar}[Z]} \tag{1.5.4}$$

Note that

- the variance and the standard deviation are "symmetric" risk measures, since both positive and negative deviations from the expected value are captured;
- the standard deviation is expressed in the same "unit" of the random amount Z; for example, if Z is expressed in Euro, the standard deviation is expressed in Euro too; conversely, the variance is expressed in the squared unit, which can be meaningless, as is the case for Euro squared.

The *variance-to-mean ratio* is defined as follows:

$$\mathrm{VMR}[Z] = \frac{\mathbb{V}\mathrm{ar}[Z]}{\mathbb{E}[Z]} \tag{1.5.5}$$

whereas the *coefficient of variation* is defined as follows:

$$\mathbb{CV}[Z] = \frac{\sigma[Z]}{\mathbb{E}[Z]} \tag{1.5.6}$$

(see also Sect. 1.4.3). The coefficient of variation, especially in the field of risk management and insurance, is also known as the *risk index*. Relevant applications will be described in Sect. 1.6.1 and 2.3.3. Note that the quantities VMR[Z] and $\mathbb{CV}[Z]$ are "relative" measures of risk; in particular, $\mathbb{CV}[Z]$ is unit-free, as both the numerator and the denominator are expressed in the same unit.

For a random amount Z with a limited interval of possible outcomes, the *range* is defined as follows:

$$\text{Range}[Z] = z_{\max} - z_{\min} \tag{1.5.7}$$

1.5.3 Downside risk measures

In order to capture only the "bad" part of a random result, risk measures other than the symmetric ones are needed. Downside risk measures can fulfill this requirement. Most of them have been proposed in the framework of portfolio management as tools for the analysis of the return. Further risk measures have been more recently proposed in the context of risk management.

First, we can focus on the possible outcomes of Z which fall below the expected value μ. To this purpose, we can use as risk measure the *semi-variance*, which captures only the negative deviations from the expected value, and is defined as follows:

$$\text{semi}\mathbb{V}\text{ar}[Z] = \mathbb{E}[(\min\{Z-\mu,0\})^2] = \int_{-\infty}^{\mu} (z-\mu)^2 f(z)\,\mathrm{d}z \tag{1.5.8}$$

Further, the *semi-standard deviation* is given by

$$\text{semi}\sigma[Z] = \sqrt{\text{semi}\mathbb{V}\text{ar}[Z]} \tag{1.5.9}$$

The idea underlying the definition of the semi-variance and the semi-standard deviation can be generalized, first by assuming as the benchmark a chosen "target" τ, instead of the expected value μ. Thus, only the negative deviations from τ are accounted for. Further, instead of considering just the second power of the deviations, we can assume the generic power k. Then, we define the *lower partial moment of degree k* as follows:

$$\text{LPM}_{\tau}^{k}[Z] = \mathbb{E}[(\min\{Z-\tau,0\})^k] = \int_{-\infty}^{\tau} (z-\tau)^k f(z)\,\mathrm{d}z \tag{1.5.10}$$

A (somewhat arbitrarily) chosen target, τ, also underpins the definition of the *shortfall risk measures*. If Z denotes a monetary result, a negative value can be chosen for the target τ; thus, the event $Z < \tau$ means a loss greater than the chosen target, namely a *tail loss*. Conversely, if Z denotes a return, the target can be positive (and small), so that focus is on the return outcomes which do not reach the stated benchmark.

The *shortfall probability*, is defined as follows

$$\mathbb{P}[Z < \tau] = \int_{-\infty}^{\tau} f(z)\,\mathrm{d}z \tag{1.5.11}$$

Referring to a monetary result, the (negative) expected value of the loss exceeding the (negative) target, conditional on exceeding the target itself, is known as the *expected shortfall*. It is given by the following expression

$$\mathrm{ES}_{\tau}[Z] = \mathbb{E}[Z|Z < \tau] = \frac{\int_{-\infty}^{\tau} z f(z)\,\mathrm{d}z}{\mathbb{P}[Z < \tau]} \tag{1.5.12}$$

The shortfall risk measures we have now defined are illustrated in Fig. 1.5.1.

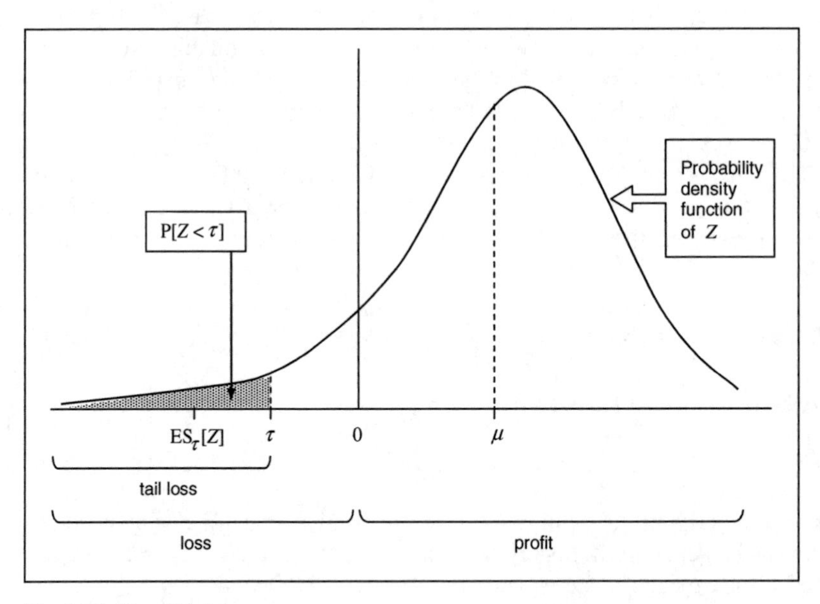

Fig. 1.5.1 Shortfall risk measures

1.5.4 Risk measures and capital requirements

Downside risk measures have been proposed, which can be interpreted as *capital requirements* aiming to "protect" a financial transaction with random result Z. In other terms, as the financial transaction can result in a loss, the agent (for example, a financial intermediary) needs to allocate a capital which can be used to cover (at least to some extent) the potential loss, so that the loss itself does not compromise other lines of business.

The *Value at Risk* (briefly, the VaR) is the (negative) amount VaR_α such that

$$\mathbb{P}[Z \leq VaR_\alpha] = \alpha \tag{1.5.13}$$

where α is a (low) probability, somewhat arbitrarily chosen, for example $\alpha = 0.01$. The probability $1 - \alpha$ is also known as the *confidence level*. See Fig. 1.5.2.

The following points should be stressed.

- The amount VaR_α is the α-percentile of the probability distribution of Z. Note that the VaR has no meaning if a probability has not been stated.
- The amount VaR_α can be interpreted as the maximum loss if an extreme event (or "tail event") does not occur. Of course, the definition of "extreme" strictly depends on the chosen probability.
- If the capital $-VaR_\alpha$ is allocated, any non-tail loss is completely funded.
- The quantity VaR_α does not provide, by itself, any information about the possible loss if a tail event occurs.

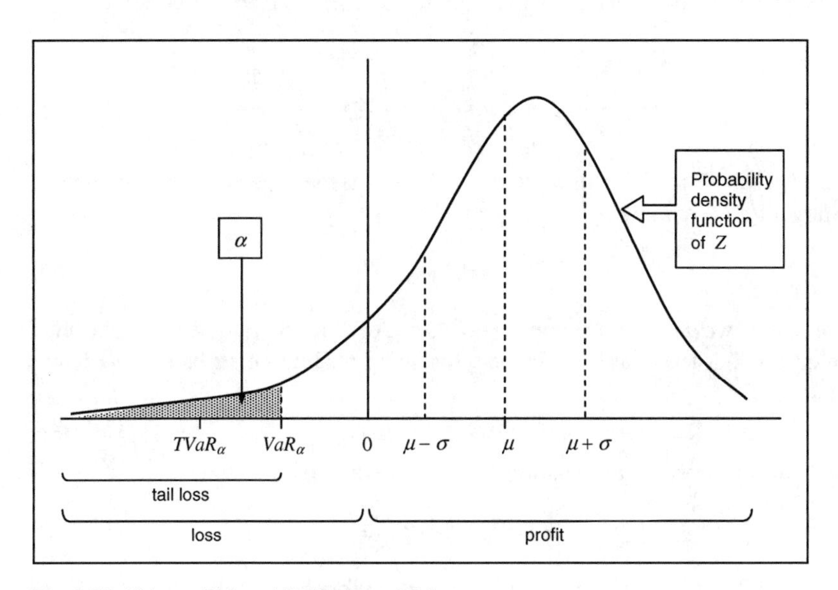

Fig. 1.5.2 Value at Risk and Tail Value at Risk

It can be useful to express VaR_α in terms of the expected value μ and the standard deviation σ of the probability distribution of Z. Let γ_α be the coefficient such that

$$\mathbb{P}[Z \leq \mu - \gamma_\alpha \sigma] = \alpha \tag{1.5.14}$$

then, we have

$$VaR_\alpha = \mu - \gamma_\alpha \sigma \tag{1.5.15}$$

Given the probability α, the coefficient γ_α can be immediately determined if, for example, we assume for Z the normal distribution, namely

$$Z \sim \mathcal{N}(\mu, \sigma) \tag{1.5.16}$$

(see Table 1.5.1).

Table 1.5.1 Coefficient γ_α (Normal distribution)

α	γ_α
0.100	1.282
0.050	1.645
0.025	1.960
0.001	3.090

The *Tail Value at Risk* (shortly, the TailVaR, or TVaR, also known as the *Conditional Tail Expectation*) is the (negative) amount $TVaR_\alpha$ defined as follows:

$$TVaR_\alpha = \mathbb{E}[Z|Z < VaR_\alpha] = \frac{\int_{-\infty}^{VaR_\alpha} z f(z)\,\mathrm{d}z}{\mathbb{P}[Z < VaR_\alpha]} \tag{1.5.17}$$

Comparing (1.5.17) to (1.5.12), we find that $TVaR_\alpha$ is the expected shortfall related to the target VaR_α, namely

$$TVaR_\alpha = \mathrm{ES}_{VaR_\alpha}[Z] \tag{1.5.18}$$

Note that, if we allocate the amount $-TVaR_\alpha$, even a loss caused by a tail event can be covered, at least partially. Indeed, for any probability distribution, we have of course

$$-TVaR_\alpha > -VaR_\alpha \tag{1.5.19}$$

For a given probability distribution of Z, we can find α' such that

$$TVaR_\alpha = VaR_{\alpha'} \tag{1.5.20}$$

Relation (1.5.20) can be useful when a procedure for the calculation of the VaR is available. Of course, $\alpha' < \alpha$. However, the exact link between the probabilities involved by VaR and TVaR depends on the probability distribution of Z (see Fig. 1.5.3). Shifting to another probability distribution, we find

$$TVaR_\alpha \approx VaR_{\alpha'} \tag{1.5.21}$$

with α' fulfilling condition (1.5.20).

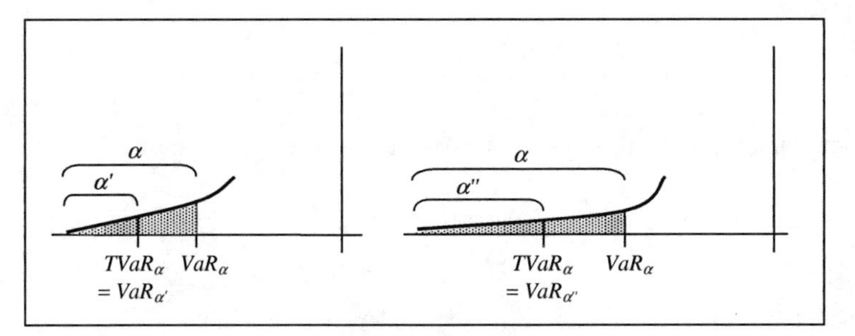

Fig. 1.5.3 Relation between VaR and TailVaR

1.6 Transferring risks

1.6.1 Building up a pool

As the pooling of risks is the rationale underlying the insurance activity, we first focus on the effects of managing jointly a number of risks of the same type (e.g. originating from fire, or third-party liability, and so on).

We refer to n individuals bearing the same type of risk. For the generic individual j, the risk implies a loss $x^{(j)}$ (a damage, a liability, and so on) if the event $\mathscr{E}^{(j)}$ occurs. Thus, we are dealing with the "basic" type of risk we have described as **Case 2** (Possible loss with fixed amount) in Sect. 1.2.3. In formal terms, the random loss, $X^{(j)}$, for $j = 1, 2, \ldots, n$, is defined as follows:

$$X^{(j)} = \begin{cases} x^{(j)} & \text{if } \mathscr{E}^{(j)} \\ 0 & \text{if } \bar{\mathscr{E}}^{(j)} \end{cases} \tag{1.6.1}$$

We assume that the events $\mathscr{E}^{(j)}$ and hence the individual losses $X^{(j)}$ are independent, and denote with $p^{(j)}$, $j = 1, 2, \ldots, n$, the probability of suffering the loss, thus

$$p^{(j)} = \mathbb{P}[\mathscr{E}^{(j)}] = \mathbb{P}[X^{(j)} = x^{(j)}] \tag{1.6.2}$$

Each individual is risk averse, and then is looking for some kind of financial protection against the potential loss. To this purpose, all the n individuals decide to set up a "pool", which will raise monies through individual *contributions*, and then will pay benefits to the individuals (members of the pool) who will have suffered a loss.

We assume that each individual benefit is equal to the loss suffered, so that the member j will receive the amount $x^{(j)}$ if she suffers the loss, 0 otherwise. In Sect. 1.6.2 we will deal with benefits in a more general context.

We now focus on some features of the risk aggregation, which originate from the construction of the pool. The total random amount $X^{[\mathrm{P}]}$, which will be paid by the pool to members suffering a loss, is defined as follows

$$X^{[P]} = \sum_{j=1}^{n} X^{(j)} \tag{1.6.3}$$

Thus, $X^{[P]}$ is the random *outgo* of the pool. Its expected value is given by

$$\mathbb{E}[X^{[P]}] = \sum_{j=1}^{n} \mathbb{E}[X^{(j)}] = \sum_{j=1}^{n} x^{(j)}\, p^{(j)} \tag{1.6.4}$$

and its variance, thanks to the hypothesis of independence, is given by

$$\mathbb{V}\mathrm{ar}[X^{[P]}] = \sum_{j=1}^{n} \mathbb{V}\mathrm{ar}[X^{(j)}] = \sum_{j=1}^{n} (x^{(j)})^2\, p^{(j)}\, (1 - p^{(j)}) \tag{1.6.5}$$

The minimum possible outcome of the random amount $X^{[P]}$ is 0, while the maximum one is $\sum_{j=1}^{n} x^{(j)}$, and hence $\mathrm{Range}[X^{[P]}] = \sum_{j=1}^{n} x^{(j)}$. The possible outcomes and the probability distribution of $X^{[P]}$ clearly depend on the values $x^{(j)}$.

To simplify the problem, we now assume that the pool is "homogeneous" in terms of both the amounts and the probabilities of loss, namely, for $j = 1, 2, \ldots, n$

$$x^{(j)} = x \tag{1.6.6}$$

$$p^{(j)} = p \tag{1.6.7}$$

It follows that, for $j = 1, 2, \ldots, n$

$$\mathbb{E}[X^{(j)}] = \mathbb{E}[X^{(1)}] = x\, p \tag{1.6.8}$$

$$\mathbb{V}\mathrm{ar}[X^{(j)}] = \mathbb{V}\mathrm{ar}[X^{(1)}] = x^2\, p\, (1 - p) \tag{1.6.9}$$

Hence, the possible outcomes of $X^{[P]}$ are

$$0, x, 2x, \ldots, nx$$

so that $\mathrm{Range}[X^{[P]}] = nx$. The expected value and variance (see (1.6.4) and (1.6.5)) then reduce to:

$$\mathbb{E}[X^{[P]}] = n\, x\, p \tag{1.6.10}$$

$$\mathbb{V}\mathrm{ar}[X^{[P]}] = n x^2\, p\, (1 - p) \tag{1.6.11}$$

The coefficient of variation (or risk index, see (1.5.6)) is given by:

$$\mathbb{CV}[X^{[P]}] = \frac{\sqrt{\mathbb{V}\mathrm{ar}[X^{[P]}]}}{\mathbb{E}[X^{[P]}]} = \frac{x\sqrt{n\, p\, (1 - p)}}{x n\, p} = \sqrt{\frac{1 - p}{n\, p}} \tag{1.6.12}$$

If we denote with K the random number of events in the pool, we have

$$X^{[P]} = K x \tag{1.6.13}$$

It follows that, in formula (1.6.10), the expected value can be read in two ways, namely:

1. $(np)x = \mathbb{E}[K]x$, that is, the expected number of losses times the amount of the individual loss;
2. $n(xp) = n\mathbb{E}[X^{(1)}]$, that is, the pool size times the individual expected loss.

Example 1.6.1. We refer to a pool of n independent risks, fulfilling assumptions (1.6.6) and (1.6.7), with $x^{(j)} = x = 1000$ and $p^{(j)} = p = 0.005$ for $j = 1, \dots, n$; then, we have:

$$\mathbb{E}[X^{(j)}] = xp = 5$$
$$\mathbb{V}\mathrm{ar}[X^{(j)}] = x^2 p(1-p) = 4975$$

Table 1.6.1 shows various results concerning the random outgo $X^{[P]}$, if $n = 100$, $n = 1000$, and $n = 10000$ respectively.

Table 1.6.1 Some typical values of the random outgo of a pool of risks

	$n = 100$	$n = 1000$	$n = 10000$
$\mathrm{Range}[X^{[P]}]$	100 000	1 000 000	10 000 000
$\mathbb{E}[X^{[P]}]$	500	5 000	50 000
$\mathbb{V}\mathrm{ar}[X^{[P]}]$	497 500	4 975 000	49 750 000
$\sqrt{\mathbb{V}\mathrm{ar}[X^{[P]}]}$	705.34	2 230.47	7 053.37
$\mathbb{CV}[X^{[P]}]$	1.411	0.446	0.141

□

As regards the effects of building-up a pool of risks, the following feature should be stressed. The variance of $X^{[P]}$ increase linearly as n increases (see Eq. (1.6.11)), whereas the standard deviation increases proportionally to $\sqrt{n}$. Hence the "absolute" riskiness increases. However, the "relative" riskiness in terms of the coefficient of variation (see (1.6.12)) decreases as the pool size increases. Similar comments have been proposed in Example 1.4.2.

In particular, we note that, for any given probability p, we have

$$\lim_{n\to\infty} \mathbb{CV}[X^{[P]}] = 0 \tag{1.6.14}$$

Example 1.6.2. We consider a pool of risks, fulfilling assumptions (1.6.6) and (1.6.7), with $p = 0.005$. Table 1.6.2 illustrates the coefficient of variation for various pool sizes.

□

The result expressed by (1.6.12) is of outstanding importance in risk theory and constitutes a kernel feature of the risk transfer process (and the insurance process in

Table 1.6.2 The coefficient of variation

n	$\mathbb{CV}[X^{[P]}]$
10	4.461
100	1.411
1000	0.446
10000	0.141
100000	0.045
...	...
∞	0.000

particular). Moreover, the result can be extended to more general pools: for example, pools which do not fulfill assumption (1.6.6), or (1.6.7). We will come back on these and related issues in Sect. 2.3.3.

1.6.2 Financing the pool

We now describe some alternative models, which aim to define possible arrangements for financing the outgo of the pool.

We still address **Case 2** (Possible loss with fixed amount). The outgo, namely the random total payment, is given by $X^{[P]} = \sum_{j=1}^{n} X^{(j)}$ (see (1.6.3)). Further, we assume the homogeneity of the pool in terms of both the amount (see (1.6.6)) and the probability (see (1.6.7)) of the individual loss.

Method 1 We assume that the total payment is to be shared equally among the members of the pool, so that the pool *income* is, by definition, equal to the outgo.

According to information available at the beginning of the period, the amount contributed by each member is of course random, and is given by $\frac{X^{[P]}}{n}$. We note that, thanks to the homogeneity hypotheses, it can also be expressed as follows (see relation (1.6.13)):

$$\frac{X^{[P]}}{n} = \frac{K x}{n} \tag{1.6.15}$$

Nonetheless, we keep the expression of the total payment as the sum of the individual losses, which is more appropriate to following developments.

The expected value of the individual contribution is

$$\mathbb{E}\left[\frac{X^{[P]}}{n}\right] = \frac{1}{n}\sum_{j=1}^{n} \mathbb{E}[X^{(j)}] = x p \tag{1.6.16}$$

and thus it turns out to be independent of the number of members of the pool. In particular, the expected value of the amount contributed is equal to the expected value of the individual loss. Thus, in terms of expected value the members do not gain any advantage by transferring the risks to the pool.

Conversely, an advantage is gained in terms of the individual riskiness, which does depend on the size of the pool. Indeed, if we assume the variance as the risk measure, we have that, for the generic j-th individual, the "original" riskiness (i.e. the riskiness before transfer to the pool) is given by

$$\mathbb{V}\mathrm{ar}[X^{(j)}] = x^2\, p\,(1-p) \tag{1.6.17}$$

whereas, for the individual as a member of the pool, the "final" riskiness, which only originates from the randomness of the contribution, is given by

$$\mathbb{V}\mathrm{ar}\left[\frac{X^{[\mathrm{P}]}}{n}\right] = \frac{1}{n^2}\sum_{j=1}^{n} \mathbb{V}\mathrm{ar}[X^{(j)}] = \frac{1}{n}x^2\, p\,(1-p) \tag{1.6.18}$$

Thus, as the size n of the pool increases, the individual riskiness (in terms of the variance) decreases.

From a theoretical point of view, the *Strong Law of Large Numbers* states that

$$\mathbb{P}\left[\lim_{n\to\infty} \frac{\sum_{j=1}^{n} X^{(j)}}{n} = \mathbb{E}[X^{(1)}]\right] = 1 \tag{1.6.19}$$

where $\mathbb{E}[X^{(1)}]$ denotes the expected value, common to all the random amounts $X^{(j)}$. According to the notation used above, we have

$$\mathbb{P}\left[\lim_{n\to\infty} \frac{X^{[\mathrm{P}]}}{n} = x\,p\right] = 1 \tag{1.6.20}$$

which means that, in the case of an "infinitely" large pool, each member's contribution is equal to her expected loss with a probability equal to one.

In conclusion, the individual contribution is random, with a riskiness decreasing as the pool size increases, whereas the coverage of the total payment is certain (of course, provided that, at the end of the period, all the members pay the contributions).

□

Method 2 We now assume that the total payment is funded in advance (that is, at the beginning of the period, and hence disregarding its actual outcome), by individual contributions to be determined according to some *calculation principle*. In general terms, we have to determine an amount which will constitute the income "facing" the outgo $X^{[\mathrm{P}]}$. As the outgo is random, we have to summarize it by using some typical values.

In particular, we assume what follows:

- the total amount of contributions, that we denote with $P^{[\mathrm{P}]}$, has to meet exactly the expected value of the total payment;

- the effect of accumulation over the year (i.e. the interest) is negligible, and hence disregarded.

Then:

$$P^{[P]} = \mathbb{E}[X^{[P]}] = n x p \tag{1.6.21}$$

So, the individual contribution is certain, and is given by

$$\frac{P^{[P]}}{n} = x p \tag{1.6.22}$$

Note that the individual contribution turns out to be the expected value of the individual loss. Thus, the individual riskiness is completely removed, as far as the amount of the contribution is concerned.

However, the outcome of the total payment $X^{[P]}$ may be greater than its expected value, and hence greater than the total amount of contributions. Therefore, it is important to focus on the event

$$X^{[P]} > P^{[P]} \tag{1.6.23}$$

which constitutes a critical point in the management of a pool of risks.

A more general setting can help us in analyzing critical aspects of the pool management. We denote with $\Pi^{[P]}$ the total amount of contributions which, however, is now assumed to be not necessarily equal to $\mathbb{E}[X^{[P]}]$, so that the individual contribution is not necessarily equal to $x p$ (but all the individual contributions are still in the same amount, namely $\frac{\Pi^{[P]}}{n}$). We consider the following situations.

- $\Pi^{[P]} < \mathbb{E}[X^{[P]}]$: in this case, the probability of covering the outgo is trivially lower than in the case $\Pi^{[P]} = P^{[P]} = \mathbb{E}[X^{[P]}]$. Anyway, since this case may be of some practical interest, we will shortly address it in terms of the consequent benefit arrangement (see Method 3).
- $\Pi^{[P]} > \mathbb{E}[X^{[P]}]$: in this case, the probability of covering the outgo is obviously higher than in the case $\Pi^{[P]} = P^{[P]} = \mathbb{E}[X^{[P]}]$. The difference $\Pi^{[P]} - P^{[P]}$ constitutes the total *safety loading* included in the amount of the contributions in order to raise the probability of covering the total payment. The assessment of appropriate safety loadings will be discussed in Sect. 2.3.5, referring to a portfolio of insured risks.

Note that, whatever the amount $\Pi^{[P]}$, when $X^{[P]} < \Pi^{[P]}$ the pool gains a profit, whilst if $X^{[P]} > \Pi^{[P]}$ the pool suffers a loss. In the former case, the profit can be (partially) redistributed to the members and (partially) accumulated, in order to increase the probability of meeting the payment in future years. In the latter case, if additional resources are not available as the result of previous accumulations, a practicable solution is given by an appropriate reduction of the payment to the members who suffered a loss. Namely, those members should receive:

$$X' = \min\left\{x, \frac{\Pi^{[P]}}{K}\right\} \tag{1.6.24}$$

Thus, the total amount available from contributions is divided equally among the K members who suffered a loss. Hence, the benefit X' is a random amount.

□

Method 3 This method can be seen as a generalization of the approach adopted in the framework of Method 2, in the case of insufficient amount of contributions (see (1.6.24)). Assume that the total amount of contributions is determined according to some rule which, for example, links the individual contribution to the annual income of each member. Hence, individual contributions are not related, at least to some extent, to the (estimated) total amount of individual losses.

Then, denoting also here with $\Pi^{[P]}$ the total amount of contributions, namely the income by the pool, relation (1.6.24) still applies to determine the individual benefit paid to members who suffered a loss. However, it is important to stress that, if compared to Method 2, Method 3 basically implies a logical "inversion", as benefits are determined, in any case, as a function of the income (and the random number K of individual losses in the pool).

□

Example 1.6.3. Refer to a pool of $n = 500$ independent risks, homogeneous in terms of both the amount of individual loss $x = 1\,000$ and the probability of loss $p = 0.01$. Note that:

$$\mathbb{E}[K] = np = 5$$
$$\mathbb{E}[X^{[P]}] = npx = 5\,000$$

Consider the following cases:

a. the number of individual losses in the pool is $K = 2$, and hence $X^{[P]} = 2\,000$;
b. the number of individual losses in the pool is $K = 6$, and hence $X^{[P]} = 6\,000$.

1. Assume that the pool is financed according to Method 1, hence the total random payment $X^{[P]}$ is to be shared equally among the members of the pool. The expected value and the variance of the amount contributed by each member are respectively given by:

$$\mathbb{E}\left[\frac{X^{[P]}}{500}\right] = xp = 10$$
$$\mathbb{V}\mathrm{ar}\left[\frac{X^{[P]}}{500}\right] = \frac{1}{500}x^2 p(1-p) = 19.80$$

Note that, conversely, the variance of the individual loss, before transfer to the pool, is given by:

$$\mathbb{V}\mathrm{ar}[X^{(j)}] = x^2 p(1-p) = 9\,900$$

In the two cases we have:

a. the outcome of the individual contribution is 4, and thus lower than its expected value;
b. the outcome of the individual contribution is 12, and thus higher than its expected value.

2. Assume that the pool is financed according to Method 2. The individual contribution is then equal to $xp = 10$ (provided that no safety loading is applied). In the two cases, we find that:

   a. the pool gains a profit, equal to $5\,000 - 2\,000 = 3\,000$;
   b. the pool suffers a loss, equal to $5\,000 - 6\,000 = -1\,000$.

3. Assume that the pool is financed according to Method 3, and that the individual contribution is still equal to 10. In the two cases, we have:

   a. the individual benefit is equal to 1 000, and the pool gains a profit, equal to $5\,000 - 2\,000 = 3\,000$;
   b. the individual benefit is equal to $\min\left\{1\,000, \frac{5\,000}{6}\right\} = 833.33$.

□

Figure 1.6.1 summarizes the relationships between contributions, losses and benefits actually paid by the pool to its members. If we compare the three methods in terms of the actual benefits maintainable by the various financing structures, we can in particular note what follows.

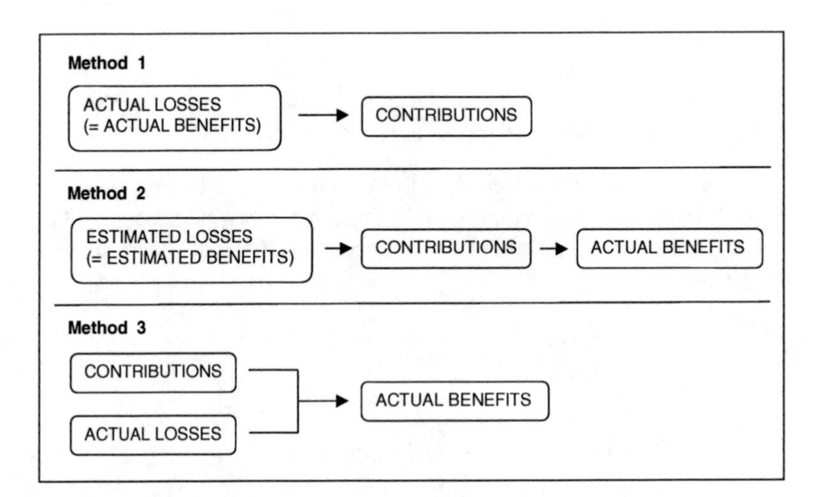

Fig. 1.6.1 Contributions and benefits in pool financing arrangements

- The construction and the management of a pool of risks are based on a mutual agreement among the members of the pool itself. The "technical equilibrium" between contributions and total payment is only guaranteed if Method 1 is adopted, because contributions are determined ex-post, as a result of the observed total amount of losses.

- Conversely, if the contributions are calculated ex-ante according to Method 2, the technical equilibrium is not guaranteed. In particular, if $X^{[P]} > \Pi^{[P]}$, and the approach (1.6.24) is adopted in determining the actual payment, benefits are no longer guaranteed. So, from a more practical point of view, Method 2 can be adopted if an intermediary intervenes in the risk pooling process, and takes the risk of paying the stated benefits even if the contributions collected do not cover the actual losses. Clearly, this role of intermediation should be taken by an insurer.
- As already noted, Method 3 implies a logical inversion in the relationship between contributions and benefits, and, by its nature, it does not provide the members of the pool with any guarantee as regards the amount of benefits. An intermediation in the pooling process is possible also in this case, but the related effect does not imply taking the risk as in Method 2, but simply managing the monetary transaction. The intermediary is, in this case, a "Mutual aid society", or "Mutual benefit society". Thus, a real transfer of the risk does not take place when Method 3 is adopted, and, for this reason, the method is outside the scope of our analysis.
- In all the three methods, the payment of benefits relies on money transfers from members who pay contributions without receiving benefits to members who pay contributions, suffer losses and then receive benefits. Such transfers constitute the so-called *mutuality* effect, which is a particular type of "cross-subsidy" in the risk transfer process. As regards mutuality in the insurance business, some examples will be presented in Sect. 1.7.4. Cross-subsidy in insurance will be discussed also in Sects. 2.2.6 and 2.2.7.

1.6.3 The role of the insurer

Assume the point of view of the individual who transfers the risk to a pool. Under her perspective, the following points are important features of a good transfer arrangement:

- the contribution to be paid to the pool is known in advance, namely at the time of transferring the risk;
- the amount paid as the benefit complies with what stated at the time of transferring the risk, whatever the number and the amounts of losses within the pool may be; in other words, the benefit is *guaranteed.*

As noted at the end of Sect. 1.6.2, such an arrangement can be realized provided that a further subject intervenes in the risk transfer process, and takes the risk of paying the guaranteed benefits, even when the contributions do not meet the total amount of individual losses. This subject is, typically, an insurance company (briefly: an *insurer*), which acts as an *intermediary* in the risk transfer process, providing the members of the pool (i.e. the *insureds*) with the guarantee of paying the benefits according to the conditions stated in the transfer arrangement (i.e. in the *insurance contract*), independently of the actual number and amounts of losses within the pool.

It should be stressed that the term "intermediary" should not be meant in an "administrative" sense only (namely, just consisting of collecting contributions, receiving the applications for benefits, paying the benefits). Besides these jobs, the insurer intermediates in a "technical" sense, by managing the mutuality within the pool, usually called the *portfolio* (of insured risks), providing the guarantee of paying the stated benefits, and hence taking the related risk.

Further, a "financial" intermediation is carried-out, when multi-year contracts are involved, which consists in managing over time the funds originated by collecting the contributions.

It is interesting to single out the nature of the risk taken by the insurer. We still denote with $\Pi^{[P]}$ the total amount of contributions, usually called *premiums*, collected by the insurer (whatever the calculation method adopted), and with $X^{[P]}$ the random amount of benefits paid. Further, we denote with $Z^{[P]}$ the net result arising from the pool, namely

$$\text{net result} = \text{income} - \text{outgo}$$

that is, in formal terms:

$$Z^{[P]} = \Pi^{[P]} - X^{[P]} \tag{1.6.25}$$

Note that, in definition (1.6.25), the time-value of money has been disregarded (i.e. an interest rate equal to 0 has been assumed). This can be reasonably accepted as we are referring to a rather short period of time (say, one year).

Clearly, the insurer gains a profit if $Z^{[P]} > 0$, whereas it suffers a loss if $Z^{[P]} < 0$. Thus, we argue that, whilst the individual risks transferred to the insurer are pure risks (see Sect. 1.2.4), the risk then borne by the insurer is a speculative risk, as it can result either in a profit or in a loss. This is the second feature of the risk transformation via pooling, provided that the pool is managed by an insurer (whereas the first feature is the decreasing relative riskiness, as expressed in particular by (1.6.12) and (1.6.14)).

Figure 1.6.2 sketches a probability distribution of the net result $Z^{[P]}$, assuming, for graphical simplicity, that the possible outcomes of $Z^{[P]}$ constitute an interval of real numbers. Of course, the (exact) distribution of $Z^{[P]}$ depends on the assumptions about the individual losses and the consequent distribution of the total payment $X^{[P]}$. From the figure we can argue what follows:

- as $\mathbb{E}[Z^{[P]}] > 0$, a total amount of contributions $\Pi^{[P]}$ greater than the expected payment $\mathbb{E}[X^{[P]}]$ has been assumed; thus, a safety loading has been included in the premiums (see Method 2 in Sect. 1.6.2);
- despite the safety loading, the probability of a loss seems to be rather high; to lower this probability, some risk management tools (besides a raise in the safety loading) are available to the insurer; this topic will be dealt with in Chap. 2.

Various risk measures can be used for capturing critical aspects of the probability distribution of the net result $Z^{[P]}$; for example, the VaR, the TailVaR, and so on. Some of these aspects will also be addressed in Chap. 2.

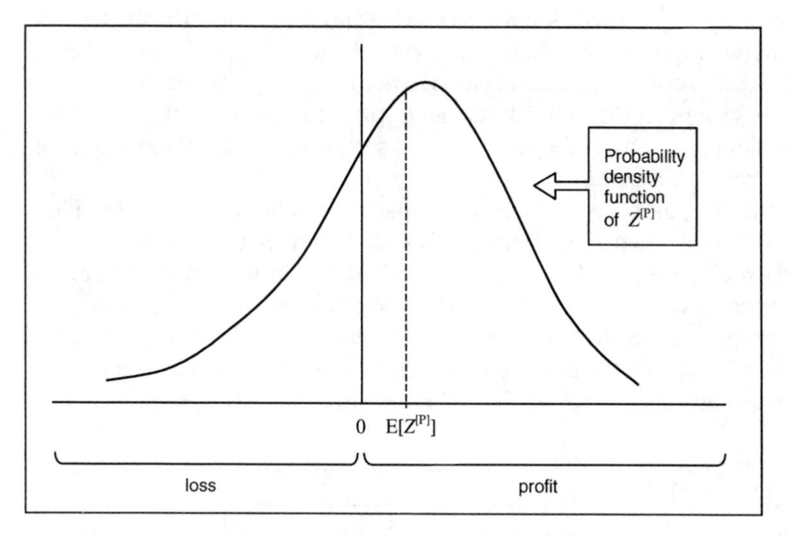

Fig. 1.6.2 Probability distribution of the net result from the pool

1.7 Insurance products

1.7.1 The insurance cover. Policy conditions

The transfer of a risk to an insurer is based, as already mentioned in Sect. 1.6.3, on the insurance contract, whose documental evidence is given by the *insurance policy*. The payment of the premium meets the benefits, which will be paid according to the *policy conditions*.

Various types of benefits can be envisaged. In particular, the benefit can consist in:

a. the reimbursement of expenses paid by the insured, for example because of third-party liability; the amount actually paid as the benefit usually depends on various policy conditions, aiming to restrict the range of amounts which can be paid by the insurer; this topic will be dealt with in Chap. 9;
b. an indemnity covering the loss suffered because of an accident, e.g. a fire; the coverage is usually partial, according to policy conditions;
c. a forfeiture amount, namely an amount stated in the insurance contract.

Benefits of type a and b are usual in non-life insurance, whereas benefits of type c are common in life insurance. Figures 1.7.1 to 1.7.3 summarize the terminology currently used in life and non-life insurance.

We first refer to non-life insurance. When some *peril* is perceived, namely some possible *event* causing a *damage*, the subject who suffers the potential damage can resort to an insurance transfer. If the event occurs, the insured applies for the benefits, namely a *claim* arises. The insurer has to *assess* the damage, and then define the amount of the benefit, that is, *settle* the claim. Finally, the *payment* follows, according to the policy conditions. A deeper analysis of some crucial steps of this sequence will be presented in Chap. 9.

In life insurance, the sequence is much simpler. The event insured can be either the death (within a stated period) or the survival of the insured (at some fixed time). In some cases also the disablement can be allowed for. In the old life insurance policies the benefit consisted in a fixed amount, stated in the policy. Conversely, in more modern policies the (initial) amount is stated in the policy, but the amount itself can vary throughout the policy duration because of linking to some index, for example expressing the inflation, or the yield from investments, and so on.

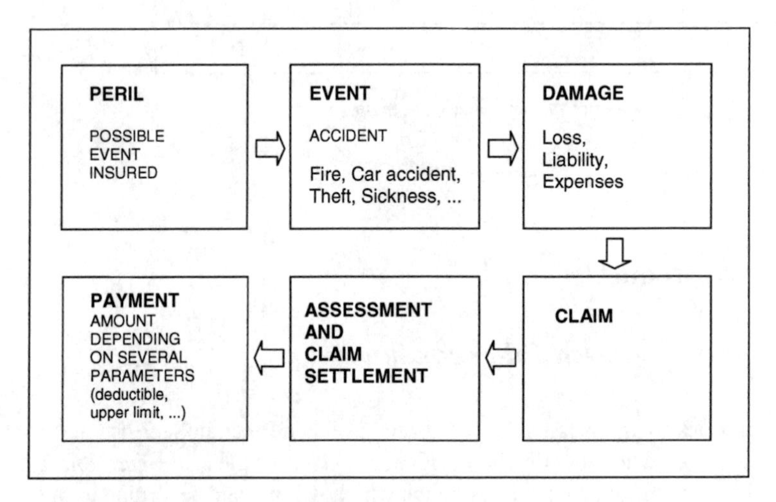

Fig. 1.7.1 Terminology in non-life insurance

1.7.2 Some examples

We refer to examples presented in Sects. 1.2.3 to 1.2.5, and discussed also in Sect. 1.4, in order to introduce various features of the insurance products covering the related risks.

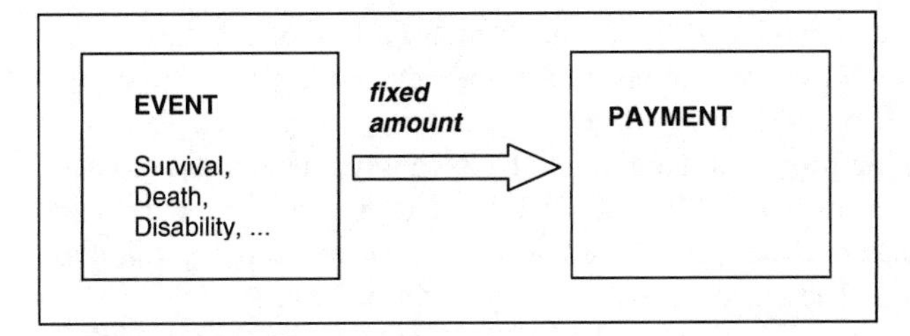

Fig. 1.7.2 Terminology in life insurance (1)

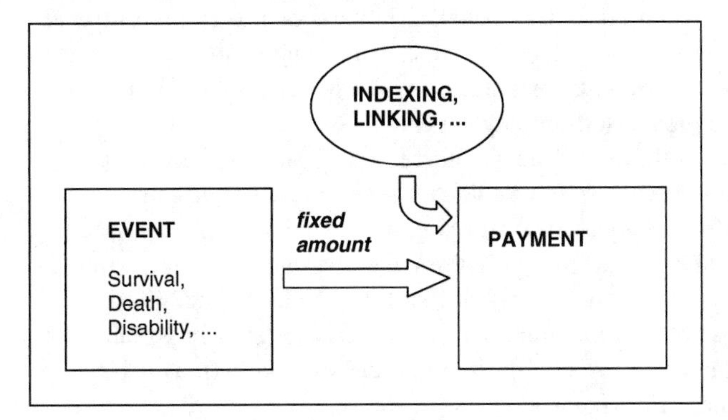

Fig. 1.7.3 Terminology in life insurance (2)

A very simple example of insurance contract is provided by **Case 2** (Possible loss with fixed amount). The insurance cover may concern the following risks (as already mentioned in Sect. 1.2.3):

1. a potential loss (e.g. a damage), which can occur just once in the period covered by the insurance contract;
2. the possible death of the insured;
3. the possible disablement (in particular resulting in a permanent disability condition) of the insured, for example because of an accident.

The insurance product covering risk 1 can be placed in the context of non-life insurance, whereas the product covering risk 2 belongs to life insurance. Risk 3 can be covered by products in the framework of life or non-life insurance, also depending on the particular legislative environment.

Insurance products covering the risk of **Case 4b** (Early death of an individual), namely the death of the insured within a stated period of r years, represent a generalization of products covering the risk 2 within Case 2. The insurance cover, clearly in the field of life insurance, is usually called *term insurance*. It should be noted that, in Case 4b time has greater importance than in Case 2, as the policy term may be rather long, say 10 years. Then, the time-value of money has to be accounted for when calculating the premium for this insurance product.

When addressing risks inherent in **Cases 3a** (Damage / loss of a cargo) to **3e** (Car driver's liability), more interesting insurance products are involved. First, the number of claims in the insured period (say, one year) may be greater than one, unlike in Case 2 (Possible loss with fixed amount). Secondly, the amount of each claim is random. Further, in many cases the benefit paid does not coincide with the amount of the damage or the liability, being lower and determined according to various policy conditions. These aspects will be dealt with in Chap. 9.

The employer can transfer to an insurer the risk inherent in **Case 3b** (Disability benefits; one-year period) and **Case 3c** (Disability benefits; multi-year period), via an insurance contract providing disability benefits. In particular, a *group insurance* can be purchased in order to transfer the set of risks pertaining to all the employees of the firm. Conversely, a disability insurance cover can be purchased also by a single individual, for example a self-employed person.

An insurance product fulfilling the needs inherent in **Case 4a** (The need for resources at retirement) is the *pure endowment insurance*. According to this contract (clearly belonging to life insurance), the insured pays a premium at policy issue (or a sequence of periodic premiums, from policy issue onwards), and will get the insured amount at the stated maturity if alive at that time. Note that nothing is paid by the insurer in the case of death before maturity. Time has a great importance (as we will see in particular when discussing the premium calculation), as the policy duration may be very long (say 10, or 20 years).

The risk of outliving the resources available at the time of retirement, which defines **Case 4c**, can be covered by purchasing a specific insurance product, called the *life annuity*. According to a life annuity contract, the insurer will pay a periodic (say, monthly or yearly) amount while the insured, in this case called the *annuitant*, is alive. So, the risk arising from the randomness of the annuitant's lifetime is borne by the insurer. Also in this case, time has a very important role in the structure of the product, as the potential duration of the life annuity (which typically starts at retirement age, say 65) can be of 25, 30 years or even more.

1.7.3 Pricing insurance products

The premiums paid by the insureds have to meet, according to a stated criterion, the benefits paid by the insurer. We now assume that the premium is paid at policy issue (thus, no splitting into a sequence of periodic premiums is allowed for), and hence just one amount, namely a *single premium*, facing future benefits has to be determined.

Although the insurance business is based on the management of pools of risks, we start by approaching premium calculation on an individual basis, namely referring to a single insured and the related insurance cover. Even though this approach might seem incomplete, as it does not explicitly allow for pooling effects, it is simple, and anyhow of great practical importance. Premium calculation in the framework of pooling features will be focussed on in Chap. 2.

The (individual) premium must rely on some "summary" of the random benefits which will be paid by the insurer. Thus, in some sense, the premium represents a *value* of the benefits. As the benefits can consist, in general, of a sequence of random amounts paid throughout the policy duration, we have to summarize:

1. with respect to time, by determining the random present value of the benefits, referred at the time of policy issue;
2. with respect to randomness, by using some typical values of the probability distribution of the random present value of the benefits, namely the expected value, the standard deviation, and so on.

Step 1 requires the choice of the annual *interest rate* for discounting benefits (or, more generally, the term structure of interest rates). It should be noted, however, that when the policy duration is short (say, one year or less), we can skip this step as time does not have a remarkable impact on the value of benefits.

Step 2 first requires appropriate *statistical bases* in order to construct the probability distribution of the random present value of the benefits, and then the choice of typical values summarizing this distribution.

So far we have only allowed for insurer's costs consisting in the payment of benefits. However, the insurer has to pay also *expenses* which are not directly connected with the payment of benefits, for example general expenses. It is common practice to charge a share of these expenses to each insurance policy, via a convenient premium increase, that is, the *expense loading*.

Finally, a further increase in the premium amount provides the insurer with a *profit* margin.

The items listed above (i.e. interest rate, statistical basis, share of insurer's expenses, profit margin) constitute the ingredients of a "recipe", called the *premium calculation principle*, whose result is the *actuarial premium*. It is worth stressing the meaning of "actuarial". The output of the procedure described above is the premium calculated according to sound actuarial (i.e. financial and statistical) principles. Nonetheless, ingredients other than those so far considered can affect the actual *price* of the insurance product. For example, competition on the insurance market could suggest to lower the premium in order to launch a more appealing product. In the following, we will not take into account these aspects.

Figure 1.7.4 summarizes the process leading to the price of an insurance product.

1.7.4 Premium calculation

We now address some of the cases already discussed, in order to illustrate premium calculations. In all the cases we disregard insurer's expenses and the related component of the premium. Thus, we focus on the calculation of the so-called *net* (or *pure*) *premiums*. As regards the profit margin, specific aspects will be discussed in the various cases.

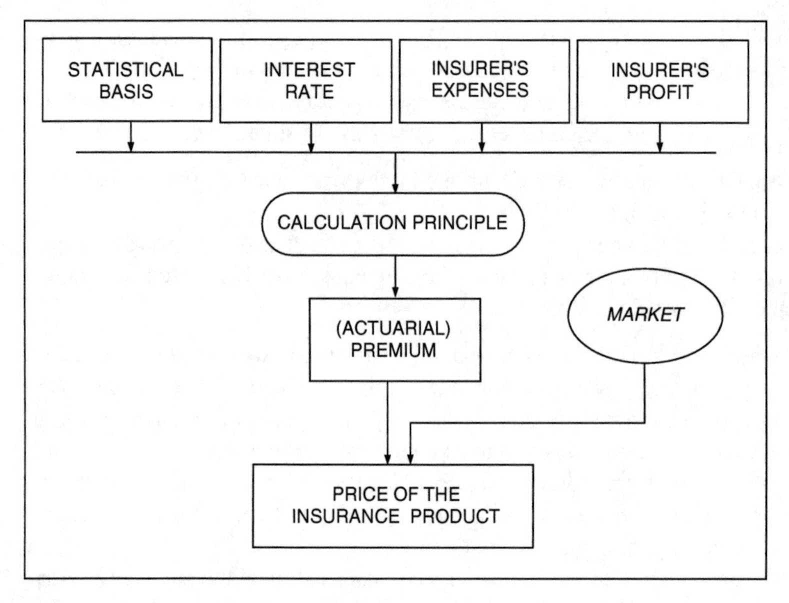

Fig. 1.7.4 Pricing an insurance product

We assume **Case 2** (Possible loss with fixed amount) as the starting point for focussing on premium calculation in a "basic" insurance cover. Since, in this case, we assume a short policy duration, time can be disregarded. The benefit is given, by definition, by the amount of the loss caused by the accident. In formal terms, the random benefit is given by:

$$X = \begin{cases} x & \text{if } \mathscr{E} \\ 0 & \text{if } \bar{\mathscr{E}} \end{cases} \tag{1.7.1}$$

that is, it coincides with the random loss defined by (1.2.1). We denote with p the probability of the event $\mathscr{E}$; hence, the expected value of the benefit is given by

$$\mathbb{E}[X] = x\,p \tag{1.7.2}$$

and thus it coincides with the expected loss as defined by (1.4.2).

We now assume (provisionally) the expected value as the premium, P, for the insurance cover, that is

$$P = x\,p \tag{1.7.3}$$

What can we expect from the application of this very simple (and simplistic) premium calculation principle? First, we note that, from the generic contract, the insurer gains a profit, equal to P, in the case of no accident, whilst suffers a loss, $P-x$, in the case of accident. In formal terms, the random result, Z, from the generic contract can be defined as follows:

$$Z = P - X \tag{1.7.4}$$

and, if the premium is given by (1.7.3), its expected value is equal to zero:

$$\mathbb{E}[Z] = P - \mathbb{E}[X] = 0 \tag{1.7.5}$$

For this reason, the premium calculation according to (1.7.3) is called the *equivalence principle*; the resulting premium can be called the *equivalence premium*.

Moving to a pool of n insured risks, namely a portfolio, claim-free contracts subsidize contracts with a claim, according to the mutuality principle (see Fig. 1.7.5), and the consequent equilibrium reflects, at the pool level, the rationale of the equivalence principle. However, in a pool of n risks, the equilibrium is achieved if and only if the actual number of claims, k, coincides with the expected number of claims, which is given by $n\,p$. Indeed, the income perfectly balances the outgo if and only if

$$nP = kx \tag{1.7.6}$$

that is

$$nxp = kx \tag{1.7.7}$$

and hence $k = n\,p$. In general, arguments presented in Sect. 1.6.2 also hold when the pool is managed by an insurer. In particular, it is worth noting that, especially when the pool size n and the probability p are small, the expected number of claims $n\,p$ could be non-integer, and thus the perfect balance could never be achieved.

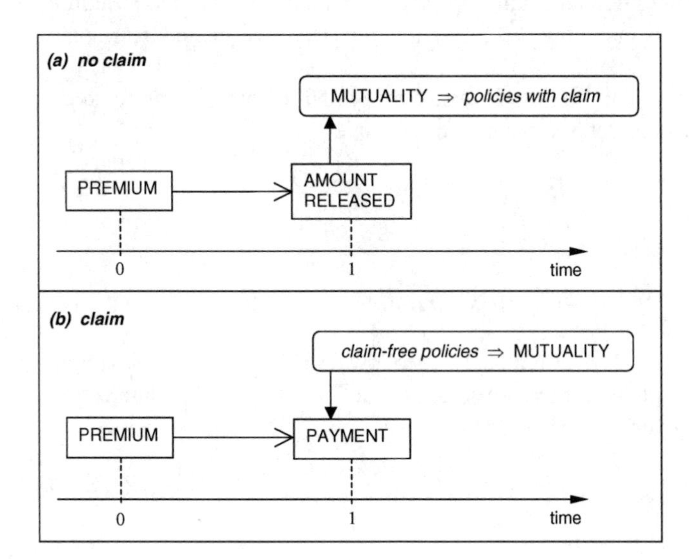

Fig. 1.7.5 Mutuality inside a pool of risks. Case 2 (Possible loss with fixed amount)

If the actual number of claims is greater than its expected value, the insurer suffers a loss. In order to keep the probability of a loss at a reasonable level, a higher premium, Π, must be charged, namely

$$\Pi = P + m \tag{1.7.8}$$

where the term m ($m > 0$) denotes a "loading" of the premium, more precisely the *safety loading*, as its purpose is to reach a higher degree of "safety" in the pool management (see also Sect. 1.6.2).

It is interesting to assess the result from the generic contract when the premium Π is charged to the contract itself. For the random result we have

$$Z = \Pi - X \tag{1.7.9}$$

so that the expected result is given by

$$\mathbb{E}[Z] = \Pi - \mathbb{E}[X] = m \tag{1.7.10}$$

Thus, the safety loading m represents the expected profit from the generic contract.

Moving again to the pool level, if the actual number of claims coincides (at least approximately) with its expected number, the total safety loading cashed by the insurer, i.e. nm, constitutes the profit produced by managing the pool. Thus, in principle the purpose of the safety loading is twofold, as it enhances the safety level and can provide the insurer with a profit margin.

As regards the magnitude of m, various formulae can be used to link the safety loading, for example, to the riskiness of the contract (or the portfolio, as we will see in Chap. 2). Indeed, premium calculation principles directly refer to the premium Π, namely the premium including the safety loading. Here we just address some aspects.

As the safety loading must be linked, in some way and to some extent, to quantitative features of the contract, we can set, for example

$$\Pi = (1 + \alpha)\,\mathbb{E}[X] \tag{1.7.11}$$

(clearly, with $\alpha > 0$) so that we have

$$m = \alpha\,\mathbb{E}[X] = \alpha\, x\, p \tag{1.7.12}$$

Although formula (1.7.11) does not explicitly allow for riskiness, this very simple premium principle is quite common in insurance practice. Moreover, an interesting interpretation can be given. From (1.7.11) we have

$$\Pi = (1 + \alpha)\, x\, p \tag{1.7.13}$$

and, setting $p' = (1 + \alpha)\, p$, we find

$$\Pi = x\, p' \tag{1.7.14}$$

It turns out that the premium Π can be interpreted as the expected value of the random loss calculated according to a risk "adjusted" probability p' ($p' > p$). So, formula (1.7.14) constitutes a straightforward application of the valuation approach based on risk-adjusted probabilities (see Sect. 1.4.7).

To conclude, we note the following points.

- If we choose a value for the loading parameter α, we are adopting an *explicit safety loading* approach, as the loading component of the premium, i.e. αP, can be recognized; see formulae (1.7.11) and (1.7.12). Of course, an explicit safety loading can also be realized by adopting formulae other than (1.7.11) and (1.7.12); examples can be derived introducing, for instance, the variance or the standard deviation of X into the calculation of Π.
- Conversely, an *implicit safety loading* approach is adopted if we directly "rise" the probability of loss. Clearly, the resulting magnitude of the safety loading, and hence the expected profit $\mathbb{E}[Z]$, can be calculated as follows:

$$m = \Pi - P = (p' - p)\,x \tag{1.7.15}$$

Case 4b (Early death of an individual) allows us to discuss the pricing of a life insurance product which is very common in all the insurance markets, namely the term insurance. We denote with C the sum assured, and assume that the sum is paid at the end of the year of death, if the insured dies within the coverage period, say r years.

We assume that the premium has to meet the expected present value of the random benefit. We can refer to the approach sketched in Sect. 1.4.5 for Case 3c (Disability benefits; multi-year period), and define the random present value, Y, of the benefit:

$$Y = \begin{cases} C\,(1+i)^{-1} & \text{if the insured dies in the first year} \\ C\,(1+i)^{-2} & \text{if the insured dies in the second year} \\ \dots & \dots \\ C\,(1+i)^{-r} & \text{if the insured dies in the } r\text{-th year} \\ 0 & \text{if the insured is alive at time } r \end{cases} \tag{1.7.16}$$

Then, we have to calculate the expected value $\mathbb{E}[Y]$. To this purpose, we need the probabilities of the events listed in (1.7.16). We assume that these probabilities can be derived from an appropriate statistical basis, and denote (according to the usual actuarial notation) with ${}_{h-1|1}q_x$ the probability that the insured, age x at policy issue, dies between time $h-1$ and h, i.e. during the h-th year. Hence, according to the equivalence principle, the premium P is given by:

$$P = \mathbb{E}[Y] = C\left((1+i)^{-1}\,{}_{0|1}q_x + (1+i)^{-2}\,{}_{1|1}q_x + \dots + (1+i)^{-r}\,{}_{r-1|1}q_x\right) \tag{1.7.17}$$

From the generic contract, the insurer gains a profit if the insured is alive at maturity (and hence no benefit is paid), whilst suffers a loss in the case of death before maturity. Inside a pool of risks, claim-free contracts subsidize contracts with claim, according to the mutuality principle. The mutuality mechanism works in a manner rather similar to that of Case 2 (Possible loss with fixed amount), with an appropriate generalization because of the duration of the contract, as shown in Fig. 1.7.6.

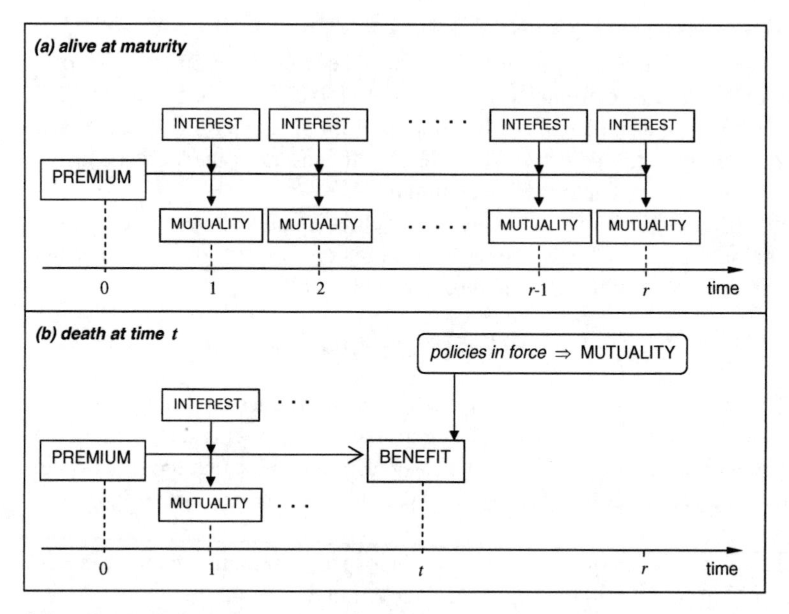

Fig. 1.7.6 Mutuality inside a pool of risks. Case 4b (Early death of an individual)

Clearly, no (explicit) safety loading appears in formula (1.7.17). However, a safety loading can be implicitly included in the premium, adopting the following procedure.

- Assume that i is a realistic estimate of the interest rate (assumed to be constant throughout the policy duration) obtained by the insurer investing the money, initially provided by the premium. Then, adopt an interest rate i' ($i' < i$) for discounting the random benefit. Equivalently, this means that an interest rate lower than that estimated on a realistic basis is credited to the policyholder.
- Assume that the ${}_{h-1|1}q_x$, $h = 1,2,\dots,r$ constitute a likely representation of the age-pattern of mortality of an insured person. Then, adopt as probabilities for the calculation of the expected value the quantities ${}_{h-1|1}q'_x$, with ${}_{h-1|1}q'_x > {}_{h-1|1}q_x$, for $h = 1,2,\dots,r$. Probabilities fulfilling this condition can be easily found, for example, referring to the mortality of a population, rather than to the mortality of a selected group of insureds.

Finally, calculate the premium as follows:

$$\Pi = C\left((1+i')^{-1}\,{}_{0|1}q'_x + (1+i')^{-2}\,{}_{1|1}q'_x + \dots + (1+i')^{-r}\,{}_{r-1|1}q'_x\right) \qquad (1.7.18)$$

Of course, we obtain $\Pi > P$.

The random present value, Z, of the result from the generic contract is given by

$$Z = \Pi - Y \tag{1.7.19}$$

and its expected value is

$$\mathbb{E}[Z] = \Pi - \mathbb{E}[Y] = \Pi - P \tag{1.7.20}$$

which also quantifies the (implicit) safety loading.

The needs inherent in **Case 4a** (The need for resources at retirement), described in Sect. 1.2.5, can be faced by a pure endowment insurance. We suppose that just one premium is paid (at time 0). Also in this case, we assume that the premium has to meet the expected value of the random benefit.

We denote with S the sum insured, with r the policy term, and with ${}_rp_x$ the probability of a person age x at policy issue being alive at time r.

The random present value, Y, of the benefit is given by

$$Y = \begin{cases} S(1+i)^{-r} & \text{if the insured is alive at time } r \\ 0 & \text{otherwise} \end{cases} \tag{1.7.21}$$

The premium, P, according to the equivalence principle, is then given by

$$P = \mathbb{E}[Y] = S(1+i)^{-r}\,{}_rp_x \tag{1.7.22}$$

Since obviously ${}_rp_x < 1$, from (1.7.22) it follows that

$$S > P(1+i)^r \tag{1.7.23}$$

Hence, the accumulation process which leads from P to the benefit S relies on the following elements:

a. the financial component, namely the (guaranteed) interest credited to the insured (with annual interest rate i);
b. the demographic component, namely the contributions from the policies which terminate because of the death of the insured before maturity, and whose amount cumulated up to the death is released and credited to policies still in-force.

Thus, the mutuality works in this insurance product according to the mechanism described under point b. Figure 1.7.7 illustrates the process leading to the sum payable at maturity, in a pool of pure endowment insurances.

Of course, no (explicit) safety loading appears in formula (1.7.22). A safety loading can be implicitly included in the premium, via a procedure similar to that adopted for the term insurance. Thus, the premium can be calculated as follows:

$$\Pi = S(1+i')^{-r}\,{}_rp'_x \tag{1.7.24}$$

where $i' < i$ and ${}_rp'_x > {}_rp_x$. Note that, as mortality of people purchasing pure endowment insurance is usually lower than that of the general population, the probability ${}_rp'_x$ cannot be drawn from a population mortality table, and then should be ad-hoc evaluated.

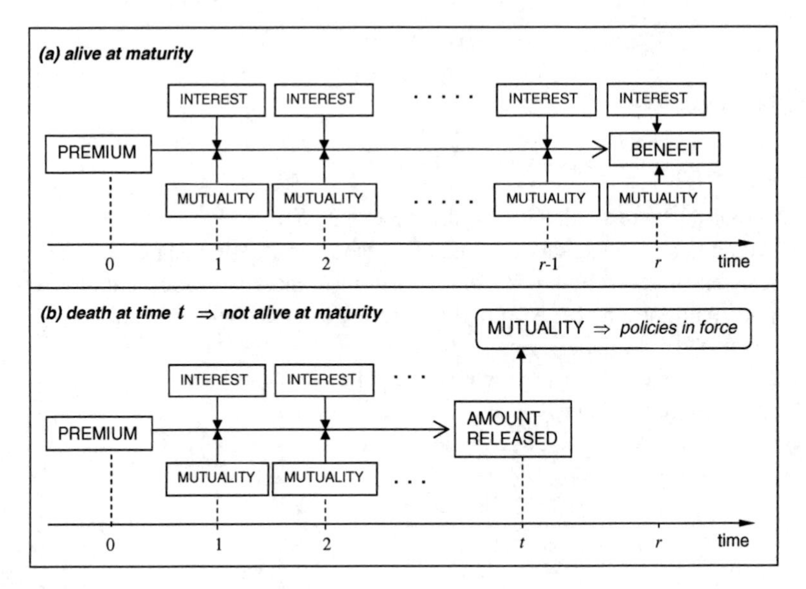

Fig. 1.7.7 Mutuality inside a pool of risks. Case 4a (The need for resources at retirement)

The random present value of the result from the generic contract is given, as in Case 4b, by formula (1.7.19), namely $Z = \Pi - Y$. Hence, when formula (1.7.24) is adopted for premium calculation, the expected present value of the result is given by

$$\mathbb{E}[Z] = \Pi - \mathbb{E}[Y] = \Pi - P = S\left((1+i')^{-r}\,{}_rp'_x - (1+i)^{-r}\,{}_rp_x\right) \tag{1.7.25}$$

1.7.5 Technical bases

In the insurance language, the expression *technical basis* denotes the set of data and assumptions which constitute the ingredients for premium calculation (as well as for other valuations, like those connected to the calculation of reserves, expected profits, and so on). As we have disregarded so far insurer's expenses and the related premium components, we only focus on the elements needed for the calculation of net premiums.

Referring to the insurance products discussed in Sect. 1.7.4, we note what follows. In the "basic" insurance product, namely the cover for Case 2 (Possible loss with fixed amount), the technical basis is simply given by the probability p. For the life insurance products, i.e. the term insurance and the pure endowment insurance, the technical basis consists of the interest rate i and the probabilities of dying or being alive, namely the table from which these probabilities can be derived.

However, the adoption of an implicit safety loading for the life insurance products implies that two different technical bases are involved, namely:

- the *pricing basis*, consisting of the interest rate i', and the probabilities q' or p', also called the *first-order basis* (or *safe-side* basis, or *prudential* basis, or *conservative* basis);
- the *scenario basis*, or *realistic basis*, which consists of the interest rate i (or a time-structure of interest rates), and the probabilities q or p, and provides a reliable description of the financial and demographical scenario; this basis is also called the *second-order basis*.

Conversely, the (usual) adoption of an explicit safety loading in non-life insurance leads to the coincidence of the pricing basis and the scenario basis.

We note that, in life insurance, premium calculation according to (1.7.18) and (1.7.24) actually relies on the equivalence principle, though implemented according to the first-order basis. Hence, the resulting premium is (formally) an equivalence premium, although calculated by adopting a technical basis other than the realistic one.

Remark It turns out that the expression "equivalence premium" is rather ambiguous. In what follows, if not specified otherwise, the expression will be referred to premiums calculated with the scenario (or realistic) basis.

Scenarios change over time, and then scenario bases should be updated. Also pricing bases should be consequently updated. However, important differences between life and non-life insurance products should be stressed, as regards the feasibility of an update in the pricing bases.

In non-life insurance, rating is usually stated on a one-year (or even shorter) basis; see, for example, insurance covers related to Cases 3a (Damage / loss of a cargo), 3d (A fire in a factory), and 3e (Car driver's liability). Hence, premiums can be based, for each insured, on the portfolio (and the market) experience. It follows that the same insured can be charged, year by year, premiums updated according to the "collective" experience.

Remark It is worth noting that, in non-life insurance, also "individual" claim experience can be taken into account in determining the annual premium for any given insured risk. See also the Remark in Sect. 2.2.6. Thus, various *experience rating systems* can be applied in pricing non-life insurance products.

As regards life insurance, contracts usually imply multi-year guarantees. Of course, new contracts can be priced according to an updated basis. Conversely, contracts already issued do not allow premium adjustments. This is obviously true in the case of single premium contracts. Further, arrangements based on periodic premiums do not allow premium updating (in particular, an increase in premiums) if, as is usual, all policy conditions are stated and guaranteed at policy issue. Hence, a risk arises from the implied use of technical bases no longer appropriate.

1.7.6 Reserving

Let us refer to the insurance products covering the risks inherent in Case 4a (The need for resources at retirement) and Case 4b (Early death of an individual). As stated in Sect. 1.7.4, we assume that, in both the insurance products, the benefits are financed by a single premium paid at the policy issue.

We note that, immediately after cashing the premium,

- the related amount (net of possible initial expenses) is available to the insurer, and has to be invested in *assets* providing the insurer with a yield, hopefully higher than the interest rate credited to the insured;
- a *liability* arises, because of the insurer's obligations to the insured (Case 4a), and to the pool of insureds (Case 4b).

Assets and liabilities generated by insurance contracts are the two aspects of the "reserving process". Thus, *reserves* (or *provisions*) must be set up because benefits are deferred with respect to premiums. These "technical" reserves should not be confused with the reserves which result from the accumulation of profits, not distributed to shareholders, and constitute a part of the shareholder's capital.

For each insurance contract, the behavior of the reserve over time strictly depends on the type of benefits (and, in general, on the premium arrangement). Referring to Case 4a and Case 4b, we can intuitively draw the time-profile of the reserve from the mechanism depicted in Fig. 1.7.7 and 1.7.6 respectively.

The interest/mutuality mechanism illustrated in Fig. 1.7.6 implies that, for each insured alive, the reserve of the term insurance, which initially (time 0) is set equal to the single premium Π, will be yearly increased by the interest credited to the reserve itself, and decreased by the amounts drawn to pay death benefits according to the mutuality principle. The annual net variation in the reserve is (usually) negative. At maturity (time r) the reserve is equal to 0, as there is no longer any obligations. Hence the behavior of the reserve (restricted to the policy anniversaries) can be represented as sketched in Fig. 1.7.8.

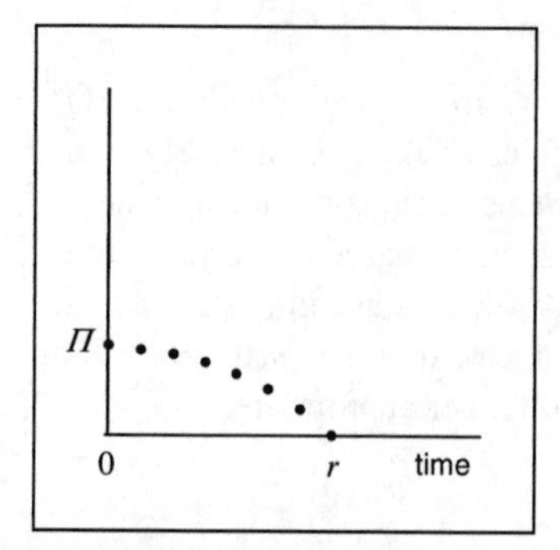

Fig. 1.7.8 Technical reserve of a term insurance

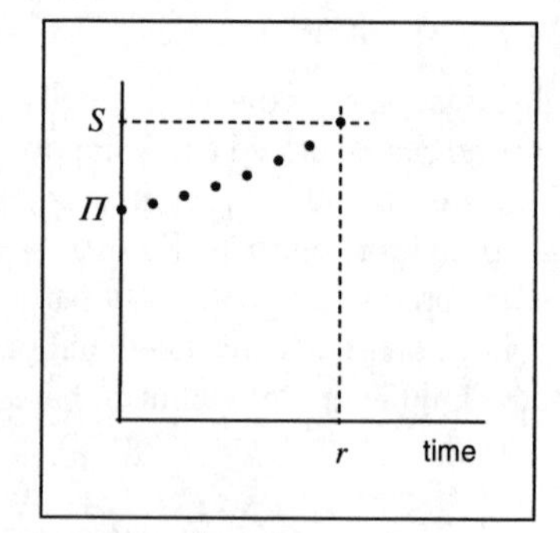

Fig. 1.7.9 Technical reserve of a pure endowment

The reserve of a pure endowment, given the interest/mutuality mechanism depicted in Fig. 1.7.7, will be yearly increased by the interest credited and the contributions from the policies terminating because of the death of the insured. At maturity, before the payment of the sum insured S, the reserve will be equal to the sum itself. Thus, the behavior of the reserve can be represented as sketched in Fig. 1.7.9.

More complex reserving processes relate to premium arrangements based on periodic premiums, instead of a single premium. Further, different problems lead to reserving in non-life insurance.

Technical reserves witness the role of an insurance company (and a life insurance company in particular) as a *financial intermediary*. The insurer brings together providers of funds (the insureds) and users of funds (private companies, public institutions, and so on): funds received while issuing insurance policies and cashing premiums are invested in financial markets in bonds and stocks issued by the users.

We recall that, together with the financial intermediation, the insurer acts as an intermediary in the risk transfer process, taking the risk of paying the guaranteed benefits even when the premiums turn out to be insufficient because of adverse deviations of the actual claim frequency from the expected one (see Sect. 1.6.3).

1.8 References and suggestions for further reading

In this Section, we only cite textbooks dealing with general aspects of risks, risk management, and insurance. Studies particularly devoted to general insurance, life insurance and post-retirement solutions will be cited in the relevant sections of the following chapters.

An effective introduction to risk and insurance is provided by [43]. The textbook [55] places emphasis on insurance products and the use of insurance within the risk management framework; life insurance, pensions, and general insurance are dealt with.

The books [30] and [57] offer complete presentations of the Risk Management process, the insurance transfers included. Practical guidelines to Risk Management in business and industry are provided by [35]. The process of analyzing and planning for both personal risks and business risks is examined by [48].

The books [38] and [14] focus on Risk Management in financial institutions. Chapter 6 in [4] deals with managing risks, also in insurance companies.

The book [34] is the classical reference on Value at Risk (VaR). The book [22] deals with risk management and the VaR approach to risk management problems, whereas [45] mainly focusses on the use of VaR in portfolio management. The book [21] is specifically devoted to the risk management process in the insurance activity, including life and non-life insurance.

The economic theory of risk and insurance and the industrial organization of insurance markets is the main focus of [51].

Readers looking for a presentation of the principles of Financial Mathematics can refer to [54], [11], and the first five chapters of [37].

Chapter 2
Managing a portfolio of risks

2.1 Introduction

Basic ideas concerning risk pooling and risk transfer, presented in Chap. 1, are progressed further in the present Chapter, mainly with the following purposes:

1. to discuss key features of premium calculation when non-homogeneous portfolios are concerned, namely portfolios consisting of risks with various claim probabilities;
2. to analyze, more deeply, the riskiness of a portfolio and the tools which can be used to face potential losses, in particular introducing the role of the shareholders' capital;
3. to illustrate the possibility, for an insurance company, to transfer, in its turn, risk of losses to another insurer, namely the possibility to resort to reinsurance;
4. to address dynamic aspects of the management of insurance portfolios.

As we will see, the actions undertaken by an insurer in order to deal with potential losses (see points 1 and 3 above) constitute important examples of risk management actions, in the specific framework of *insurance risk management*.

The "basic" insurance cover, namely the cover related to Case 2 (Possible loss with fixed amount) widely used in Chap. 1, will still be addressed while dealing with the issues mentioned above, in order to keep the presentation at an acceptable level of complexity.

A. Olivieri, E. Pitacco, *Introduction to Insurance Mathematics*,
DOI 10.1007/978-3-642-16029-5_2, © Springer-Verlag Berlin Heidelberg 2011

2.2 Rating: the basics

2.2.1 Some preliminary ideas

We refer to a portfolio of "basic" insurance covers, as defined in Chap. 1 (see, in particular Case 2 in Sects. 1.2.3, 1.4.2, 1.6.1, and 1.7.2), and we focus on the calculation of net premiums (i.e. not including loadings for expenses).

We assume that, for each risk, the premium is proportional to the benefit (that we also call the "sum insured") paid in the case of a claim. Denoting (as in Chap. 1) with x the benefit for the generic risk, the premium is then given by $x\bar{p}$, where the quantity $\bar{p}$ represents the premium for one monetary unit of benefit. In the insurance language, $\bar{p}$ is commonly called the *premium rate*.

The following are natural choices:

a. set $\bar{p}$ equal to the probability of a claim, p, as implied by the equivalence principle (see, for example, Sect. 1.7.4 and formula (1.7.3) in particular);
b. set $\bar{p}$ equal to the adjusted probability of a claim, p', so that riskiness is accounted for via an implicit safety loading (see formula (1.7.14) in particular).

Although we now do not deal with implicit safety loadings, the first choice is not the only feasible one, as we will see in the next sections. Anyhow, the premium rate should reflect, at least to some extent, the probability of a claim. As a consequence, a number of premium rates, $\bar{p}_1, \bar{p}_2, \ldots$, should be used for calculating the premiums for risks with various claim probabilities. The set of rules which link the premium rates to the claim probabilities constitutes a *rating system*. The rating system is the basis underlying the construction of an *insurance tariff* (which also includes loading for expenses, possible discounts, and so on).

2.2.2 Homogeneous risks

First, we assume that the n risks, which constitute the portfolio, are homogeneous in probability. As usual, we denote with $x^{(j)}$ the potential loss and hence the benefit for the j-th risk, and with p the probability of loss for each of the insured risks.

According to the equivalence principle, the net premium for the j-th risk, $P^{(j)}$, is then given by

$$P^{(j)} = x^{(j)}\, p \tag{2.2.1}$$

Thus, the premium rate is given by the probability p.

At the portfolio level, the premiums expressed by (2.2.1) lead to the so-called *technical equilibrium* (clearly, in terms of expected value). Indeed, we have

$$\sum_{j=1}^{n} P^{(j)} = \sum_{j=1}^{n} x^{(j)}\, p = \mathbb{E}[X^{[\mathrm{P}]}] \tag{2.2.2}$$

where $X^{[P]}$ denotes the portfolio random outgo. Thus

$$\text{TOTAL INCOME} = \text{TOTAL EXPECTED OUTGO} \tag{2.2.3}$$

and hence the expected portfolio result, defined as follows

$$\text{TOTAL EXPECTED RESULT} = \text{TOTAL INCOME} - \text{TOTAL EXPECTED OUTGO} \tag{2.2.4}$$

is equal to 0. Equation (2.2.3) expresses the *equivalence principle at the portfolio level.*

2.2.3 Non-homogeneous risks

We now shift to non-homogeneous portfolios, namely portfolios consisting of risks with various claim probabilities. For simplicity, we refer to a portfolio which consists of n_1 risks with claim probability p_1, and n_2 risks with claim probability p_2. Let $n = n_1 + n_2$. Thus, the portfolio can be split into two homogeneous *sub-portfolios*. Without loss of generality, we assume $p_1 < p_2$.

The obvious choice for premium calculation consists in charging each risk with a premium calculated according to the related claim probability. This means that we set:

- in the first sub-portfolio, i.e. for $j = 1, 2, \ldots, n_1$:

$$P^{(j)} = x^{(j)} p_1 \tag{2.2.5}$$

- in the second sub-portfolio, i.e. for $j = n_1 + 1, n_1 + 2, \ldots, n_1 + n_2$:

$$P^{(j)} = x^{(j)} p_2 \tag{2.2.6}$$

The premiums defined by (2.2.5) and (2.2.6) ensure the technical equilibrium, as expressed by (2.2.3), in both the first and the second sub-portfolio, and hence, of course, in the whole portfolio.

The technical equilibrium within each sub-portfolio is the natural consequence of adopting the equivalence principle (and implementing this principle with the appropriate claim probabilities). Conversely, the target of achieving the technical equilibrium within each sub-portfolio can be interpreted as a constraint in the premium calculation, and, as such, can be "relaxed", or replaced by weaker constraints.

In particular, we can assume that our aim is charging all the risks with the same premium rate, $\bar{p}$. This premium rate cannot ensure the equilibrium in each sub-portfolio; hence, the target is now the equilibrium within the whole portfolio, as expressed by (2.2.3). Clearly, we will have $p_1 < \bar{p} < p_2$.

Possible aims of such a rating system are the following ones:

- simplify the insurance tariff;

- charge "reasonable" premiums to risks with a high claim probability, transferring part of the cost to risks with a low claim probability.

 The premium for the j-th risk, $j = 1,2,\dots,n$, is then given by

$$P^{(j)} = x^{(j)}\,\bar{p} \tag{2.2.7}$$

and the premium rate $\bar{p}$ must fulfil the equivalence principle at the portfolio level (see (2.2.3)), namely

$$\sum_{j=1}^{n} x^{(j)}\,\bar{p} = \sum_{j=1}^{n_1} x^{(j)}\,p_1 + \sum_{j=n_1+1}^{n_1+n_2} x^{(j)}\,p_2 \tag{2.2.8}$$

from which we find

$$\bar{p} = \frac{\sum_{j=1}^{n_1} x^{(j)}\,p_1 + \sum_{j=n_1+1}^{n_1+n_2} x^{(j)}\,p_2}{\sum_{j=1}^{n} x^{(j)}} \tag{2.2.9}$$

Hence, the premium rate $\bar{p}$ is the arithmetic weighted average of the probabilities p_1 and p_2, and the weights are given by the totals of sums insured in the first and second sub-portfolio respectively.

It is interesting to note that, if all the sums insured are equal to x, formula (2.2.9) reduces to

$$\bar{p} = \frac{n_1\,p_1 + n_2\,p_2}{n} \tag{2.2.10}$$

Thus, the premium rate $\bar{p}$ is the arithmetic weighted average of the probabilities p_1 and p_2, weighted by the sub-portfolio sizes.

2.2.4 A more general rating system

Rating systems defined by formulae (2.2.5), (2.2.6) and, respectively, (2.2.7) constitute particular cases of a more general structure.

In order to define a rather general rating system, let $\bar{p}_1$, $\bar{p}_2$ denote two premium rates, charged to risks with claim probability p_1, p_2 respectively. Premiums are then given by the following formulae:

- in the first sub-portfolio, i.e. for $j = 1,2,\dots,n_1$:

$$P^{(j)} = x^{(j)}\,\bar{p}_1 \tag{2.2.11}$$

- in the second sub-portfolio, i.e. for $j = n_1+1, n_1+2,\dots,n_1+n_2$:

$$P^{(j)} = x^{(j)}\,\bar{p}_2 \tag{2.2.12}$$

 Let the following inequalities hold:

$$p_1 \le \bar{p}_1 \le \bar{p}_2 \le p_2 \tag{2.2.13}$$

Assume that the premium rates $\bar{p}_1$ and $\bar{p}_2$ ensure the technical equilibrium at the portfolio level, and consider the following cases:

- setting $\bar{p}_1 = p_1$ and $\bar{p}_2 = p_2$, we find the "natural" rating system, with premiums differentiated according to the claim probabilities (see (2.2.5) and (2.2.6));
- setting $\bar{p}_1 = \bar{p}_2$, we find the system with just one premium rate (see (2.2.7));
- to find new rating systems, we restrict to the cases such that

$$p_1 < \bar{p}_1 < \bar{p}_2 < p_2 \tag{2.2.14}$$

Let us define a rating system satisfying inequalities (2.2.14), and such that the equilibrium at the portfolio level (as expressed by (2.2.3)) is achieved. Thus

$$\sum_{j=1}^{n_1} x^{(j)}\,\bar{p}_1 + \sum_{j=n_1+1}^{n_1+n_2} x^{(j)}\,\bar{p}_2 = \sum_{j=1}^{n_1} x^{(j)}\,p_1 + \sum_{j=n_1+1}^{n_1+n_2} x^{(j)}\,p_2 \tag{2.2.15}$$

In particular, if all the sums insured are equal to x, Eq. (2.2.15) reduces to

$$n_1\,\bar{p}_1 + n_2\,\bar{p}_2 = n_1\,p_1 + n_2\,p_2 \tag{2.2.16}$$

We note, from Eqs. (2.2.15) and (2.2.16), that the unknowns $\bar{p}_1$ and $\bar{p}_2$ cannot be univocally determined. Then, an additional condition is required, for example $\bar{p}_1 = \alpha\,\bar{p}_2$, or $\bar{p}_1 = \bar{p}_2 - \beta$, with α and β such that inequalities (2.2.14) are anyway fulfilled.

Clearly, the aim of such a rating system is to keep premium rates differentiated, while charging a "reasonable" premium to risks with a higher claim probability, and then transferring part of the cost to risks with a lower probability.

Remark Although inequalities (2.2.14) are quite reasonable, in principle we could also assume

$$\bar{p}_1 < p_1 < p_2 < \bar{p}_2 \tag{2.2.17}$$

that is, aiming to "reward" risks with a low probability, while "penalizing" risks with a high probability.

2.2.5 Rating systems and technical equilibrium

When rating systems other than those constructed by setting the premium rates equal to the claim probabilities are adopted, problems concerning the technical equilibrium may arise. To discuss such problems, we refer, for simplicity, to a portfolio in which all the sums insured are equal to x, and just one premium rate is adopted.

Let us turn back to formula (2.2.10), which expresses the premium rate $\bar{p}$ assuming the equilibrium only at the portfolio level as the target. We note that $\bar{p}$ is a function of the sizes, n_1 and n_2, of the two sub-portfolios. Clearly, when the premiums, based on the premium rate $\bar{p}$, are charged to a group of new applicants for the insurance cover, the actual sizes of the sub-groups of risks with claim probability p_1 and p_2 respectively are unknown. Thus, n_1 and n_2 should be understood only as estimates of the actual numbers of applicants.

We denote with n_1^*, n_2^* the actual sizes of the sub-groups. If

$$\frac{n_1^*}{n_1^*+n_2^*} = \frac{n_1}{n_1+n_2} \tag{2.2.18}$$

the technical equilibrium is ensured, as the relative sizes of the actual groups coincide with the estimated relative sizes. Conversely, assume that

$$\frac{n_1^*}{n_1^*+n_2^*} < \frac{n_1}{n_1+n_2} \tag{2.2.19}$$

In this case, the technical equilibrium is not achieved. Indeed, in the actual situation the premium $\bar{p}^*$, given by

$$\bar{p}^* = \frac{n_1^* p_1 + n_2^* p_2}{n_1^*+n_2^*} \tag{2.2.20}$$

should be applied; that is, a lower weight should be attributed to the lower probability, i.e. to p_1 because of (2.2.19). Clearly, $\bar{p}^* > \bar{p}$, and thus an expected loss follows.

Similar problems arise when the rating system is based on two premium rates, as defined by formulae (2.2.15) and (2.2.16).

Example 2.2.1. Two different rating systems, A and B, are defined. Both the systems are constructed by assuming that the number of risks with the lower probability, p_1, is twice the number of risks with the higher probability, p_2, that is, $n_1 = 2n_2$; see Table 2.2.1.

Table 2.2.2 shows the expected payment, the premium income, and the expected portfolio result, referred to two actual portfolios, the first one leading to an equilibrium situation, whilst the second one (for which inequality (2.2.19) holds) implies an expected payment greater than the premium income, whatever the rating system adopted, and hence a negative expected result. As regards the portfolio leading to a non-equilibrium situation, the system A obviously implies a higher loss.

Table 2.2.1 Claim probabilities and premium rates

n_1	n_2	p_1	p_2	Rating A	Rating B	
				$\bar{p}$	$\bar{p}_1$	$\bar{p}_2$
4000	2000	0.005	0.008	0.006	0.0055	0.007

□

Table 2.2.2 Expected outgo, premium income, and expected portfolio result

n_1^*	n_2^*	Expected outgo	Premium income		Expected result	
			Rating A	Rating B	Rating A	Rating B
8 000	4 000	72 000	72 000	72 000	0	0
3 000	3 000	39 000	36 000	37 500	−3 000	−1 500

A practical problem is as follows: is the situation described by inequality (2.2.19) a likely one? The following points provide an answer to this critical question.

- The (expected) equilibrium at the portfolio level is based on a transfer of money (shares of premiums) from insureds charged with a premium higher than their "true" premium, i.e. the premium resulting from the probability of a claim, to insureds charged with a premium lower than their "true" premium. In the technical language, such a transfer of money is called *solidarity* (among the insureds). In particular, the generic insured with claim probability p_1 transfers to the pool the amount
$$S_1^{(j)} = x^{(j)}\,\bar{p} - x^{(j)}\,p_1 > 0 \tag{2.2.21}$$
whereas the pool transfers to the generic insured with claim probability p_2 the amount
$$S_2^{(j)} = x^{(j)}\,\bar{p} - x^{(j)}\,p_2 < 0 \tag{2.2.22}$$
The amounts $S_1^{(j)}$ and $S_2^{(j)}$ are usually called *solidarity premiums* (positive and negative, respectively).
- Rating systems based on solidarity may cause *self-selection*, as individuals forced to provide solidarity to other individuals can reject the policy, moving to other insurance solutions (or, more generally, risk management actions). The resulting effect is a portfolio with a (relative) prevalence of risks with the higher claim probability. Thus, from the insurer's point of view, self-selection is *adverse selection*.
- The severity of this self-selection phenomenon depends on how people perceive the solidarity mechanism, as well as on the premium systems adopted by competitors in the insurance market.
- So, in practice, solidarity mechanisms can work provided that they are compulsory (for example, imposed by insurance regulation) or they constitute a common market practice.

2.2.6 From risk factors to rating classes

The rating system defined by formulae (2.2.7) to (2.2.10) adopts one premium rate $\bar{p}$ versus two claim probabilities p_1, p_2. The underlying rationale can be extended to more general situations.

When we define a “population”, we have to adopt a rigorous criterion to decide whether a given “individual” belongs to the population (i.e. is a “member” of the population) or not. For example, the population can be defined as consisting of all males currently alive, born in Italy in the period 1950-1970. Although the definition is rigorous, we are aware that the population consequently defined is rather heterogeneous, in particular with regard to the risk of death. Indeed, individuals can have various ages, can be more or less healthy, can have a more or less risky occupation, etc. Thus, we can recognize various *risk factors* (age, current health conditions, occupation, and so on), which should be taken into account when stating, for example, the probability of dying within one year.

Problems concerning heterogeneity and the use of risk factors in life and non-life insurance calculations will be specifically addressed in Chaps. 3, 4, and 9. Now, we just provide a first insight into the role of risk factors in the pricing procedures.

We assume that each risk factor can take one out of a given (integer) number of “values”, either scalar (e.g. the age), or nominal (e.g. the gender). Figure 2.2.1 refers to a population for which three risk factors have been initially recognized, with 4, 3 and 2 values respectively (each factor is represented by a coordinate). Thus, the population has been split into $4 \times 3 \times 2 = 24$ *risk classes* (see panel (a)).

In principle, a specific claim probability, and hence a specific premium rate, should be determined for each risk class. However, the resulting tariff structure could be considered too complex, or some premium rates too high. Then, a first simplification could be obtained disregarding one of the risk factors; see Fig. 2.2.1, panel (b), which shows that risk factor 3 has been disregarded. A further grouping of risk classes is illustrated by panel (c), in which we see the grouping of some values of risk factors 1 and 2. As the final result, the population is split into $3 \times 2 = 6$ *rating classes*.

When two or more risk classes are aggregated into one rating class, some insureds pay a premium higher than their “true” premium, i.e. the premium resulting from the risk classification, while other insureds pay a premium lower than their “true” premium. Thus, the equilibrium inside a rating class relies on a money transfer among individuals belonging to different risk classes. As mentioned above, this transfer is usually called solidarity (among the insureds).

When the rating classes coincide with the risk classes, the rating system is “tailored” on the features of each insured risk (at least to the extent these features can be detected), and no solidarity transfer works. Conversely, the solidarity effect is stronger when the number of rating classes is smaller, compared with the number of risk classes.

Remark Even if the rating classes coincide with the risk classes, a “residual” heterogeneity still affects the insured risks inside each rating class, because of the presence of *unobservable risk factors*; for example: genetic characteristics as regards mortality, personal attitude to cause accidents in car insurance, and so on. Thus, an unavoidable degree of solidarity among insured risks is implied by unobservable risk factors, whatever the number of rating classes.
The residual heterogeneity (and hence the solidarity) can be reduced if the individual claim experience allows the insurer to “learn” about the features of each insured risk. In particular, in non-life insurance rating classes can be defined accounting, for example, for the numbers of claims experienced in previous years. So, an *individual experience rating* (also called *merit rating* in car

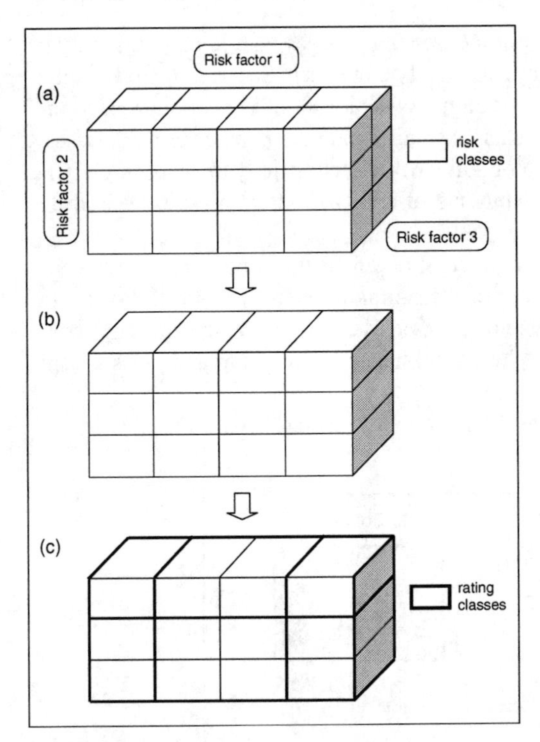

Fig. 2.2.1 From risk factors to rating classes

insurance) determines an *a-posteriori risk classification*, whereas an *a-priori risk classification*, based on rating factors known in advance, works at policy issue. This topic will be specifically dealt with in Chap. 9.

In the field of private insurance, an extreme case is achieved when one rating class only relates to a large number of underlying risk classes. Outside the area of private insurance, the solidarity principle is commonly applied in social security. In this field, the extreme case arises when the whole national population contribute to fund the benefits, even if only a part of the population itself is eligible to receive benefits; so, the burden of insurance is shared among the community.

2.2.7 Cross-subsidy: mutuality and solidarity

Mutuality and solidarity constitute two forms of *cross-subsidy* among the insureds (or, in general, among the members of a pool). However, some important points should be stressed in order to single out the different features of these forms of cross-subsidy.

First, mutuality is an implication of the pooling process (and in particular of the risk transfer to an insurance company), as clearly emerges in Sect. 1.7.4. Conversely, solidarity among the insureds is the straight consequence of the adoption of a rating system with a number of rating classes smaller than the number of risk classes. So, the presence and the magnitude of solidarity effects strictly depend on the tariff structure (see, in particular, the amounts of solidarity premiums, expressed by formulae (2.2.21) and (2.2.22)).

Secondly, it is worth noting that the mutuality affects the benefit payment phase, so that the "direction" and "measure" of the mutuality effect in a portfolio (or, in general, in a pool of risks) are only known ex-post. Conversely, the solidarity (possibly) affects the premium income phase, and hence its direction and measure are known ex-ante.

Figure 2.2.2 illustrates cross-subsidy in a pool of insured risks.

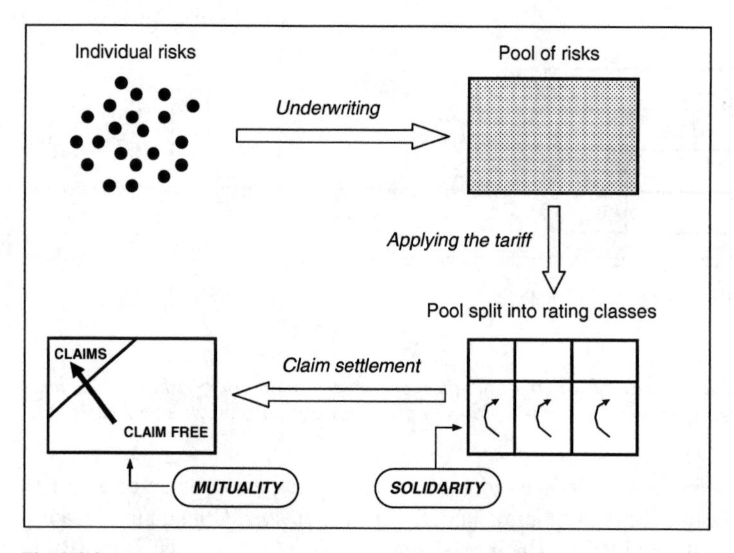

Fig. 2.2.2 Mutuality and solidarity in a pool of insured risks

2.3 Facing portfolio riskiness

Risks inherent in the results obtained by managing a pool of risks have been already discussed in Chap. 1 (see Sects. 1.6.2 and 1.6.3). We now turn back on these issues, referring to a portfolio of insured risks. In particular, we focus on the following aspects:

- what are the "components" of the risk inherent in portfolio outgoes (and hence in portfolio results);

- what are the elements of an appropriate tool-kit for managing this risk.

2.3.1 Expected outgo versus actual outgo

We consider a portfolio of n basic insurance covers (see Case 2 in Sects. 1.7.2 and 1.7.4), in which all the sums insured are equal to x, and we assume that the portfolio is homogeneous with respect to the claim probability; we denote with p this probability.

Let f denote the observed relative claim frequency, i.e. $f = \frac{k}{n}$, where k is the observed number of claims. If $f = p$, the equilibrium is actually achieved (of course, provided that the premium rate is set equal to p), as the actual outgo, given by $n x f$, is equal to the expected outgo, $n x p$, and hence to the premium income. Indeed, we have

$$\sum_{j=1}^{n} P^{(j)} = n x p = n x f \tag{2.3.1}$$

Conversely, we may find that $f \neq p$, and clearly our concern is for the case $f > p$. Figure 2.3.1 sketches three portfolio stories in which we find that, in various years, $f \neq p$. Reasons underlying this inequality may be quite different in the three stories.

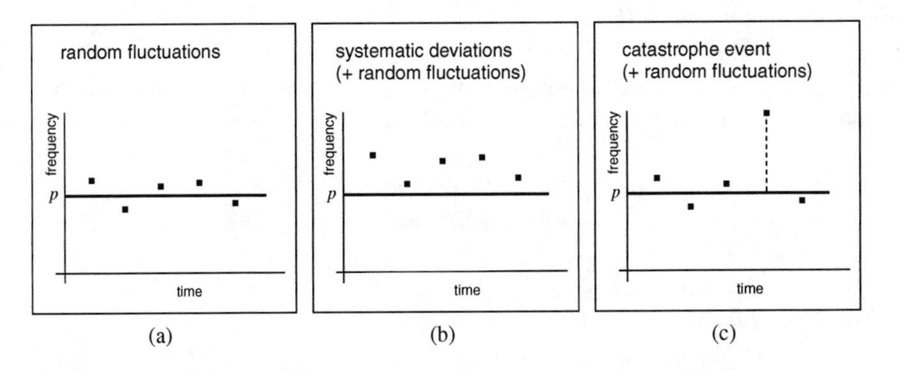

Fig. 2.3.1 Observed frequency versus probability

- In Fig. 2.3.1a, we see that the observed claim frequency randomly fluctuates around the probability, namely around the expected frequency. This possibility is usually denoted as the *risk of random fluctuations*, or the *process risk*.
- On the contrary, Fig. 2.3.1b depicts a situation in which, besides random fluctuations, we see "systematic" deviations from the expected frequency; likely, this occurs because the assessment of the probability p does not capture the true nature of the insured risks. This possibility is usually called the *risk of systematic*

deviations, or the *uncertainty risk*, referring to the uncertainty in the assessment of the expected frequency.
- In Fig. 2.3.1c, the effect of a "catastrophe", which causes a huge number of claims in a given year, clearly appears. This possibility is commonly known as the *catastrophe risk*.

Thus, three *risk components* have been singled out. All the components impact on the monetary results of the portfolio. However, the severity of the impact strongly depends on the portfolio structure, and the portfolio size in particular.

- The severity of the risk of random fluctuations decreases, in relative terms, as the portfolio size increases. This feature is the direct consequence of the risk pooling (see Sect. 1.6.1), and thus is commonly known as the *pooling effect*.
- The severity of the risk of systematic deviations is independent, in relative terms, of the portfolio size (as we will see in Sect. 2.3.10). Indeed, the systematic deviations affect the pool as an aggregate. Conversely, the total impact on portfolio results increases as the portfolio size increases.
- The severity of the catastrophe risk can be higher due, for example, to a high concentration of insured risks within a geographic area.

In the following sections we focus on the risk of random fluctuations.

2.3.2 Risk assessment

We still refer to a portfolio of n basic insurance covers; for the generic cover, the insurer's random payment is given by

$$X^{(j)} = \begin{cases} x^{(j)} & \text{in the case of claim} \\ 0 & \text{otherwise} \end{cases} \tag{2.3.2}$$

where $x^{(j)}$ is the sum insured.

We assume that:

- the portfolio is homogeneous with respect to the claim probability, denoted with p;
- claims and hence random numbers $X^{(j)}$ are independent each other.

Let $P^{(j)}$ denote the expected value of $X^{(j)}$ (namely, the equivalence premium according to the realistic basis), thus

$$P^{(j)} = \mathbb{E}[X^{(j)}] = x^{(j)}\, p \tag{2.3.3}$$

Moving to the portfolio level, we denote with $X^{[\mathrm{P}]}$ the total payment

$$X^{[\mathrm{P}]} = \sum_{j=1}^{n} X^{(j)} \tag{2.3.4}$$

whose expected value, denoted by $P^{[\mathrm{P}]}$, is given by

$$P^{[\mathrm{P}]} = \mathbb{E}[X^{[\mathrm{P}]}] = p \sum_{j=1}^{n} x^{(j)} = \sum_{j=1}^{n} P^{(j)} \tag{2.3.5}$$

Our first aim is to quantify the portfolio riskiness, in order to determine an appropriate safety loading. In general, a basic information about riskiness is obviously provided by the variance of the total payment.

For the generic insured risk, the variance of the random payment is given by

$$\mathbb{V}\mathrm{ar}[X^{(j)}] = (x^{(j)})^2 p (1-p) \tag{2.3.6}$$

Then, for the total payment, thanks to the independence assumption, we find

$$\mathbb{V}\mathrm{ar}[X^{[\mathrm{P}]}] = \sum_{j=1}^{n} \mathbb{V}\mathrm{ar}[X^{(j)}] = p(1-p) \sum_{j=1}^{n} (x^{(j)})^2 \tag{2.3.7}$$

It is interesting to analyze the link between the variance of the total payment and the structure of the portfolio itself, in terms of the sums insured. We denote with $\bar{x}$ the average sum insured, namely

$$\bar{x} = \frac{1}{n} \sum_{j=1}^{n} x^{(j)} \tag{2.3.8}$$

and with $\bar{x}^{(2)}$ the second moment of the distribution of the sums insured, that is

$$\bar{x}^{(2)} = \frac{1}{n} \sum_{j=1}^{n} (x^{(j)})^2 \tag{2.3.9}$$

Finally, we denote with v the variance of the distribution of the sums insured

$$v = \frac{1}{n} \sum_{j=1}^{n} (x^{(j)} - \bar{x})^2 \tag{2.3.10}$$

which can also be expressed as follows:

$$v = \bar{x}^{(2)} - (\bar{x})^2 \tag{2.3.11}$$

From relations (2.3.7) to (2.3.11), it follows that

$$\mathbb{V}\mathrm{ar}[X^{[\mathrm{P}]}] = n p (1-p) \bar{x}^{(2)} = n p (1-p) \left(v + (\bar{x})^2\right) \tag{2.3.12}$$

Thus, for a given portfolio size n and a given average sum insured $\bar{x}$ (and hence a given value of $(\bar{x})^2$), the variance of the total payment is lower when the variance of the sums insured, v, is lower. In particular, we find the minimum variance $\mathbb{V}\mathrm{ar}[X^{[\mathrm{P}]}]$ when $v = 0$, that is, when all the policies have the same sum insured. Note that, in this case, the actual total payment (and hence the actual portfolio result) only depends on the number of claims in the portfolio, whilst it does not depend on which policies are affected by claims.

2.3.3 The risk index

As shown in Sect. 1.6.1, an interesting insight into the riskiness of a pool of risks (and thus a portfolio of insured risks, in particular) is given by the coefficient of variation of the total payment, $X^{[P]}$. The coefficient of variation provides a measure of relative riskiness, i.e. riskiness related to the expected value of the total payment. The coefficient of variation of $X^{[P]}$ is also called, in the actuarial literature, the *risk index* of the portfolio. We will denote it with ρ (reference to the portfolio payment $X^{[P]}$ is understood). Hence,

$$\rho = \mathbb{CV}[X^{[P]}] = \frac{\sqrt{\mathbb{V}\mathrm{ar}[X^{[P]}]}}{\mathbb{E}[X^{[P]}]} = \frac{\sigma^{[P]}}{P^{[P]}} \tag{2.3.13}$$

where $\sigma^{[P]}$ denotes the standard deviation of the total payment.

We now analyze some aspects of the link between the risk index and the portfolio structure. We still refer to the portfolio defined in Sect. 2.3.2.

From Eqs. (2.3.5) and (2.3.7), we find

$$\rho = \sqrt{\frac{1-p}{p}} \frac{\sqrt{\sum_{j=1}^{n} (x^{(j)})^2}}{\sum_{j=1}^{n} x^{(j)}} = \sqrt{\frac{1-p}{n\,p}} \frac{\sqrt{\bar{x}^{(2)}}}{\bar{x}} \tag{2.3.14}$$

From (2.3.14) we note that, for a given portfolio size n and a given average sum insured $\bar{x}$, the risk index ρ is higher when $\bar{x}^{(2)}$ is higher, and thus the variance v of the distribution of the sums insured is higher (see the conclusions after formula (2.3.12)).

Example 2.3.1. Tables 2.3.1 to 2.3.3 refer to three portfolios, all with the same average sum insured, $\bar{x} = 1000$; in all the portfolios, the claim probability is $p = 0.005$. However, the three portfolios have different sizes, or structures in terms of sums insured. Various typical values (among which the risk index) summarize the total payment and the inherent risk.

By comparing the results in Table 2.3.1 to those in Table 2.3.2, we clearly perceive the magnitude of the pooling effect. Conversely, by comparing results in Table 2.3.1 to those in Table 2.3.3, we can see the effect of heterogeneity in the sums insured.

□

What can we say, in general terms, about the range of values of the risk index ρ, for a given portfolio size n and a given claim probability p ? First, it can be proved

Table 2.3.1 Portfolio A

Number of policies	Sum insured
100 000	1 000
Typical values:	$\bar{x} = 1\,000$ $v = 0$ $P^{[P]} = \mathbb{E}[X^{[P]}] = 500\,000$ $\sigma^{[P]} = \sqrt{\mathbb{V}\mathrm{ar}[X^{[P]}]} = 22\,304$ $\rho = \dfrac{\sigma^{[P]}}{P^{[P]}} = 0.0446$

Table 2.3.2 Portfolio B

Number of policies	Sum insured
10 000	1 000
Typical values:	$\bar{x} = 1\,000$ $v = 0$ $P^{[P]} = \mathbb{E}[X^{[P]}] = 50\,000$ $\sigma^{[P]} = \sqrt{\mathbb{V}\mathrm{ar}[X^{[P]}]} = 7\,053$ $\rho = \dfrac{\sigma^{[P]}}{P^{[P]}} = 0.1411$

Table 2.3.3 Portfolio C

Number of policies	Sum insured
70 000	500
25 000	1 000
5 000	8 000
Typical values:	$\bar{x} = 1\,000$ $v = 2\,625\,000$ $P^{[P]} = \mathbb{E}[X^{[P]}] = 500\,000$ $\sigma^{[P]} = \sqrt{\mathbb{V}\mathrm{ar}[X^{[P]}]} = 42\,467$ $\rho = \dfrac{\sigma^{[P]}}{P^{[P]}} = 0.0849$

that

$$\frac{\sqrt{n}}{n} \leq \frac{\sqrt{\sum_{j=1}^{n} (x^{(j)})^2}}{\sum_{j=1}^{n} x^{(j)}} \leq 1 \tag{2.3.15}$$

Then, from these inequalities, it follows that

$$\sqrt{\frac{1-p}{np}} \leq \rho \leq \sqrt{\frac{1-p}{p}} \tag{2.3.16}$$

As regards the lower bound, we have already shown that it is actually reached if (and only if) all sums insured are equal (see Sect. 1.6.1). As regards the upper bound, note that, if one sum insured "diverges" (ceteris paribus), we have:

$$\frac{\sqrt{\sum_{j=1}^{n}(x^{(j)})^2}}{\sum_{j=1}^{n}x^{(j)}} \to 1 \tag{2.3.17}$$

and hence

$$\rho \to \sqrt{\frac{1-p}{p}} \tag{2.3.18}$$

In more practical terms, when just one sum insured is extremely high if compared to the other sums, the advantage provided by the portfolio size vanishes, so that the riskiness of the portfolio is roughly equal to the riskiness of a portfolio consisting of just one policy.

Hence, we can conclude stating that the relative riskiness reduces as the portfolio size increases, provided that each individual position (and the related contribution to the riskiness) becomes negligible in respect of the overall portfolio.

2.3.4 The probability distribution of the total payment

More information about the riskiness of a portfolio can be achieved via the probability distribution of the total payment $X^{[P]}$. Deriving this probability distribution is, in general, a rather complex problem. Then, we restrict our attention to a particular case, and to the use of approximations.

We assume that our portfolio, which consists of n independent risks, is homogeneous with respect to both the probability, p, and the sum insured, x. Hence, the random total payment can be expressed as follows:

$$X^{[P]} = Kx \tag{2.3.19}$$

where K denotes the random number of claims in the portfolio.

Thanks to the hypothesis of independence, K has a binomial distribution, thus

$$K \sim \mathrm{Bin}(n,p) \tag{2.3.20}$$

and hence

$$\mathbb{P}[X^{[P]} = kx] = \mathbb{P}[K = k] = \binom{n}{k} p^k (1-p)^{n-k};\ k = 0,1,\dots,n \tag{2.3.21}$$

In order to get more tractable calculation procedures, various approximations to the binomial distribution can be used. In particular, for a large size n and small probability p, the Poisson distribution can be adopted. Thus, we can assume

$$K \sim \text{Pois}(\lambda) \tag{2.3.22}$$

and hence

$$\mathbb{P}[X^{[P]} = kx] = e^{-\lambda}\,\frac{\lambda^k}{k!};\;\; k = 0, 1, \ldots \tag{2.3.23}$$

with

$$\lambda = np \quad (= \text{expected number of claims in the portfolio}) \tag{2.3.24}$$

Further, the normal distribution provides an approximation, which relies on the Central Limit Theorem. Then,

$$X^{[P]} \sim \mathcal{N}(P^{[P]}, \sigma^{[P]}) \tag{2.3.25}$$

where

$$P^{[P]} = \mathbb{E}[X^{[P]}] = nxp \tag{2.3.26}$$

$$\sigma^{[P]} = \sqrt{\mathbb{V}\text{ar}[X^{[P]}]} = x\sqrt{np(1-p)} \tag{2.3.27}$$

Hence

$$\frac{X^{[P]} - P^{[P]}}{\sigma^{[P]}} \sim \mathcal{N}(0,1) \tag{2.3.28}$$

So, we have for example

$$\mathbb{P}\left[z_1 < \frac{X^{[P]} - P^{[P]}}{\sigma^{[P]}} \le z_2\right] = \Phi_{\mathcal{N}(0,1)}(z_2) - \Phi_{\mathcal{N}(0,1)}(z_1) \tag{2.3.29}$$

where $\Phi_{\mathcal{N}(0,1)}(z)$ denotes the cumulative distribution function, namely

$$\Phi_{\mathcal{N}(0,1)}(z) = \frac{1}{\sqrt{2\pi}} \int_{-\infty}^{z} e^{-\frac{u^2}{2}}\,\mathrm{d}u \tag{2.3.30}$$

The normal approximation can also be adopted in more general cases, e.g. for portfolios of insured risks with various sums insured and/or various probabilities of claim.

The goodness of some approximations is briefly discussed, via numerical examples, in Appendix 2.A.

Some interesting results can be achieved looking at how the risk index enters probabilities concerning the total payment $X^{[P]}$. For example, consider the following probability:

$$\psi_\delta = \mathbb{P}\left[(1-\delta)\,P^{[P]} < X^{[P]} \le (1+\delta)\,P^{[P]}\right] \tag{2.3.31}$$

(see Fig. 2.3.2). The probability on the right-hand side of (2.3.31) can be expressed in terms of the risk index ρ. Indeed, we find

$$\psi_\delta = \mathbb{P}\left[-\delta\,\frac{1}{\rho} < \frac{X^{[P]} - P^{[P]}}{\sigma^{[P]}} \le \delta\,\frac{1}{\rho}\right] \tag{2.3.32}$$

and then:

$$\psi_\delta = \Phi\left(\delta\,\frac{1}{\rho}\right) - \Phi\left(-\delta\,\frac{1}{\rho}\right) \tag{2.3.33}$$

where Φ denotes the cumulative distribution function of the standardized random variable $\frac{X^{[P]}-P^{[P]}}{\sigma^{[P]}}$. From (2.3.33) we argue that, for any given value of δ, the lower is ρ the higher is ψ_δ. Thus, the "concentration" increases as the risk index decreases, e.g. because the size of the portfolio increases (see also Table 2.3.4 in Example 2.3.2).

Focussing on "downside" payments is clearly of great interest when assessing the riskiness of a portfolio. To this purpose, probabilities like

$$\pi(t) = \mathbb{P}\left[X^{[P]} > P^{[P]} + t\right] \tag{2.3.34}$$

should be addressed; t represents a critical "threshold", which expresses the insurer's capability to meet the total payment. For example, consider the probability ν_δ defined as follows:

$$\nu_\delta = \pi\left(\delta\, P^{[P]}\right) = \mathbb{P}\left[X^{[P]} > (1+\delta)\, P^{[P]}\right] \tag{2.3.35}$$

in which the threshold t is expressed in terms of the expected value $P^{[P]}$ (see Fig. 2.3.2). We find:

$$\nu_\delta = \mathbb{P}\left[\frac{X^{[P]} - P^{[P]}}{\sigma^{[P]}} > \delta\,\frac{1}{\rho}\right] = 1 - \Phi\left(\delta\,\frac{1}{\rho}\right) \tag{2.3.36}$$

It is easy to understand that, for any given δ, the probability ν_δ decreases as ρ decreases, e.g. because the size of the pool increases.

If we assume, in particular, the normal approximation to the distribution of $X^{[P]}$, we find

$$\nu_\delta = \frac{1 - \psi_\delta}{2} \tag{2.3.37}$$

(See Table 2.3.5 in Example 2.3.2 for a numerical illustration).

Example 2.3.2. We refer to a portfolio, which consists of n independent risks, homogeneous with respect to both the sum insured and the claim probability p. We assume $p = 0.005$. The normal approximation has been used for the numerical evaluations. Table 2.3.4 illustrates the concentration, in terms of the probability (2.3.31), for some values of δ and various pool sizes. On the other hand, Table 2.3.5 shows the probability of "downside" payments.

□

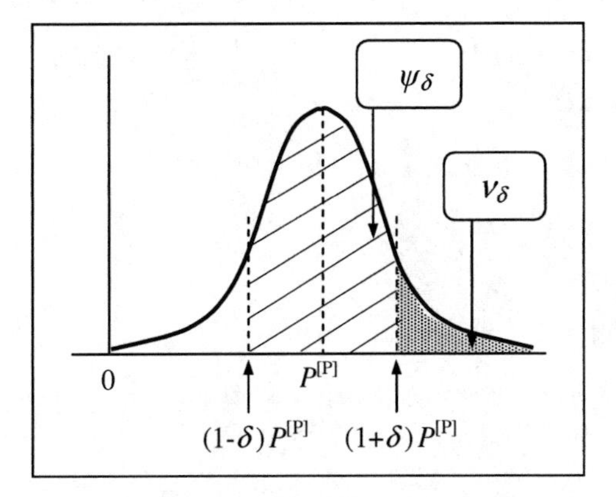

Fig. 2.3.2 Probability distribution of the random payment $X^{[P]}$

Table 2.3.4 The concentration around the expected value

	ψ_δ		
n	$\delta = 0.10$	$\delta = 0.05$	$\delta = 0.01$
100	0.06983	0.03495	0.00699
1000	0.17737	0.08924	0.01788
10000	0.61920	0.33877	0.06984

Table 2.3.5 The probability of "downside" payments

	ν_δ		
n	$\delta = 0.10$	$\delta = 0.05$	$\delta = 0.01$
100	0.46509	0.48253	0.49651
1000	0.41132	0.45538	0.49106
10000	0.19040	0.33062	0.46508

2.3.5 The safety loading

In this Section we show how to calculate the safety loading consistently with the portfolio riskiness. So, a practical feature of the risk index will clearly emerge.

Refer to the portfolio of n basic insurance covers, described in Sect. 2.3.2. Let $m^{(j)}$ denote the (explicit) safety loading for risk j, and $\Pi^{(j)}$ the premium including the safety loading, that is

$$\Pi^{(j)} = P^{(j)} + m^{(j)} \tag{2.3.38}$$

where $P^{(j)} = x^{(j)}\,p$ (see Eq. (2.3.3)).

Moving to the portfolio level, let $\Pi^{[P]}$ denote the total premium income

$$\Pi^{[P]} = \sum_{j=1}^{n} \Pi^{(j)} \tag{2.3.39}$$

which can also be expressed as

$$\Pi^{[P]} = P^{[P]} + m^{[P]} \tag{2.3.40}$$

with obvious meaning of the symbol $m^{[P]}$.

The portfolio result, $Z^{[P]}$, is then defined as follows:

$$Z^{[P]} = \Pi^{[P]} - X^{[P]} \tag{2.3.41}$$

We obviously have:

$$\mathbb{E}[Z^{[P]}] = m^{[P]} \tag{2.3.42}$$

$$\mathbb{V}\text{ar}[Z^{[P]}] = \mathbb{V}\text{ar}[X^{[P]}] \tag{2.3.43}$$

We consider the event $Z^{[P]} < 0$, that is the event $X^{[P]} > P^{[P]} + m^{[P]}$. According to the notation defined by (2.3.34), the probability of this event, namely the *probability of loss*, is denoted as follows:

$$\pi(m^{[P]}) = \mathbb{P}\left[X^{[P]} > P^{[P]} + m^{[P]}\right] \tag{2.3.44}$$

Clearly, the probability of loss should be kept reasonably low, via an appropriate choice of the (total) safety loading $m^{[P]}$.

Figures 2.3.3 and 2.3.4 show the probability distributions of the random payment $X^{[P]}$ and the portfolio result $Z^{[P]}$, respectively (the probability distributions are assumed to be continuous, so that the behavior of the density functions is displayed).

Note that, in the present setting of the problem, the safety loading $m^{[P]}$ is the only parameter whose value can be chosen to lower the probability of a loss (i.e. a negative value of $Z^{[P]}$). Clearly, the effect of a change in this parameter (see Fig. 2.3.5) is a shift in the probability distribution of $Z^{[P]}$ (see Fig. 2.3.6).

From (2.3.44), we have

$$\pi(m^{[P]}) = \mathbb{P}\left[\frac{X^{[P]} - P^{[P]}}{\sigma^{[P]}} > \frac{m^{[P]}}{\sigma^{[P]}}\right] = 1 - \Phi\left(\frac{m^{[P]}}{\sigma^{[P]}}\right) \tag{2.3.45}$$

where Φ denotes the cumulative distribution function of the random number $\frac{X^{[P]}-P^{[P]}}{\sigma^{[P]}}$, with expected value equal to 0 and standard deviation equal to 1.

Let ε denote the accepted probability of loss. We want to find $m^{[P]}$ such that

$$\pi(m^{[P]}) = \varepsilon \tag{2.3.46}$$

that is

$$1 - \Phi\left(\frac{m^{[P]}}{\sigma^{[P]}}\right) = \varepsilon \tag{2.3.47}$$

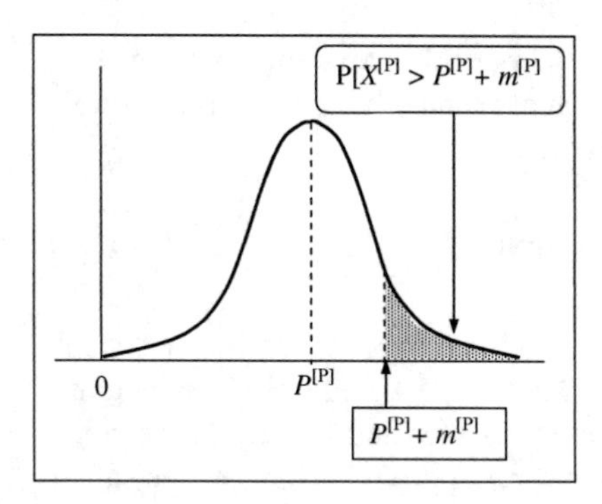

Fig. 2.3.3 The probability distribution of the random payment $X^{[P]}$

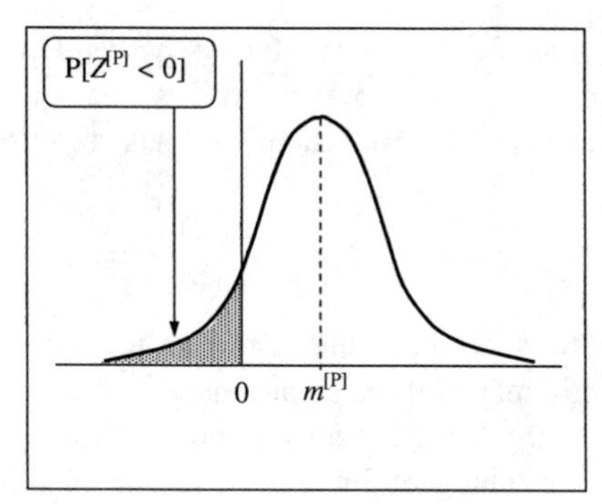

Fig. 2.3.4 The probability distribution of the random result $Z^{[P]}$

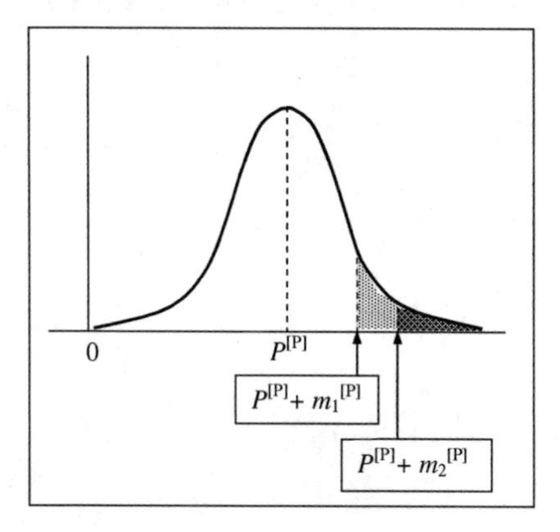

Fig. 2.3.5 The probability distribution of $X^{[P]}$: probability of exceeding two different levels of safety loading

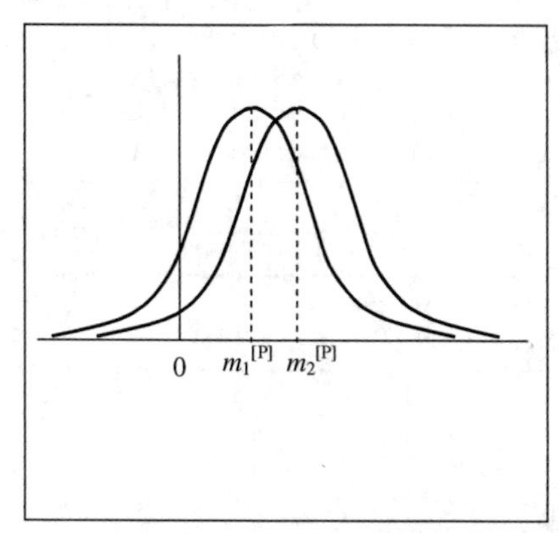

Fig. 2.3.6 The probability distribution of $Z^{[P]}$: safety loading as a shift parameter of the random result

and then

$$m^{[P]} = \sigma^{[P]} \Phi^{-1}(1-\varepsilon) \tag{2.3.48}$$

Finally, we find that the required safety loading per unit of expected value, namely the *safety loading rate*, is given by

$$\frac{m^{[P]}}{P^{[P]}} = \frac{\sigma^{[P]}}{P^{[P]}} \Phi^{-1}(1-\varepsilon) \tag{2.3.49}$$

that is

$$\frac{m^{[P]}}{P^{[P]}} = \rho\, \Phi^{-1}(1-\varepsilon) \tag{2.3.50}$$

Thus, for a given accepted probability ε, the lower is the risk index ρ, the lower is the safety loading rate.

Example 2.3.3. Tables 2.3.6 to 2.3.8 refer to the portfolio structures described by Tables 2.3.1 to 2.3.3, respectively. The normal approximation has been used to evaluate the probabilities, namely it has been assumed:

$$\frac{X^{[P]} - P^{[P]}}{\sigma^{[P]}} \sim \mathcal{N}(0,1) \tag{2.3.51}$$

The analysis of the results in the three tables leads, of course, to conclusions strictly related to those presented in Example 2.3.1. Now, the effect of risk pooling (compare Table 2.3.6 to Table 2.3.7) and the effect of heterogeneity in the sums insured (compare Table 2.3.6 to Table 2.3.8) clearly appears in terms of the safety loading rate $\frac{m^{[P]}}{P^{[P]}}$. Note, in particular, the huge values of this rate in Portfolio B when a very low probability of loss is assumed as the target. So, the need for tools other than the safety loading clearly emerges.

Table 2.3.6 Safety loading - Portfolio A

$m^{[P]}$	$\frac{m^{[P]}}{P^{[P]}}$	$\frac{m^{[P]}}{\sigma^{[P]}}$	$\pi(m^{[P]})$
64 200	0.1284	2.880	0.002
57 550	0.1151	2.580	0.005

Table 2.3.7 Safety loading - Portfolio B

$m^{[P]}$	$\frac{m^{[P]}}{P^{[P]}}$	$\frac{m^{[P]}}{\sigma^{[P]}}$	$\pi(m^{[P]})$
20 312	0.4063	2.880	0.002
18 195	0.3639	2.580	0.005
6420	0.1284	0.910	0.181

Table 2.3.8 Safety loading - Portfolio C

$m^{[P]}$	$\frac{m^{[P]}}{P^{[P]}}$	$\frac{m^{[P]}}{\sigma^{[P]}}$	$\pi(m^{[P]})$
122 300	0.2446	2.880	0.002
109 550	0.2191	2.580	0.005
64 200	0.1284	1.512	0.065

□

2.3.6 Capital allocation and beyond

The outcome of the total payment $X^{[P]}$ can be higher than the amount of premiums, even when these include an appropriate safety loading. In order to manage this risk, the insurer can assign to the portfolio a fund which consists of shareholders' capital (and, as such, may derive from previous profits, or from the issue of shares). This action is usually referred to as the *capital allocation*. Hence, the purpose of the allocation is to protect the insurance company against possible negative results produced by the portfolio.

Let M denote the amount of capital allocated to the portfolio. Figure 2.3.7 illustrates the use of resources available to the insurer, in order to face the portfolio total payment, and the results corresponding to the possible outcomes of the payment itself.

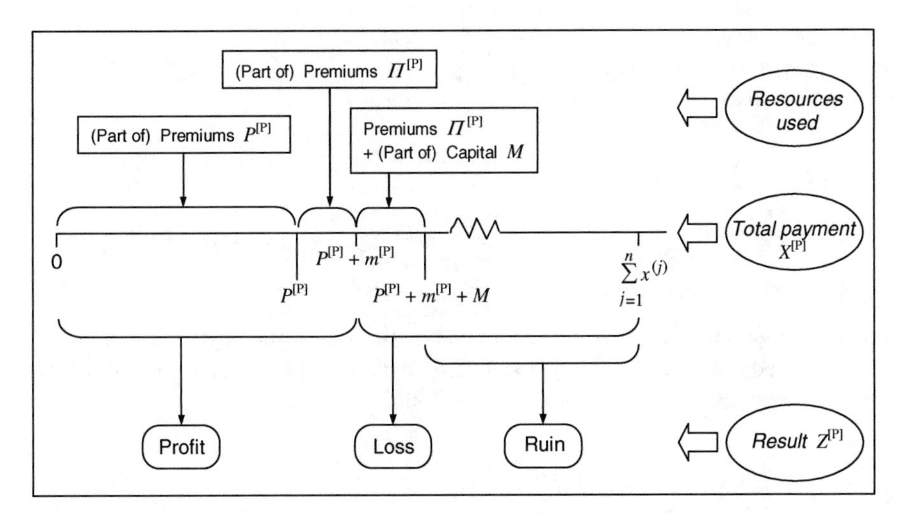

Fig. 2.3.7 Facing the total payment

In particular, the event $X^{[P]} > P^{[P]} + m^{[P]} + M$ means the *portfolio default*, or *ruin*. We note that both the safety loading $m^{[P]}$ and the capital M are variables whose values can be chosen to lower the *probability of default*, namely the probability:

$$\pi(m^{[P]} + M) = \mathbb{P}[X^{[P]} > P^{[P]} + m^{[P]} + M] = \mathbb{P}[Z^{[P]} < -M] \tag{2.3.52}$$

If the total safety loading $m^{[P]}$ has been already stated, the following problem should be considered: find the amount M such that

$$\pi(m^{[P]} + M) = \alpha \tag{2.3.53}$$

where α is an assigned, low probability (see Fig. 2.3.9). Of course, we have:

$$M = -VaR_{\alpha} \tag{2.3.54}$$

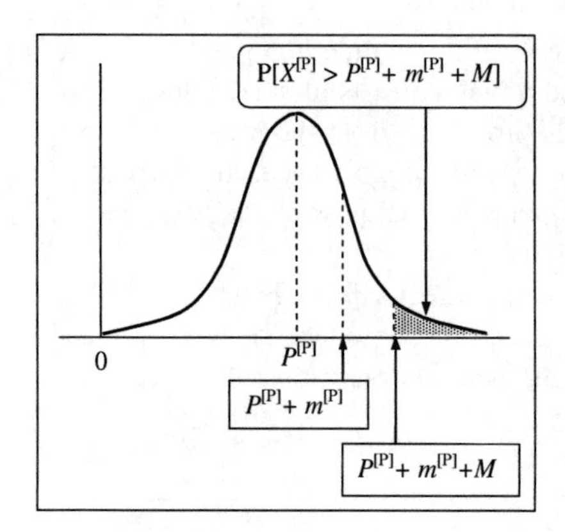

Fig. 2.3.8 The probability distribution of the random payment $X^{[P]}$

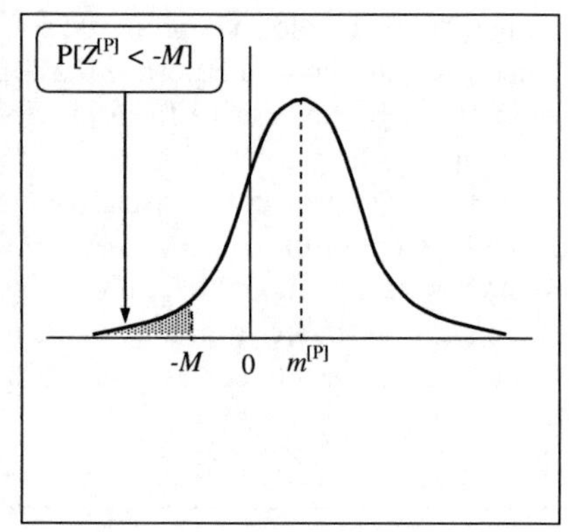

Fig. 2.3.9 The probability distribution of the random result $Z^{[P]}$

From (2.3.52) we have:

$$\pi(m^{[P]} + M) = \mathbb{P}\Big[\frac{X^{[P]} - P^{[P]}}{\sigma^{[P]}} > \frac{m^{[P]} + M}{\sigma^{[P]}}\Big] = 1 - \Phi\Big(\frac{m^{[P]} + M}{\sigma^{[P]}}\Big) \tag{2.3.55}$$

where Φ denotes the cumulative distribution function of the random number $\frac{X^{[P]} - P^{[P]}}{\sigma^{[P]}}$, with expected value equal to 0 and standard deviation equal to 1. Thus, the target expressed by (2.3.53) can also be written as follows:

$$1 - \Phi\Big(\frac{m^{[P]} + M}{\sigma^{[P]}}\Big) = \alpha \tag{2.3.56}$$

and hence

$$\frac{m^{[P]} + M}{\sigma^{[P]}} = \Phi^{-1}(1 - \alpha) \tag{2.3.57}$$

(see Fig. 2.3.10).

For a given probability α and a given standard deviation $\sigma^{[P]}$ (which is univocally determined by the portfolio structure), Eq. (2.3.57) can be solved with respect to the total amount $m^{[P]} + M$. In other terms, if the safety loading is not yet stated, both the amounts $m^{[P]}$ and M can be chosen in order to achieve the target probability.

The unit-free index

$$s = \frac{m^{[P]} + M}{\sigma^{[P]}} \tag{2.3.58}$$

is sometimes called the *relative stability index*. From (2.3.55), we see that the higher is s, the lower is the ruin probability. To raise s, the following actions can be taken:

1. raise the safety loading $m^{[P]}$;

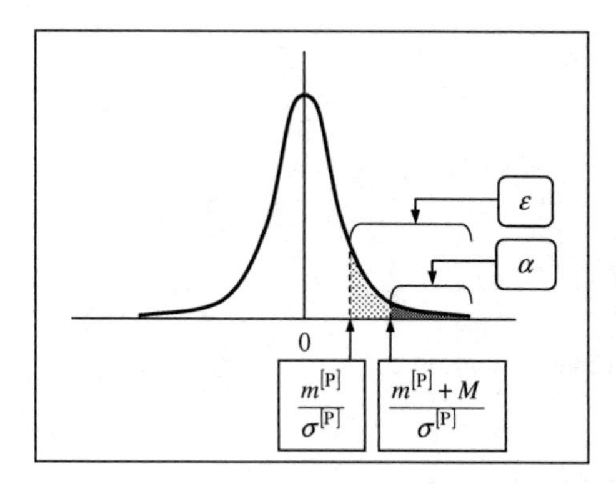

Fig. 2.3.10 The standardized probability distribution of the random payment

2. raise the allocated capital M;
3. reduce $\sigma^{[P]}$, via appropriate reinsurance arrangements (thus affecting the portfolio structure, in terms of sums insured), and, in particular, by choosing the "retention" level (we will deal with these concepts in Sects. 2.4 and 2.5).

As the insurer can choose (at least in principle) the safety loading, the amount of allocated capital, and the retention level, these quantities are called *decision variables*. However the following aspects should be stressed. Action 1 affects the premiums, and hence is bounded by market constraints. Conversely, action 2 has constraints at the company level, because capital is a limited resource.

As regards action 3, whatever reinsurance arrangements may be chosen, the related cost obviously affects the resources available to the portfolio, in particular reducing the expected profit $m^{[P]}$. As both numerator and denominator of the stability index are affected (see (2.3.58)), the effect is not univocally determined in general.

Example 2.3.4. Tables 2.3.9 to 2.3.11 refer to the portfolio structures described by Tables 2.3.1 to 2.3.3, respectively.

In particular, from Tables 2.3.10 and 2.3.11 the important role of the capital allocation clearly appears, especially when very high safety loading rates should otherwise be applied, because of either the size of the portfolio or its structure, in order to keep low the probability of default.

Table 2.3.9 Capital allocation and safety loading - Portfolio A

M	$m^{[P]}$	$\frac{m^{[P]}}{P^{[P]}}$	s	$\pi(m^{[P]}+M)$
10000	50000	0.100	2.6901	0.0036
15000	50000	0.100	2.9143	0.0018
20000	50000	0.100	3.1385	0.0009

Table 2.3.10 Capital allocation and safety loading - Portfolio B

M	$m^{[P]}$	$\frac{m^{[P]}}{P^{[P]}}$	s	$\pi(m^{[P]}+M)$
10000	5000	0.100	2.1268	0.0167
13200	5000	0.100	2.5805	0.0050
10000	8200	0.164	2.5805	0.0050

Table 2.3.11 Capital allocation and safety loading - Portfolio C

M	$m^{[P]}$	$\frac{m^{[P]}}{P^{[P]}}$	s	$\pi(m^{[P]}+M)$
10000	50000	0.100	1.4129	0.0788
60000	50000	0.100	2.5902	0.0048
35000	75000	0.150	2.5902	0.0048

□

2.3.7 Solvency

As seen above, the event $Z^{[P]} < -M$ represents the portfolio default. Conversely, when $M + Z^{[P]} \geq 0$ the insurer is able to meet the total payment by using the premiums and, possibly, (part of) the allocated capital, that is, the insurer is *solvent*. Hence, a *solvency requirement* can be expressed as follows:

$$\mathbb{P}[M + Z^{[P]} \geq 0] = 1 - \alpha \tag{2.3.59}$$

where α is the accepted default probability (see Eq. (2.3.53)).

Equation (2.3.59) can be solved with respect to M. The solution (see (2.3.57), for given values of $m^{[P]}$ and $\sigma^{[P]}$) provides the capital requirement for solvency purposes.

It is worth noting that, in the ordinary language, the term "solvency" is often used in a not well defined sense. Commonly, it is used to denote the capability of an agent to pay the amounts when these fall due. It is apparent that this definition does not fit obvious actuarial requirements. Indeed, in the insurance activity, the capability cannot be meant in a deterministic sense (which leads to the concept of "absolute solvency"): actually, the total amount due could be equal to the sum of all sums insured with the policies in-force at a given time, if all the insureds claim at that time. Hence, insurance business needs a definition of solvency in a probabilistic sense, as witnessed in particular by Eq. (2.3.59).

2.3.8 Creating value

We now return to the choice between action 1 (raise $m^{[P]}$) and 2 (raise M), aiming to lower the probability of default (or to achieve an assigned target probability α). First, we note that allocating capital implies a cost to the shareholders, whereas raising the safety loading leads to a higher cost to the policyholders.

Let r denote the (annual) rate which quantifies the opportunity cost of the shareholders' capital. Thus, the cost of allocating the amount M is given by rM. However, the common definition of a profit (or a loss) is only based on the comparison between actual incomes and costs. Thus, for the portfolio we are addressing, we have

$$\begin{aligned} \Pi^{[P]} < X^{[P]} &\Rightarrow \text{loss} \\ \Pi^{[P]} > X^{[P]} &\Rightarrow \text{profit} \end{aligned}$$

(note that the only cost accounted for is given by the payment for claims, $X^{[P]}$, as, in our setting, expenses are disregarded). Conversely, if we want to assess the portfolio result also allowing for the cost of capital allocation, $\Pi^{[P]}$ has to be compared to $X^{[P]} + rM$. A new concept then arises, namely the *creation of value*. Thus, for our portfolio we have:

$$\begin{aligned} \Pi^{[P]} < X^{[P]} &\Rightarrow \text{loss and value destruction} \\ X^{[P]} < \Pi^{[P]} < X^{[P]} + rM &\Rightarrow \text{profit and value destruction} \\ \Pi^{[P]} = X^{[P]} + rM &\Rightarrow \text{profit and no value} \\ \Pi^{[P]} > X^{[P]} + rM &\Rightarrow \text{profit and value creation} \end{aligned}$$

In order to compare strategies which consist in mixing action 1 and action 2, we have to move to expected values. Thus, we have to replace $X^{[P]}$ with its expected value $\mathbb{E}[X^{[P]}] = P^{[P]}$. Noting that $\Pi^{[P]} = P^{[P]} + m^{[P]}$, we find, in terms of expected values:

$$m^{[P]} < rM \Rightarrow \text{value destruction} \tag{2.3.60}$$

$$m^{[P]} > rM \Rightarrow \text{value creation} \tag{2.3.61}$$

Example 2.3.5. We refer to portfolio B, and assume $\alpha = 0.005$ as the target probability; hence, an amount $M + m^{[P]} = 18\,200$ is required (see Table 2.3.10). Further, we assume $r = 0.08$. Table 2.3.12 illustrates some situations of value creation (Value > 0), value destruction (Value < 0), and "equilibrium" (Value $= 0$), respectively.

□

Whatever the target probability, the equation

$$m^{[P]} = rM \tag{2.3.62}$$

Table 2.3.12 Value creation versus value destruction - Portfolio B

M	$m^{[P]}$	rM	Value $m^{[P]} - rM$
10000	8200	800	7400
15000	3200	1200	2000
16852	1348	1348	0
18000	200	1440	−1240

defines the border line between value creation and value destruction. Conversely, for a given target probability, we have

$$m^{[P]} + M = \text{const.} \tag{2.3.63}$$

(represented by a "level line") as it results from Eq. (2.3.57). See Fig. 2.3.11.

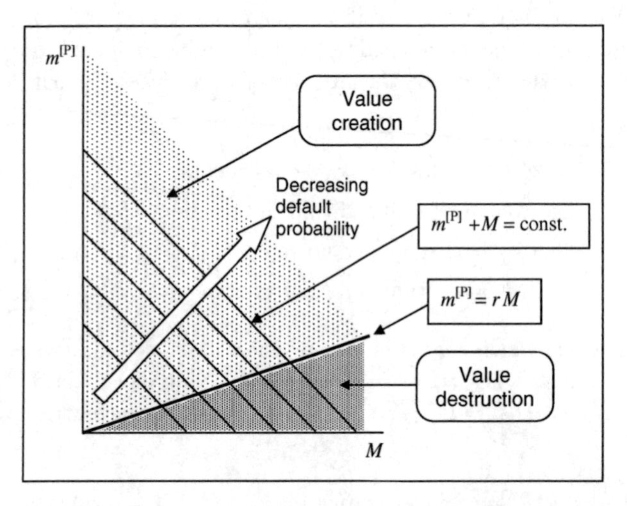

Fig. 2.3.11 Capital allocation; value creation versus value destruction

It is worth stressing that both value creation and solvency are two important goals for any insurance business. Clearly, for any given portfolio (and a given amount of safety loading), the two goals require opposite actions: a higher amount of capital improves the solvency level, while reducing the value creation.

Example 2.3.6. We still refer to portfolio B, and assume 5% as the safety loading rate, so that $m^{[P]} = 2500$. The opportunity cost of capital is $r = 0.08$. Table 2.3.13 illustrates value creation and default probability as functions of the capital allocation.

□

Table 2.3.13 Value creation and default probability - Portfolio B

M	rM	Value $m^{[P]} - rM$	Default prob. $\pi(m^{[P]} + M)$
10000	800	1700	0.063
15700	1256	1244	0.005
20000	1600	900	0.001
35000	2800	−300	≈ 0

2.3.9 Risk management and risk analysis: some remarks

Various issues dealt with in the previous sections of this Chapter can be properly placed in the framework of insurance risk management, and in particular can be interpreted as risk management actions.

Pricing the insurance product, which in our setting simply reduces to calculate an appropriate safety loading, aims at loss prevention and loss reduction (see Sect. 1.3.3). In a more general setting, also product design (and, in particular, the design of various policy conditions, see Sect. 1.7.1) contributes to loss prevention and loss reduction.

Capital allocation is the action aiming at loss financing via retention (see Sect. 1.3.3). More precisely, the shareholders' capital allocated to a portfolio constitutes the tool for funding possible future losses.

Like other business entities, insurers can finance potential losses via risk transfer. In the following sections, we will first focus on traditional risk transfer, namely via reinsurance arrangements (Sects. 2.4 and 2.5). Then, alternative risk transfers (Sect. 2.6), and in particular the transfer to capital markets, will be analyzed in the framework of loss financing actions.

Risk Management, as a methodological framework, has provided important contributions to risk analysis and risk assessment. Nevertheless, it should be stressed that the earliest contribution to risk quantification can be traced back to the 18th century. In 1786 Johannes Tetens first addressed the analysis of the process risk inherent in a life insurance portfolio. Tetens showed that the risk in absolute terms increases as the portfolio size n increases, whereas the risk in respect of each insured decreases in proportion to $\sqrt{n}$. This feature of the risk pooling process has been described in Sect. 1.6.1 (in particular, see Examples 1.6.1 and 1.6.2), and Sect. 2.3.3 (see Example 2.3.1).

In a modern theoretical perspective, Tetens' ideas constitute a pioneering contribution to the *individual risk theory*. Note that the term "individual" recalls the nature of the approach, which starts from the description of the individual risks $X^{(j)}$ (in Case 2, the amount $x^{(j)}$ of the potential loss, and the relevant probability $p^{(j)}$), and leads to the construction of the probability distribution (or, at least, some typical values) of the total payment $X^{[P]}$. According to the terminology commonly used in the Risk Management context, the adoption of this method is called the *bottom-up approach*.

The *collective risk theory*, whose origin can be traced back to the seminal contribution by Filip Lundberg, dated 1909, directly focusses on the characteristics of the total payment $X^{[P]}$. In the Risk Management context, this approach is usually called the *top-down approach*. Well-known implementations lead, for instance, to the calculation of the VaR and the TailVaR (see Sect. 1.5.4), and to various solvency requirements according to a dynamic perspective (as we will see in Sect. 2.7).

2.3.10 The "uncertainty risk"

We refer, as in Sects. 2.3.2 and 2.3.3, to a portfolio of n basic insurance covers, all with the same probability of claim. Further, we assume that all the policies have the same sum insured x. We denote simply with X the random payment for the generic policy.

Unlike the previous sections, we now suppose that p does not necessarily represent the "correct" estimate of the claim probability. If p is not a correct estimate of this probability, situations like the one displayed in Fig. 2.3.1b, and thus involving systematic deviations, can occur.

To make explicit our awareness, we can express uncertainty about the estimate of the claim probability through a random quantity $\tilde{p}$, to which a probability distribution should be assigned. We now denote with p the generic outcome of the random quantity $\tilde{p}$.

As regards the probability distribution of $\tilde{p}$, we can, for example, choose a Beta distribution (see Fig. 2.3.12), the parameters of which are usually denoted with α, β. Thus,

$$\tilde{p} \sim \text{Beta}(\alpha, \beta) \tag{2.3.64}$$

Hence, for the random quantity $\tilde{p}$, we have:

$$\mathbb{E}[\tilde{p}] = \frac{\alpha}{\alpha + \beta} \tag{2.3.65}$$

$$\mathbb{V}\text{ar}[\tilde{p}] = \frac{\alpha \beta}{(\alpha + \beta)^2 (\alpha + \beta + 1)} \tag{2.3.66}$$

When uncertainty about the claim probability is accounted for, the expected value of X, conditional on any value p of $\tilde{p}$ is given by

$$\mathbb{E}[X|p] = x\,p \tag{2.3.67}$$

Conversely, the quantity

$$\mathbb{E}[X|\tilde{p}] = x\,\tilde{p} \tag{2.3.68}$$

is a random amount, as it is a function of $\tilde{p}$. Its expectation, according to the Beta distribution assigned to $\tilde{p}$, is given by

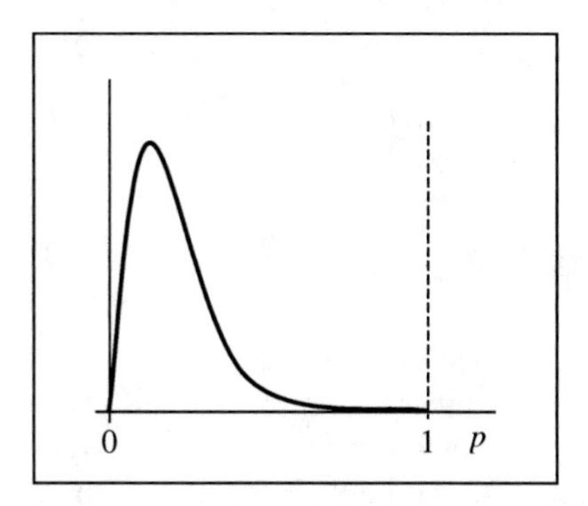

Fig. 2.3.12 The pdf of a Beta distribution

$$\mathbb{E}_{\text{Beta}}[\mathbb{E}[X|\tilde{p}]] = \mathbb{E}_{\text{Beta}}[x\,\tilde{p}] = x\,\frac{\alpha}{\alpha+\beta} \tag{2.3.69}$$

Note that, in the uncertainty framework, formula (2.3.69) expresses the unconditional expected value, namely $\mathbb{E}[X]$. For the variance of the random amount $\mathbb{E}[X|\tilde{p}]$ we find

$$\mathbb{V}\text{ar}_{\text{Beta}}[\mathbb{E}[X|\tilde{p}]] = \mathbb{V}\text{ar}_{\text{Beta}}[x\,\tilde{p}] = x^2\,\frac{\alpha\,\beta}{(\alpha+\beta)^2\,(\alpha+\beta+1)} \tag{2.3.70}$$

In the presence of uncertainty, the variance of X, conditional on any value p of $\tilde{p}$, is given by

$$\mathbb{V}\text{ar}[X|p] = x^2\,p\,(1-p) \tag{2.3.71}$$

while

$$\mathbb{V}\text{ar}[X|\tilde{p}] = x^2\,\tilde{p}\,(1-\tilde{p}) \tag{2.3.72}$$

is a random quantity. Its expectation, according to the Beta distribution assigned to $\tilde{p}$, is given by

$$\mathbb{E}_{\text{Beta}}[\mathbb{V}\text{ar}[X|\tilde{p}]] = x^2\,\mathbb{E}_{\text{Beta}}[\tilde{p}\,(1-\tilde{p})] \tag{2.3.73}$$

It can be proved that $\mathbb{E}_{\text{Beta}}[\tilde{p}\,(1-\tilde{p})] = \frac{\alpha\,\beta}{(\alpha+\beta)(\alpha+\beta+1)}$, so that

$$\mathbb{E}_{\text{Beta}}[\mathbb{V}\text{ar}[X|\tilde{p}]] = x^2\,\frac{\alpha\,\beta}{(\alpha+\beta)(\alpha+\beta+1)} \tag{2.3.74}$$

Moving to the portfolio level, we now address the total payment $X^{[P]}$. When uncertainty in the claim probability is allowed for, we have that the expected value $\mathbb{E}[X^{[P]}|p]$ and the variance $\mathbb{V}\text{ar}[X^{[P]}|p]$ must be meant as conditional on the generic value p of the random quantity $\tilde{p}$, as for the corresponding typical values of X. Further, we have:

$$\mathbb{E}[X^{[P]}|\tilde{p}] = n\,\mathbb{E}[X|\tilde{p}] \tag{2.3.75}$$

and for the variance

$$\mathbb{V}\text{ar}[X^{[P]}|\tilde{p}] = n\mathbb{V}\text{ar}[X|\tilde{p}] \tag{2.3.76}$$

Expected value and variance, as given by (2.3.75) and (2.3.76) respectively, are random quantities. We have:

$$\mathbb{E}[X^{[P]}] = \mathbb{E}_{\text{Beta}}[n\mathbb{E}[X|\tilde{p}]] = nx\frac{\alpha}{\alpha+\beta} \tag{2.3.77}$$

Note that (2.3.77) expresses the unconditional expected value of $X^{[P]}$.

As regards the variance of $X^{[P]}$, first it should be stressed that the independence among the individual random claims must be meant only conditional on any given value of the probability p. Then, in the presence of uncertainty about this probability, namely when the random quantity $\tilde{p}$ is addressed, the unconditional variance of $X^{[P]}$ cannot be expressed as the sum of the individual unconditional variances. Conversely, it can be proved that the unconditional variance of $X^{[P]}$ can be expressed as follows:

$$\begin{aligned}\mathbb{V}\text{ar}[X^{[P]}] &= \mathbb{V}\text{ar}_{\text{Beta}}[\mathbb{E}[X^{[P]}|\tilde{p}]] + \mathbb{E}_{\text{Beta}}[\mathbb{V}\text{ar}[X^{[P]}|\tilde{p}]] \\ &= \mathbb{V}\text{ar}_{\text{Beta}}[n\mathbb{E}[X|\tilde{p}]] + \mathbb{E}_{\text{Beta}}[n\,\mathbb{V}\text{ar}[X|\tilde{p}]]\end{aligned} \tag{2.3.78}$$

Hence, from (2.3.70) and (2.3.74) we have:

$$\mathbb{V}\text{ar}[X^{[P]}] = n^2x^2\frac{\alpha\beta}{(\alpha+\beta)^2(\alpha+\beta+1)} + nx^2\frac{\alpha\beta}{(\alpha+\beta)(\alpha+\beta+1)} \tag{2.3.79}$$

Finally, for the (unconditional) coefficient of variation, namely the risk index, after some manipulations we find

$$\mathbb{CV}[X^{[P]}] = \frac{\sqrt{\mathbb{V}\text{ar}[X^{[P]}]}}{\mathbb{E}[X^{[P]}]} = \sqrt{\frac{\beta}{\alpha(\alpha+\beta+1)} + \frac{1}{n}\frac{\beta(\alpha+\beta)}{\alpha(\alpha+\beta+1)}} \tag{2.3.80}$$

Hence, we have

$$\lim_{n\to\infty}\mathbb{CV}[X^{[P]}] = \sqrt{\frac{\beta}{\alpha(\alpha+\beta+1)}} > 0 \tag{2.3.81}$$

Note that, on the contrary, when no uncertainty is allowed for, the risk index tends to 0 when the pool size n diverges (see (1.6.14)). In more practical terms, this means that:

- the process risk (namely, the risk of random fluctuations) is a *diversifiable* risk, and the diversification is achieved by increasing the portfolio size;
- the uncertainty risk (namely, the risk of systematic deviations) is an *undiversifiable* risk, as its (relative) magnitude is independent of the portfolio size.

(see also Sect. 2.3.1).

Example 2.3.7. We assume, for the random quantity $\tilde{p}$, the Beta distribution with the following parameters:

$$\alpha = 4; \ \beta = 796 \tag{2.3.82}$$

Hence, from (2.3.65) and (2.3.66), we find:

$$\mathbb{E}[\tilde{p}] = 0.005$$
$$\mathbb{V}\text{ar}[\tilde{p}] = 7.754 \times 10^{-9}$$

Let us now assume the following parameters

$$\alpha = 2; \ \beta = 398 \tag{2.3.83}$$

In this case, we have:

$$\mathbb{E}[\tilde{p}] = 0.005$$
$$\mathbb{V}\text{ar}[\tilde{p}] = 3.094 \times 10^{-8}$$

Note that, while keeping the same expected value, we now have a higher variance, which clearly expresses a higher degree of uncertainty about the claim probability.

Table 2.3.14 shows the behavior of the risk index, namely $\mathbb{CV}[X^{[P]}]$, for various portfolio sizes n; the cases of no uncertainty (i.e. a fixed value of p) and uncertainty expressed by the parameters specified by (2.3.82) and (2.3.83) respectively, are considered. The results are self-evident: the undiversifiable part of the risk clearly appears when uncertainty is explicitly introduced into the valuations.

Table 2.3.14 The coefficient of variation $\mathbb{CV}[X^{[P]}]$

n	$p = 0.005$	$\alpha = 4, \beta = 796$	$\alpha = 2, \beta = 398$
10	4.461	4.486	4.511
100	1.411	1.495	1.575
1 000	0.446	0.669	0.834
10 000	0.141	0.518	0.718
...	...	...	...
∞	0.000	0.498	0.704

□

2.4 Reinsurance: the basics

2.4.1 General aspects

The reinsurance is the traditional risk transfer from an insurer (the *cedant*) to another insurer (the *reinsurer*). From a technical point of view, the main aim of the reinsurance transfer is to find protection against the portfolio ruin (and the insurer's ruin, as well). Further aims of reinsurance will be addressed in Sect. 2.5.4.

The basic idea underlying any *reinsurance form* (or *arrangement*) is to split the portfolio random payment, $X^{[\mathrm{P}]}$, as follows:

$$X^{[\mathrm{P}]} = X^{[\mathrm{ret}]} + X^{[\mathrm{ced}]} \tag{2.4.1}$$

where:

- the random amount $X^{[\mathrm{ced}]}$ is the *ceded* part of the total payment; this amount will be paid by the reinsurer to the cedant;
- the random amount $X^{[\mathrm{ret}]}$ is the *retained* part of the total payment, hence it is the net payment of the cedant.

A *reinsurance premium* is paid by the cedant to the reinsurer, as the price of the possible reinsurer's intervention.

How to define the two terms on the right-hand side of (2.4.1)? The two following approaches can be adopted.

1. In principle, the simplest way to define the splitting consists in assigning a retention function Γ, which works at the portfolio level, so that

$$X^{[\mathrm{ret}]} = \Gamma(X^{[\mathrm{P}]}) \tag{2.4.2}$$

In some cases, the retained payment can also depend on other quantities, e.g. the total number of claims, K, in the portfolio, thus

$$X^{[\mathrm{ret}]} = \Gamma(X^{[\mathrm{P}]}, K) \tag{2.4.3}$$

Anyway, this approach relies on the definition of the splitting on a *portfolio basis*, and then leads to a *global reinsurance* arrangement.

2. As the random payment is the sum of the payments related to the various risks, namely $X^{[\mathrm{P}]} = \sum_{j=1}^{n} X^{(j)}$, we can split each $X^{(j)}$ by defining a retention function γ, so that

$$X^{(j)[\mathrm{ret}]} = \gamma(X^{(j)}) \tag{2.4.4}$$

Then, the retained total payment is given by

$$X^{[\mathrm{ret}]} = \sum_{j=1}^{n} X^{(j)[\mathrm{ret}]} \tag{2.4.5}$$

In some cases, a set of retention functions $\gamma^{(j)}$, $j = 1, 2, \ldots, n$, must be defined, instead of a single function γ. Anyhow, this approach requires the splitting on an *policy basis*, hence leading to an *individual reinsurance* arrangement.

We now describe an implementation of approach 1. Another implementation of this approach will be presented in Sect. 2.5.3.

2.4.2 Stop-loss reinsurance

Stop-loss reinsurance provides a "direct" protection against the portfolio ruin, as it directly refers to the portfolio total payment. The reinsurer gets the reinsurance premium $\Pi^{[\text{reins}]}$ and pays the part of $X^{[\text{P}]}$ which exceeds a stated amount, Λ, the *stop-loss retention*, or *priority*. The priority is commonly expressed in terms of the total premium income $\Pi^{[\text{P}]}$ (and usually $\Lambda > \Pi^{[\text{P}]}$).

The cedant's retention and the reinsurer's payment are then given by:

$$X^{[\text{ret}]} = \begin{cases} X^{[\text{P}]} & \text{if } X^{[\text{P}]} \leq \Lambda \\ \Lambda & \text{if } X^{[\text{P}]} > \Lambda \end{cases} \tag{2.4.6a}$$

$$X^{[\text{ced}]} = \begin{cases} 0 & \text{if } X^{[\text{P}]} \leq \Lambda \\ X^{[\text{P}]} - \Lambda & \text{if } X^{[\text{P}]} > \Lambda \end{cases} \tag{2.4.6b}$$

Figure 2.4.1(a) shows the reinsurer's intervention.

An *upper limit*, Θ, to reinsurer's intervention can be stated. In this case, the cedant's retention and the reinsurer's payment are respectively given by:

$$X^{[\text{ret}]} = \begin{cases} X^{[\text{P}]} & \text{if } X^{[\text{P}]} \leq \Lambda \\ \Lambda & \text{if } \Lambda < X^{[\text{P}]} < \Lambda + \Theta \\ X^{[\text{P}]} - \Theta & \text{if } X^{[\text{P}]} \geq \Lambda + \Theta \end{cases} \tag{2.4.7a}$$

$$X^{[\text{ced}]} = \begin{cases} 0 & \text{if } X^{[\text{P}]} \leq \Lambda \\ X^{[\text{P}]} - \Lambda & \text{if } \Lambda < X^{[\text{P}]} < \Lambda + \Theta \\ \Theta & \text{if } X^{[\text{P}]} \geq \Lambda + \Theta \end{cases} \tag{2.4.7b}$$

Figure 2.4.1(b) shows the reinsurer's intervention.

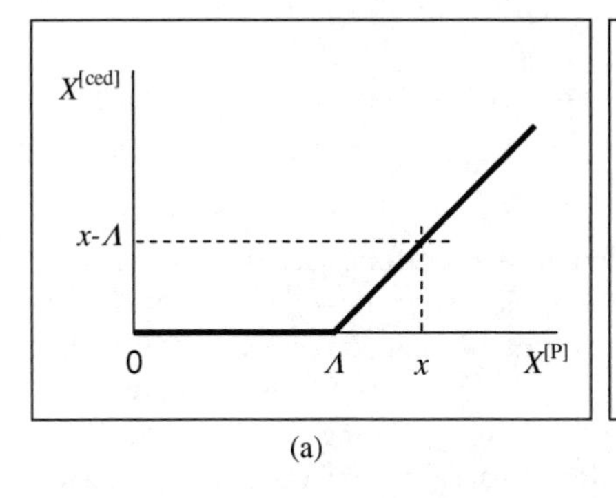

(a)

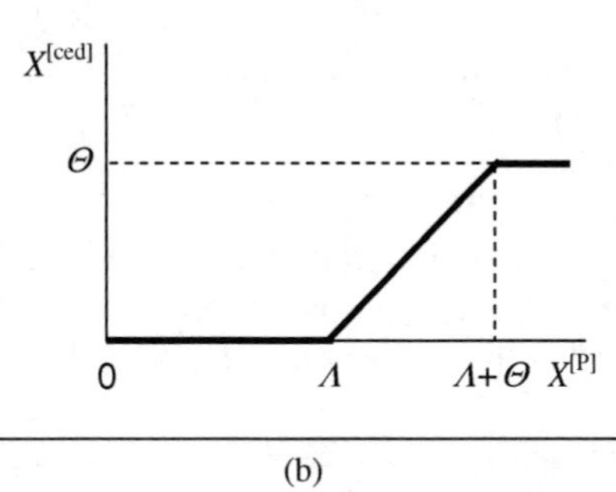

(b)

Fig. 2.4.1 The reinsurer's payment

Note that Eqs. (2.4.6) and (2.4.7) constitute two implementations of the general scheme expressed by Eqs. (2.4.1) and (2.4.2).

When dealing with reinsurance arrangements, the portfolio loss, L, rather than the portfolio result $Z^{[P]}$, is often referred to. The loss of the cedant is given, in the absence of reinsurance, by:

$$L = X^{[P]} - \Pi^{[P]} \tag{2.4.8}$$

Clearly, $L = -Z^{[P]}$.

If a stop-loss reinsurance works (without an upper limit, and hence with $X^{[\text{ced}]}$ defined by Eqs. (2.4.6)), the loss, $L^{[\text{SL}]}$, is given by:

$$L^{[\text{SL}]} = X^{[P]} - \Pi^{[P]} + \Pi^{[\text{reins}]} - X^{[\text{ced}]} = \begin{cases} L + \Pi^{[\text{reins}]} & \text{if } X^{[P]} \leq \Lambda \\ \Lambda - \Pi^{[P]} + \Pi^{[\text{reins}]} & \text{if } X^{[P]} > \Lambda \end{cases} \tag{2.4.9}$$

(see Fig. 2.4.2). Note that, in the presence of reinsurance, the portfolio outgo also includes the reinsurance premium, and thus is given by $X^{[P]} + \Pi^{[\text{reins}]}$, whereas the income is given by $\Pi^{[P]} + X^{[\text{ced}]}$.

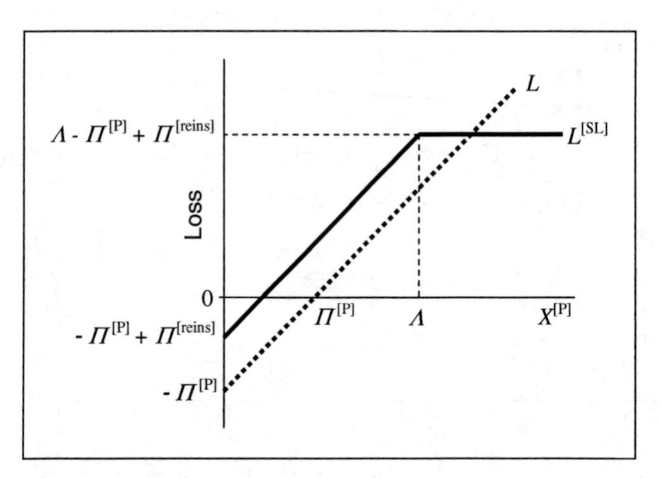

Fig. 2.4.2 The cedant's loss

As the stop-loss reinsurance directly refers to the portfolio loss, it represents in theory the best solution to portfolio protection. However, in practice, it should be noted that this reinsurance form implies a potentially dangerous exposure of the reinsurer, related to the tail of the probability distribution of $X^{[P]}$ (especially if no upper limit is stated). This means that a very high safety loading should be included into the premium $\Pi^{[\text{reins}]}$, possibly making this reinsurance cover extremely expensive. Hence, it is mainly used as an ingredient in a reinsurance programme (see Sect. 2.5.6), after other reinsurance covers have been implemented to protect the portfolio.

2.4.3 From portfolios to contracts

We now move to individual reinsurance arrangements, whose parameters are thus defined at a contract level (rather than a portfolio level), still referring to the "basic" insurance cover.

A *reinsurance policy* at a contract level is defined as

$$\underline{a} = (a^{(1)}, a^{(2)}, \dots, a^{(n)}) \tag{2.4.10}$$

where $a^{(j)}$ $(0 < a^{(j)} \leq 1)$ is the share retained of the j-th contract, i.e. the *retained proportion.*

For any given reinsurance policy $\underline{a}$, relation (2.4.4) becomes:

$$X^{(j)[\text{ret}]} = a^{(j)} X^{(j)} = \begin{cases} a^{(j)} x^{(j)} & \text{in the case of claim} \\ 0 & \text{otherwise} \end{cases} \tag{2.4.11}$$

and hence we have:

$$\mathbb{E}[X^{(j)[\text{ret}]}] = a^{(j)} \mathbb{E}[X^{(j)}] = a^{(j)} P^{(j)} \tag{2.4.12}$$

$$\mathbb{V}\text{ar}[X^{(j)[\text{ret}]}] = (a^{(j)})^2 \mathbb{V}\text{ar}[X^{(j)}] \leq \mathbb{V}\text{ar}[X^{(j)}] \tag{2.4.13}$$

where $P^{(j)}$ denotes the equivalence premium (relying on a realistic basis).

Shares of premiums and, hence, safety loadings (namely, expected profits) are ceded to the reinsurer. For $j = 1, 2, \dots, n$, let $\Pi^{(j)[\text{ret}]}$ and $m^{(j)[\text{ret}]}$ denote the retained share of premium (including the safety loading) and safety loading respectively. Clearly,

$$m^{(j)[\text{ret}]} = \Pi^{(j)[\text{ret}]} - a^{(j)} P^{(j)} \tag{2.4.14}$$

In particular, if

$$\Pi^{(j)[\text{ret}]} = a^{(j)} \Pi^{(j)} \tag{2.4.15}$$

it follows that

$$m^{(j)[\text{ret}]} = a^{(j)} \Pi^{(j)} - a^{(j)} P^{(j)} = a^{(j)} m^{(j)} \tag{2.4.16}$$

However, the ceded share can be different from $(1 - a^{(j)})m^{(j)}$, and, in particular:

- it can be lower, if
  - the reinsurer grants a reward to the cedant for the underwriting work (namely, a *reinsurance commission*);
  - the reinsurer accepts a lower safety loading thanks to a larger portfolio size;
- it can be either lower or higher because the reinsurer adopts a technical basis different from the one adopted by the ceding company, and hence a different premium.

Example 2.4.1. Assume that, for the policy 1 in the portfolio, the sum insured is $x^{(1)} = 1\,000$, and the probability of claim (assessed by the cedant) is $p^{(1)} = 0.01$;

the safety loading is 10% of the equivalence premium $P^{(1)} = 10$, and thus $m^{(1)} = 1$. Hence, $\Pi^{(1)} = 11$. Let $a^{(1)} = 0.70$ be the retained share of the risk.

First, assume that the reinsurer agrees on the technical basis, i.e. on $p^{(1)} = 0.01$, and 10% as the safety loading, and is willing to obtain a proportional share of the safety loading. Thus, for the ceding company we have $m^{(1)[\text{ret}]} = 0.7$, so that $\Pi^{(1)[\text{ret}]} = 7.7$, thus resulting proportional to $\Pi^{(1)}$ according to the retention share.

Secondly, suppose that the reinsurer still agrees on the technical basis, but is willing to leave to the cedant a share of the safety loading higher than 70%, say 80%. Hence, we find

$$\Pi^{(1)[\text{ret}]} = 0.70 P^{(1)} + 0.80 m^{(1)} = 7.8$$

Finally, assume that the reinsurer does not agree on the technical basis. In particular, he accepts a safety loading equal to 10% of the equivalence premium, whilst evaluates the claim probability as $\tilde{p}^{(1)} = 0.012$. Thus, according to the reinsurer's judgement, the equivalence premium should be $\tilde{P}^{(1)} = 12$, and the premium including the safety loading should be $\tilde{\Pi}^{(1)} = 13.2$. If the reinsurer is willing to obtain a proportional share of $\tilde{\Pi}^{(1)}$, namely $0.30 \times 13.2 = 3.96$, the cedant retains

$$\Pi^{(1)[\text{ret}]} = \Pi^{(1)} - 0.30 \tilde{\Pi}^{(1)} = 11 - 3.96 = 7.04$$

and thus

$$m^{(1)[\text{ret}]} = \Pi^{(1)[\text{ret}]} - 0.70 P^{(1)} = 7.04 - 7 = 0.04$$

□

To assess the effect of reinsurance on the portfolio riskiness, we have to look at the retained total payment, $X^{[\text{ret}]}$, and some related typical values, in particular the index defined by (2.3.58).

The retained total payment is defined by (2.4.5). Then, we have

$$\mathbb{E}[X^{[\text{ret}]}] = \mathbb{E}\left[\sum_{j=1}^{n} X^{(j)[\text{ret}]}\right] = \sum_{j=1}^{n} a^{(j)} P^{(j)} \tag{2.4.17}$$

and (assuming the independence among the insured risks)

$$\mathbb{V}\text{ar}[X^{[\text{ret}]}] = \sum_{j=1}^{n} \mathbb{V}\text{ar}[X^{(j)[\text{ret}]}] = \sum_{j=1}^{n} (a^{(j)})^2 \, \mathbb{V}\text{ar}[X^{(j)}] \tag{2.4.18}$$

Let $\sigma^{[\text{ret}]}$ denote the standard deviation of the total payment, that is

$$\sigma^{[\text{ret}]} = \sqrt{\mathbb{V}\text{ar}[X^{[\text{ret}]}]} \tag{2.4.19}$$

Further, let $m^{[\text{ret}]}$ denote the retained safety loading (and hence the retained expected profit):

$$m^{[\text{ret}]} = \sum_{j=1}^{n} m^{(j)[\text{ret}]} \tag{2.4.20}$$

Then, we have:

$$s^{[\text{ret}]} = \frac{m^{[\text{ret}]} + M}{\sigma^{[\text{ret}]}} \tag{2.4.21}$$

From (2.4.21) we can argue that, in the presence of a reinsurance arrangement, the probability of default depends on:

- the effect of reinsurance on the variability of the total payout, expressed by $\sigma^{[\text{ret}]}$;
- the retained share of the total expected profit, expressed by $m^{[\text{ret}]}$.

Note that, in particular, we have:

$$m^{[\text{ret}]} < m \quad \text{and} \quad \sigma^{[\text{ret}]} < \sigma \tag{2.4.22}$$

The probability of default, $\pi(m^{[\text{ret}]} + M)$, is then given by:

$$\pi(m^{[\text{ret}]} + M) = \mathbb{P}[X^{[\text{ret}]} > P^{[\text{ret}]} + m^{[\text{ret}]} + M] = 1 - \Phi\left(\frac{m^{[\text{ret}]} + M}{\sigma^{[\text{ret}]}}\right) = 1 - \Phi(s^{[\text{ret}]}) \tag{2.4.23}$$

(see Eq. (2.3.55))

To quantify the probability of default, and then to determine an appropriate capital allocation, we need to refer to specific reinsurance policies $\underline{a} = (a^{(1)}, a^{(2)}, \dots, a^{(n)})$, and to the rules adopted for splitting the safety loading (see Example 2.4.1 in particular).

2.4.4 Two reinsurance arrangements

The *quota-share reinsurance* is defined by the following policy:

$$\underline{a} = (a, a, \dots, a); \quad 0 < a < 1 \tag{2.4.24}$$

namely, the same retention share is applied to all the individual risks. The effect on the sums insured is illustrated by Fig. 2.4.3 which shows that, in relative terms, all the sums insured are reduced in the same proportion.

For the standard deviation of the portfolio payment, we immediately find:

$$\sigma^{[\text{ret}]} = a\sigma \tag{2.4.25}$$

whereas the retained profit is given by

$$m^{[\text{ret}]} = am \tag{2.4.26}$$

if the reinsurer and the cedant agree on a proportional sharing.

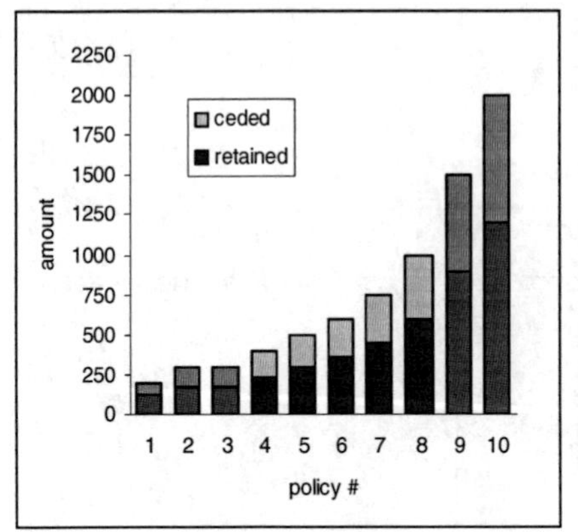

Fig. 2.4.3 Quota-share reinsurance

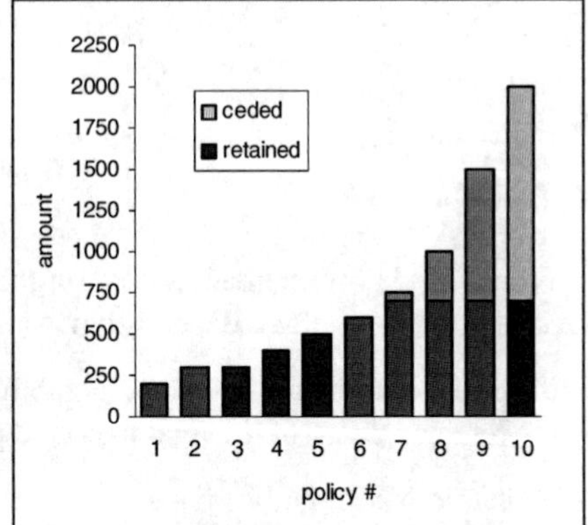

Fig. 2.4.4 Surplus reinsurance

A *surplus reinsurance* arrangement is defined by the retention, $x^{[\text{ret}]}$, in terms of the sum insured. The amount $x^{[\text{ret}]}$ is commonly called the *retention line*. For the generic risk, whose sum insured is $x^{(j)}$, the splitting (see (2.4.4)) is determined as follows:

- the amount $\min\{x^{(j)}, x^{[\text{ret}]}\}$ is retained;
- the amount $\max\{0, x^{(j)} - x^{[\text{ret}]}\}$, i.e. the *surplus*, is ceded.

Hence, the reinsurance policy $\underline{a}$ is defined as follows:

$$a^{(j)} = \frac{\min\{x^{(j)}, x^{[\text{ret}]}\}}{x^{(j)}} = \min\left\{1, \frac{x^{[\text{ret}]}}{x^{(j)}}\right\}; \quad j = 1, 2, \ldots, n \tag{2.4.27}$$

Figure 2.4.4 illustrates the effect of the surplus reinsurance, namely the "leveling" of sums insured.

Intuitively, a higher efficiency is expected from surplus reinsurance, thanks to the leveling effect. It is worth recalling (see Sect. 2.3.3, and formula (2.3.17) in particular) that, as a consequence of a huge sum insured, the diversification via pooling tends to disappear. Clearly, the surplus reinsurance can mitigate this dangerous effect, by leveling (at least to some extent) the sums insured. On the contrary, according to the quota-share arrangement there is no leveling, as all the sums insured are reduced in the same proportion.

Remark We note that, comparing the effects of quota-share and surplus reinsurance is, to some extent, similar to comparing the effects of fixed-percentage deductible and fixed-amount deductible, discussed in Sect. 1.3.4.

2.4.5 Examples

We address the following aspects of reinsurance policies, by using numerical examples:

- first, we discuss the effects of quota-share and surplus reinsurance, in terms of the retained expected profit, the standard deviation of the portfolio payment, and the resulting probability of default $\pi(m^{[\text{ret}]}+M)$ (as given by formula (2.4.23)); see Example 2.4.2;
- then, we compare various combinations of surplus reinsurance and capital allocation, in terms of the retained expected profit and the standard deviation of the portfolio payment, for a fixed level of probability of default; see Example 2.4.3.

Example 2.4.2. We refer to portfolio C, described in Example 2.3.1 (see Table 2.3.3). We assume what follows:

- safety loading rate $\frac{m^{[\text{P}]}}{P^{[\text{P}]}}=0.10$;
- allocated capital $M=10\,000$;
- retained share of premiums (and hence expected profit) equal to retained share of sums insured.

Some comments can help in understanding the higher effectiveness of the surplus reinsurance compared to the quota-share arrangement.

The same amount of retained expected profit, namely $m^{[\text{ret}]}=45\,000$, is achieved with $a=0.90$ and $x^{[\text{ret}]}=6\,000$; however, in the quota-share reinsurance the standard deviation is higher ($\sigma^{[\text{ret}]}=38\,220$ versus $\sigma^{[\text{ret}]}=33\,271$), and hence the probability of default is higher ($\pi(m^{[\text{ret}]}+M)=0.075$ versus $\pi(m^{[\text{ret}]}+M)=0.049$). A similar situation holds for $a=0.75$ and $x^{[\text{ret}]}=3\,000$.

Finally, we note that the same probability of default, $\pi(m^{[\text{ret}]}+M)=0.004$, is achieved in the quota-share with $a=0.157$, and the surplus reinsurance with $x^{[\text{ret}]}=1\,500$; however, the latter arrangement leaves a much higher expected profit ($m^{[\text{ret}]}=33\,750$ versus $m^{[\text{ret}]}=7\,865$).

Table 2.4.1 Quota-share reinsurance - Portfolio C

a	$m^{[\text{ret}]}$	$\sigma^{[\text{ret}]}$	$s^{[\text{ret}]}$	$\pi(m^{[\text{ret}]}+M)$
1.000	50 000	42 467	1.413	0.079
0.900	45 000	38 220	1.439	0.075
0.750	37 500	31 850	1.491	0.068
0.157	7 865	6 680	2.674	0.004

□

Example 2.4.3. We refer to portfolio B, described in Example 2.3.1 (see Table 2.3.2), which consists of 10 000 risks, all with $x=1\,000$ as the sum insured and $p=0.005$

Table 2.4.2 Surplus reinsurance - Portfolio C

$x^{[\text{ret}]}$	$m^{[\text{ret}]}$	$\sigma^{[\text{ret}]}$	$s^{[\text{ret}]}$	$\pi(m^{[\text{ret}]}+M)$
≥ 8000	50000	42467	1.413	0.079
6000	45000	33271	1.653	0.049
5000	42500	28867	1.819	0.034
3000	37500	20864	2.277	0.011
1500	33750	16353	2.675	0.004

as the claim probability. We focus on some combinations of retention line $x^{[\text{ret}]}$ and allocated capital M, leading to the the same probability of default $\pi(m^{[\text{ret}]}+M) = 0.005$, and hence to the same value $s^{[\text{ret}]} = 2.5805$. Thus,

$$\frac{m^{[\text{ret}]}+M}{\sigma^{[\text{ret}]}} = 2.5805$$

We assume that the safety loading rate $\frac{m^{[\text{P}]}}{P^{[\text{P}]}} = 0.10$ is adopted, which leads to $m^{[\text{P}]} = 5000$ (see Table 2.3.10). Then, we find:

$$m^{[\text{ret}]} = \begin{cases} m^{[\text{P}]}\,\dfrac{x^{[\text{ret}]}}{x} = 5\,x^{[\text{ret}]} & \text{for } x^{[\text{ret}]} < 1000 \\ m^{[\text{P}]} = 5000 & \text{for } x^{[\text{ret}]} \geq 1000 \end{cases}$$

Further, we have:

$$\sigma^{[\text{ret}]} = \sqrt{10000\,(x^{[\text{ret}]})^2\,p\,(1-p)} = 100\,x^{[\text{ret}]}\,\sqrt{p\,(1-p)} = 7.053\,x^{[\text{ret}]}$$

so that we find:

$$M = 13.2\,x^{[\text{ret}]}$$

This formula can be generalized (although referring still to the particular portfolio structure) as follows

$$M = \kappa x^{[\text{ret}]} \tag{2.4.28}$$

where the coefficient κ depends, in particular, on the target probability of default. Figure 2.4.5 illustrates the linear relation (2.4.28), for various target probabilities.

Table 2.4.3 illustrates the effects of some choices of retention line and capital allocation (all the combinations leading to the same result in terms of the probability of default, that is 0.005).

□

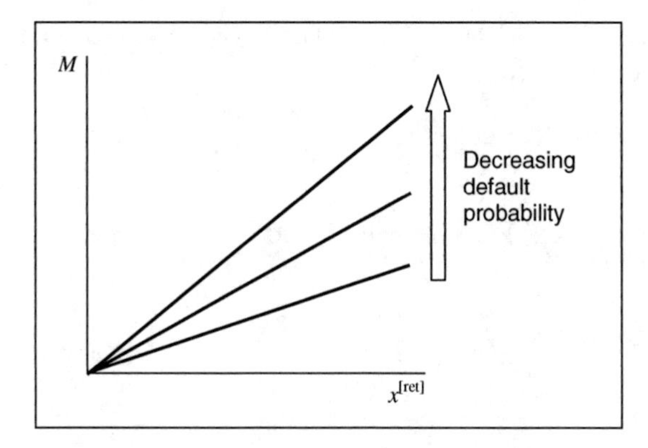

Fig. 2.4.5 Capital allocation versus surplus reinsurance

Table 2.4.3 Capital allocation versus surplus reinsurance - Portfolio B

M	$x^{[ret]}$	$m^{[ret]}$	$\sigma^{[ret]}$
13 200	$\geq$ 1 000	5 000	7 053
6 600	500	2 500	3 527
2 640	200	1 000	1 411
1 320	100	500	705

2.4.6 Optimal reinsurance policy

We consider the following problem: find the reinsurance policy

$$\underline{a} = (a^{(1)}, a^{(2)}, \dots, a^{(n)})$$

which implies the lowest probability of default, out of the set of reinsurance policies leading to the same amount of retained expected profit $m^{[ret]}$. It is worth noting that the results reported below hold in general situations, namely are not restricted to the "basic" insurance cover we have so far addressed.

The problem we are attacking is a problem of constrained optimization. In formal terms, let $\hat{m}^{(j)}$ denote the safety loading of the j-th risk ceded in the case of zero retention (that is, if $a^{(j)} = 0$). As seen in Sect. 2.4.3, we can have $\hat{m}^{(j)} \lesseqgtr m^{(j)}$. Assume that, for any value of $a^{(j)}$ ($0 \leq a^{(j)} \leq 1$), the ceded safety loading is $(1 - a^{(j)})\,\hat{m}^{(j)}$. Then, we have

$$m^{(j)[ret]} = m^{(j)} - (1 - a^{(j)})\,\hat{m}^{(j)} \tag{2.4.29}$$

and, for the total retained safety loading:

$$m^{[ret]} = m^{[P]} - \sum_{j=1}^{n} (1 - a^{(j)})\,\hat{m}^{(j)} \tag{2.4.30}$$

Consider the index $s^{[\text{ret}]}$, defined by (2.4.21), and the probability of default, given by (2.4.23). Note that, under the constraint

$$m^{[\text{ret}]} + M = \text{constant} \tag{2.4.31}$$

we have

$$\min_{\underline{a}}\{\sigma^{[\text{ret}]}\} \Rightarrow \max_{\underline{a}}\{s^{[\text{ret}]}\} \Rightarrow \min_{\underline{a}}\{\pi(m^{[\text{ret}]} + M)\} \tag{2.4.32}$$

where

$$\sigma^{[\text{ret}]} = \sqrt{\sum_{j=1}^{n} (a^{(j)})^2 (\sigma^{(j)})^2} \tag{2.4.33}$$

with $(\sigma^{(j)})^2 = \mathbb{V}\text{ar}[X^{(j)}]$

Hence, the optimization problem is as follows:

$$\min_{\underline{a}} \sum_{j=1}^{n} (a^{(j)})^2 (\sigma^{(j)})^2 \tag{2.4.34}$$

subject to:

$$\begin{cases} \sum_{j=1}^{n} (1 - a^{(j)})\, \hat{m}^{(j)} = A \\ 0 \le a^{(j)} \le 1;\ j = 1, 2, \dots, n \end{cases}$$

We note that the optimization problem is parametric, as its solution depends on the parameter A.

It is possible to prove that the optimal solution is given by:

$$a^{(j)} = \min\left\{1, B \frac{\hat{m}^{(j)}}{(\sigma^{(j)})^2}\right\} \tag{2.4.35}$$

where the parameter B depends, in particular, on the value assigned to A, and hence on the amount of ceded expected profit: the lower is the ceded expected profit, the higher is B and then the retention.

We now return to the "basic" insurance cover, and assume the same claim probability p for all the n risks. Hence, for $j = 1, 2, \dots, n$, we have

$$(\sigma^{(j)})^2 = (x^{(j)})^2 p (1 - p) \tag{2.4.36}$$

Moreover, we assume that, for $j = 1, 2, \dots, n$, the quantity $\hat{m}^{(j)}$ is proportional to the sum insured $x^{(j)}$:

$$\hat{m}^{(j)} = \alpha x^{(j)} \tag{2.4.37}$$

Note that relation (2.4.37) holds, in particular, if:

1. $m^{(j)} = \beta P^{(j)} = \beta p x^{(j)}$,
   and
2. $\hat{m}^{(j)} = m^{(j)}$.

From (2.4.35) it follows that

$$a^{(j)} = \min\left\{1, B\frac{\alpha}{x^{(j)}\,p(1-p)}\right\} \tag{2.4.38}$$

and, in monetary terms:

$$a^{(j)}\,x^{(j)} = \min\left\{x^{(j)}, B\frac{\alpha}{p(1-p)}\right\} \tag{2.4.39}$$

The amount $B\frac{\alpha}{p(1-p)}$ is independent of j, so that we can write:

$$a^{(j)}\,x^{(j)} = \min\left\{x^{(j)}, x^{[\mathrm{ret}]}\right\} \tag{2.4.40}$$

Hence, the solution of the constrained optimization problem (2.4.34) is given by the surplus reinsurance.

It is worth noting that, conversely, if a surplus reinsurance arrangement is adopted, the probability of default is minimized, subject to the loss of expected profit related to the value of A implied by the retention level $x^{[\mathrm{ret}]}$.

2.5 Reinsurance: further aspects

2.5.1 Reinsurance arrangements

Reinsurance arrangements can be classified according to several criteria. In particular, the classification into global reinsurance arrangements (that is, on a portfolio basis) and individual arrangements (on a policy basis) has been mentioned in Sect. 2.4 (see also Fig. 2.5.1).

When a reinsurance arrangement is defined on a *policy basis*, the relevant parameters concern the individual risks (for example: the share a in the quota-share reinsurance, the retained line $x^{[\mathrm{ret}]}$ in the surplus reinsurance). Another reinsurance arrangement belonging to this category, the so-called Excess-of-loss reinsurance, will be described in Sect. 2.5.2.

The parameters of reinsurance arrangements defined on a *portfolio basis* relate to quantities concerning the portfolio total payment (for example, the priority Λ and the upper limit Θ in the stop-loss reinsurance). Another reinsurance arrangement belonging to this category, the so-called catastrophe reinsurance, will be described in Sect. 2.5.3.

According to another criterion, reinsurance arrangements can be classified into proportional and non-proportional arrangements (see Fig. 2.5.1).

In a *proportional* reinsurance arrangement, claims and premiums are divided between the cedant and the reinsurer in the ratio of their shares in the reinsurance contract. Hence, the sharing of claims is determined when the reinsurance arrangement is defined. Quota-share and surplus reinsurance belong to this category.

In a *non-proportional* reinsurance arrangement, the rule for the sharing of claims is stated when the reinsurance contract is defined, but the actual sharing of claims is determined depending on the severity of each claim, or the number of claims in the portfolio, or the total portfolio payment. Examples are given by the stop-loss, the catastrophe and the XL reinsurance.

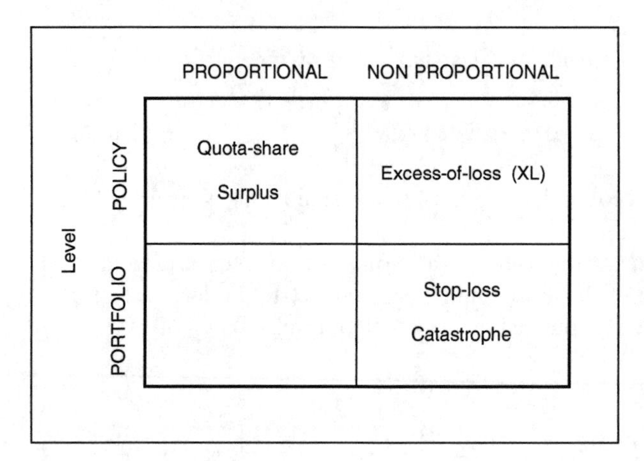

Fig. 2.5.1 Reinsurance arrangements

2.5.2 *Random claim sizes. XL reinsurance*

Other features of the reinsurance arrangements we have already dealt with, namely quota-share and surplus reinsurance, emerge when moving to individual risks more general than those related to the basic insurance cover, in particular by allowing for random claim sizes. For example, we can refer to risks described as Cases 3d (A fire in a factory) and 3e (Car driver's liability) in Sect. 1.2.4. Further, the specific role of the Excess-of-Loss reinsurance emerges if we allow for random claim sizes.

Let us refer to the j-th risk in the portfolio. An example of the (continuous) probability distribution of the generic k-th claim, $X_k^{(j)}$, is provided, in terms of the related density function, by Fig. 2.5.2; $x_{\max}^{(j)}$ represents the maximum possible outcome.

In a quota-share arrangement, with *retention share* a for all the risks in the portfolio, the retained amount is defined as follows:

$$X_k^{(j)[\mathrm{ret}]} = a X_k^{(j)} \tag{2.5.1}$$

In a surplus reinsurance, with $x^{[\mathrm{ret}]}$ as the retained line, we have:

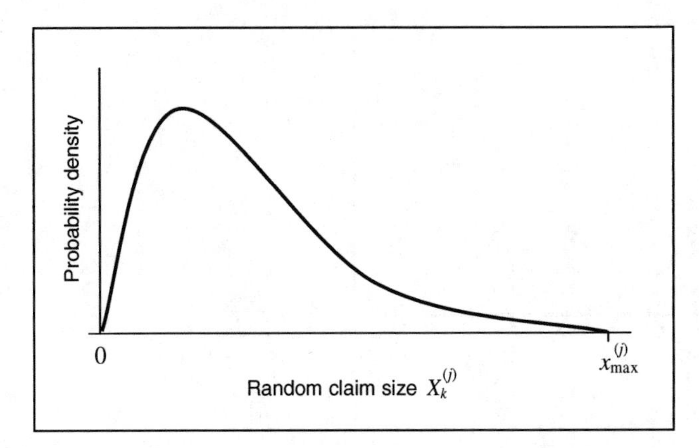

Fig. 2.5.2 Probability density of the random payment in a claim

$$X_k^{(j)[\mathrm{ret}]} = \frac{x^{[\mathrm{ret}]}}{x_{\max}^{(j)}} X_k^{(j)} \tag{2.5.2}$$

We note that, while in a quota-share arrangement the retained share is trivially equal to a for all the risks in the portfolio, according to the surplus reinsurance the retained share is $\frac{x^{[\mathrm{ret}]}}{x_{\max}^{(j)}}$, and hence depends on $x_{\max}^{(j)}$ which is specific to each insured risk.

The retention and the reinsurer's intervention in the *Excess-of-Loss* reinsurance (briefly, XL reinsurance) are defined as follows:

$$X_k^{(j)[\mathrm{ret}]} = \min\{X_k^{(j)}, \Lambda\} \tag{2.5.3a}$$

$$X_k^{(j)[\mathrm{ced}]} = \max\{X_k^{(j)} - \Lambda, 0\} \tag{2.5.3b}$$

where Λ denotes the *deductible*. The analogy with the deductible in a generic risk transfer is apparent (see Sect. 1.3.4, and Eqs. (1.3.4)).

We note that, in this simple XL arrangement, the reinsurer pays the whole amount beyond the deductible, net of the deductible itself, namely no upper-limit has been stated. The retained share decreases as the claim size $X_k^{(j)}$ increases; see Fig. 2.5.3. Indeed, from (2.5.3a) we have:

$$\frac{X_k^{(j)[\mathrm{ret}]}}{X_k^{(j)}} = \min\left\{1, \frac{\Lambda}{X_k^{(j)}}\right\} \tag{2.5.4}$$

Assume, conversely, that the upper limit of the reinsurance cover is set to $h\Lambda$ (with h an integer number, $h \geq 2$). For a generic claim with random size $X_k^{(j)}$, possible situations are as follows:

1. if $X_k^{(j)} \leq \Lambda$, then the insurer totally retains the claim amount;
2. if $\Lambda < X_k^{(j)} \leq h\Lambda$, then the XL cover exhausts the cession;

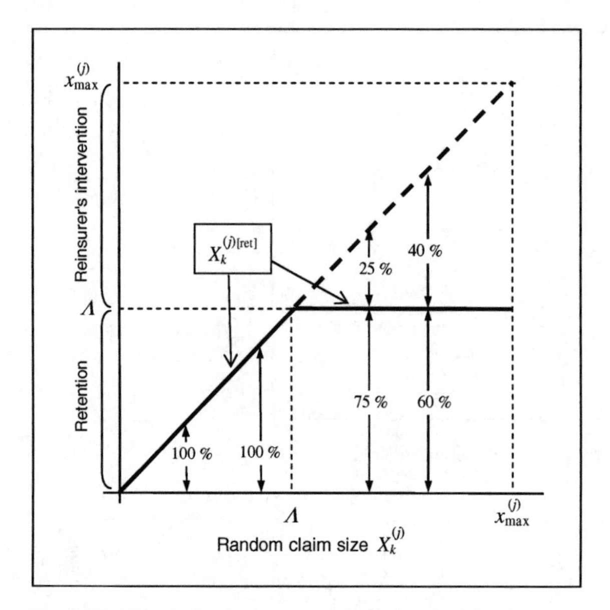

Fig. 2.5.3 The retained payment of the cedant in XL reinsurance (no upper limit)

3. if $X_k^{(j)} > h\Lambda$, then the insurer has still to cede $X_k^{(j)} - h\Lambda$, through a second XL cover (or possibly more XL covers), with another reinsurer (or even with the first reinsurer, however according to a technical basis usually different from the one used in the first cover).

Hence, the cession is split into two (or more) *layers*: the first layer covers the interval $(\Lambda, h\Lambda)$, whereas the interval $(h\Lambda, X_k^{(j)})$ can be covered by a further XL reinsurance (or more than one XL). See Fig. 2.5.4, where it has been assumed $h = 3$.

2.5.3 Catastrophe reinsurance

The *Catastrophe reinsurance* (briefly, *Cat-XL*) is a non-proportional reinsurance arrangement, at a portfolio level. Its aim is to protect the portfolio (and the insurance company) against the risk that a single accident (that is, a "catastrophe") causes a huge number of claims in the portfolio itself. For example:

- in a generic portfolio, a high number of claims can occur because of a disaster (hurricane, earthquake, and so on);
- in "a group insurance", a number of insureds can suffer body injuries owing to a single accident in the workplace (explosion, fire, collapse, and so on); see, for example, Cases 3b (Disability benefits; one-year period) and 3c (Disability benefits; multi-year period) in Sect. 1.7.2.

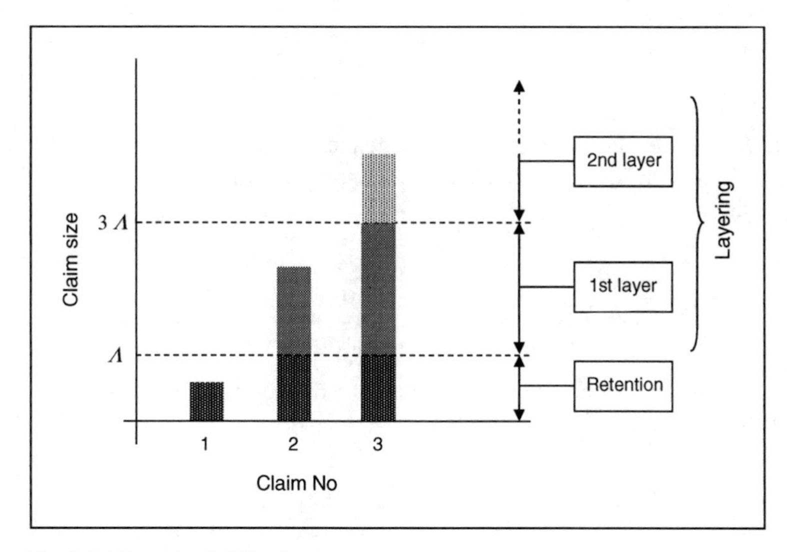

Fig. 2.5.4 Layering in XL reinsurance

A *catastrophe* is usually defined in term of a given (minimum) number of claims, c, within a time interval of a given (maximum) duration, for example 48 hours. In formal terms, let K denote the random number of claims, X the consequent total payment (before reinsurer's intervention); the reinsurer will intervene only if $K \geq c$.

There are various definitions of the Cat XL structure. We just focus on the two following definitions.

First, the Cat XL arrangement can be defined on a claim-number basis. Let $k^{[\mathrm{ret}]}$ denote the deductible in terms of number of claims. Then, the cedant's retention and the reinsurer's intervention are respectively given by:

$$X^{[\mathrm{ret}]} = \min\left\{X, \frac{k^{[\mathrm{ret}]}}{K} X\right\} \tag{2.5.5a}$$

$$X^{[\mathrm{ced}]} = \max\left\{0, \frac{K - k^{[\mathrm{ret}]}}{K} X\right\} \tag{2.5.5b}$$

Note that, according to this definition, if X is large then $X^{[\mathrm{ret}]}$ is large. Thus, the reinsurance arrangement is effective if individual claims have approximately the same amount, and hence the total payment X mainly depends on the number of claims. Otherwise, effectiveness can be gained via a preliminary surplus or XL reinsurance.

Another definition of the Cat XL arrangement is based on the amount X of the total payment. Let $x^{[\mathrm{ret}]}$ denote the deductible (in monetary terms). Then:

$$X^{[\text{ret}]} = \min\{X, x^{[\text{ret}]}\} \tag{2.5.6a}$$

$$X^{[\text{ced}]} = \max\{X - x^{[\text{ret}]}, 0\} \tag{2.5.6b}$$

Example 2.5.1. Consider the Cat XL reinsurance defined by Eqs. (2.5.5), with $c = 5$, and $k^{[\text{ret}]} = 8$. According to the outcome of the number of claims, K, we have the following situations:

$$K = \underbrace{\overbrace{1,2,3,4}^{\text{no cat}},\ 5,6,7,8,}_{\text{no reinsurer's intervention}} \quad \underbrace{9,10,11,\dots}_{\text{reinsurer's intervention}}$$

Move to the Cat XL arrangement defined by Eqs. (2.5.6), still with $c = 5$, and with $x^{[\text{ret}]} = 1200$. Then, we have

$$K = \overbrace{1,2,3,4}^{\text{no cat}}, \qquad \underbrace{5,6,7,8,9,10,11,\dots}_{\text{possible reinsurer's intervention, depending on } X}$$

Consider the following cases:

(a) $K = 10,\ X = 1\,000$;
(b) $K = 10,\ X = 5\,000$.

In case (a), according to the first Cat XL arrangement we have:

$$X^{[\text{ret}]} = \frac{8}{10} 1\,000 = 800, \quad X^{[\text{ced}]} = \frac{2}{10} 1\,000 = 200$$

whereas the second arrangement yields:

$$X^{[\text{ret}]} = 1\,000, \quad X^{[\text{ced}]} = 0$$

In case (b), the first arrangement leads to:

$$X^{[\text{ret}]} = \frac{8}{10} 5\,000 = 4\,000, \quad X^{[\text{ced}]} = \frac{2}{10} 5\,000 = 1\,000$$

while the second yields

$$X^{[\text{ret}]} = 1\,200, \quad X^{[\text{ced}]} = 3\,800$$

□

2.5.4 Purposes of reinsurance

Although, from a strictly actuarial point of view, it is apparent that reinsurance arrangements aim to keep the portfolio riskiness at a level acceptable by the insurance company, resorting to reinsurance can have various purposes. Some considerations follow.

1. As regards the reduction of the portfolio riskiness, it should be noted that reinsurance arrangements mainly aim at reducing the impact of random fluctuations and catastrophic events. In fact, the reinsurance company is willing to take the ceded risks as it can achieve a higher pooling effect and an improved diversification of risks (see Sect. 2.3.1). From the point of view of the cedant, more insurance implies:
   - a lower capital allocation;
   - an increased underwriting capacity.

   Conversely, risks affected by possible systematic deviations could be rejected by reinsurers, as these deviations affect the pool as an aggregate, and the total impact on portfolio results increases as the portfolio size increases. Notwithstanding, the reinsurer can take the risk of systematic deviations, with the proviso that a further transfer of this risk can be worked out. We will address this issue in Sect. 2.6.
2. The cedant company can benefit from technical advise provided by the reinsurer. In particular:
   - the reinsurer, thanks to specific experience, can suggest statistical bases and inform about market features for new insurance products;
   - as regards in-force portfolios, the reinsurer can provide the cedant with an update of statistical bases (which is more effective if a quota-share arrangement works, as this allows the reinsurer to monitor all claims pertaining to the reinsured portfolio).
3. Reinsurance can have a "financing" role, thanks to a sharing of policy and portfolio expenses between the cedant and the reinsurer.

2.5.5 Insurance-reinsurance networks

Figure 2.5.5 illustrates an *insurance-reinsurance network*. Following the paths marked by solid arrows, we firstly find an example of *direct insurance* (or *primary insurance*): insurer X directly takes risks from clients A1, A2, ..., An. Hence, X works in the insurance market. Then, we find examples of *cession*: insurer X cedes risks to Y and Z; for example, policies implying a huge exposure are only partially accepted by Y, so that the residual portions are ceded to Z. Thus, companies Y and Z provide company X with *reinsurance*. Finally, company Y cedes to W part of the risks taken from X; this reinsurance transaction is called *retrocession*.

Further examples can be found following the paths marked by dashed arrows. First, we find another example of direct insurance: insurer Y directly takes risks from clients B1, B2, ..., Bm. Note that company Y works both in primary insurance and in reinsurance as well, as it takes risks ceded by company X. The relationship between X and Y is twofold, as Y also cedes risks to company Y. Finally, we note that companies X and Y share a risk ceded by client B1, and this constitutes an example of *coinsurance*.

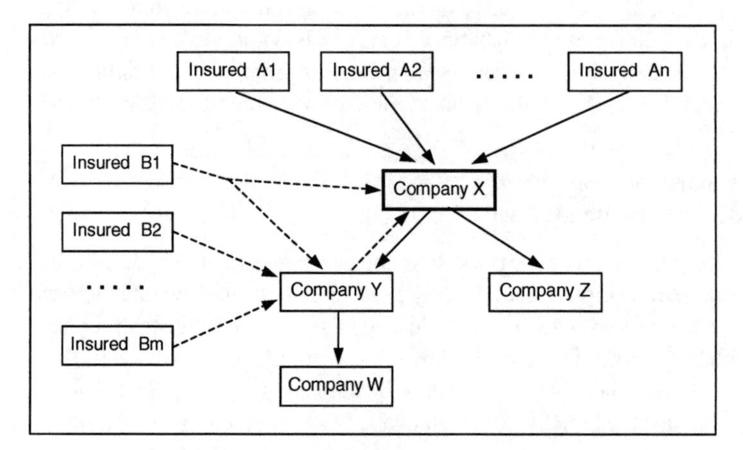

Fig. 2.5.5 Insurance, coinsurance, and reinsurance: a network

Reinsurance arrangements can be stated on various bases, for the cedant and the reinsurer respectively:

1. facultative / facultative (briefly, *facultative*);
2. obligatory / obligatory (briefly, *obligatory*);
3. facultative / obligatory (briefly, *facob*).

In arrangement type 1, an insurer can cede a risk to an insurer, and the reinsurer can accept the risk itself. Usually, this arrangement concerns the cession of single risks, in particular those involving huge exposures.

Types 2 and 3 require that a reinsurance contract, usually called a *treaty*, has been written by the cedant and the reinsurer. In particular, in an arrangement of type 2 the insurer is obliged to cede portions (as defined in the treaty) of the risks underwritten, and the reinsurer is obliged to accept them. In type 3, the insurer can decide to cede risks and, if so, the reinsurer is obliged to take them.

2.5.6 Reinsurance treaties. Reinsurance programmes

A *reinsurance treaty* concerns all the aspects of a reinsurance arrangement, in particular:

- the time interval of the reinsurance cover;
- the reinsurance form (stop-loss, quota-share, XL, and so on);

- the *limitations* of the reinsurance cover (priority, upper limit, deductibles, retention lines, and so on);
- the technical bases for the calculation of the reinsurance premiums, and the conditions concerning the premium payment.

Limitations to a reinsurance cover can be classified into "vertical" and "horizontal" limitations. *Horizontal limitations* refer to the total reinsurer's payment related to the cover interval; an example is provided by the upper limit in the stop-loss reinsurance (see Sect. 2.4.2).

Vertical limitations concern the reinsurer's payment related either to each single claim or to each single policy. An example of vertical limitations concerning each single claim is provided by the *layering* in the XL arrangement (see Sect. 2.5.2).

A *reinsurance programme* combines several reinsurance treaties, possibly supplemented by facultative reinsurance when needed (for example, in relation to single huge exposures), and can involve various reinsurers. Resorting to reinsurance programmes is more common in non-life insurance, because of the random size of the claims and, hence, the higher riskiness.

Usually, reinsurance programmes are designed on a class-by-class basis, namely separate reinsurance programmes concern, for example, fire insurance, third party liability, domestic property, and so on. Notwithstanding, reinsurance programmes can include special treaties arranged to cover risks, although belonging to various classes, in specific geographic areas, for example exposed to the risk of hurricanes, or earthquakes.

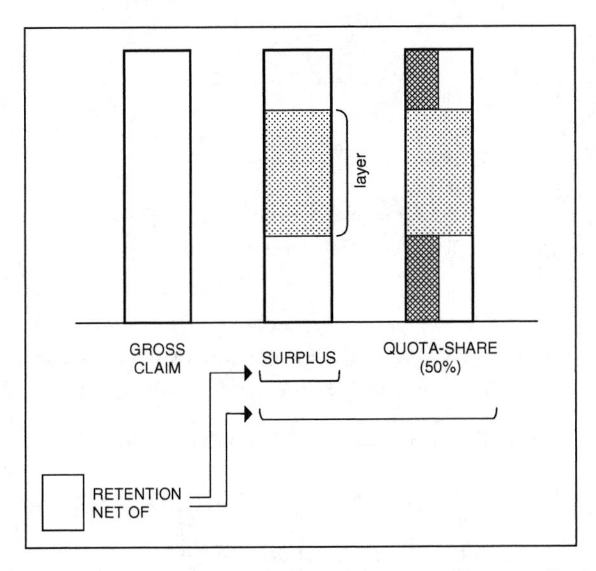

Fig. 2.5.6 Applying a reinsurance programme; effects at policy level

Applying a reinsurance programme to each individual risk within a portfolio determines a progressive reduction of the cedant's exposure, and hence of the default probability. Figure 2.5.6 illustrates the effect, at a policy level, of a surplus reinsurance followed by a quota-share reinsurance. Figure 2.5.7, conversely, illustrates the effects on the portfolio exposure, for which a stop-loss arrangement supplements the proportional reinsurance covers. At a portfolio level, a reinsurance programme can be formally represented by a sequence of m mathematical operators, each one corresponding to a reinsurance form:

$$X^{[\mathrm{ret}]} = f_m\left[\cdots f_3\left[f_2\left[f_1[X^{(1)}, X^{(2)}, \ldots, X^{(n)}]\right]\right]\right] \tag{2.5.7}$$

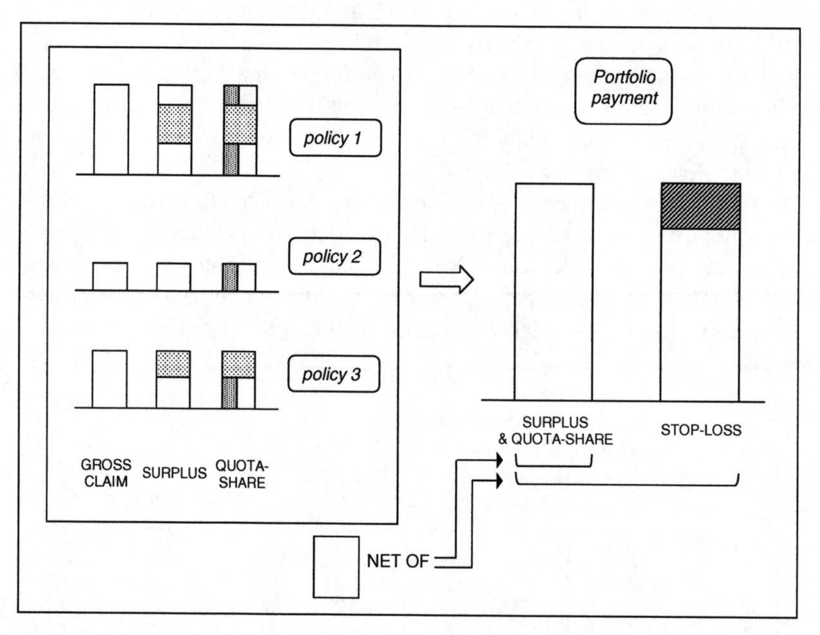

Fig. 2.5.7 Applying a reinsurance programme; effects at portfolio level

Combining quota-share and surplus arrangements provides basic examples of reinsurance programmes. Assume the retention share a for the quota-share, and the retention line $x^{[\mathrm{ret}]}$ for the surplus. We have, for the j-th risk, the following results:

- a quota-share "followed" by a surplus reinsurance leads to the retention

$$x^{(j)[\mathrm{ret1}]} = \min\{ax^{(j)}, x^{[\mathrm{ret}]}\} \tag{2.5.8}$$

- a surplus "followed" by a quota-share leads to the retention

$$x^{(j)[\mathrm{ret2}]} = a\min\{x^{(j)}, x^{[\mathrm{ret}]}\} \tag{2.5.9}$$

2.6 Alternative risk transfers

2.6.1 Some preliminary ideas

The (traditional) *insurance - reinsurance process* can be split into two basic steps (see Fig. 2.6.1):

1. the *insurance step*, which consists in transferring risks from individuals (families, firms, institutions, and so on) to an insurance company, and whose effects are
   a. building-up a pool;
   b. reducing the relative riskiness (caused by random fluctuations);
2. the *reinsurance step*, which consists in transferring risks from the insurance company (the cedant) to the reinsurer, and whose effects are
   a. building-up larger pools;
   b. a further reduction of the relative riskiness (caused by random fluctuations).

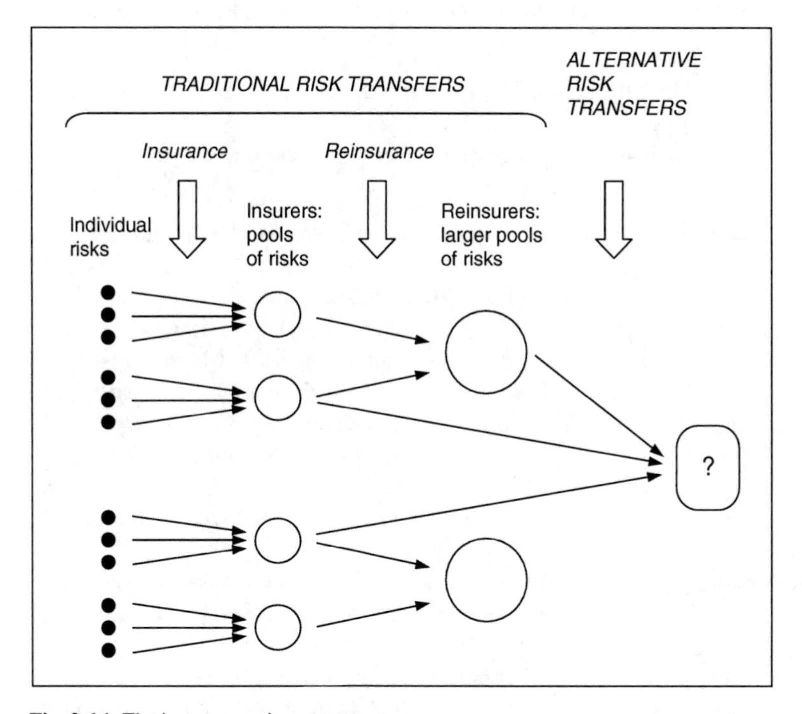

Fig. 2.6.1 The insurance-reinsurance process

However, risk components other than random fluctuations can affect insurers' and reinsurers' results, namely systematic deviations and catastrophic events. As regards the latter, larger pools can improve diversification, for instance thanks to an increased variety of geographical locations of insured risks. As regards the former, the relative impact of systematic deviations is independent of the pool size (and the absolute impact increases as the pool size increases). Thus, risk transfer arrangements other than the traditional reinsurance, namely Alternative Risk Transfers (*ART*),

- are needed for transferring (at least to some extent) the risk of systematic deviations;
- can help in managing the catastrophe risk (lowering the cost of reinsurance, and / or the need for capital allocation).

In the following sections we will focus on ART in life insurance and reinsurance.

2.6.2 Securitization and the role of capital markets

Securitization consists in packaging a pool of assets or, more generally, a cash-flow stream into securities traded on the market. The aims of a securitization transaction can be:

- to raise liquidity by selling future flows (such as recovery of acquisition costs or expected profits);
- to transfer risks whenever contingent payments or random cash-flows are involved.

Since new securities are issued, a counter-party risk arises for the investor (see below).

The organizational aspects of a securitization transaction are rather complex. Figure 2.6.2 sketches a simple design for a life insurance deal, focussing on the main agents involved. The transaction starts in the insurance market, where policies underwritten give rise to the cash-flows which are securitized (at least in part). The insurer then sells the right to some cash-flows to a *Special Purpose Vehicle* (*SPV*), which is a financial entity established to link the insurer to the capital market. Securities backed by the chosen cash-flows are issued by the SPV, which raises money from the capital market. Such funds are (at least partially) acknowledged to the insurer.

According to the specific features of the transaction, further items may be added to the structure. For example, a fixed interest rate could be paid to investors, so that intervention by a Swap counter-party is required; see Fig. 2.6.3.

As it has been pointed out above, some counterpart risk is originated by the securitization transaction. This is due to a possible default of the insurer with respect to the obligations assumed against the SPV, as well as of policyholders in respect of the insurer, for example in the form of lapses which affect the securitized cash-flow stream. To reduce such default risks, some form of credit enhancement may be introduced, both internal (e.g. transferring to the SPV a higher value of cash-flows

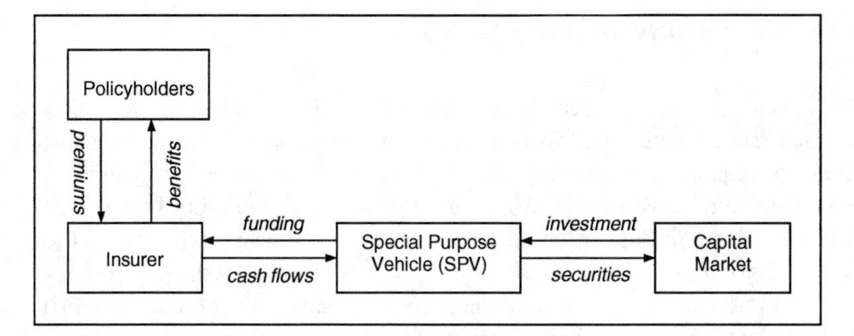

Fig. 2.6.2 The securitization process in life insurance: a basic structure

than those required by the actual size of the securities) and external, through intervention of a specific entity (issuing, for example, credit insurance, letters of credit, and so on); see again Fig. 2.6.3. Further counterpart risk emerges from the other parties involved, similarly to any financial transaction. Note that intervention by a third financial institution may anyhow result in an increase of the rating of the securities.

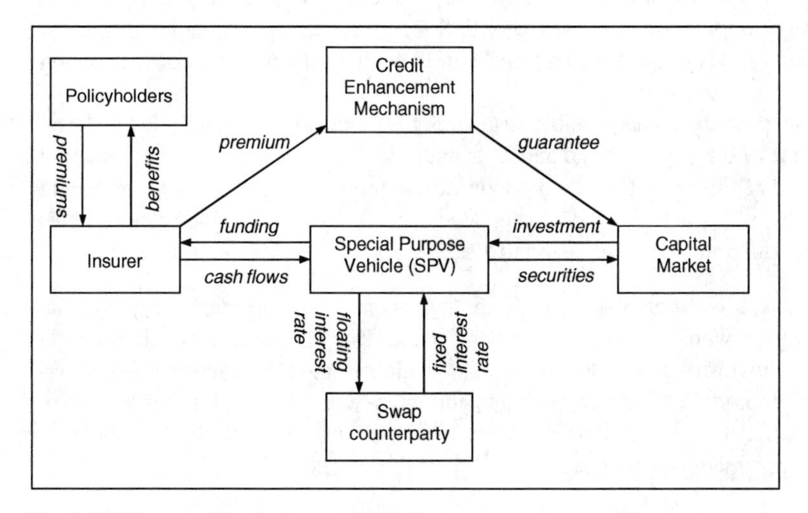

Fig. 2.6.3 The securitization process in life insurance: a more complex structure

Further details of the securitization transaction concern services for payments provided by external bodies, investment banks trading the securities on the market, and so on. Since we are interested on the main technical aspects of the securitization process, we do not go deeper into these topics (which, anyhow, do play an important role for the success of the overall transaction).

2.6.3 An example: the mortality bonds

Mortality-linked securities are securities whose pay-off is contingent on mortality in a given *reference population*; this is obtained, in particular, by embedding some derivatives whose underlying is a mortality index assessed on the given population. These securities may serve two opposite purposes: to hedge mortality higher than expected, or survivorship higher than expected. In the former case, we will refer to them as *mortality bonds*, in the latter as *longevity bonds*. We restrict the terminology to "bond", without making explicit reference (in the name) to the derivative which is included in the security (which could be option-like, swap-like or other) because we are more interested on hedging opportunities rather than on the organizational aspects of the deal (of course, we are anyhow aware of the importance that such aspects play from a practical point of view, but their discussion goes beyond the aim of this Section).

The purpose of mortality bonds is to provide liquidity in the case of mortality in excess of what expected, possibly owing to epidemics or natural disasters. So typically a short position on the bond may hedge liabilities of an insurer/reinsurer dealing with life insurances. Mortality bonds are typically short term (3-5 years) and they are linked to a mortality index expressing the frequency of death observed in the reference population in a given period. Some thresholds are set at bond issue. If the mortality index outperforms a threshold, then either the principal or the coupon are reduced.

We now describe some possible structures for mortality bonds. In what follows, 0 is the time of issue of the bond and T its maturity. Further, S_t denotes the principal of the bond at time t, and C_t the coupon due at time t. Finally, with I_t we denote the mortality index at time t years from bond issue ($t = 0, 1, \ldots, T$). Some examples will be provided in Example 2.6.1 and 2.6.2.

Example 2.6.1. The bond aims at protecting against high mortality experienced throughout the whole lifetime of the bond itself. This is obtained by reducing the principal at maturity. Albeit just some ages could be considered in detecting situations of high mortality, it is reasonable to address a range of ages. Further, the index should account for mortality over the whole lifetime of the bond. So the following quantities represent possible examples of mortality index:

$$I_T = \max_{t=1,2,\ldots,T} \{q(t)\} \tag{2.6.1}$$

$$I_T = \frac{\sum_{t=1}^{T} q(t)}{T} \tag{2.6.2}$$

where $q(t)$ is the observed annual mortality rate averaged over the reference population in year t.

At maturity the principal paid-back to investors is

$$S_T = S_0 \times \begin{cases} 1 & \text{if } I_T \le \lambda' I_0 \\ \Phi(I_T) & \text{if } \lambda' I_0 < I_T \le \lambda'' I_0 \\ 0 & \text{if } I_T > \lambda'' I_0 \end{cases} \tag{2.6.3}$$

where $I_0 = q(0)$, λ' and λ'' are two parameters (stated under bond conditions), with $1 \le \lambda' < \lambda''$, and $\Phi(I_T)$ is a proper decreasing function, such that $\Phi(\lambda' I_0) = 1$ and $\Phi(\lambda'' I_0) = 0$. For example

$$\Phi(I_T) = \frac{\lambda'' I_0 - I_T}{(\lambda'' - \lambda') I_0} \tag{2.6.4}$$

The coupon is independent of the experienced mortality. In particular, it can be given by

$$C_t = S_0 (i_t + r) \tag{2.6.5}$$

where i_t is the market interest rate at time t, and r is an extra-yield rewarding investors for taking the mortality risk.

□

While the cash-flows related to the bond described in Example 2.6.1 try to match the flows in the life insurance portfolio just at the end of a period of some years, an alternative design of the mortality bond can be conceived to provide a match on a yearly basis.

Example 2.6.2. Assume that the coupon is given by

$$C_t = S_0 \times \begin{cases} i_t + r & \text{if } I_t \le \Lambda'_t \\ (i_t + r)\,\Psi(I_t) & \text{if } \Lambda'_t < I_t \le \Lambda''_t \\ 0 & \text{if } I_t > \Lambda''_t \end{cases} \tag{2.6.6}$$

where Λ'_t, Λ''_t set two mortality thresholds. For example,

$$\Lambda'_t = \lambda' \mathbb{E}[D_t] \tag{2.6.7}$$

$$\Lambda''_t = \lambda'' \mathbb{E}[D_t] \tag{2.6.8}$$

where $1 \le \lambda' < \lambda''$, and $\mathbb{E}[D_t]$ is the expected number of deaths in the reference population (according to a specified mortality assumption). In this structure, the mortality index I_t should express the number of deaths in year $(t-1,t)$. The function $\Psi(I_t)$ should then be decreasing; for example:

$$\Psi(I_t) = \frac{\Lambda''_t - I_t}{\Lambda''_t - \Lambda'_t} \tag{2.6.9}$$

As in (2.6.5), the rate r in (2.6.6) is the extra-yield rewarding investors for the mortality risk inherent in the pay-off of the bond. Note that, in this structure, the principal at maturity can be assumed independent of the experienced mortality, for example

$$S_T = S_0 \tag{2.6.10}$$

□

2.7 The time dimension

2.7.1 General aspects

Insurance contracts with durations longer than one year have been addressed in Chap. 1; see, for example, Cases 4a (The need for resources at retirement), and 4b (Early death of an individual) in Sect. 1.7.4. Nonetheless, in the present Chapter, for the sake of simplicity, we have mainly focussed on one-year insurance covers; see, for example, Sects. 2.3 and 2.4, in which the "basic" insurance cover, namely the Case 2 (Possible loss with fixed amount), has been referred to.

However, a one-year (or, more in general, a one-period) insight of the management of an insurance portfolio, whatever the policy term, can provide us just with a static perspective. Conversely, a number of problems of practical interest can be properly defined and solved only allowing for a sequence of periods, that is, according to a dynamic perspective. The evolution throughout time of the portfolio fund, which originates from premium income and claim payment, and the related capital allocation policies constitute important examples of a perspective involving the "time dimension".

When defining a multi-period analysis of a portfolio (or an insurance company), various approaches are available. For simplicity, we assume that all the policies in the portfolio have the same policy term r. In Figs. 2.7.1 to 2.7.3 various policy generations are represented with the aid of a coordinate system that has the calendar time as abscissa and the duration as ordinate. The solid part of each line represents the part of the related generation accounted for according to the various approaches.

A *run-off* analysis only addresses the "in-force" portfolio, namely the policies already written. Thus, the portfolio is assumed to be "closed" to new entries, and hence no future business is accounted for. See Fig. 2.7.1.

Conversely, according to a *going-concern* approach the portfolio is assumed "open", and hence also future business is allowed for. See Fig. 2.7.2. Of course, such an approach requires an estimate of the numbers of policies written in the future years.

The *break-up* (or *wind-up*) approach, on the contrary, consists in analyzing the insurer's capability of meeting all the obligations assuming that the insurance company has to stop all business within a very short period (say, one year). Figure 2.7.3 refers to this approach.

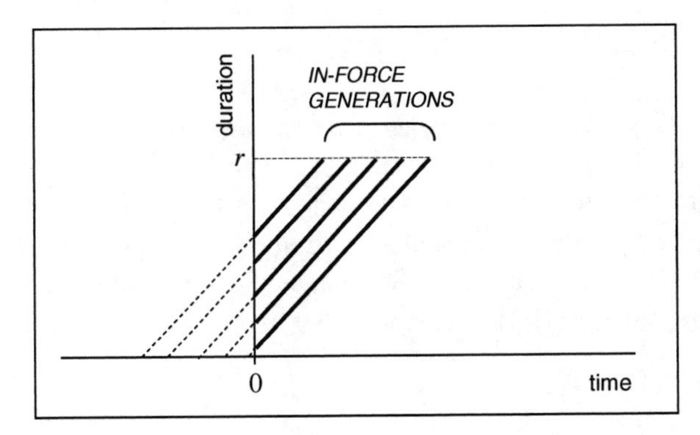

Fig. 2.7.1 *Run-off* of a portfolio

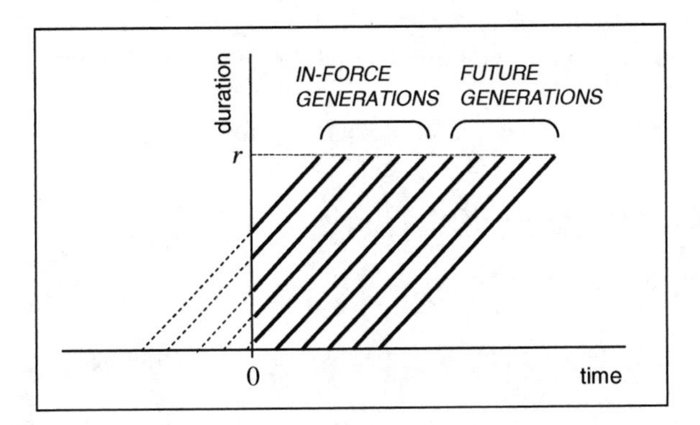

Fig. 2.7.2 A *going-concern* portfolio

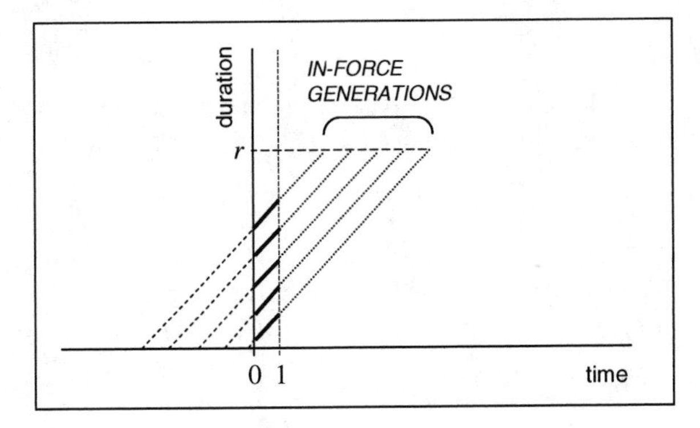

Fig. 2.7.3 *Break-up* of a portfolio

2.7.2 Premiums, payments, portfolio fund

Consider a portfolio consisting of n one-year policies providing the "basic" insurance cover, namely the cover related to Case 2 (Possible loss with fixed amount). According to a going-concern approach, we assume a time horizon of T years.

As regards the first year, let $\Pi_0^{[P]}$ denote the premium income (including safety loading) at the beginning of the year, i.e. at time 0. Such amount is assumed to be known. Further, let $X_1^{[P]}$ denote the total random payment, that is

$$X_1^{[P]} = \sum_{j=1}^{n} X^{(j)} \tag{2.7.1}$$

We assume that, at the beginning of each future year, the insurer underwrites new policies, which constitute a generation of the same type of the first one (possibly, however, with a variable size).

So, we generalize the one-year portfolio model by defining, for $t = 1,2,\dots,T$, the following quantities:

$\Pi_{t-1}^{[P]}$ = premium income at time $t-1$, i.e. at the beginning of year t
$X_t^{[P]}$ = total payment in year t

The annual portfolio result, $Z_t^{[P]}$, referred at the end of the year, can be defined as follows:

- if we disregard the time-value of the money, we have

$$Z_t^{[P]} = \Pi_{t-1}^{[P]} - X_t^{[P]} \tag{2.7.2}$$

- conversely, if we assume that all the claims are paid at the end of the year of occurrence, and that i is the yield on investment, we have

$$Z_t^{[P]} = \Pi_{t-1}^{[P]}\,(1+i) - X_t^{[P]} \tag{2.7.3}$$

According to the second assumption, the *portfolio fund* (or *surplus*), $F_t^{[P]}$, $t = 1,2,\dots$, is defined as follows:

$$F_t^{[P]} = \sum_{h=1}^{t} Z_h^{[P]}\,(1+i)^{t-h} = \sum_{h=0}^{t-1} \Pi_h^{[P]}(1+i)^{t-h} - \sum_{h=1}^{t} X_h^{[P]}\,(1+i)^{t-h} \tag{2.7.4}$$

With the (provisional) assumption

$$F_0^{[P]} = 0 \tag{2.7.5}$$

we then find:

$$F_t^{[P]} = F_{t-1}^{[P]}\,(1+i) + Z_t^{[P]};\; t = 1,2,\dots \tag{2.7.6}$$

namely

$$F_t^{[P]} = (F_{t-1}^{[P]} + \Pi_{t-1}^{[P]})(1+i) - X_t^{[P]}; \; t = 1,2,\ldots \tag{2.7.7}$$

From recursion (2.7.6), it clearly appears that, as regards the annual results, the hypothesis underlying the definition of $F_t^{[P]}$ is the accumulation of profits (and possibly losses) in the portfolio fund.

If the portfolio fund takes, for some t, a negative value, a *default* (or *ruin*) situation occurs. To lower the probability of such an event, shareholders' capital should be allocated to the portfolio, in particular at time $t = 0$. If M_0 denotes the (initial) allocation, the portfolio fund process must be redefined as follows:

$$F_t^{[P]} = M_0 \, (1+i)^t + \sum_{h=1}^{t} Z_h^{[P]} \, (1+i)^{t-h} \tag{2.7.8}$$

which implies

$$F_0^{[P]} = M_0 \tag{2.7.9}$$

in recursions (2.7.6) and (2.7.7).

2.7.3 Solvency and capital requirements

As seen in Sect. 2.3.7, the insurer's solvency should be meant in a probabilistic sense, namely as the capability of meeting, with an assigned (high) probability, the random payments as described by a probabilistic model (which specifies the claim probability and, as regards more general insurance covers, the probability distribution of the claim size, interest rates, expenses, and so on).

The following quantities must be stated:

- the probability of meeting the random payments (say 0.99, or 0.995, ...);
- the quantity representing the insurer's solvency level; for example, the portfolio fund $F_t^{[P]}$ can be addressed; if, at time t, we have $F_t^{[P]} < 0$, then the portfolio is in the default state;
- the time horizon which the concept of solvency is referred to (say 2 years, or 5 years, ...).

Note that the time horizon must be chosen, as we are working in a multi-year framework.

In formal terms, the following equation expresses the solvency requirement, when the fund $F_t^{[P]}$ is addressed to check the solvency:

$$\mathbb{P}[F_1^{[P]} \geq 0 \cap F_2^{[P]} \geq 0 \cap \cdots \cap F_T^{[P]} \geq 0] = 1 - \alpha \tag{2.7.10}$$

where $1 - \alpha$ denotes the stated probability of meeting the random payments (and hence α denotes the accepted default probability).

In order to achieve the stated probability $1-\alpha$, Eq. (2.7.10) has to be solved with respect to M_0, which enters the definition of the portfolio fund $F_t^{[P]}$ via Eq. (2.7.8).

The following equation represents an alternative solvency requirement:

$$\mathbb{P}[F_T^{[P]} \geq 0] = 1-\alpha \tag{2.7.11}$$

For a given probability α, Eq. (2.7.11) expresses a requirement weaker than that expressed by (2.7.10) (trivially, if $T > 1$). Note, however, that temporary negative values of the portfolio fund $F_t^{[P]}$ are feasible only if capital outside the portfolio is available and can be used for an immediate reinstatement of the fund. Thus, requirement (2.7.11) should not be adopted when referring to the whole insurance company.

Remark Solvency concepts described above generalize ideas presented in Sect. 2.3.7, referring to the one-period model. In particular, we note that, given the expression (2.7.8), requirements expressed by (2.7.10) and (2.7.11) can be interpreted as generalizations of the solvency requirement (2.3.59).

To achieve a required degree of solvency $1-\alpha$, Eq. (2.7.10), or (2.7.11) must be solved with respect to capital allocation M_0. In practice, numerical methods based on Montecarlo simulation must be adopted to solve those equations. The simulation procedure consists in generating a sample of paths of $F_t^{[P]}$, for $t = 1, 2, \ldots, T$. Then, the probability $\mathbb{P}[F_1^{[P]} \geq 0 \cap F_2^{[P]} \geq 0 \cap \cdots \cap F_T^{[P]} \geq 0]$ can be estimated via the sample frequency

$$\frac{\text{number of paths with } F_t^{[P]} \geq 0 \text{ for } t = 0, 1, \ldots, T}{\text{number of simulations}} \tag{2.7.12}$$

whereas the probability $\mathbb{P}[F_T^{[P]} \geq 0]$ can be estimated via

$$\frac{\text{number of paths with } F_T^{[P]} \geq 0}{\text{number of simulations}} \tag{2.7.13}$$

Example 2.7.1. We refer to a portfolio initially consisting of $n = 10000$ one-year policies, all with sum insured $x = 1000$, and claim probability $p = 0.005$. Assuming a safety loading rate equal to 10%, we have a premium income $\Pi_0^{[P]} = 55000$. Further, we assume a time horizon of $T = 5$ years, and suppose that at the beginning of each future year a new generation, with the same size and structure of the first one, enters the portfolio. Finally, we assume an initial capital allocation $M_0 = 10000$.

Figure 2.7.4 illustrates 50 paths of the portfolio fund. It has been assumed that times of claim occurrence and payment are uniformly distributed over each year. Time-value of the money has been disregarded (that is, setting $i = 0$). Moreover, the construction of the statistical distribution of the portfolio fund $F_5^{[P]}$, relying on the simulated paths, is sketched.

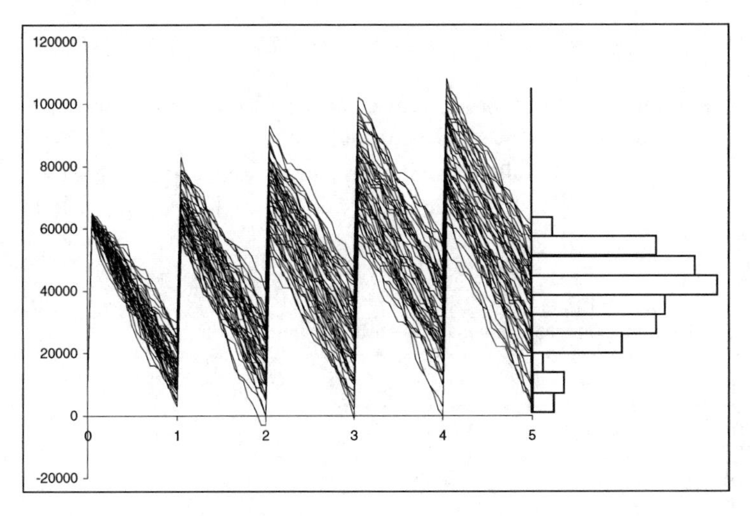

Fig. 2.7.4 50 paths of the portfolio fund

Finally, Figs. 2.7.5 and 2.7.6 show the statistical distribution of the fund $F_1^{[P]}$ and $F_5^{[P]}$ respectively. In particular, it is interesting to note the higher dispersion of the fund at time $t = 5$. Further, both statistical distributions reveal a positive frequency of negative values of the portfolio fund. Clearly, risk management actions should be taken (e.g. a higher capital allocation) if these frequencies seem to be to high.

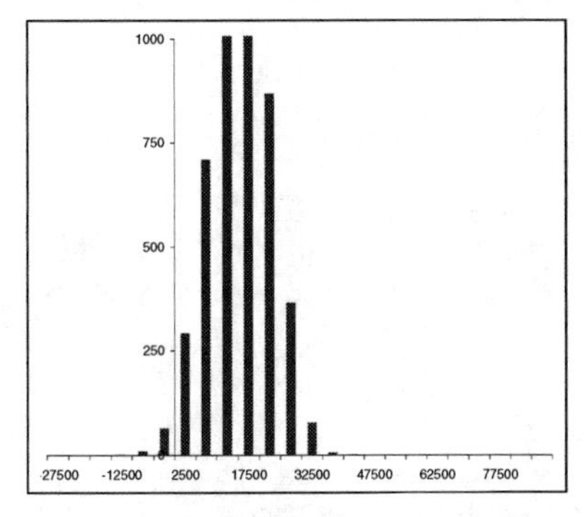

Fig. 2.7.5 Statistical distribution of $F_1^{[P]}$ (5000 simulations)

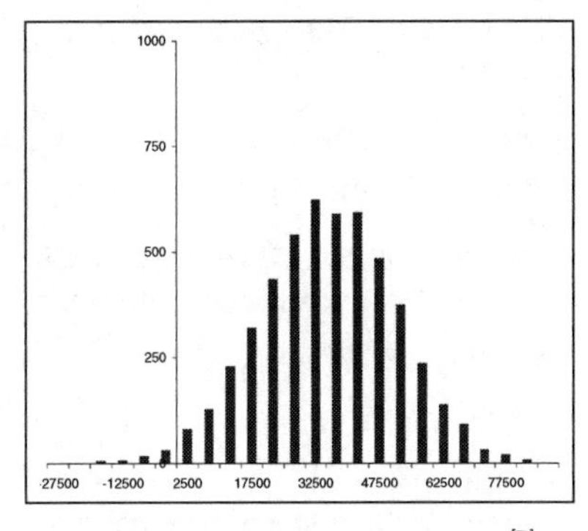

Fig. 2.7.6 Statistical distribution of $F_5^{[P]}$ (5000 simulations)

□

Example 2.7.2. To provide an example of capital allocation effects on the solvency degree, we still refer to the portfolio described in Example 2.7.1. Table 2.7.1 shows some probabilities related to the behavior of the portfolio fund $F_t^{[P]}$. Of course, all the probabilities depend on the initial capital allocation M_0, and, in particular, increase as M_0 increases.

If we choose, according to the criterion expressed by Eq. (2.7.11), a solvency degree $1-\alpha = 0.99$, the required capital allocation is $M_0 = 10000$: indeed $\mathbb{P}[F_5^{[P]} \geq 0] \approx 0.99$. Conversely, this allocation implies a lower solvency degree if the criterion expressed by (2.7.10) is adopted: in fact, we find $\mathbb{P}[F_1^{[P]} \geq 0 \cap F_2^{[P]} \geq 0 \cap \dots \cap F_5^{[P]} \geq 0] \approx 0.95$.

Table 2.7.1 Probabilities concerning the non-negativity of the portfolio fund

M_0	$\mathbb{P}[F_1^{[P]} \geq 0]$	$\mathbb{P}[F_5^{[P]} \geq 0]$	$\mathbb{P}[F_1^{[P]} \geq 0 \cap \dots \cap F_5^{[P]} \geq 0]$
0	0.7844	0.9454	0.6954
5000	0.9264	0.9710	0.8606
10000	0.9848	0.9894	0.9518
14000	0.9970	0.9928	0.9788

Finally, we note that if M_0 is equal to 0, or anyhow is small, compensations among period results are possible, as we can realize by comparing $\mathbb{P}[F_1^{[P]} \geq 0]$ to $\mathbb{P}[F_5^{[P]} \geq 0]$.
□

2.7.4 Generalizing the model

The model described above can be generalized in various ways. We just outline some ideas. For example, we can assume that:

1. policies are issued throughout each year according to a time-uniform stream; this implies a time-continuous premium income; the premium income cumulated up to time t, $\Pi^{[P]}(t)$, is given by
$$\Pi^{[P]}(t) = \Pi^{[P]} t \tag{2.7.14}$$
where $\Pi^{[P]}$ denotes the annual income, assumed constant over time;
2. each (one-year) policy can claim more times over the year;
3. each claim has a random size.

Note that, thanks to assumptions 2 and 3 a more realistic representation of claims in a portfolio is achieved. In the time-continuous setting, it is usual to define, for any real t ($t \geq 0$), the following quantities:

$$\begin{aligned} K(t) &= \text{number of claims up to time } t \\ X^{[P]}(t) &= \text{total payment cumulated up to time } t \end{aligned}$$

The quantity $K(t)$, as a function of t, is called the *claim number process*, whereas $X^{[P]}(t)$ is called the *aggregate claim process* (see Figs. 2.7.7 and 2.7.8).

If we disregard the time-value of the money (namely, if we assume $i = 0$), the *portfolio fund process*, $F^{[P]}(t)$, can be defined as follows:

$$F^{[P]}(t) = M_0 + \Pi^{[P]}(t) - X^{[P]}(t) \tag{2.7.15}$$

(see Fig. 2.7.9).

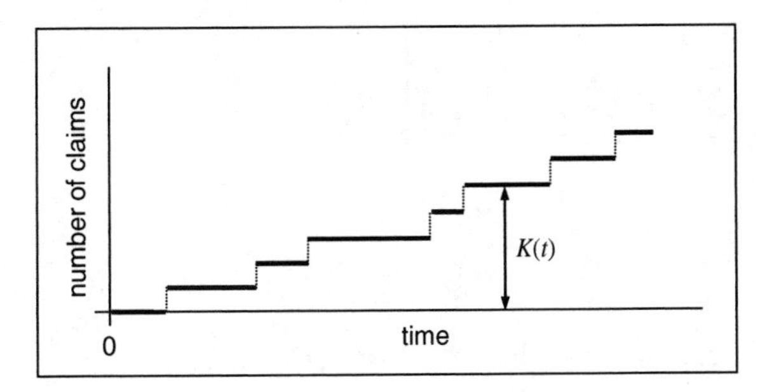

Fig. 2.7.7 The claim number process

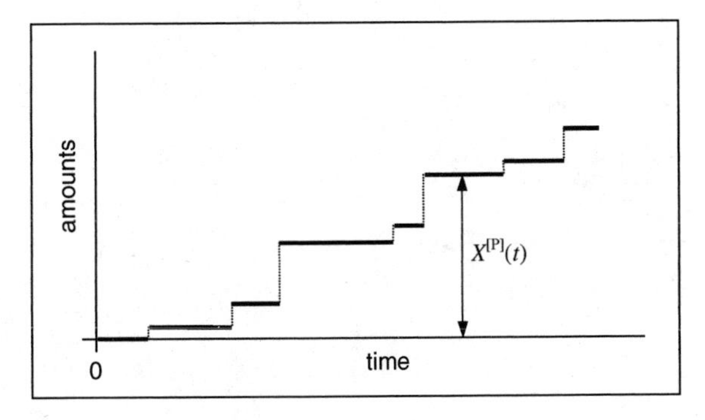

Fig. 2.7.8 The aggregate claim process

2.7.5 Solvency and capital flows

Also capital allocation strategies, aiming at solvency, can be redesigned in a more general context. We still assume that the amount M_0 represents the initial capital allocation. Then, we assume that capital flows can take place in various anniversaries, with the following purposes:

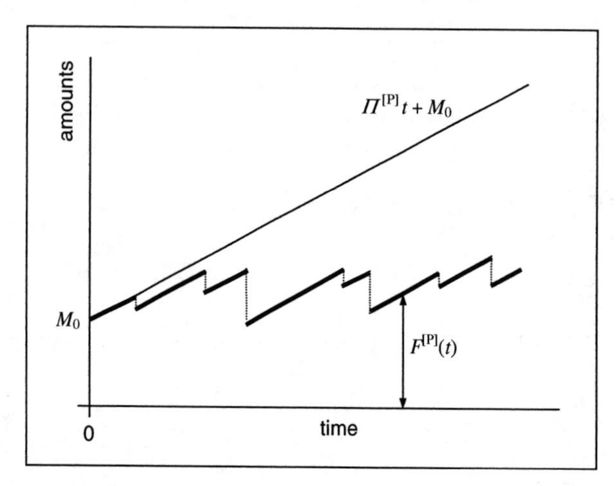

Fig. 2.7.9 The portfolio fund process

- to protect the portfolio against possible default (see Figs. 2.7.10 and 2.7.11);
- to release capital exceeding a reasonable solvency target (see Fig. 2.7.12).

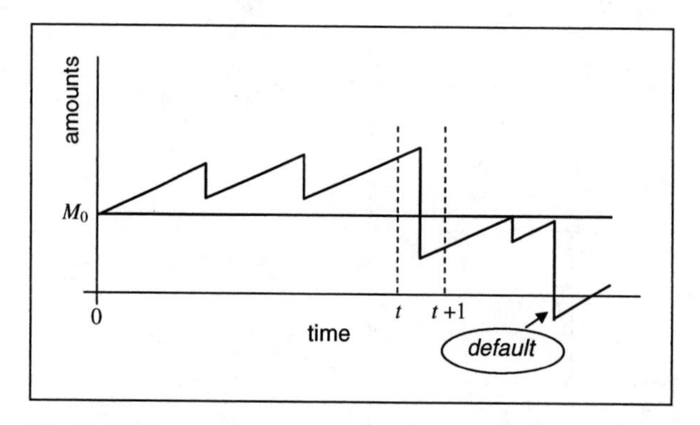

Fig. 2.7.10 Portfolio fund process incurring in default

Note that this more general setting can be properly represented in terms of a *barrier model*: the two barriers provide thresholds which suggest capital release and, respectively, capital allocation to reinstate the portfolio solvency.

Remark Simulations of real-world portfolios require a significant computation time, especially when a multi-year framework is involved. Hence, alternative approaches leading to feasible formulae, which can approximate the relevant results, can be very useful in insurance practice. In particular, the so-called *short-cut formulae* express the required capital, for example M_0, as a function of some known quantities (e.g. the total amount of insured benefits, the total amount of premiums, etc.) and a set of parameters which should reflect the risk profile of the portfolio (or the insurance company). Formulae of this type are proposed, for instance, by the supervisory authorities.

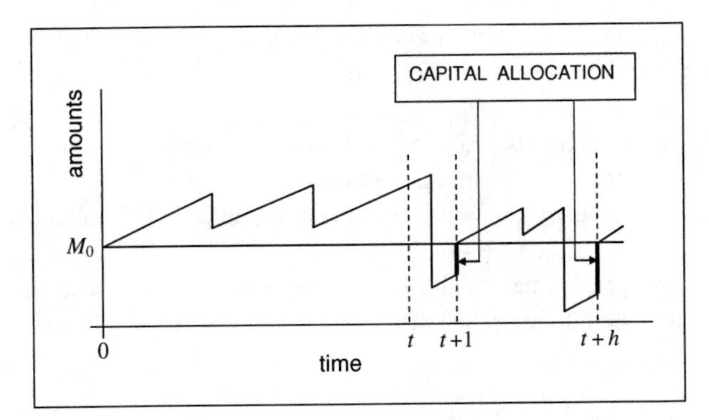

Fig. 2.7.11 Portfolio fund process with further capital allocations

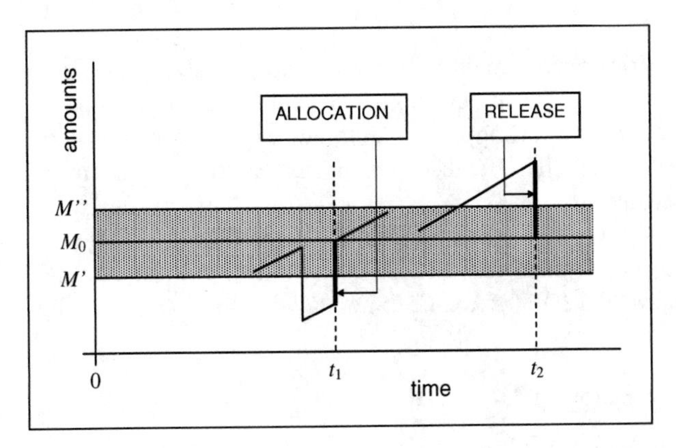

Fig. 2.7.12 Portfolio fund process with capital flows according to a "barrier" model

2.8 References and suggestions for further reading

Also in this Section, as in Sect. 1.8, we only cite textbooks dealing with general aspects of risks and insurance. Studies specifically devoted to non-life insurance, life insurance and post-retirement solutions will be cited in the relevant sections of the following chapters.

Chapters 6, 9 and 15 in [4] focus on managing risks, the need for capital and solvency issues, respectively.

The textbook [9] deals with various technical and financial aspects of life and non-life insurance and pension funds. All the important topics of risk theory are presented in [17], which provides a significant bridge between theory and insurance practice.

Quantitative tools, and in particular statistical models, used in non-life insurance are described in [33].

The object of [13] is to explain the fundamental principles and practice of non-life reinsurance. A more technical presentation of reinsurance issues is provided by [17].

The transfer of risks to capital markets via insurance-linked securities is dealt with by [3]. In [2], longevity bonds are in particular addressed.

An extensive presentation of solvency issues, with specific reference to a number of supervisory systems, is given by [50].

Finally, [27] provides extensive information about the early history of risk theory and insurance mathematics and technique up to 1919.

2.A Appendix

As noted in Sect. 2.3.4, various approximations to the (exact) probability distribution of the total random payment $X^{[P]}$ can be adopted. Whatever the approximating distribution may be, the goodness of the approximation must be carefully assessed, especially with regard to the right tail of the distribution itself, as this tail quantifies the probability of large payments.

The following examples can provide some ideas about the degree of approximation obtained by using the Poisson (see (2.3.22) to (2.3.24)) and the Normal approximation (see (2.3.25) to (2.3.30)) to the binomial distribution (given by (2.3.21)).

Example 2.A.1. Assume the following data:

- individual loss: $x^{(j)} = 1$, for $j = 1, \dots, n$;
- probability: $p = 0.005$;
- pool sizes: $n = 100$, $n = 500$, $n = 5000$.

The (exact) binomial distribution and the normal approximation have been adopted for $n = 500$ and $n = 5000$; the (exact) binomial distribution and the Poisson approximation have been used for $n = 100$. Tables 2.A.1 to 2.A.3 and Figures 2.A.1 and 2.A.2 show numerical results.

The following aspects should be stressed. In relation to portfolio sizes $n = 500$ and $n = 5000$, the normal approximation tends to underestimate the right tail of the payment distribution (see Table 2.A.1). Conversely, the Poisson distribution provides a good approximation to the exact distribution, also for $n = 100$ (see Tables 2.A.2 and 2.A.3); unlike the normal approximation, the Poisson model tends to overestimate the right tail, so that a prudential assessment of the payment follows.
□

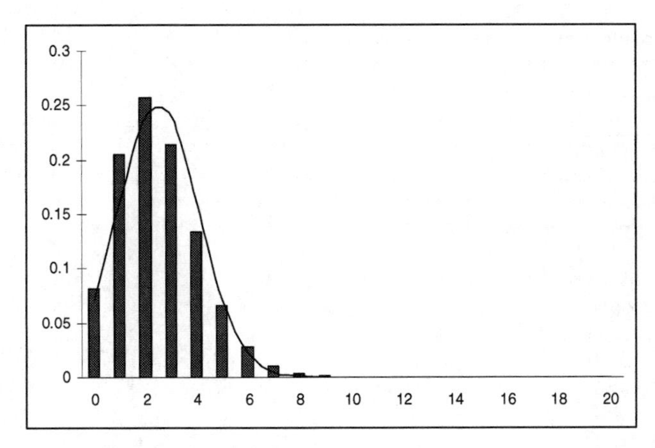

Fig. 2.A.1 Probability distribution of the random payment ($n = 500$). Binomial distribution and Normal approximation

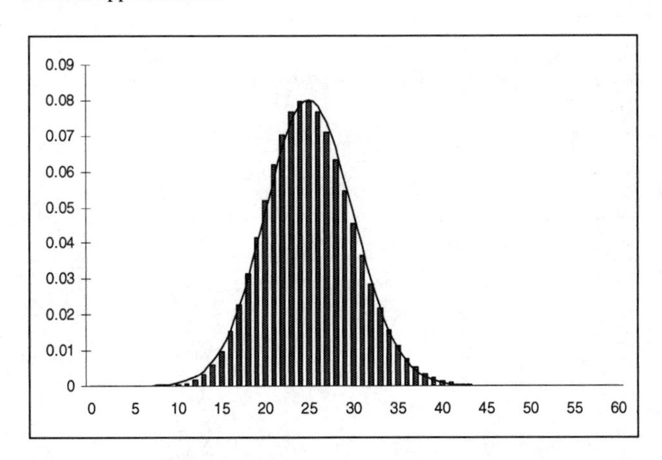

Fig. 2.A.2 Probability distribution of the random payment ($n = 5\,000$). Binomial distribution and Normal approximation

Table 2.A.1 Right tails of Binomial distribution and Normal approximation

	$n = 500$ $\mathbb{P}[X^{[P]} > k]$			$n = 5000$ $\mathbb{P}[X^{[P]} > k]$	
k	Binomial	Normal	k	Binomial	Normal
5	0.04160282	0.056471062	30	0.136121887	0.158048811
6	0.013944069	0.013238288	35	0.022173757	0.022480517
7	0.004135437	0.002164124	40	0.001983179	0.001316908
8	0.001097966	0.000244022	45	0.000101743	3.03545E-05
9	0.000263551	1.88389E-05	50	3.13201E-06	2.68571E-07
10	5.76731E-05	9.90663E-07	55	6.02879E-08	8.9912E-10
...	...	...	...	...	...

Table 2.A.2 Binomial distribution and Poisson approximation

	$n=100$	
	$\mathbb{P}[X^{[P]}=k]$	
k	Binomial	Normal
0	0.605770436	0.60653066
1	0.304407255	0.30326533
2	0.075719392	0.075816332
3	0.012429649	0.012636055
4	0.001514668	0.001579507
5	0.000146139	0.000157951
6	1.16275E-05	1.31626E-05
7	7.84624E-07	9.40183E-07
8	4.58355E-08	5.87614E-08
9	2.35447E-09	3.26452E-09
10	1.07667E-10	1.63226E-10
...	...	...

Table 2.A.3 Right tails of Binomial distribution and Poisson approximation

	$n=100$	
	$\mathbb{P}[X^{[P]}>k]$	
k	Binomial	Normal
3	0.001673268	0.001752
4	0.000158599	0.000172
5	1.24604E-05	1.42E-05
6	8.32926E-07	1.00E-06
7	4.83022E-08	6.22E-08
...	...	...

Chapter 3
Life insurance: modeling the lifetime

3.1 Introduction

When writing insurance contracts, the insurer takes risks originating from various causes. In life insurance, causes of risk relate to financial aspects (e.g. investment yield, inflation, etc.), demographical aspects (e.g. lifetimes of policyholders, lapses and surrenders, etc.), expenses. In this Chapter we deal with demographical aspects only, focussing on policyholders' lifetimes, which in turn determine the frequency of death in a portfolio.

A number of risk factors affect individual mortality. Important risk factors are age, gender, health status, profession, smoking habits, etc. So, formulae used to calculate premiums and reserves for life insurance and annuity products should allow for various risk factors. In particular, the insured's age enters formulae via the age-pattern of mortality, that is a structure linking probabilities of survival and death to the attained age.

The age-pattern of mortality can be specified, in quantitative terms, by using various "tools". A common choice, rather usual in actuarial practice, consists in taking the so-called life table as the basis for premium and reserve calculation.

Remark Mortality data and mortality assumptions constitute a critical issue in life insurance technique. However, need for mortality data and models also arise in a number of other fields, for example: social security, pension funds, health care (both public and private), and so on.

3.2 Life tables

3.2.1 Elements of a life table

The expression *life table* is commonly used to denote a set of sequences, like those represented in Table 3.2.1. The first column indicates the age, denoted by x. In the second column, the l_x's represent the estimated (rounded) numbers of people alive

A. Olivieri, E. Pitacco, *Introduction to Insurance Mathematics*,
DOI 10.1007/978-3-642-16029-5_3, © Springer-Verlag Berlin Heidelberg 2011

at age x in a properly defined population. The exact meaning of the l_x's will be explained after discussing two approaches to the calculation of these numbers. Whatever the exact meaning, the numbers $l_0, l_1, l_2, \ldots$, constitute a decreasing sequence. Note that, in Table 3.2.1, $l_{109} \approx 0$; thus, 108 represents the maximum attainable age, or *limiting age*. This age is usually denoted by ω; hence, $l_\omega > 0$ whilst $l_{\omega+1} = 0$.

The sequences of d_x's and q_x's are strictly related to the l_x's. In particular, d_x denotes the number of deaths between exact age x and $x+1$; thus

$$d_x = l_x - l_{x+1} \tag{3.2.1}$$

Note that

$$\sum_{x=0}^{\omega} d_x = l_0 \tag{3.2.2}$$

The quantity q_x is the probability of an individual aged x dying within 1 year, and can be expressed as follows

$$q_x = \frac{d_x}{l_x} \tag{3.2.3}$$

Expression (3.2.3) will be discussed further in the following sections.

The graphs obtained by plotting the l_x's and the d_x's against age x are usually called the *survival curve* and the *curve of deaths* respectively; see Example 3.2.1.

Table 3.2.1 A life table

x	l_x	d_x	$1000 q_x$
0	100 000	879	8.788
1	99 121	46	0.461
2	99 076	33	0.332
...	...	...	...
50	93 016	426	4.582
51	92 590	459	4.961
...	...	...	...
108	1	1	1 000.000
109	≈ 0	≈ 0	–

3.2.2 *Cohort tables and period tables*

Assume that the sequence $l_0, l_1, \ldots, l_\omega$ is directly provided by statistical evidence, that is by a *longitudinal mortality observation* of the actual numbers of individuals alive at age $1, 2, \ldots, \omega$, out of a given initial *cohort* consisting of l_0 newborns. Thus, the observation is *by year of birth*. The sequence $l_0, l_1, \ldots, l_\omega$ is called a *cohort life table*.

If ω is the limiting age, then the construction of the cohort table requires $\omega + 1$ years.

Assume, conversely, that the statistical evidence consists of the frequency of death at the various ages, observed throughout a given period, for example one year. Thus, the mortality observation is *by year of death*. Further, assume that the frequency of death at age x (possibly after a graduation with respect to x) is an estimate of the probability q_x.

Then, for $x = 0, 1, \ldots, \omega - 1$, define

$$l_{x+1} = l_x (1 - q_x) \tag{3.2.4}$$

with l_0 (the *radix*) assigned (e.g. $l_0 = 100000$), and ω denoting, as previously, the age such that $l_\omega > 0$ and $l_{\omega+1} = 0$ (or $l_{\omega+1} \approx 0$). Hence, l_x is the expected number of survivors out of a notional cohort (also called a *synthetic cohort*) initially consisting of l_0 individuals. The sequence $l_0, l_1, \ldots, l_\omega$, defined by recursion (3.2.4), is called a *period life table*, as it is derived from period mortality observations.

Period observations are also called *cross-sectional observations*, because they analyze (in terms of the frequency of death) an existing population "across" the various ages. Note, in particular, that the q_x's derive from the observed mortality of people born $\omega, \omega - 1, \ldots, x, \ldots, 1, 0$ years before the observation year.

An important hypothesis underlying recursion (3.2.4) should be stressed. As the q_x's are assumed to be estimated from mortality experience in a given period (say, one year), the calculation of the l_x's relies on the assumption that the mortality pattern does not change in the future.

Statistical evidence shows that human mortality, in many countries, has declined over the 20th century, and in particular over its last decades (for more details, see Sect. 3.8.1). So, the hypothesis of a "static" mortality cannot be assumed in principle, at least when long periods of time are referred to. Hence, in life insurance applications, the use of period life tables should be restricted to products involving short or medium durations (5 to 10 years, say), like the term insurance and the endowment insurance, whilst it should be avoided when dealing with life annuities and pension plans. Conversely, life annuities and pensions require life tables which allow for the anticipated future mortality trend, namely *projected life tables* constructed on the basis of the experienced mortality trend. This topic will be dealt with in Sects. 3.8.2 to 3.8.5.

Example 3.2.1. In Fig. 3.2.1 a survival curve is plotted. The l_x's are calculated starting from a period mortality observation. The related d_x's, which constitute the curve of deaths, are plotted in Fig. 3.2.2.

Some features, which are shared by most life tables, clearly emerge (in particular looking at the curve of deaths):

1. the *infant mortality*;
2. the *mortality hump* at young-adult ages, mainly due to accidental deaths;
3. the *age of maximum mortality* (at old ages).

Note that the point of highest mortality (at old ages) in the curve of deaths corresponds to the inflexion point in the survival curve.

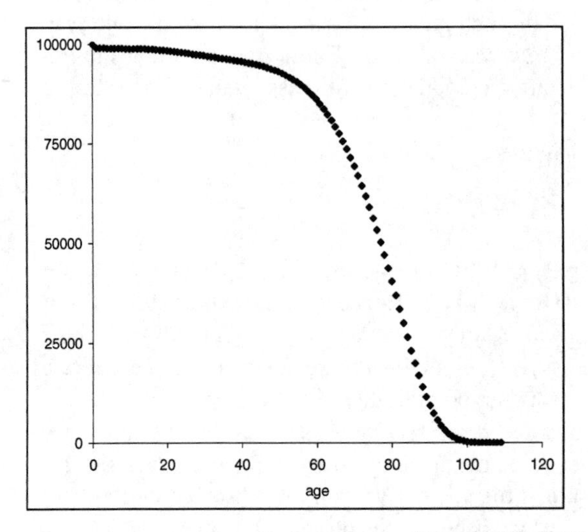

Fig. 3.2.1 l_x in the Italian male population - 1992 (source: ISTAT)

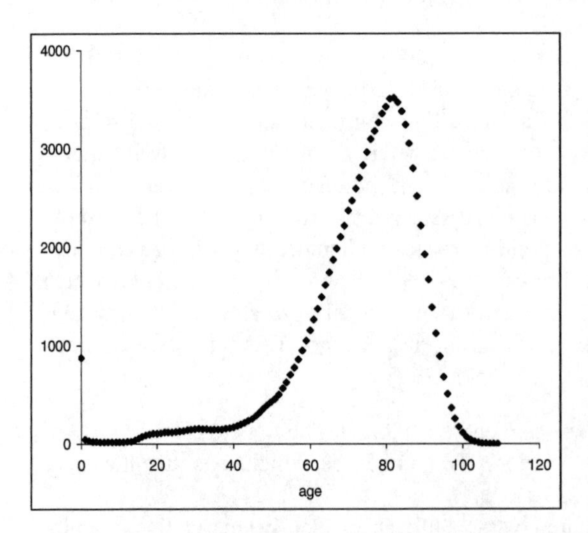

Fig. 3.2.2 d_x in the Italian male population - 1992 (source: ISTAT)

□

3.2.3 Construction of a period life table

Several methods can be adopted for constructing a period life table. As a detailed discussion of this topic is beyond the scope of this book, we just mention a method, which can be implemented in order to obtain a numerical assessment of the one-year probabilities of dying.

We denote with θ_x the observed number of deaths between age x and $x+1$, and with ETR_x, the number of individuals *exposed to risk*, i.e. "generating" the θ_x deaths. The number ETR_x can be estimated according to various approaches. Here, we briefly describe the so-called *census method*. We assume one year as the observation period.

Let $P_x(0)$ denote the size of the population aged between x and $x+1$ at the beginning of the year (i.e. at time 0), and $P_x(1)$ the size of the population aged between x and $x+1$ at the end of the year (time 1). In Fig. 3.2.3 the numbers $P_x(t)$, for $x = 0, 1, \dots$, referred to a generic time t, are depicted, separately for males ([M]) and females ([F]). Lower ages are in the bottom part of the graph. The resulting graph describes the structure of a population by age and gender, and is usually called in Demography the *age-gender pyramid* (or the *population pyramid*). Note that the shape of the "pyramid" reflects the evolution of a population over time: for example, a small size in the low age classes, compared to the size in the medium and high age classes, denotes an ageing population.

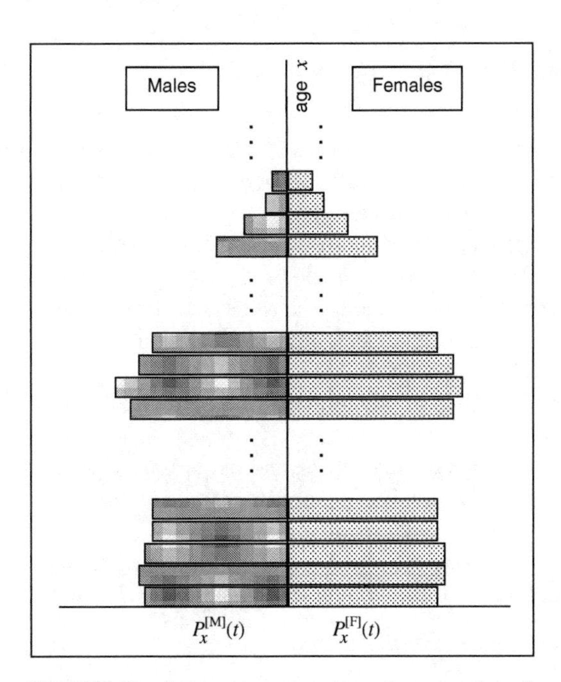

Fig. 3.2.3 Population structure at time t, by age and gender

We define the number of individuals exposed to risk as follows:

$$\mathrm{ETR}_x = \frac{P_x(0) + P_x(1) + \theta_x}{2} \tag{3.2.5}$$

Thus, we take as exposed to risk first the average between the population at the beginning of the year and the population at the end of the year, both populations consisting of individuals aged between x and $x+1$; further, we add to this quantity one half of the number of people dying in the year, as we assume that the deaths occur on average at the mid of the year.

The frequency of death at age x, denoted with $\hat{q}_x$, is then calculated as follows:

$$\hat{q}_x = \frac{\theta_x}{\mathrm{ETR}_x} \tag{3.2.6}$$

The $\hat{q}_x$'s, which result from statistical observation, are called *raw mortality rates*. As they may have an erratic behavior, for example because of very small population sizes at very old ages, whereas previous experience and intuition suggest a smooth progression, a *graduation* procedure is usually applied to the sequence of $\hat{q}_x$. Graduated period mortality rates should exhibit a progressive change over a set of ages, without sudden and/or huge jumps, which cannot be explained by intuition nor supported by past experience.

Various approaches to graduation can be adopted. In particular, two broad categories can be recognized:

- *parametric* graduation, involving the use of *mortality laws*;
- *non-parametric* graduation.

According to a parametric approach, a functional form is chosen (some examples will be presented in Sects. 3.4.2 and 3.9.5), and the relevant parameters are estimated in order to find the parameter values which provide the best fit to the observed mortality rates. Various fitting criteria can be adopted for parameter estimation, for example maximum likelihood.

The choice of a particular functional form is avoided when a non-parametric graduation method is adopted. Traditional methods in this category are, for example, the weighted moving average methods. In what follows, we simply assume that some graduation procedure has been applied to raw mortality data, providing as its output a set of graduated values.

We denote with q_x, $x = 0, 1, 2, \dots$ the graduated values, and we assume q_x as the probability of an individual age x dying within one year, thus before reaching age $x+1$. Hence, the q_x's are the *annual* (or *one-year*) *probabilities of death*.

Finally, the l_x's can be calculated by using the relation (3.2.4). Figure 3.2.4 summarizes the procedure which, starting from the population structure, leads to the sequence of l_x.

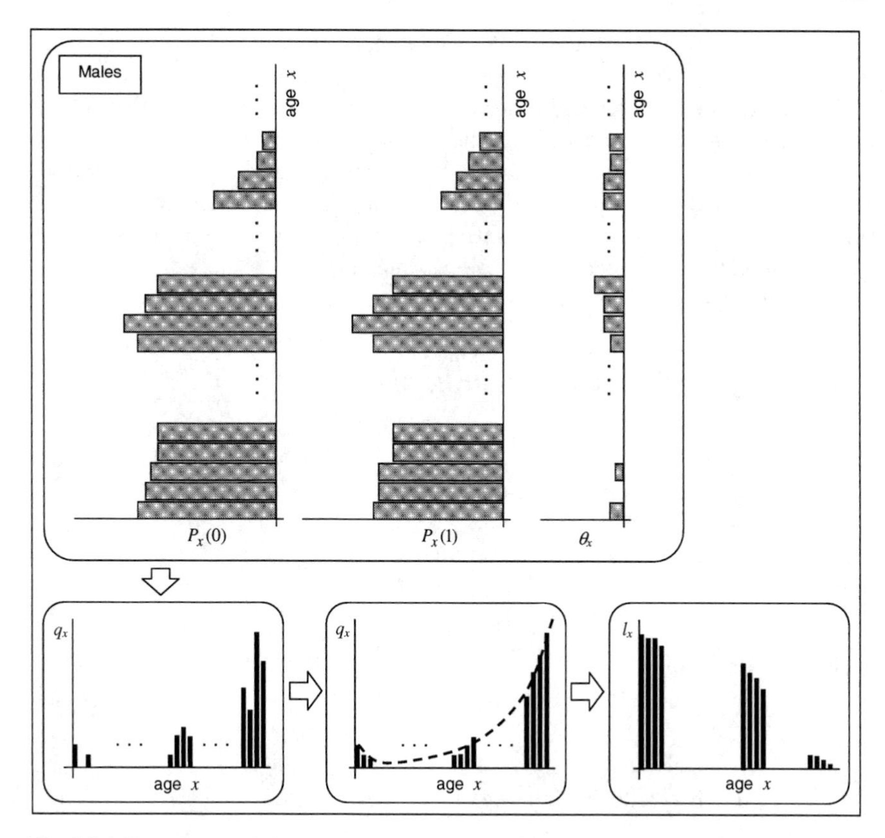

Fig. 3.2.4 From the population structure by age to the life table

3.2.4 "Population" tables versus "market" tables

Mortality data, and hence life tables, can originate from observations concerning a whole national population, a specific part of a population (for example, retired workers, disabled people, etc.), an insurer's portfolio, a pension plan, and so on.

Life tables constructed on the basis of observations involving a whole national population (usually split into females and males) are commonly referred to as *population life tables*.

Market life tables are constructed using mortality data arising from a collection of insurance portfolios and/or pension plans. Usually, distinct tables are constructed for insurance products providing death benefits (for example term insurances), life annuities purchased on an individual basis, pensions (namely annuities paid to the members of a pension plan).

The rationale for distinct market tables lies in the fact that mortality levels may significantly differ as we move from one type of insurance product to another. This aspect will be discussed in Sect. 3.6.2.

Market tables provide experience-based data for premium and reserve calculations and for the assessment of expected profits. Population tables can provide a starting point when market tables are not available. Moreover, population tables usually reveal mortality levels higher than those expressed by market tables and hence are likely to constitute a *prudential* (or *conservative*, or *on the safe-side*) assessment of mortality in portfolios of insurance products providing death benefits. Thus, population tables can be used when pricing such products, in order to include a profit margin (or an implicit safety loading) into the premiums.

3.2.5 The life table as a probabilistic model

We now assume that the sequence $l_0, l_1, \ldots, l_x, \ldots, l_\omega$ constitutes our data base, and define various probabilities, useful in life insurance calculations, taking this sequence as the starting point.

We denote by p_x the probability of an individual age x being alive at age $x+1$. Clearly

$$p_x = 1 - q_x \tag{3.2.7}$$

and hence (see Eq. (3.2.4))

$$p_x = \frac{l_{x+1}}{l_x} \tag{3.2.8}$$

Further, we denote by ${}_hp_x$ the probability that an individual age x is alive at age $x+h$. This event can be expressed in terms of one-year events concerning a given individual, namely:

- the individual age x is alive at age $x+1$;
- the individual age $x+1$ is alive at age $x+2$;
- ...
- the individual age $x+h-1$ is alive at age $x+h$.

Hence

$${}_hp_x = p_x\, p_{x+1} \cdots p_{x+h-1} \tag{3.2.9}$$

and then, using (3.2.8), we find

$${}_hp_x = \frac{l_{x+1}}{l_x}\frac{l_{x+2}}{l_{x+1}} \cdots \frac{l_{x+h}}{l_{x+h-1}} = \frac{l_{x+h}}{l_x} \tag{3.2.10}$$

Note that, clearly, ${}_0p_x = 1$. Conversely, ${}_1p_x = p_x$. The following relation is useful in a number of actuarial calculations:

$${}_{h+k}p_x = {}_hp_x\; {}_kp_{x+h} \tag{3.2.11}$$

We denote by ${}_hq_x$ the probability that an individual age x dies before attaining age $x+h$. We have

$$
{}_hq_x = 1 - {}_hp_x = \frac{l_x - l_{x+h}}{l_x} \tag{3.2.12}
$$

Of course, ${}_0q_x = 0$, whereas

$$
{}_1q_x = q_x = \frac{l_x - l_{x+1}}{l_x} = \frac{d_x}{l_x} \tag{3.2.13}
$$

Remark Sometimes the one-year probabilities q_x and p_x are called *mortality rates* and *survival rates* respectively. We prefer to avoid these expressions to denote probability of death and survival, as the term "rate" should be referred to a counter expressing the number of events per unit of time.

The probability of a person age x dying between age $x+h$ and $x+h+k$ is denoted with ${}_{h|k}q_x$. Referring to a given individual, this event can be split as follows:

- the individual age x is alive at age $x+h$;
- the individual age $x+h$ dies before age $x+h+k$

Hence, we have

$$
{}_{h|k}q_x = {}_hp_x \, {}_kq_{x+h} = \frac{l_{x+h} - l_{x+h+k}}{l_x} \tag{3.2.14}
$$

The probability ${}_{h|k}q_x$ is usually called "deferred" probability of dying, the deferment being the period of h years.

The following relations can be easily interpreted and proved by using the formulae presented above:

$$
{}_{h|k}q_x = {}_{h+k}q_x - {}_hq_x = {}_hp_x - {}_{h+k}p_x \tag{3.2.15}
$$

3.2.6 One-year measures of mortality

Consider the probability defined in (3.2.14). In particular, with $k = 1$ we find

$$
{}_{h|1}q_x = {}_hp_x \, q_{x+h} = \frac{l_{x+h} - l_{x+h+1}}{l_x} = \frac{d_{x+h}}{l_x} \tag{3.2.16}
$$

Referring to a newborn, namely setting $x = 0$, we have

$$
{}_{h|1}q_0 = \frac{d_h}{l_0} \tag{3.2.17}
$$

We note that

$$
\sum_{h=0}^{\omega} {}_{h|1}q_0 = \frac{1}{l_0} \sum_{h=0}^{\omega} d_h = 1 \tag{3.2.18}
$$

Actually, the ${}_{h|1}q_0$'s constitute the probability distribution of the lifetime of a newborn (with integer outcomes $0, 1, \ldots, \omega$; see Sect. 3.2.7). In particular, ${}_{0|1}q_0$ is the probability of death during the first year of life, ${}_{1|1}q_0$ is the probability of death during the second year of life, and so on. Further, for all integer k, we have:

$$\sum_{h=k}^{\omega} {}_{h|1}q_0 = \frac{1}{l_0} \sum_{h=k}^{\omega} d_h = \frac{l_k}{l_0} = {}_k p_0 \tag{3.2.19}$$

Consider the following probabilities:

- q_x, expressed by (3.2.13);
- ${}_{x|1}q_0$, expressed by (3.2.17) replacing h with x.

Both the probabilities quantify one-year mortality, namely between age x and $x+1$. Figure 3.2.5 illustrates the behavior of the two probabilities as functions of age x (assuming, for simplicity, that x can take all real values). We note that q_x (see Fig. 3.2.5a) refers to an individual alive at age x, whereas ${}_{x|1}q_0$ (see Fig. 3.2.5b) refers to a newborn. The different behavior is easily explained looking at the definitions of the two one-year probabilities, i.e. (3.2.13) and (3.2.17) respectively, and noting that l_x decreases as x increases. In particular, when l_x is close to l_0, the two graphs are quite similar, whereas as l_x strongly decreases, q_x definitely increases. Note also that the behavior of the ${}_{x|1}q_0$ trivially reflects the behavior of the d_x (for example, see Fig. 3.2.2).

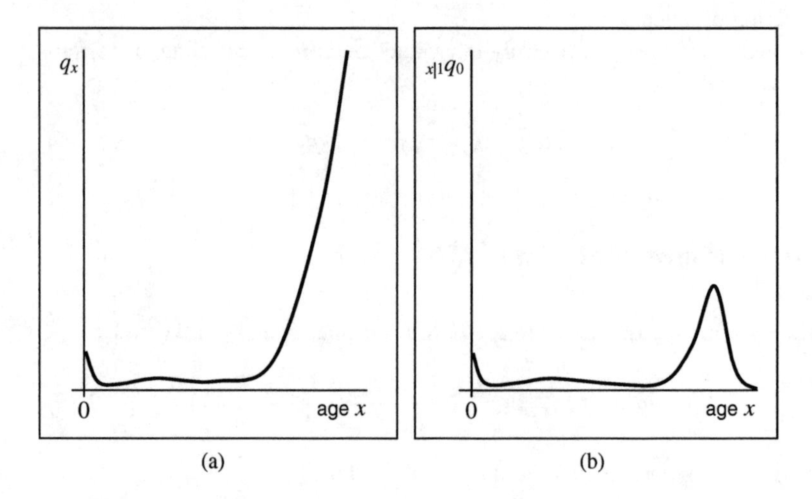

Fig. 3.2.5 One-year probabilities of death

Another one-year measure of mortality can be defined, namely the quantity

$$m_x = \frac{d_x}{\frac{l_x + l_{x+1}}{2}} \tag{3.2.20}$$

usually called the *central mortality rate*.

It is interesting to compare m_x to the probability q_x (see (3.2.13)). Both the quantities relate the expected number, d_x, of people dying between age x and $x+1$ to an expected number of "exposed to risk". The latter takes as the number of exposed to risk the quantity l_x, namely the "initial" number of people in the age interval $(x, x+1)$, whereas the former relates the numerator to the average number $\frac{l_x+l_{x+1}}{2}$, i.e. the "central" number of people in the same age interval.

Table 3.2.2 One-year measures of mortality

definition	usual name	age referred to	exposed to risk
$q_x = \frac{d_x}{l_x}$	(initial) mortality rate	x	l_x
$_{x\|1}q_0 = \frac{d_x}{l_0}$	deferred mortality rate	0	l_0
$m_x = \frac{d_x}{\frac{l_x+l_{x+1}}{2}}$	central mortality rate	x	$\frac{l_x+l_{x+1}}{2}$
$\phi_x = \frac{q_x}{1-q_x}$	mortality odds	x	l_x

Finally, we introduce a further quantity related to one-year mortality. When q_x can be expressed as $q_x = \frac{\phi_x}{1+\phi_x}$, the function ϕ_x represents the so-called *mortality odds*, namely

$$\phi_x = \frac{q_x}{1-q_x} \tag{3.2.21}$$

From $0 < q_x < 1$ (for $x < \omega$), it follows that $\phi_x > 0$. Thus, focussing on the odds, rather than the annual probabilities of dying, can make easier the choice of a mathematical formula fitting the age-pattern of mortality (see Sect. 3.4.2), as the only constraint is the positivity of the odds.

Table 3.2.2 summarizes these one-year measures of mortality. As regards the use of the term "rate" (common in actuarial practice) also to denote the probabilities q_x and $_{x|1}q_0$, see the Remark in Sect. 3.2.5.

Example 3.2.2. A life table includes the elements represented in Table 3.2.3.

For example, the following probabilities can be calculated:

1. the probability of a newborn (i.e. age 0) dying between age 65 and 66,

$$_{65|1}q_0 = \frac{l_{65} - l_{66}}{l_0} = 0.004$$

2. the probability of a person age 43 being alive at age 65

Table 3.2.3 Expected number of survivors in a life table

x	l_x
0	100000
...	...
40	98000
41	97920
42	97800
43	97650
44	97450
45	97220
...	...
65	85000
66	84600
...	...

$$_{22}p_{43} = \frac{l_{65}}{l_{43}} = 0.87046$$

3. the probability of a person age 40 dying between age 42 and 45

$$_{2|3}q_{40} = \frac{l_{42} - l_{45}}{l_{40}} = 0.00592$$

4. the probability of a person age 42 dying between age 42 and 45

$$_3q_{42} = \frac{l_{42} - l_{45}}{l_{42}} = 0.00593$$

□

Example 3.2.3. Probabilities 1 and 2 in Example 3.2.2 involve very long time intervals (65 years as the deferred period in probability 1, and 22 years in probability 2). If the l_x's are drawn from a period life table, these probabilities (although formally correct) can be affected by severe errors in the presence of a mortality trend.

Conversely, the other two probabilities involve short intervals, and thus can be reasonably accepted. Hence, an appropriate use of a period life table should be restricted to rather short interval, say 10 years at most. For example, refer to an insured age 40 at policy issue; the following probabilities can be used for a five-year term insurance (see Case 4b in Sect. 1.7.4):

$$_1q_{40} = \frac{l_{40} - l_{41}}{l_{40}}, \; _{1|1}q_{40} = \frac{l_{41} - l_{42}}{l_{40}}, \; \dots, \; _{4|1}q_{40} = \frac{l_{44} - l_{45}}{l_{40}}$$

□

3.2.7 A more formal setting: the random lifetime

A more formal setting can be defined if we refer our probabilistic model to the *remaining lifetime* of an individual age x. We denote by T_x this lifetime, which is clearly a random variable. Whatever its outcome may be, the (random) age at death is given by $T_x + x$. The possible outcomes of T_x are the positive real numbers; however, it is rather usual to take $\omega - x$ as the maximum possible outcome.

In particular, T_0 represents the *total lifetime* of an individual age 0, namely a newborn. Of course, we have

$$T_x = T_0 - x \,|\, T_0 > x \tag{3.2.22}$$

In life insurance calculations, probabilities like $\mathbb{P}[T_x > h]$, $\mathbb{P}[h < T_x \leq h+k]$, and so on, are needed. When a life table is available, those probabilities can be immediately derived from the life table itself, provided that the ages and durations are integers. Thus, we have for example

$$\mathbb{P}[T_x > h] = {}_h p_x = \frac{l_{x+h}}{l_x} \tag{3.2.23}$$

$$\mathbb{P}[T_x \leq h] = {}_h q_x = 1 - {}_h p_x = \frac{l_x - l_{x+h}}{l_x} \tag{3.2.24}$$

$$\mathbb{P}[h < T_x \leq h+k] = {}_{h|k} q_x = \frac{l_{x+h} - l_{x+h+k}}{l_x} \tag{3.2.25}$$

If we have to calculate probabilities like (3.2.23), (3.2.24) and (3.2.25) when ages or durations are real number, then an extension of the probabilistic model is needed. We will address this topic in Sect. 3.9.

The *curtate remaining lifetime*, usually denoted with K_x is defined as the integer part of T_x. Thus, the possible outcomes of K_x are $0, 1, 2, \ldots$, according to the following scheme:

$$\begin{array}{ccc} 0 < T_x < 1 & \Leftrightarrow & K_x = 0 \\ 1 \leq T_x < 2 & \Leftrightarrow & K_x = 1 \\ 2 \leq T_x < 3 & \Leftrightarrow & K_x = 2 \\ \ldots & & \ldots \end{array}$$

A similar definition applies in particular to the random variable T_0, leading to the *curtate total lifetime* K_0. Note that the probability distribution of K_0 is given by

$$_{0|1}q_0,\; _{1|1}q_0,\; \ldots, _{x|1}q_0, \ldots, _{\omega|1}q_0 \tag{3.2.26}$$

Indeed, we have

$$_{x|1}q_0 = \mathbb{P}[x < T_0 \leq x+1] = \mathbb{P}[K_0 = x];\;\; x = 0, 1, \ldots, \omega \tag{3.2.27}$$

3.3 Summarizing a life table

Age-specific functions are usually needed in actuarial calculations. For example, in the age-discrete context functions like l_x, q_x, etc. are commonly used in order to calculate premiums, reserves, and so on.

Nevertheless, the role of single-figure indices, also called *markers*, which summarize the life table and hence the lifetime probability distribution, should not be underestimated. In particular, important features of past mortality trends can be singled out by focussing on the behavior of some indices over time, as we will see in Sect. 3.8.1.

3.3.1 The life expectancy

The expected value of the random lifetime is a typical marker. More precisely, consider the curtate total lifetime K_0, and then the random variable $K_0 + \frac{1}{2}$. Hence, calculate the expected value $\mathbb{E}[K_0 + \frac{1}{2}]$, which is given by

$$\mathbb{E}[K_0 + \tfrac{1}{2}] = \sum_{h=0}^{\omega} \left(h + \tfrac{1}{2}\right) {}_{h|1}q_0 \tag{3.3.1}$$

The quantity expressed by (3.3.1) is called *expected total lifetime* (or *life expectancy at the birth*), and is usually denoted with $\overset{\circ}{e}_0$. Note that, using (3.2.19), after a little algebra we obtain:

$$\overset{\circ}{e}_0 = \tfrac{1}{2} + \sum_{k=1}^{\omega} {}_k p_0 \tag{3.3.2}$$

Referring to an individual age x, the *expected remaining lifetime* (or *life expectancy at age x*), usually denoted with $\overset{\circ}{e}_x$, is defined as follows

$$\overset{\circ}{e}_x = \mathbb{E}[K_x + \tfrac{1}{2}] = \sum_{h=0}^{\omega-x} \left(h + \tfrac{1}{2}\right) {}_{h|1}q_x \tag{3.3.3}$$

Using a relation similar to (3.2.19), we find:

$$\overset{\circ}{e}_x = \tfrac{1}{2} + \sum_{k=1}^{\omega-x} {}_k p_x \tag{3.3.4}$$

The *expected age at death* (that is, the expected total lifetime for an individual alive at age x) is then given by $x + \overset{\circ}{e}_x$.

The probabilities adopted in the previous formulae are commonly provided by cross-sectional observations. Then, the expected values $\overset{\circ}{e}_0$ and $\overset{\circ}{e}_x$ represent *period life expectancies*, and hence rely on the hypothesis of static mortality (see Sect. 3.2.2). Expected values calculated accounting for future mortality trend will be introduced in Sect. 3.8.3.

Example 3.3.1. Assume that the life table constructed via a period observation of mortality in population A leads to the expected total lifetime $\overset{\circ}{e}_0^{[A]}$. An analogous observation concerning population B leads to $\overset{\circ}{e}_0^{[B]}$. Suppose, for example, that we find

$$\overset{\circ}{e}_0^{[A]} = 74$$
$$\overset{\circ}{e}_0^{[B]} = 76$$

How can we interpret the difference $\overset{\circ}{e}_0^{[B]} - \overset{\circ}{e}_0^{[A]} = 2$? What can we say about the impact of this difference, for instance, on the costs related to the payment of pensions and life annuities ?

Consider the following statement: "The higher expected total lifetime implies that the costs for paying pensions to population B are higher then the costs concerning population A, as people in B receive on average two annual payments more". This statement may be wrong. Let's try to understand why.

The expected total lifetime $\overset{\circ}{e}_0$ depends on the whole probability distribution of the random variable K_0 (see definition (3.3.1)), hence including, in particular, the infant mortality and the young-adult mortality hump (see Fig. 3.2.5b). So, the higher value of $\overset{\circ}{e}_0^{[B]}$ can be explained, in particular, in terms of

1. a lower infant mortality;
2. a lower mortality at young-adult ages;
3. a longer life expectation for people who reach, for example, age 50.

Clearly, items 1 and 2 cannot support the above statement. Conversely, item 3 does support the statement itself, and, at the same time, stresses an interesting aspect. When pension problems are dealt with, a useful information is provided by the expected remaining lifetime at a given adult age, say 50 or 60. So, if we find $\overset{\circ}{e}_{60}^{[B]} > \overset{\circ}{e}_{60}^{[A]}$, then we can state that the costs for paying pensions to population B are likely to be higher then the costs concerning population A.

□

By using formula (3.3.3), it is easy to prove that

$$1 + \overset{\circ}{e}_x = \frac{1}{p_{x-1}} \overset{\circ}{e}_{x-1} \tag{3.3.5}$$

and hence

$$x+\overset{\circ}{e}_x > x-1+\overset{\circ}{e}_{x-1} \tag{3.3.6}$$

From (3.3.6), it follows in particular that

$$x+\overset{\circ}{e}_x > \overset{\circ}{e}_0 \tag{3.3.7}$$

Inequalities (3.3.6) and (3.3.7) are self-evident: the expected total lifetime increases as the attained age increases, because the individual has overcome the risk of dying in the past years.

3.3.2 Other markers

A number of markers, other than the expected total lifetime (or the expected remaining lifetime at some given age), can be adopted to summarize a life table. Some examples follow.

- The *Lexis point* is the modal value, at old ages, of the probability distribution of the total lifetime, namely the (old) age with the highest mortality, i.e. the highest d_x (and hence the highest ${}_{x|1}q_0$).
- The *variance* of the probability distribution of the total lifetime (or its *standard deviation*) is a traditional variability measure.
- The probability that a newborn dies before a given age x', namely ${}_{x'}q_0$, provides, for x' small (say 1, or 5), a measure of the infant mortality.

Although these and other markers, which summarize the probability distribution of the lifetime, are of great interest in demographical studies, their use is quite limited in the actuarial field. Actually, life insurance calculations require working with functions of the random lifetime, rather than directly with the random lifetime itself.

3.4 A mortality "law"

3.4.1 From tables to parameters

Since the earliest attempt to describe in analytical terms a mortality schedule (due to A. De Moivre and dating back to 1725), great effort has been devoted by demographers and actuaries to the construction of analytical formulae (or *mortality laws*) that fit the age-pattern of mortality. When a mortality law is used to fit observed data, namely a parametric graduation is chosen (see Sect. 3.2.3), the age-pattern of mortality is summarized by a small number of parameters (two to ten, say, in the mortality laws commonly used in actuarial and demographical applications). Thus, we can replace the 110, say, items of a life table by a small number of parameters without sacrificing much information.

Many mortality laws have been proposed in the age-continuous context. Some of these laws will be presented and discussed in Sect. 3.9.5. Here we focus on one type of mortality law only, namely the Heligman-Pollard formula, which, although defined for any real age x, expresses the one-year probability of death q_x and the mortality odds $\frac{q_x}{1-q_x}$, and hence can perfectly work in a framework in which ages and durations are integers.

3.4.2 The Heligman-Pollard law

Heligman and Pollard proposed in 1980 a class of formulae which aim to represent the age-pattern of mortality over the whole span of life. The *first Heligman-Pollard law*, expressed in terms of the odds, is

$$\phi_x = A^{(x+B)^C} + D\mathrm{e}^{-E(\ln x - \ln F)^2} + GH^x \tag{3.4.1}$$

while the *second Heligman-Pollard law*, in terms of q_x, is given by

$$q_x = A^{(x+B)^C} + D\mathrm{e}^{-E(\ln x - \ln F)^2} + \frac{GH^x}{1+GH^x} \tag{3.4.2}$$

Note that, in both cases, at high ages we have

$$q_x \approx \frac{GH^x}{1+GH^x} \tag{3.4.3}$$

Formula (3.4.3) can be used as an approximation when calculating values related to life annuities and pensions, e.g. for $x \geq 65$.

Two other laws, generalizing the second Heligman-Pollard law, were proposed; however, a deep analysis of such a topic is beyond the scope of this book.

Example 3.4.1. Assume the following values for the parameters of the first Heligman-Pollard law (see (3.4.1)):

$$\begin{aligned} A &= 0.000544 & B &= 0.017 \\ C &= 0.101 & D &= 0.000158 \\ E &= 10.72 & F &= 18.67 \\ G &= 0.0000183 & H &= 1.11 \end{aligned}$$

These parameters have been estimated on the basis of a UK mortality experience. [1] Figure 3.4.1 illustrates the age-pattern of mortality in terms of the probabilities q_x.

[1] See: Dellaportas P., Smith A.F.M., Stavropoulos P. (2001), Bayesian Analysis of Mortality Data, *Journal of the Royal Statistical Society*. Series A, vol. 164 (2), pp. 275-291.

A better representation, because of the range of values, is provided by the graph of the logarithms $\ln q_x$; see Fig. 3.4.2. Finally, the probabilities ${}_{x|1}q_0$ are depicted in Fig. 3.4.3.

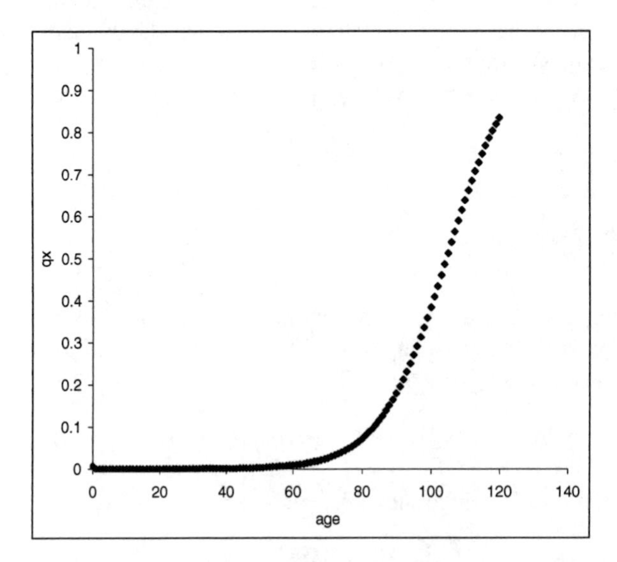

Fig. 3.4.1 The first Heligman-Pollard law: q_x

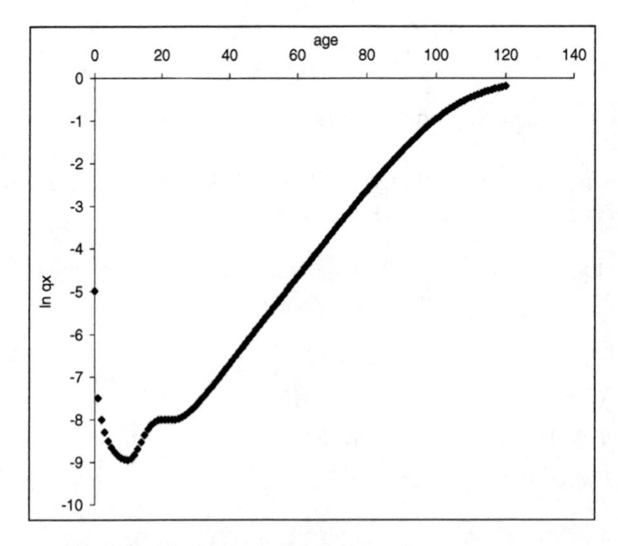

Fig. 3.4.2 The first Heligman-Pollard law: $\ln q_x$

□

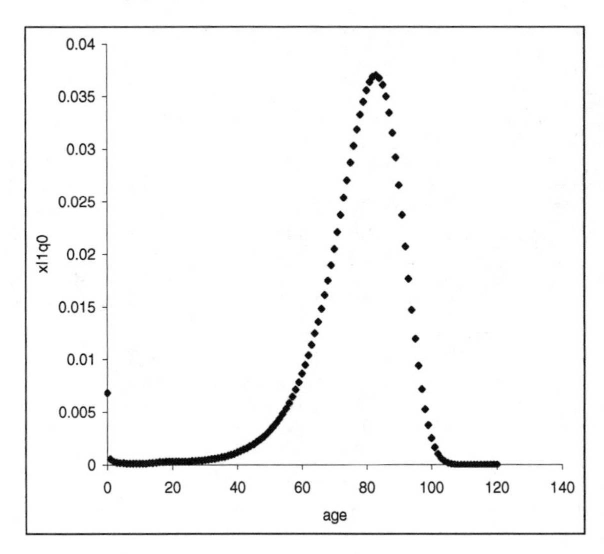

Fig. 3.4.3 The first Heligman-Pollard law: ${}_{x|1}q_0$

3.5 From the basic model to more general models

The model we have so far dealt with can be considered a "basic" one, as only the attained age is accounted for in assigning the probability of an individual dying within one year (or being alive after one year, or after a given number of years, and so on).

However, statistical experience and, at least to some extent, intuition suggest that, in many applications among which the life insurance and pension business, more complex models are needed, for example allowing for heterogeneity (inside a population) in respect of mortality, for future mortality trends, for the effect of medical ascertainment in the underwriting process, and so on.

Figure 3.5.1 illustrates the main directions along which we will now move, in order to build up more general models to be used in life insurance and pension calculations. The various terms used in the blocks of the figure will be explained in the next sections.

3.6 Heterogeneity

3.6.1 Some preliminary ideas

Any given population is affected by some degree of heterogeneity, as far as individual mortality is concerned. Heterogeneity in populations should be approached addressing two main issues:

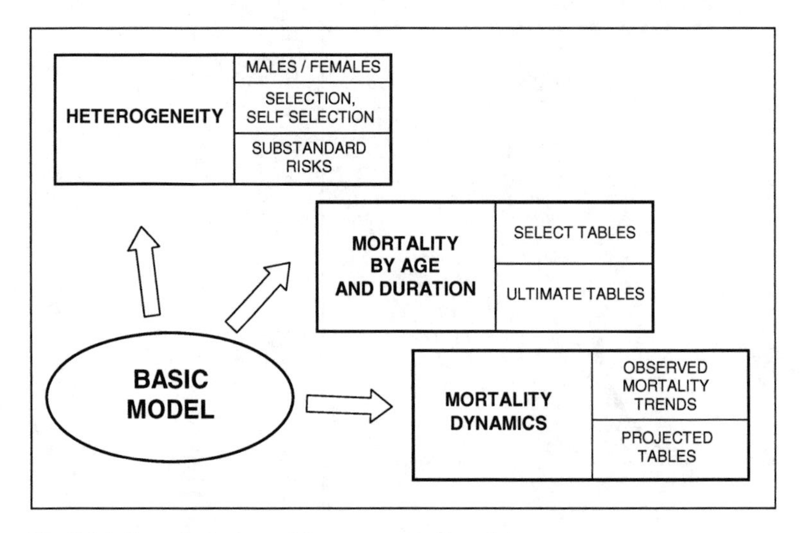

Fig. 3.5.1 From the basic model to more general models

- detecting and modeling observable heterogeneity factors (e.g. age, gender, occupation, etc.);
- allowing for unobservable heterogeneity factors.

In the insurance framework, heterogeneity factors are also called *risk factors* (see Sect. 2.2.6). As regards *observable heterogeneity factors*, mortality depends on:

1. biological and physiological factors, such as age, gender, genotype;
2. features of the living environment; in particular: climate and pollution, nutritional standards (mainly with reference to excesses and deficiencies in diet), population density, hygienic and sanitary conditions;
3. occupation, in particular in relation to possible professional disability or exposure to injury, and educational attainment;
4. individual lifestyle, in particular with regard to nutrition, alcohol and drug consumption, smoking, physical activities and pastimes;
5. current health conditions, personal and/or family medical history, civil status, and so on.

Item 2 affects the overall mortality of a population. That is why mortality tables are typically considered specifically for a given geographic area. The remaining items concern the individual and, when dealing with life insurance, they can be observed at policy issue. Their assessment is performed through appropriate questions in the application form and, as to health conditions, possibly through a medical examination. The specific items considered for insurance rating depend on the types of benefits provided by the insurance contract (see Sect. 3.6.2).

Differences among the individuals can also be attributed to *unobservable heterogeneity factors*. Examples of unobservable factors are the individual's attitude towards health, and some congenital personal characteristics.

When allowing for unobservable heterogeneity factors, various approaches can be adopted. However, the basic idea is that the population life table, or the population mortality law, should be interpreted as a mixture of a set of tables or laws, each one expressing a specific level of mortality. We do not deal with these aspects, which are beyond the scope of this book.

3.6.2 Rating classes

The observable risk factors lead to a partitioning of the insured population into *risk classes*. However, for various reasons, not all the risk factors are allowed for when pricing an insurance product (and hence a solidarity effect is introduced in the premium system). Risk factors accounted for in the pricing (or "rating") procedure are called *rating factors*; consequently, the insured population is split into *rating classes* (see also Sect. 2.2.6).

The rating procedure should be organized, for any given insurance product, as follows.

1. An appropriate choice of the rating factors should aim at grouping people in classes within which insured lives bear an analogous expected mortality profile.
2. For each individual applying for insurance, a *selection process* should be performed, whose aim is to assign the applicant to her proper rating class.

When defining a rating procedure, possible *adverse selection* (or *anti-selection*) should be taken into account. This expression denotes a higher propensity to buy insurance in people bearing a worse risk profile.

The specific risk factors considered for life insurance rating depend, to some extent, on the types of benefits provided by the insurance contract. *Age* is always considered, due to the apparent variability of mortality in this regard. *Gender* is usually accounted for, especially when living benefit are involved, given that females on average live longer than males. This difference clearly appears in Figs. 3.6.1 and 3.6.2, in terms of the curves of death and the survival curves respectively.

As far as *genetic aspects* are concerned, the evolving knowledge in this area has raised a lively debate (which is still running) on whether it is legitimate for insurance companies to resort to genetic tests for underwriting purposes.

Applicants for life annuities are usually in good health, so a medical examination is not necessary; on the contrary, a proper investigation is needed for those who buy death benefits, given that people in poorer health conditions may be more interested in them and hence more likely to buy such benefits.

When death benefits are dealt with, *health conditions*, *occupation* and *smoking status* can be taken as rating factors. These lead to a classification into *standard* and *substandard risks* in life insurance. For the latter (also referred to as *impaired lives*), a higher premium level is adopted in order to avoid adverse selection, given that they bear a higher probability to become eligible for the benefit. In some markets, standard risks are further split into *regular* and *preferred risks*, the latter having a

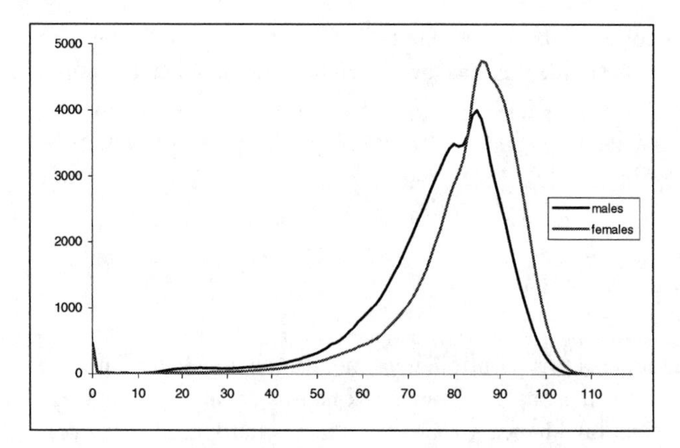

Fig. 3.6.1 Curves of death in the Italian male and female populations - 2002 (source: ISTAT)

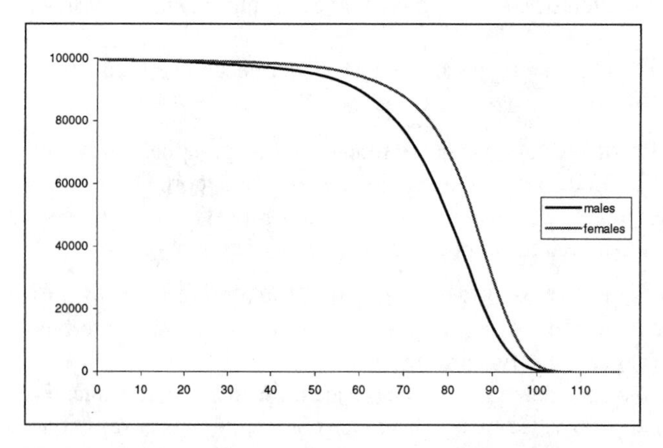

Fig. 3.6.2 Survival curves in the Italian male and female populations - 2002 (source: ISTAT)

better profile than the former (for example, because they never smoked); as such, they are allowed to pay a reduced premium rate.

Mortality for people in poorer or better conditions than the average is usually expressed in relation to average (or standard) mortality. This allows us to deal only with one life table (or one mortality law), properly adjusted when substandard or preferred risks are dealt with. Thus, if q_x denotes the annual probability of death in the age-pattern of mortality taken as the standard, the adjusted probability of death, $q_x^{[\mathrm{adj}]}$, is assumed to be expressed as follows:

$$q_x^{[\mathrm{adj}]} = \Phi(q_x) \tag{3.6.1}$$

where Φ denotes an appropriate function.

3.6.3 Substandard risks

In this Section we introduce some models which can be adopted to express the age-pattern of mortality for substandard risks, as a "transform" of the standard mortality.

We denote by x the age at policy issue, and by m the policy term. Further, we denote by q_{x+t} $(0 \leq t \leq m)$ the one-year probability of dying according to the life table (or the mortality law) adopted for expressing the age-pattern of mortality of standard risks.

A rather general transform is provided by the *linear model*, that is

$$q_{x+t}^{[L]} = (1+\beta)\, q_{x+t} + \alpha; \;\; 0 \leq t \leq m \tag{3.6.2}$$

From this model, more specific transforms can be derived. The *additive model* (see Fig. 3.6.3) is defined by setting $\beta = 0$ and $\alpha > 0$ in (3.6.2). Thus:

$$q_{x+t}^{[A]} = q_{x+t} + \alpha; \;\; 0 \leq t \leq m \tag{3.6.3}$$

Note that the additive model implies an extra-mortality, given by α, which is constant and independent of the initial age. Such a model is consistent, for example, with extra-mortality due to accidents (related either to occupation or to extreme sports).

A slight modification of model (3.6.3) allows us to express a constant extra-mortality which, however, depends on the age x at policy issue via the probability of death q_x:

$$q_{x+t}^{[A]} = q_{x+t} + \alpha' q_x; \;\; 0 \leq t \leq m \tag{3.6.4}$$

Conversely, setting $\alpha = 0$ and $\beta > 0$ in (3.6.2), we obtain the *multiplicative model* (see Fig. 3.6.4):

$$q_{x+t}^{[M]} = (1+\beta)\, q_{x+t}; \;\; 0 \leq t \leq m \tag{3.6.5}$$

In this model, the extra-mortality is given by $\beta\, q_{x+t}$. In the age intervals of interest, q_{x+t} increases as the attained age $x+t$ increases. Hence, the multiplicative model implies an increasing extra-mortality.

The evolution of some diseases, which either lead to an early death or have a short recovery time, suggests the adoption of models implying a *decreasing extra-mortality*. An example is provided by the following model:

$$q_{x+t}^{[D]} = \begin{cases} (1+\beta)\, q_{x+t} + \alpha; & 0 \leq t \leq r \\ q_{x+t}; & r < t \leq m \end{cases} \tag{3.6.6}$$

with $0 < r < m$, and α, β such that

$$q_x^{[D]} = q_x + \text{initial extra-mortality} \tag{3.6.7}$$

$$q_{x+r}^{[D]} = q_{x+r} \tag{3.6.8}$$

We note that, according to model (3.6.6), the extra-mortality extinguishes within a period of r years (see Fig. 3.6.5).

The mortality pattern of substandard risks can be assumed, at least approximately, as equal to the standard mortality pattern referred to an older individual. The *age-shift model* (see Fig. 3.6.6) implements this idea, and can be considered as an approximation to the multiplicative model. It is defined as follows:

$$q_{x+t}^{[S]} = q_{x+t+s}; \; 0 \leq t \leq m, \; s > 0 \tag{3.6.9}$$

A higher increment s in the insured's age expresses a higher extra-mortality (corresponding to a higher value for the parameter β in the multiplicative model (3.6.5)).

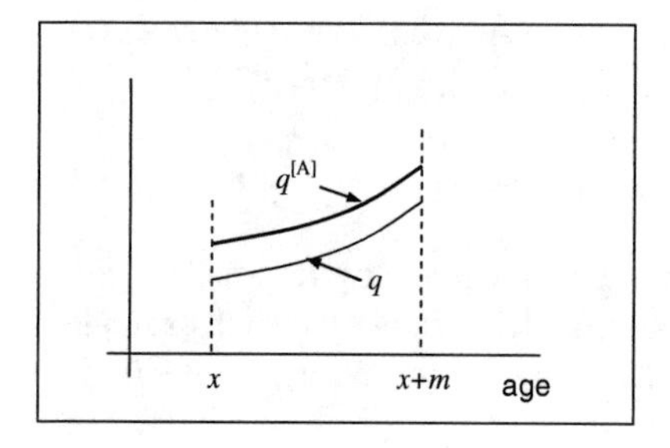

Fig. 3.6.3 The additive model

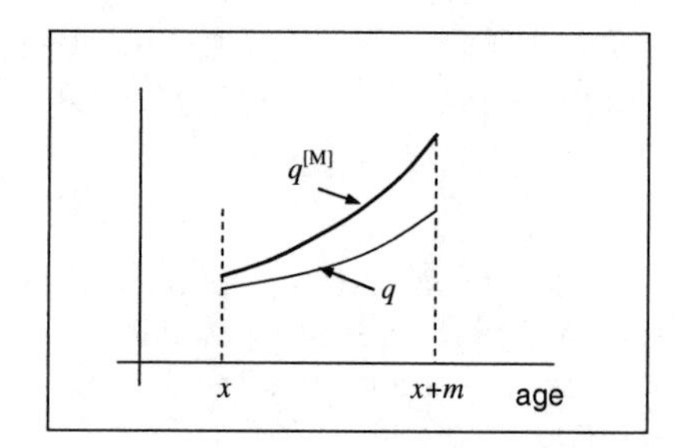

Fig. 3.6.4 The multiplicative model

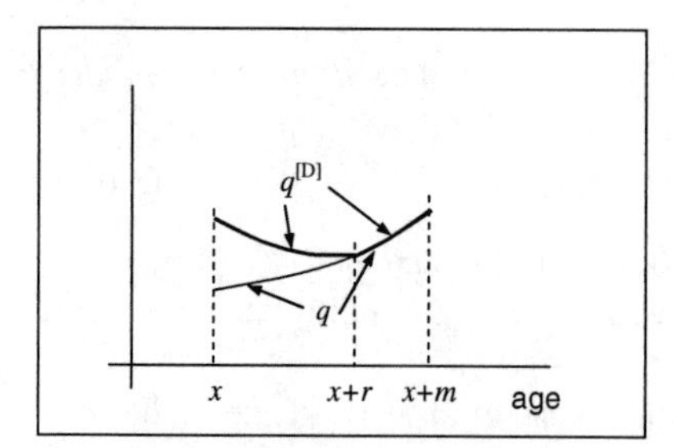

Fig. 3.6.5 Decreasing extra-mortality

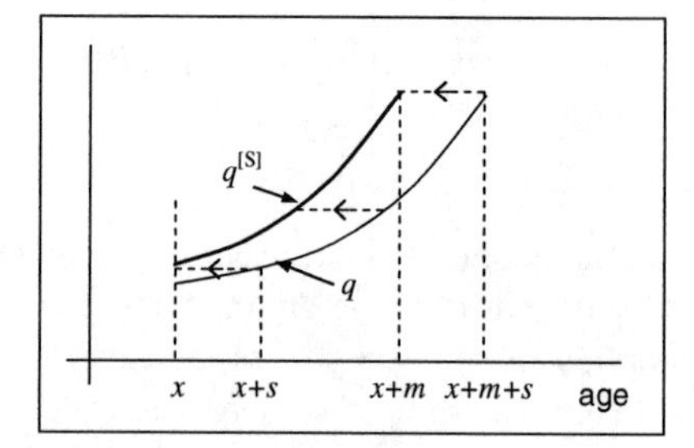

Fig. 3.6.6 The age-shift model

3.6.4 The "factor formula"

Model (3.6.5) can be used not only to represent a mortality higher than the standard one. Indeed, setting $-1 < \beta < 0$ a mortality lower than the standard is expressed. So, the model can be adopted to represent a wide range of mortality patterns in terms of the standard one given by the q_{x+t}'s.

An interesting example is provided by the so-called *numerical rating system*, introduced in 1919 by New York Life Insurance and still adopted by many insurers. A set of k rating factors is referred to. The annual specific probability of death of a given individual currently age $x+t$, $q_{x+t}^{[\mathrm{spec}]}$, is expressed by the following formula (also called the *factor formula*):

$$q_{x+t}^{[\mathrm{spec}]} = q_{x+t}\left(1+\sum_{h=1}^{k}\gamma_h\right) \tag{3.6.10}$$

The parameters γ_h lead to a higher or lower death probability for the individual in relation to the values assumed by the chosen rating factors. Clearly, the following constraint must be fulfilled for all ages $x+t$:

$$-1 < \sum_{h=1}^{k}\gamma_h < \frac{1}{q_{x+t}} - 1 \tag{3.6.11}$$

Note that an additive effect of each of the rating factors is assumed.

Remark In insurance practice, a mortality different from the standard one is frequently accounted for by adjusting directly the premium rates, rather than the probabilities of death. For example, this may be the case for the age-shifting, or the factor formula. Although the results may be quite similar, at least over some age ranges and for some insurance products, the approach is not correct, as in premium calculation elements other than the demographic one are included, e.g. expenses, financial aspects summarized by the technical rate of interest, etc.

3.7 Mortality by age and duration

3.7.1 Some preliminary ideas

Consider, for example, a group of insureds, all age 45, deriving from a population whose mortality can be described by a given life table. Is q_{45} (drawn from the assumed life table) a reasonable assessment of the one-year probability of dying for each insured in the group?

In order to answer this question, the following points should be addressed.

1. When starting a life insurance policy with an insurance company, an individual may be subject to medical screening and, possibly, to a medical examination (see Sect. 3.6.2).
2. It has been observed that the mortality experienced by policyholders recently accepted (as standard risks) is lower than the mortality experienced by policyholders (of the same age) with a longer duration since policy issue.

So, the answer to the above question is negative if the insureds have entered insurance in different years: it is reasonable to expect that an individual, who has just bought insurance, will be of better health than an individual who bought insurance several years ago, and whose health conditions could have worsened over those years.

In order to express the dependence of the probability of death on the time elapsed since policy issue, the attained age (45, in the example) should be split as follows:

$$\text{attained age} = \text{age at entry} + \text{time since policy issue}$$

The following notation is usually adopted to address the annual probabilities of death for an insured currently age 45:

$$q_{[45]}, q_{[44]+1}, \dots, q_{[40]+5}, \dots$$

where the number in square brackets denotes the age at policy issue, whereas the second number denotes the time since policy issue. In general, $q_{[x]+u}$ denotes the probability that an individual currently aged $x+u$, who bought insurance at age x, dies within one year.

According to point 2 above, it is usual to assume:

$$q_{[45]} < q_{[44]+1} < \dots < q_{[40]+5} < \dots$$

3.7.2 Select tables and ultimate tables

Allowing for the dependence of the probability of death on the time elapsed since policy issue requires the use of life tables in which probabilities are functions of age x at entry and time u since policy issue.

We look at the life table in terms of the probabilities of death. We assume that the generic row of the table contains the following elements:

$$q_{[x]},\ q_{[x]+1},\ q_{[x]+2},\ \dots,\ q_{[x]+u},\ \dots \tag{3.7.1}$$

We denote by $x_{\min}$ and $x_{\max}$ the minimum and respectively the maximum age at entry (for example, $x_{\min} = 20$, and $x_{\max} = 70$ if death benefits are involved). The set of sequences (3.7.1), for $x = x_{\min}, x_{\min+1}, \dots, x_{\max}$, is called a *select life table*.

However, experience shows that it is reasonable to assume that the selection effect vanishes after some years, say r years after policy issue. Hence, we can assume:

$$q_{[x]} < q_{[x-1]+1} < \dots < q_{[x-r]+r} = q_{[x-r-1]+r+1} = \dots = \bar{q}_x \tag{3.7.2}$$

where $\bar{q}_x$ denotes the probability that an individual currently age x, who bought insurance more than r years ago, dies within one year. The period r is called the *select period*.

Assuming, for example, a select period of $r = 3$ years, the following probabilities should be used (rather than those in (3.7.1)) for an individual entering insurance at age x:

$$q_{[x]}, q_{[x]+1}, q_{[x]+2}, \bar{q}_{x+3}, \bar{q}_{x+4}, \dots \tag{3.7.3}$$

The set of sequences (3.7.3), for $x = x_{\min}, x_{\min+1}, \dots, x_{\max}$, is called a *select-ultimate table*. In particular, the table used after the select period, namely the sequence

$$\bar{q}_{x_{\min}+r}, \ \bar{q}_{x_{\min}+r+1}, \ \dots, \ \bar{q}_z, \ \dots \tag{3.7.4}$$

(where z denotes a generic age) is called the *ultimate life table*.

Life tables in which mortality is assumed to depend on attained age only (as is the case for the life tables described in Sect. 3.2.2) are called *aggregate life tables*. Clearly, the ultimate life table is an aggregate life table.

Figure 3.7.1 illustrates a likely behavior of one-year probabilities of death, in the select part and the ultimate part of a select-ultimate life table.

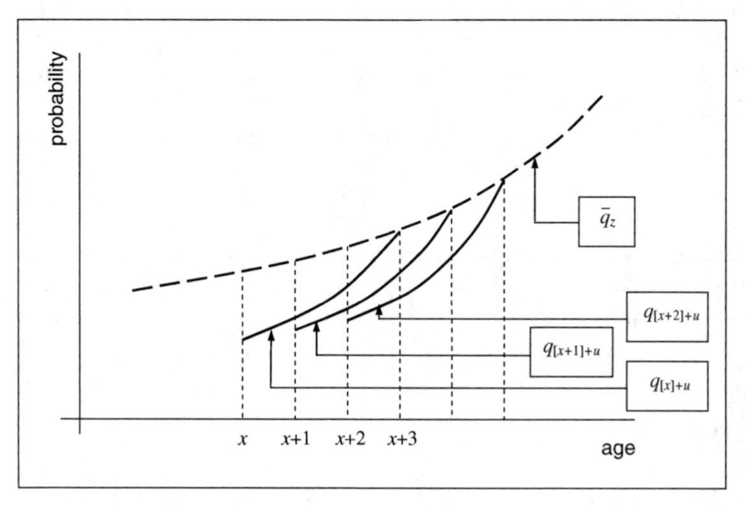

Fig. 3.7.1 Select and ultimate probabilities ($r = 3$)

Remark The selection effect, due to medical ascertainment (in the case of insurances with death benefit) or self-selection (in the case of life annuities), operates during the first years after policy issue, and the related age-pattern of mortality is often called *issue-select*. Another type of selection is allowed for, when some contingency can adversely affect the individual mortality. For example, in actuarial calculations regarding insurance benefits in the case of disability, the mortality of disabled policyholders is usually considered to be dependent on the time elapsed since the time of disablement inception (as well as on the attained age). In this case, the mortality is called *inception-select*.

3.7.3 *A practical issue*

Select probabilities $q_{[x]+u}$ should be estimated from observations of insureds' mortality. However, this requires the splitting of the insured population into a high number of "cells", as age at policy issue and duration since policy issue should be separately accounted for. Likely, such an estimation would be based on small numbers of individuals in each cell, then leading to a poor reliability of the resulting estimate.

Assume, conversely, that just the ultimate mortality is estimated (that is, irrespective of the time since policy issue, provided that this time is greater than the select period), leading to the probabilities $\bar{q}_z$ which are functions of the attained age z only. Then, the selection effect can be expressed by using appropriate reduction factors.

Trivially, select probabilities can be formally expressed as follows:

$$q_{[x]+u} = \bar{q}_{x+u}\,\rho_x(u); \quad \text{for } u = 0, 1, \ldots, r-1 \tag{3.7.5}$$

where the factor $\rho_x(u)$ depends on both the age at entry x and the time u. However, the use of factors $\rho_x(u)$ does not reduce the dimension of the estimation problem. Instead of (3.7.5), we can then assume the (approximate) relation

$$q_{[x]+u} = \bar{q}_{x+u}\,\rho(u); \quad \text{for } u = 0, 1, \ldots, r-1 \tag{3.7.6}$$

where the factor $\rho(u)$ ($\rho(u) < 1$) only depends on the time since policy issue u (or, at least, can be assumed to be independent of age x for wide age ranges, say 20 to 40, 41 to 60, etc.).

3.8 Mortality dynamics

3.8.1 *Mortality trends*

In many countries, mortality experience over the last decades shows some aspects affecting the shape of curves representing the mortality as a function of the attained age. Figures 3.8.1 and 3.8.2 illustrate the moving mortality scenario referring to the Italian male population, in terms of survival curves, i.e. in terms of l_x, and curves of deaths, i.e. in terms of d_x. Survival curves and curves of deaths relate to various period mortality observations from 1881 to 2002 ("SIM t" refers to period observations on Italian males centered on calendar year t).

Obviously, experienced trends also affect the behavior of other quantities expressing the mortality pattern, such as the life expectancy and the mortality rates. In Fig. 3.8.3, referring to Italian males, the life expectancy at the birth, the life expectancy at age 65, and the mode of the curve of deaths (i.e. the Lexis point) are compared in their evolution.

Finally, Figs. 3.8.4 and 3.8.5 concern the behavior of mortality rates. In Fig. 3.8.4 mortality rates q_x referring to various life tables are plotted against the age x, while

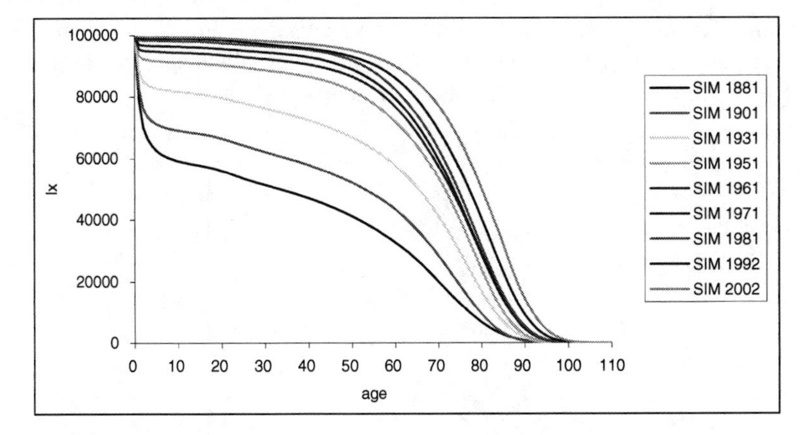

Fig. 3.8.1 Survival curves

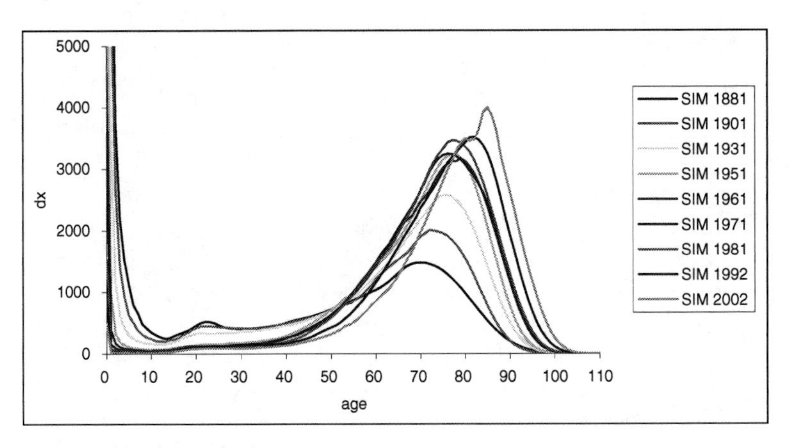

Fig. 3.8.2 Curves of deaths

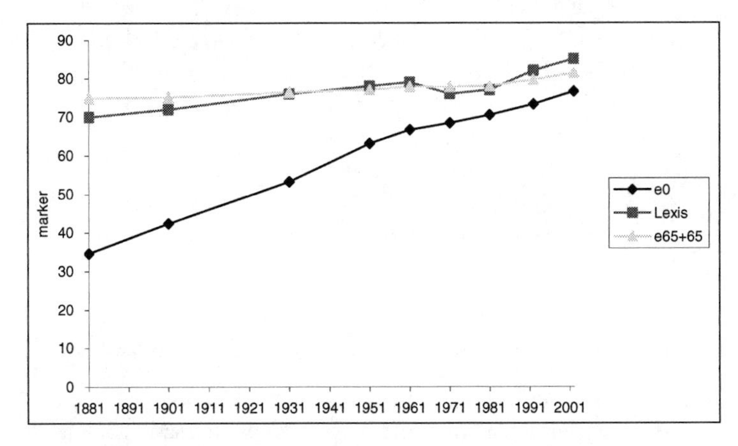

Fig. 3.8.3 Life expectancy and Lexis point

Fig. 3.8.5 shows the so-called mortality profiles at various age x in relative terms, namely the mortality rates $q_x(t)$ as functions of calendar year t divided by the mortality rate $q_x(1881)$ referring to the oldest table considered.

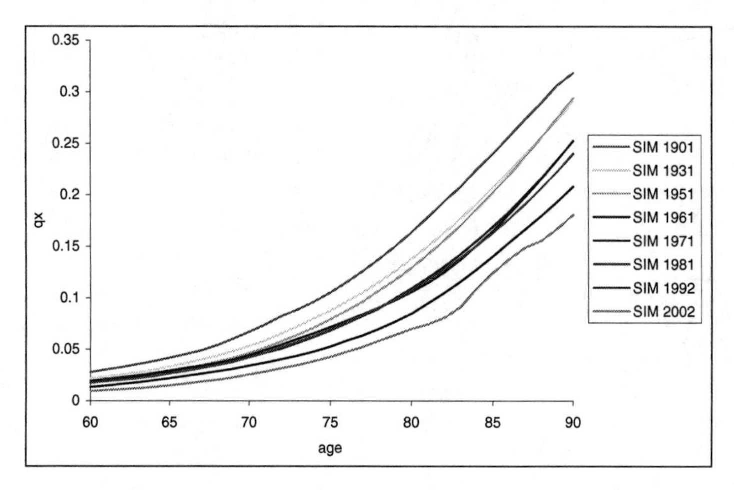

Fig. 3.8.4 Mortality rates

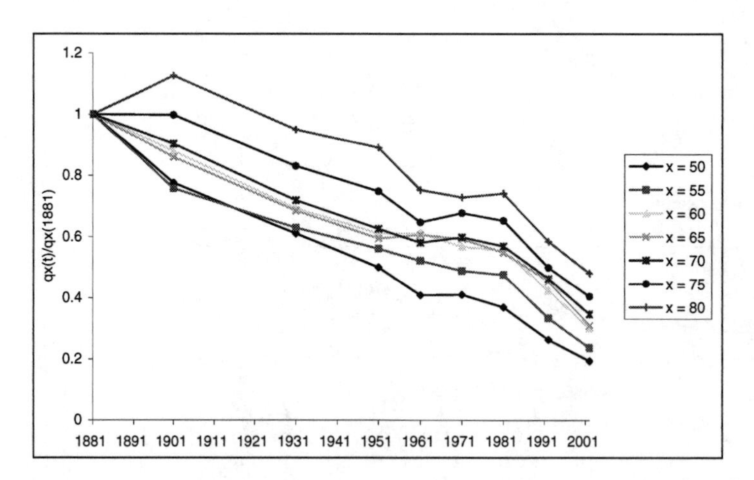

Fig. 3.8.5 Mortality profiles

Results are self-evident. In particular the following aspects can be pointed out:

1. an increase in the life expectancy (at the birth as well as at old ages);
2. a decrease in the infant mortality, and in mortality rates in particular at adult and old ages.

Turning back to the shape of the survival function and the curve of deaths, the following aspects of mortality in many countries can be singled out:

3. an increasing concentration of deaths around the mode (at old ages) of the curve of deaths is evident; so the survival function moves towards a rectangular shape, whence the term *rectangularization* to denote this aspect (see Fig. 3.8.6a);
4. the mode of the curve of deaths (which, because of the rectangularization, tends to coincide with the maximum age ω) moves towards very old ages; this aspect is called the *expansion* of the survival function (see Fig. 3.8.6b);
5. higher levels and a larger dispersion of accidental deaths at young ages (the so-called young mortality hump) have been more recently observed.

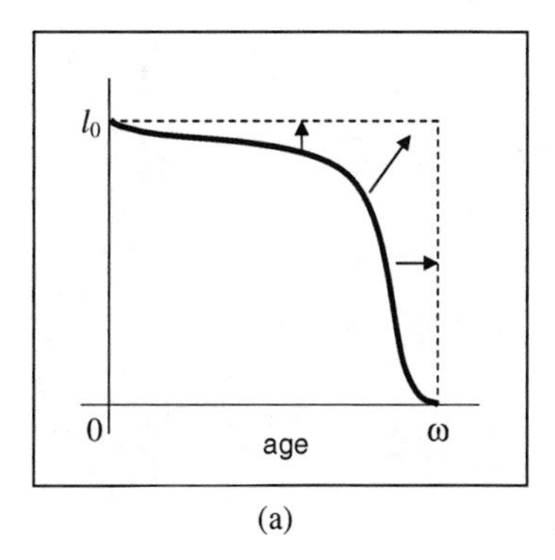

(a)

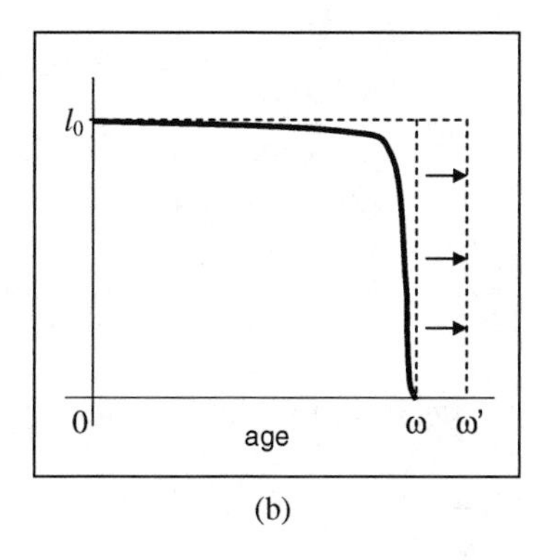

(b)

Fig. 3.8.6 Mortality trends in terms of the survival function

3.8.2 Representing mortality dynamics

The progressive decline of human mortality, witnessed by a number of population statistics, leads to the rejection of the hypothesis of "static" mortality, which would lead to biased actuarial evaluations. Trends in mortality imply the use of "projected" survival models for several purposes in life insurance and annuity calculations.

A dynamic approach to mortality underpins mortality forecasts or projections. When working in a dynamic context, the basic idea is to express mortality as a function of the (future) calendar year t. As in actuarial calculations age-specific measures of mortality are usually needed, in a dynamic context mortality is assumed to be a function of both age x and calendar year t.

In particular, we now focus on one-year probabilities of death. We denote by $q_x(t)$ the probability that a person age x in the calendar year t dies within one year. A matrix of one-year probabilities of death is represented in Table 3.8.1.

The probabilities in Table 3.8.1 can be read according to three arrangements:

Table 3.8.1 Annual probabilities of death in a dynamic context

	...	$t-1$	t	$t+1$	...
0	...	$q_0(t-1)$	$q_0(t)$	$q_0(t+1)$	...
1	...	$q_1(t-1)$	$q_1(t)$	$q_1(t+1)$	...
...	...	...	...	...	...
x	...	$q_x(t-1)$	$q_x(t)$	$q_x(t+1)$	...
$x+1$	...	$q_{x+1}(t-1)$	$q_{x+1}(t)$	$q_{x+1}(t+1)$	...
...	...	...	...	...	...
ω	...	$q_\omega(t-1)$	$q_\omega(t)$	$q_\omega(t+1)$	...

1. a *vertical* arrangement (i.e. by columns),

$$q_0(t),\ q_1(t), \ldots,\ q_x(t), \ldots \tag{3.8.1}$$

   corresponding to a sequence of *period life tables*, with each table referring to people living in a given calendar year t;
2. a *diagonal* arrangement,

$$q_0(t),\ q_1(t+1), \ldots,\ q_x(t+x), \ldots \tag{3.8.2}$$

   corresponding to a sequence of *cohort life tables*, with each table referring to the cohort born in year t;
3. a *horizontal* arrangement (i.e. by rows),

$$\ldots, q_x(t-1),\ q_x(t),\ q_x(t+1), \ldots \tag{3.8.3}$$

   yielding the *mortality profiles*, with each profile expressing the mortality trend at a given age x.

In general, the matrix in Table 3.8.1 contains elements referring to past years (and possibly originating from mortality observations) and elements referring to future years. Let t' denote the current calendar year, or possibly the year for which the most recent (reliable) period life table is available. Thus, probabilities $q_x(t)$ for $t > t'$ refer to future years, or years for which a life table is not yet available. Hence, these probabilities should be estimated by using a projection procedure.

For a given year t' and a given maximum year t^* (time horizon), the *projected life table* consists of the submatrix

$$\{q_x(t)\}; \quad x = 0, 1, \ldots \omega; \ t = t'+1, t'+2, \ldots, t^* \tag{3.8.4}$$

(see Fig. 3.8.7).

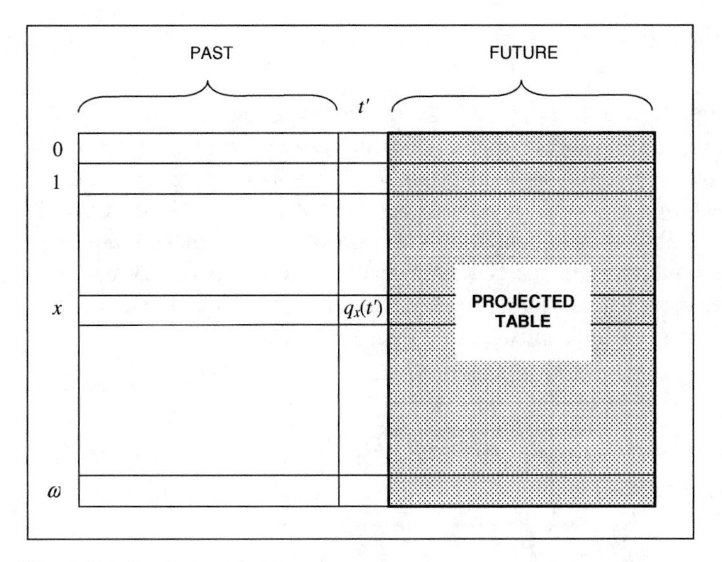

Fig. 3.8.7 The projected table

3.8.3 Probabilities and life expectancy in a dynamic context

The appropriate use of the one-year probabilities in Table 3.8.1 requires that, in each year t, probabilities concerning the lifetime of a person age x in that year are derived from the diagonal

$$q_x(t), q_{x+1}(t+1), \dots \tag{3.8.5}$$

that is, from the relevant cohort table. Then, the probability of a person age x in year t being alive at age $x+h$ is given by:

$${}_hp_x(t) = \big(1-q_x(t)\big)\big(1-q_{x+1}(t+1)\big)\dots\big(1-q_{x+h-1}(t+h-1)\big) \tag{3.8.6}$$

From probabilities (3.8.6), we can derive the following probabilities of dying:

$${}_{h|1}q_x(t) = {}_hp_x(t)\, q_{x+h}(t+h) \tag{3.8.7}$$

and then the *cohort life expectancy at age x* (namely, for an individual age x in year t):

$$\mathring{e}_x(t) = \tfrac{1}{2}\,{}_{0|1}q_x(t) + (1+\tfrac{1}{2})\,{}_{1|1}q_x(t) + \dots \tag{3.8.8}$$

Note that, in a dynamic context, formula (3.8.8) should be used instead of (3.3.3), in order to evaluate the expected lifetime allowing for future mortality trends.

3.8.4 Approaches to mortality forecasts

A number of approaches can be adopted to mortality projection in order to obtain forecasts of future mortality. Whatever the approach may be, an important role is obviously played by the mortality experienced in the past, which constitutes the data base for the projection procedures (see Fig. 3.8.8). Usually the data base consists of period tables, possibly complemented by segments of cohort tables. According to some approaches, mortality forecasts are only based on mortality observed in the past, whereas other approaches require further inputs. We just mention the following approaches, which can provide an insight into forecasting methods.

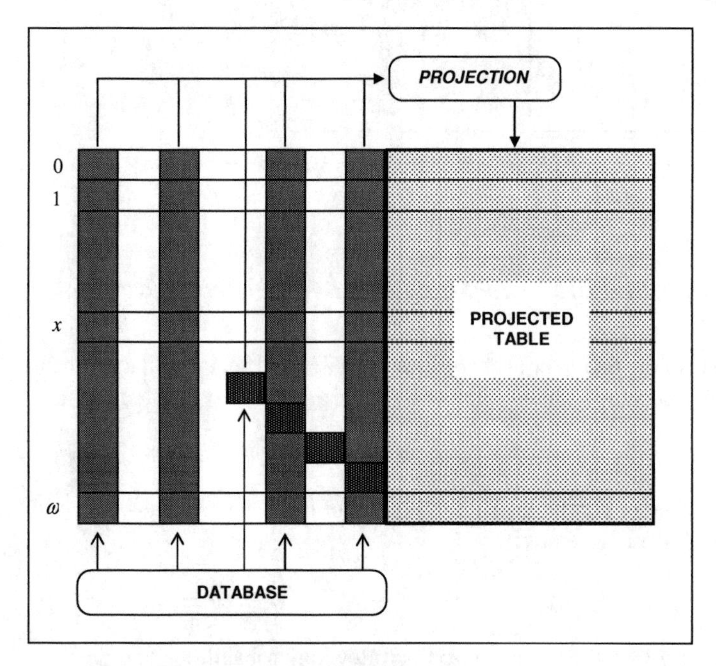

Fig. 3.8.8 Construction of a projected table

1. The analysis of mortality profiles for each age x, namely the sequences (3.8.3) for $t \leq t'$, can suggest a likely behavior of $q_x(t)$ for $t > t'$. Basically, the projection procedure consists in the *graduation* of the observed mortality rates with respect to time, and the consequent *extrapolation* to obtain the one-year probability of dying in future years. In the framework of graduation-extrapolation procedures, an important point should be addressed, namely: how are the items in the database interpreted? Depending on the answer, two classes of projection procedures are defined.

    a. If the answer is "data are simply numbers", then the extrapolation procedure does not allow for any statistical feature of the information available, as, for example, the reliability of the data. In this case, the output of the procedure is just a point estimate of future mortality (see Fig. 3.8.9).

b. Conversely, when the data are interpreted as the outcomes of random variables (namely, random frequencies of death), the extrapolation procedure must rely on sound statistical assumptions and, as a consequence, future mortality can be represented in terms of both point estimates and interval estimates.

2. When projecting mortality, the collateral information available to the forecaster can be allowed for. Information may concern a wide range of trends and events, for example trends in smoking habits, trends in prevalence of some illness, improvements in medical knowledge and surgery, etc. Thus, projections can be performed according to an assumed *scenario*. The introduction of relationships between events (e.g. advances in medical science) and effects (mortality improvements) underpins mortality projections which are carried out according to assumed scenarios. Obviously, some degree of arbitrariness follows, affecting the results.
3. Both extrapolation procedures and scenarios can be used to project *mortality by different causes* separately, instead of projecting mortality in "aggregate" terms. Projections by cause of death offer a useful insight into the changing incidence of the various causes. Conversely, some important problems arise when this type of projection is adopted. In particular, it should be stressed that complex interrelationships exist among causes of death, whilst the classic assumption of independence is commonly accepted. For example, mortality from heart diseases and lung cancer are positively correlated, as both are linked to smoking habits. A further problem concerns the difficult identification of the cause of death for elderly people.

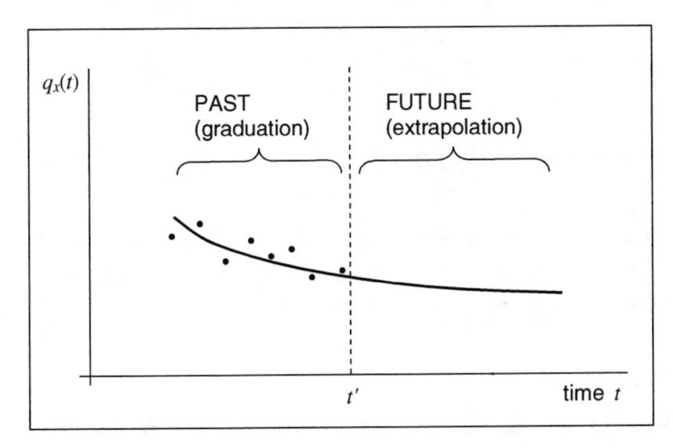

Fig. 3.8.9 Extrapolation of the $q_x(t)$'s

Remark Graduation - extrapolation methods rely on the assumption that the observed trend continues in future years. Even if a very long sequence of observations is available (throughout a time interval of, say, more than 50 years), the past trend addressed in the graduation procedure should be restricted to rather recent observations, in order to avoid the inclusion of causes of mortality improvements whose effect should be considered already extinguished. Figure 3.8.10 illustrates a possible overestimation of future mortality improvements, due to a too long period assumed as the basis of the graduation.

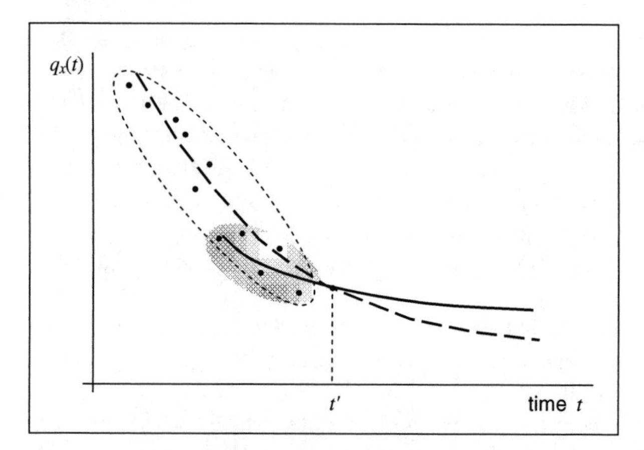

Fig. 3.8.10 Extrapolation results depending on the graduation period

The existence of various approaches to mortality forecasts witnesses that this topic is very complex. A deeper analysis of these issues is beyond the scope of this book. Hence, in Sect. 3.8.5 we just mention a simple extrapolation method, which can be placed in the framework of approach 1(a). Despite the lack of a rigorous statistical support, such a method is still widely used in current actuarial practice. Basic ideas underpinning approach 1(b) are presented in Sect. 3.8.6.

3.8.5 Extrapolation via exponential formulae

Let us assume that several period observations are available for a given population. Each observation consists of the age-pattern of mortality for a given set of ages, say $x_{\min}, x_{\min}+1, \dots, x_{\max}$. The observation referred to calendar year t, $t = t_1, t_2, \dots, t_n$, is expressed by:

$$q_{x_{\min}}(t), q_{x_{\min}+1}(t), \dots, q_{x_{\max}}(t) \tag{3.8.9}$$

that is, a (part of a) column of the matrix in Table 3.8.1. Note that, for each x, the sequence

$$q_x(t_1), q_x(t_2), \dots, q_x(t_n) \tag{3.8.10}$$

represents the observed mortality profile at age x (namely, along a row of the matrix in Table 3.8.1).

Assume that the trend observed in past years can be graduated via an exponential function. Further, suppose that the observed trend will continue in future years. Then, future mortality can be estimated extrapolating the trend itself. In formal terms, we assume:

$$q_x(t) = q_x(t')\, r_x^{t-t'} \tag{3.8.11}$$

where t' is the base year, and r_x is the mortality (annual) *variation factor* (*reduction factor* if $r_x < 1$) at age x, estimated on the basis of the observed mortality profile (3.8.10).

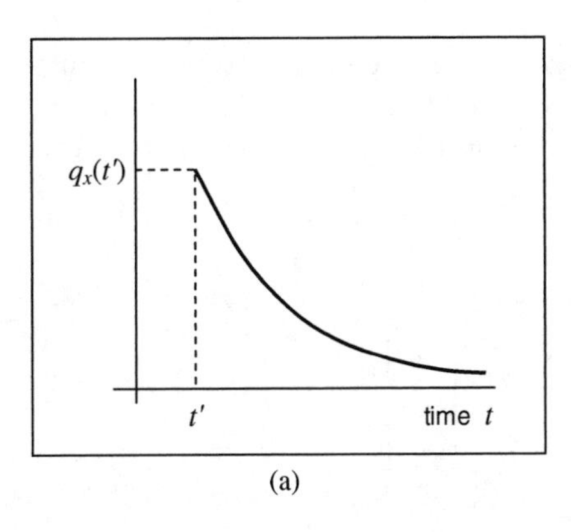

(a)

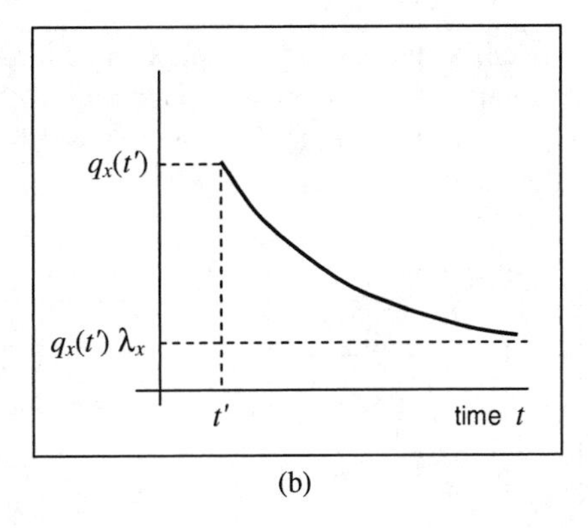

(b)

Fig. 3.8.11 Exponential models

From formula (3.8.11) it follows that, if $r_x < 1$ then

$$\lim_{t \to +\infty} q_x(t) = 0 \tag{3.8.12}$$

(see Fig. 3.8.11a). Although the validity of mortality forecasts should be restricted to a limited time interval, it is more realistic to assign a positive limit to the mortality at any age x. To this purpose, the following formula with an assigned (positive) *asymptotic mortality* can be adopted:

$$q_x(t) = q_x(t') \left(\lambda_x + (1 - \lambda_x)\, r_x^{t-t'} \right) \tag{3.8.13}$$

where $\lambda_x \geq 0$ for all x. Thus, the asymptotic mortality at age x is given by

$$\lim_{t \to +\infty} q_x(t) = q_x(t')\, \lambda_x \tag{3.8.14}$$

(see Fig. 3.8.11b).

3.8.6 Mortality forecasts allowing for random fluctuations

A rigorous approach to mortality forecasts should take into account the stochastic nature of mortality. In particular, the following points should underpin a stochastic projection model:

- observed mortality rates are outcomes of random variables representing past mortality;
- forecasted mortality rates are estimates of random variables representing future mortality.

Hence, stochastic assumptions about mortality are required, namely probability distributions for the random numbers of death (see Sect. 3.10.1 and 3.10.3), and a statistical structure linking forecasts to observations must be specified (as sketched in Fig. 3.8.12).

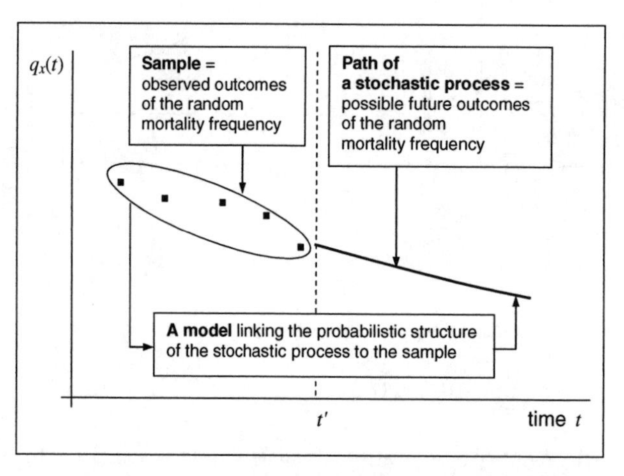

Fig. 3.8.12 A statistical approach in the graduation - extrapolation procedure

In a stochastic framework, the results of projection procedures consist in both point estimates and interval estimates of future mortality rates (see Fig. 3.8.13) and other life table functions. Clearly, traditional graduation - extrapolation procedures, which do not explicitly allow for randomness in mortality, produce just one numerical value for each future mortality rate. Moreover, such values can be hardly interpreted as point estimates, because of the lack of an appropriate statistical structure and model.

An effective graphical representation of randomness in future mortality is given by the so-called *fan charts*; see Fig. 3.8.14, which refers to the projection of the expected lifetime. The fan chart shows a "central projection" together with some "prediction intervals". The narrowest interval, namely the one with the darkest shading, correspond to a low probability prediction, say 10%, and is included in prediction intervals with higher probabilities (say 25%, 50%, etc.).

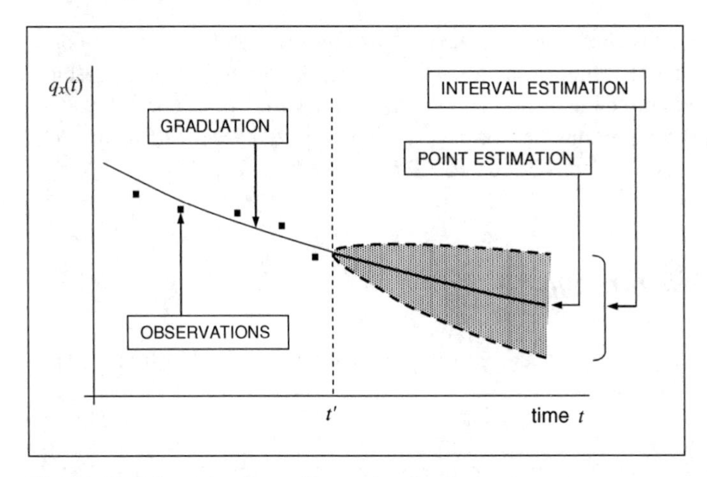

Fig. 3.8.13 Point estimation and interval estimation

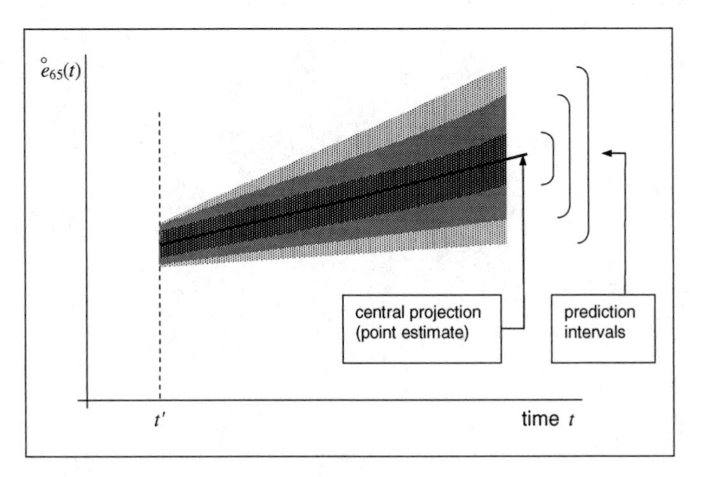

Fig. 3.8.14 Fan chart of the expected lifetime

The Lee-Carter method (proposed in 1992) represents a significant example of the stochastic approach to mortality forecasts, and constitutes one of the most influential proposals in recent times.

3.9 Moving to a time-continuous context

As seen in Sect. 3.2.7, if we want to evaluate probabilities like (3.2.23), (3.2.24) and (3.2.25) when ages or durations are real numbers, tools other than the life table are needed. In this Section we describe some tools which allows us to extend the calculation of probabilities of survival and death to a time-continuous context.

Although in the following chapters calculations concerning life insurance contracts will be presented in a time-discrete context, some important issues suggest us to extend the survival model to a time-continuous framework. An important example is provided by the expression of mortality assumptions via the so-called force of mortality, as we will see in Sects. 3.9.3 and 3.9.5.

3.9.1 The survival function

Assume that the function $S(t)$, called the *survival function* and defined for $t \geq 0$ as follows

$$S(t) = \mathbb{P}[T_0 > t] \tag{3.9.1}$$

has been assigned. Of course, T_0 denotes the random lifetime for a newborn.

Consider the probability (3.2.23), which can be expressed as follows:

$$\mathbb{P}[T_x > h] = \mathbb{P}[T_0 > x+h \,|\, T_0 > x] = \frac{\mathbb{P}[T_0 > x+h]}{\mathbb{P}[T_0 > x]} \tag{3.9.2}$$

From relation (3.2.22), we then find

$${}_hp_x = \frac{S(x+h)}{S(x)} \tag{3.9.3}$$

For probability (3.2.24) we obtain

$${}_hq_x = \frac{S(x) - S(x+h)}{S(x)} \tag{3.9.4}$$

The same reasoning leads to

$${}_{h|k}q_x = \frac{S(x+h) - S(x+h+k)}{S(x)} \tag{3.9.5}$$

The survival function and the life table are strictly related each other. We note that, since l_x is the expected number of people alive out of a cohort initially consisting of l_0 individuals, we have:

$$l_x = l_0 \, \mathbb{P}[T_0 > x] \tag{3.9.6}$$

and, in terms of the survival function,

$$l_x = l_0 \, S(x) \tag{3.9.7}$$

(provided that all the individuals in the cohort have the same mortality pattern, described by $S(x)$). Thus, the l_x's are proportional to the values the survival function takes on integer ages x, and hence the life table can be interpreted as a tabulation of the survival function.

The typical shape of the *survival curve*, namely the graph of the survival function, is illustrated in Fig. 3.9.1. The analogy with the behavior of the l_x's is apparent (see, for example, Fig. 3.2.1), and is justified by relation (3.9.7).

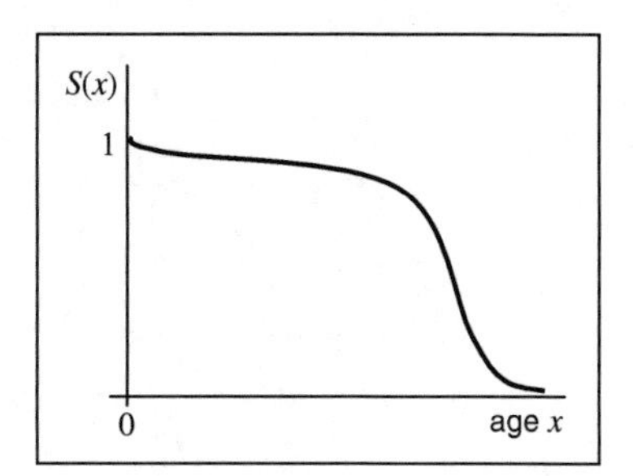

Fig. 3.9.1 The survival function

We now assume that a life table is available, for example thanks to a period observation providing an estimate of the mortality rates, from which the l_x's are calculated, for $x = 0, 1, 2, \ldots$, according to the procedure described in Sect. 3.2.3. How to obtain the survival function for all real ages x?

Relation (3.9.7) suggests a practicable approach. First, for $x = 0, 1, \ldots, \omega$, set

$$S(x) = \frac{l_x}{l_0} \tag{3.9.8}$$

using the available life table. Then, for $x = 0, 1, \ldots, \omega$ and $0 < t < 1$, define

$$S(x+t) = (1-t)\, S(x) + t\, S(x+1) \tag{3.9.9}$$

and assume $S(x) = 0$ for $x > \omega$. Hence, a piece-wise linear function is obtained.

Graduation models other than the linear model used in (3.9.9) can be adopted. Moreover, the values of $S(x)$ can be fitted using some mathematical formula; however, the use of formulae for representing the age-continuous mortality pattern can be better placed in the framework we will describe in Sect. 3.9.3.

3.9.2 Other related functions

Other functions can be involved in age-continuous actuarial calculations. The most important is the force of mortality (or mortality intensity), dealt with in Sect. 3.9.3. In the present Section we introduce the *probability density function* (pdf) and the *distribution function* of the random variable T_x, $x \geq 0$.

First, we focus on the random lifetime T_0. Let $f_0(x)$ and $F_0(x)$ denote, respectively, the pdf and the distribution function of T_0. In particular, $F_0(x)$ expresses, by definition, the probability of a newborn dying within x years. Hence,

$$F_0(x) = \mathbb{P}[T_0 \leq x] \tag{3.9.10}$$

or, according to the usual notation,

$$F_0(x) = {}_xq_0 \tag{3.9.11}$$

Of course, we have

$$F_0(x) = 1 - S(x) \tag{3.9.12}$$

The following relation holds between the pdf $f_0(x)$ and the distribution function $F_0(x)$:

$$F_0(x) = \int_0^x f_0(t)\,\mathrm{d}t \tag{3.9.13}$$

Usually it is assumed that, for $x > 0$, the pdf $f_0(x)$ is a continuous function. Then, we have

$$f_0(x) = \frac{\mathrm{d}F_0(x)}{\mathrm{d}x} = -\frac{\mathrm{d}S(x)}{\mathrm{d}x} \tag{3.9.14}$$

The graph of the pdf $f_0(x)$ is frequently called the *curve of deaths* (see also Sect. 3.2.1).

Figure 3.9.2 illustrates the typical behavior of the pdf $f_0(x)$. Equation (3.9.14) justifies the relation between the curve of deaths and the survival curve (see Fig. 3.9.1). In particular, we note that the point of maximum downward slope in the survival curve corresponds to the modal point (at adult-old ages) in the curve of deaths.

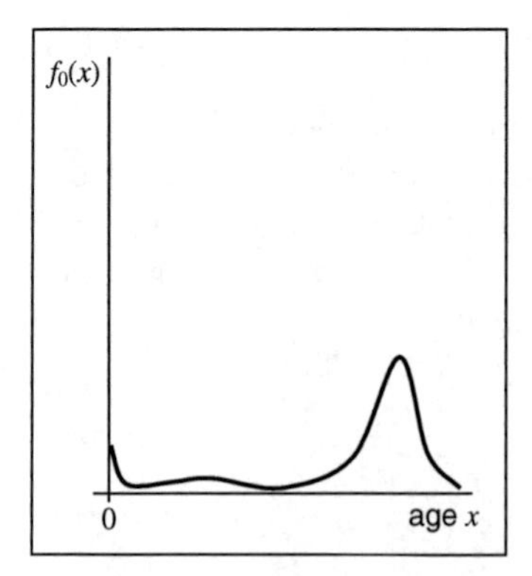

Fig. 3.9.2 The probability density function

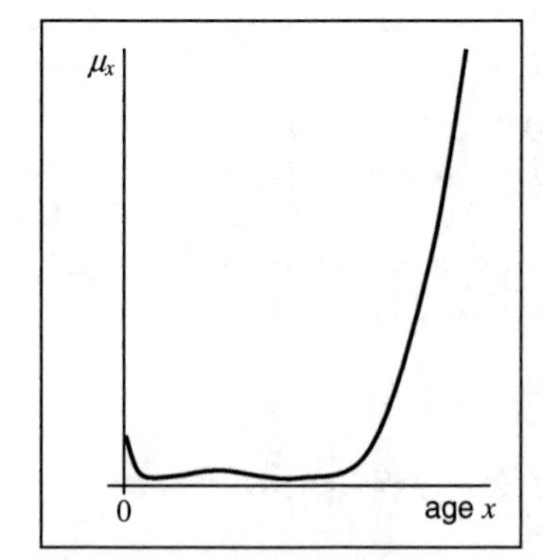

Fig. 3.9.3 The force of mortality

Moving to the remaining lifetime at age x, T_x ($x > 0$), the following relations link the distribution function and the pdf of T_x with the analogous functions relating to T_0:

$$F_x(t) = \mathbb{P}[T_x \leq t] = \frac{\mathbb{P}[x < T_0 \leq x+t]}{\mathbb{P}[T_0 > x]} = \frac{F_0(x+t) - F_0(x)}{S(x)} \tag{3.9.15}$$

$$f_x(t) = \frac{\mathrm{d}F_x(t)}{\mathrm{d}t} = \frac{\dfrac{\mathrm{d}F_0(x+t)}{\mathrm{d}t}}{S(x)} = \frac{f_0(x+t)}{S(x)} \tag{3.9.16}$$

From functions $F_x(t)$ and $f_x(t)$ (and in particular, via (3.9.15) and (3.9.16), from $F_0(t)$ and $f_0(t)$), all of the probabilities involved in actuarial calculations can be derived. For example:

$${}_tp_x = 1 - F_x(t) = \int_t^{+\infty} f_x(u)\,du = \frac{1}{S(x)} \int_t^{+\infty} f_0(x+u)\,du \tag{3.9.17}$$

3.9.3 *The force of mortality*

Consider the function μ_x, defined for all $x \geq 0$ as follows:

$$\mu_x = \lim_{t \to 0} \frac{{}_tq_x}{t} \tag{3.9.18}$$

The function μ_x is called the *force of mortality* (or the *mortality intensity*, or the *hazard function*). It can be estimated, for example for $x = 0, 1, \ldots$, using period mortality observations. Then, the estimated values can be graduated, in particular using a *mathematical mortality law*. A number of laws have been proposed in actuarial and demographical literature, and are used in actuarial practice. Some important examples are presented in Sect. 3.9.5.

Figure 3.9.3 shows the typical behavior of the force of mortality. The relation between its graph and the curve of deaths can be explained thanks to relation (3.9.21).

An interesting relation links the survival function to the force of mortality. From definition (3.9.18), by using (3.9.4) we obtain

$$\mu_x = \lim_{t \to 0} \frac{S(x) - S(x+t)}{t\,S(x)} \tag{3.9.19}$$

and then

$$\mu_x = \frac{-\dfrac{\mathrm{d}S(x)}{\mathrm{d}x}}{S(x)} \tag{3.9.20}$$

and also (see (3.9.14)):

$$\mu_x = \frac{f_0(x)}{S(x)} \tag{3.9.21}$$

Hence, once the survival function $S(x)$ has been assigned, the force of mortality can be derived. Thus, the force of mortality does not add any information concerning the age-pattern of mortality, provided that this has been described in terms of $S(x)$.

The role of the force of mortality is to provide a tool for a fundamental statement of assumptions about the behavior of individual mortality as a function of the attained age. The Gompertz law for the force of mortality (see Sect. 3.9.5) provides an excellent example. Indeed, when μ_x has been assigned, relation (3.9.20) is a differential equation. Solving with respect to $S(x)$ (with the obvious boundary condition $S(0) = 1$) leads to:

$$S(x) = \mathrm{e}^{-\int_0^x \mu_t \, \mathrm{d}t} \tag{3.9.22}$$

Clearly, the possibility of finding a "closed" form for $S(x)$ strictly depends on the mathematical structure of μ_x.

Once the survival function has been obtained, then all survival and death probabilities can be derived (see equations (3.9.3) to (3.9.5), with x, h and k positive real numbers). In particular, for example

$$q_x = 1 - p_x = 1 - \frac{S(x+1)}{S(x)} = 1 - \mathrm{e}^{-\int_x^{x+1} \mu_t \, \mathrm{d}t} \tag{3.9.23}$$

Remark Functions of age x, like l_x, q_x, d_x, etc. in the age-discrete context, and $S(x)$, $f_0(x)$, μ_x, etc. in the age-continuous context, constitute examples of *biometric functions* (other biometric functions relate to disability, mortality of disabled people, and so on). In the age-discrete context, they are also named *life table functions*.
We recall that, once one of these functions has been assigned, the other functions (in the same context) can be derived. For example, in age-discrete calculations from the l_x values we can derive the functions q_x, d_x, etc.; in the age-continuous framework, from the force of mortality μ_x the survival function can be calculated and then all of the probabilities of interest.
Links between quantities used in an age-discrete context (like l_x, d_x, etc.) and quantities used in age-continuous circumstances (like $S(x)$, $f_0(x)$, etc.) may be of interest, especially when comparing and interpreting graphical representations of data provided by statistical experiences. The analogy between l_x and $S(x)$ immediately emerges from (3.9.7). As regards d_x (see Eq. (3.2.1)), the analogy with the pdf $f_0(x)$ follows from the fact that the former is minus the first-order difference of the function l_x, while the latter is minus the derivative of the survival function $S(x)$. Further, thanks to relation (3.2.17), also the link between ${}_{x|1}q_0$ and $f_0(x)$ emerges.

3.9.4 Markers

Single-figure indices, namely markers, summarizing the lifetime probability distribution can be defined also in a time-continuous context.

The *expected total lifetime* (or *life expectancy at the birth*) is defined as follows:

$$\bar{e}_0 = \mathbb{E}[T_0] = \int_0^{+\infty} t\, f_0(t)\, \mathrm{d}t \tag{3.9.24}$$

Integrating by parts, we can express $\bar{e}_0$ in terms of the survival function:

$$\bar{e}_0 = \int_0^{+\infty} S(t)\, \mathrm{d}t \tag{3.9.25}$$

The definition can be extended to all (real) ages x. So, the *expected remaining lifetime at age x* (or *life expectancy at age x*) is given by

$$\bar{e}_x = \mathbb{E}[T_x] = \int_0^{+\infty} t\, f_x(t)\, \mathrm{d}t = \frac{1}{S(x)} \int_0^{+\infty} S(x+t)\, \mathrm{d}t \tag{3.9.26}$$

For an individual age x, the expected age at death is clearly given by:

$$x + \mathbb{E}[T_x] = \bar{e}_x \tag{3.9.27}$$

It is possible to prove that the expected values $\mathring{e}_0$ and $\mathring{e}_x$ (see Eqs. (3.3.1) to (3.3.4)) are approximations to the expected values $\bar{e}_0$ and $\bar{e}_x$ respectively, by applying the trapezoidal rule to integrals in Eqs. (3.9.25) and (3.9.26).

Another location index is provided by the *Lexis point* which, in a time-continuous context, is defined as the (old) age $x^{(\mathrm{L})}$ such that

$$f_0(x^{(\mathrm{L})}) = \max_x \{f_0(x)\} \tag{3.9.28}$$

A traditional variability measure is provided by the *variance of the random lifetime*:

$$\mathbb{V}\mathrm{ar}[T_0] = \int_0^{+\infty} (t - \bar{e}_0)^2 f_0(t)\, \mathrm{d}t \tag{3.9.29}$$

The *interquartile range* provides another variability measure. It is defined as follows:

$$\mathrm{IQR}[T_0] = x^{(75)} - x^{(25)} \tag{3.9.30}$$

where $x^{(25)}$ and $x^{(75)}$ are respectively the first quartile (the 25-th percentile) and the third quartile (the 75-th percentile) of the probability distribution of T_0, namely the ages such that $S(x^{(25)}) = 0.75$ and $S(x^{(75)}) = 0.25$. Note that IQR decreases as the lifetime distribution becomes less dispersed.

The 10-th percentile of the probability distribution of T_0, $x^{(10)}$, is usually called the *endurance*; thus, $S(x^{(10)}) = 0.90$.

The probability of a new born dying before a given age x',

$${}_{x'}q_0 = 1 - S(x') = \int_0^{x'} f_0(t)\, \mathrm{d}t \tag{3.9.31}$$

for x' small (say 1, or 5), provides a measure of infant mortality.

Figure 3.9.4 illustrates some markers of practical interest.

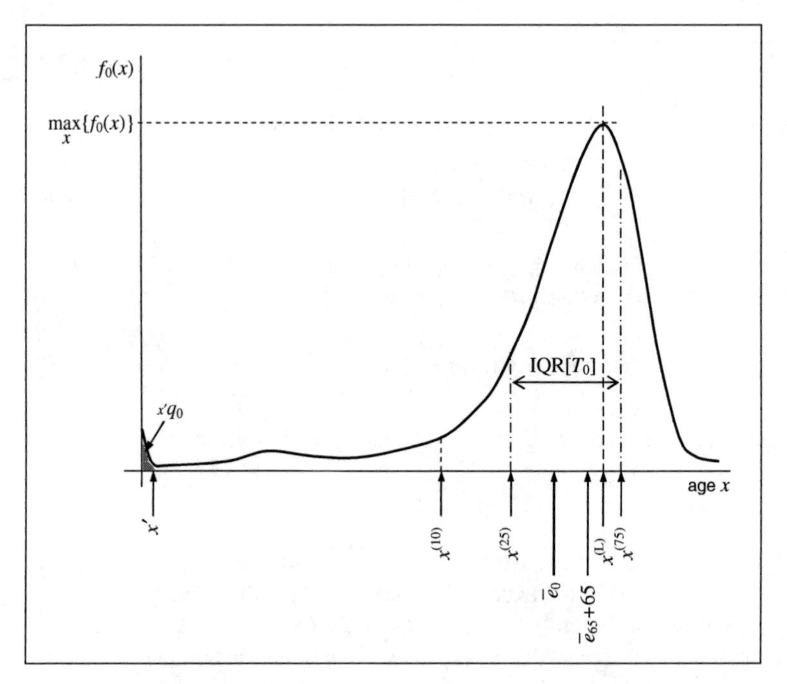

Fig. 3.9.4 Markers in a time-continuous context

3.9.5 Parametric models

Parametric models, i.e. mortality laws, have been proposed in relation to various functions expressing the age-pattern of mortality, in both the age-discrete and the age-continuous context. An important example, namely the Heligman-Pollard laws which focus on the odds and the one-year probabilities of death, have been presented in Sect. 3.4.2. In the age-continuous context, a number of mortality laws refer to the force of mortality, μ_x, although some of them have been originally proposed in different terms.

The *Gompertz law*, proposed in 1825, is as follows:

$$\mu_x = B\,c^x \tag{3.9.32}$$

with $B, c > 0$. Sometimes the following equivalent notation is used

$$\mu_x = \alpha\,e^{\beta x} \tag{3.9.33}$$

It is interesting to look at the hypothesis underlying the Gompertz law. Assume that, moving from age x to age $x + \Delta x$, the increment of the mortality intensity is proportional to its initial value, μ_x, and to the length of the interval, Δx; thus

$$\Delta\mu_x = \beta\,\mu_x\,\Delta x \tag{3.9.34}$$

with $\beta > 0$. This assumption leads to the differential equation

$$\frac{\mathrm{d}\mu_x}{\mathrm{d}x} = \beta\,\mu_x, \tag{3.9.35}$$

and finally to (3.9.33), with $\alpha > 0$. The Gompertz law is used to represent the age progression of mortality at the old ages, namely the senescent mortality.

The *(first) Makeham law*, proposed in 1867, is a generalization of the Gompertz law, namely

$$\mu_x = A + B\,c^x \tag{3.9.36}$$

where the term $A \geq 0$ (independent of age) represents non-senescent mortality, e.g. because of accidents. The following equivalent notation is also used:

$$\mu_x = \gamma + \alpha\,\mathrm{e}^{\beta x} \tag{3.9.37}$$

By using (3.9.22), from (3.9.36) we obtain:

$$S(x) = \exp\left(-A\,x - \frac{B}{\log c}(c^x - 1)\right) \tag{3.9.38}$$

In particular, setting $A = 0$ we find the survival function of the Gompertz law.

The *second Makeham law*, proposed in 1890, is as follows

$$\mu_x = A + H\,x + B\,c^x \tag{3.9.39}$$

and hence constitutes a further generalization of the Gompertz law.

The *Thiele law*, proposed in 1871, can represent the age-pattern of mortality over the whole life span (see Fig. 3.9.3):

$$\mu_x = A\,\mathrm{e}^{-Bx} + C\,\mathrm{e}^{-D(x-E)^2} + F\,G^x \tag{3.9.40}$$

where all the parameters are positive real numbers. The first terms decreases as the age increases and represents the infant mortality. The second term, which has a “Gaussian” shape, represents the mortality hump (mainly due to accidents) at young-adult ages. Finally, the third term (of a Gompertz type) represents the senescent mortality.

It is worth noting that the structure of the first Heligman-Pollard law (as well as its aim, namely to represent the age-pattern of mortality over the whole life span) is analogous to the structure of Thiele’s law.

In 1932 Perks proposed two mortality laws. The *first Perks law* is as follows:

$$\mu_x = \frac{\alpha\,\mathrm{e}^{\beta x} + \gamma}{\delta\,\mathrm{e}^{\beta x} + 1} \tag{3.9.41}$$

Conversely, the *second Perks law* has the following more general structure

$$\mu_x = \frac{\alpha\,e^{\beta x} + \gamma}{\delta\,e^{\beta x} + \varepsilon\,e^{-\beta x} + 1} \tag{3.9.42}$$

Perks' laws, whose graphs have a logistic shape, play an important role in representing the mortality pattern at very old ages (say, beyond 80). Actually, recent statistical observations show that the force of mortality is slowly increasing at very old ages, approaching a rather flat shape. This fact leads to the rejection of the exponential increase (implied by the previous models).

3.10 Stochastic mortality

3.10.1 Number of people alive in a cohort

Assume that, at time $t = 0$, a "group" (for example a pension fund, or a portfolio of life insurance contracts) consists of n_0 initial individuals. Further, assume that all the members of this group are aged x initially, and that no other individual will enter the group in future years. Thus, the group is a *cohort*. Finally, assume that the only cause of exit is the death.

The number of people alive at time t, $t = 1,2,\ldots$, is a random number, which we denote with N_t. Any sequence of integers $n_1, n_2, \ldots$, such that

$$n_0 \geq n_1 \geq n_2 \geq \ldots \tag{3.10.1}$$

is a possible outcome of the random sequence

$$N_1,\ N_2,\ \ldots \tag{3.10.2}$$

Of course, any single outcome of the random sequence (3.10.2) does not provide, by itself, significant information about the reasonable evolution of the cohort. Conversely, the meaning of "reasonable" can be specified as soon as a probabilistic structure describing the lifetimes of the cohort members has been assigned.

3.10.2 Deterministic models versus stochastic models

We assume that the individuals in the cohort are analogous in respect of the age-pattern of mortality, thus for all the individuals we assume the same life table (or survival function). Hence, the probability of being alive at time t is given, for any member of the cohort, by ${}_tp_x = \frac{l_{x+t}}{l_x}$.

It follows that the expected number of individuals alive at time t, out of the initial n_0 members, is given by

$$\mathbb{E}[N_t] = n_0\,{}_tp_x; \quad t = 1,2,\ldots \tag{3.10.3}$$

It should be noted that, although formula (3.10.3) involves probabilities, the model built up so far is a deterministic model, as probabilities are only used to determine expected values and the probabilities themselves are assumed to be known. A first step towards stochastic models follows.

We assume that the random lifetimes of the individuals in the cohort are independent. For any given t and for $j = 1,2,\dots,n_0$, we denote by $\mathscr{E}_t^{(j)}$ the event "the member j is alive at time t". Of course, $\mathbb{P}[\mathscr{E}_t^{(j)}] = {}_tp_x$ for all j. From the independence of the lifetimes, the independence of the events $\mathscr{E}_t^{(j)}$, $j = 1,2,\dots,n_0$ follows. We note that N_t can be defined as the random number of true events out of the n_0 events defined above; hence, N_t has a binomial distribution, with parameters n_0, ${}_tp_x$. Thus

$$\mathbb{P}[N_t = k] = \binom{n_0}{k} ({}_tp_x)^k (1 - {}_tp_x)^{n_0-k}; \quad k = 0,1,\dots,n_0 \tag{3.10.4}$$

In particular, the variance of N_t is given by

$$\mathbb{V}\mathrm{ar}[N_t] = n_0\, {}_tp_x\, (1 - {}_tp_x) \tag{3.10.5}$$

Example 3.10.1. We consider a cohort of n_0 individuals, all age $x = 40$ initially. We assume that the age-pattern of mortality is described by the first Heligman-Pollard law, with the parameters specified in the Example 3.4.1.

Figures 3.10.1 and 3.10.2 refer to a cohort initially consisting of $n_0 = 500$ individuals. The probability distributions of N_5 and N_{10}, respectively, are depicted. In particular, we have

$$\mathbb{E}[N_5] = 496.269;\ \mathbb{V}\mathrm{ar}[N_5] = 3.703;\ \mathbb{CV}[N_5] = 0.003878$$
$$\mathbb{E}[N_{10}] = 490.083;\ \mathbb{V}\mathrm{ar}[N_{10}] = 9.720;\ \mathbb{CV}[N_{10}] = 0.006362$$

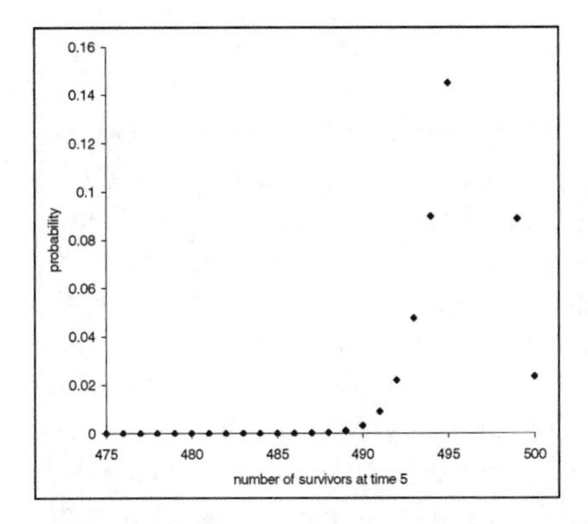

Fig. 3.10.1 Probability distribution of N_5 ($n_0 = 500$)

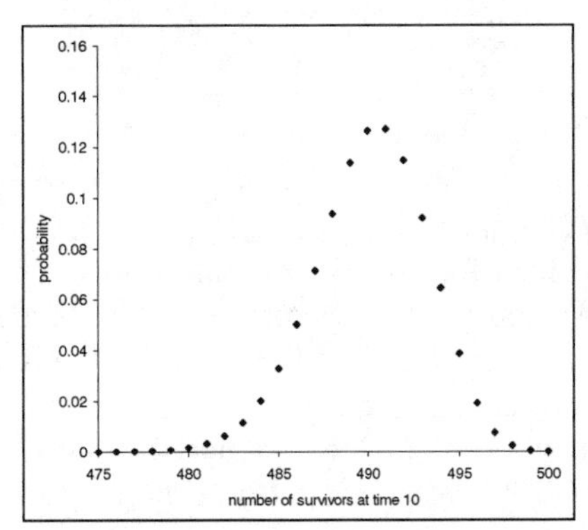

Fig. 3.10.2 Probability distribution of N_{10} ($n_0 = 500$)

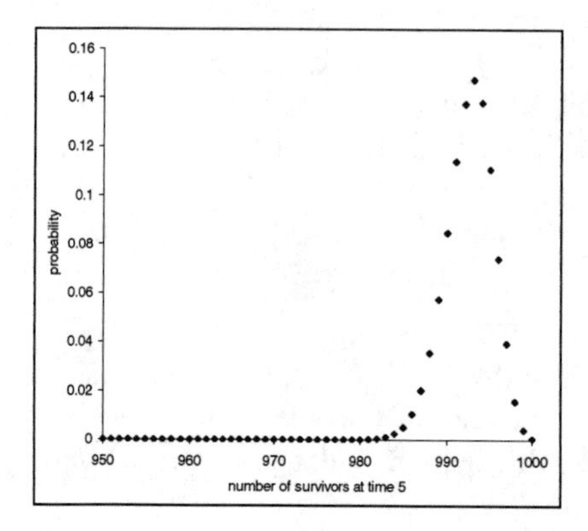

Fig. 3.10.3 Probability distribution of N_5 ($n_0 = 1\,000$)

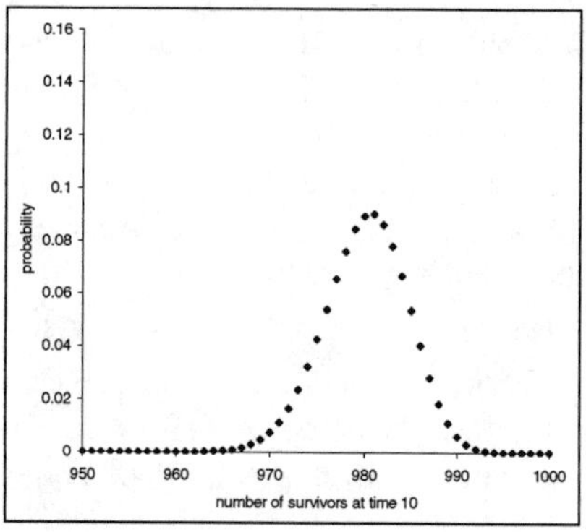

Fig. 3.10.4 Probability distribution of N_{10} ($n_0 = 1\,000$)

In Figs. 3.10.3 and 3.10.4 the probability distributions of N_5 and N_{10} are respectively illustrated, referring to a cohort initially consisting of $n_0 = 1\,000$ individuals. In particular, we have

$$\mathbb{E}[N_5] = 992.538; \;\; \mathbb{V}\mathrm{ar}[N_5] = 7.406; \;\; \mathbb{CV}[N_5] = 0.002741;$$
$$\mathbb{E}[N_{10}] = 980.166; \; \mathbb{V}\mathrm{ar}[N_{10}] = 19.441; \; \mathbb{CV}[N_{10}] = 0.004498$$

The effect of the portfolio size on the variability of the results (namely, the pooling effect) is evident: in particular the relative variability, which is expressed by the coefficient of variation (see Sect. 1.5.2), decreases, at any time, as the portfolio size increases. The effect of the portfolio size also emerges if we compare the graphs in Figs. 3.10.1 and 3.10.2, on the one hand, to those in Figs. 3.10.3 and 3.10.4 on the other, of course taking into account the different scales adopted for the axes. Note that, on the contrary, for any given portfolio size the variability (in both absolute and relative terms) increases as the time increases.
□

The probability distribution of N_t witnesses the presence of random fluctuations in the number of survivors around its expected value $\mathbb{E}[N_t]$. As seen in Sect. 2.3.1, random fluctuations are the consequence of the *process risk*, which, in the demographical framework, constitutes one of the components of the mortality / longevity risk. Conversely, systematic deviations are the consequence of the *uncertainty risk*, which constitutes another component of the mortality / longevity risk.

According to recent glossary standards, in the following we will use the term *mortality risk* to denote a mortality higher than that expected, when this generates negative consequences (for example, for an insurer dealing with death benefits). Conversely, the term *longevity risk* will denote a mortality lower than expected, when this originates negative consequences (for example, for an insurer dealing with life annuities). The expression mortality / longevity risk will be used to generically denote risks arising from lifetimes. Note that the mortality and the longevity risk may consist of random fluctuations as well as systematic deviations.

Remark The mortality and the longevity risks belong to the class of *biometric risks*, which include all the risks related to human life conditions. Thus, besides mortality and longevity risks, also risks arising from the behavior of disability, natality, and so on, fall in the class of biometric risks.

3.10.3 Random fluctuations in mortality

Further insights into the process risk can be obtained looking at the random behavior of the number of survivors in the cohort over time. As life insurance is, typically, a medium-long term business, the features of this activity can be better perceived in a dynamic perspective.

To this purpose, we can implement a simulation procedure, based on the generation of (pseudo-) random numbers. The procedure can be as follows:

1. simulate the random lifetime (i.e. the age at death) for each member of the cohort;
2. given the simulated values of the n_0 lifetimes, calculate the numbers of individuals alive at times $1,2,\dots$, namely the simulated outcome $n_1,n_2,\dots$ of the random sequence $N_1,N_2,\dots$;
3. repeat steps 1 and 2, for a given number s of times.

The output of this procedure is a (simulated) sample consisting of s outcomes, or *paths*, of the random sequence $N_1,N_2,\dots$.

Example 3.10.2. We consider a cohort of $n_0 = 100$ individuals, all age $x = 40$ initially. As in Example 3.10.1, we assume the age-pattern of mortality described by the first Heligman-Pollard law, with the parameters specified in Example 3.4.1.

Figure 3.10.5 illustrates the behavior of 50 paths of the random sequence

$$N_1,N_2,\dots,N_{10}$$

namely limited to the first 10 years. The dashed line represents the sequence of expected values

$$\mathbb{E}[N_1],\mathbb{E}[N_2],\dots,\mathbb{E}[N_{10}]$$

around which the simulated paths develop.

□

For any given time t, information about the distribution of the random number N_t can be obtained looking at the simulated outcomes of N_t, namely by constructing the statistical distribution of N_t.

However, it is worth noting that, when just one cohort consisting of individuals with the same age-pattern of mortality is involved, probability distributions of the random numbers of survivors can be found via analytical formulae, as seen in

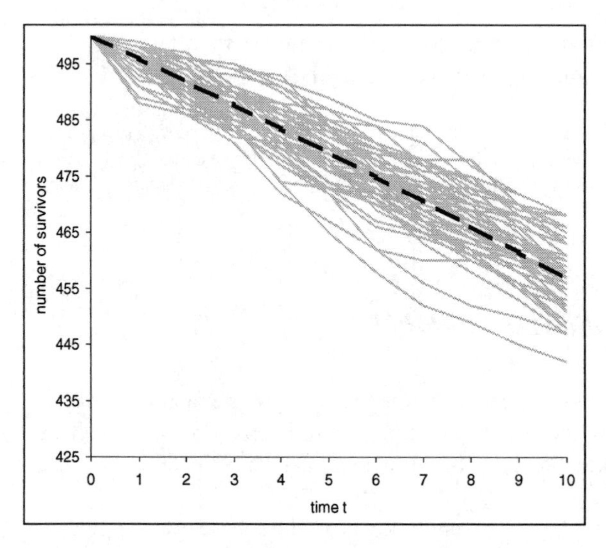

Fig. 3.10.5 Simulated number of survivors: random fluctuations

Sect. 3.10.2. Further, approximations to the probability distribution of the numbers of people dying in the various years can be adopted, when several initial ages and thus several mortality patterns are involved.

Conversely, simulation procedures are useful, even when the structure by age of the cohort is very simple, when we have to analyze the behavior of quantities depending on the random numbers of people alive or dying. Important examples are given by the cash-flows in life insurance portfolios. So, the simulation procedure we have described should be meant as the starting point for building up more complex models involving, for example, incomes and outflows. Examples will be provided in the following chapters.

3.10.4 Systematic deviations in mortality

In order to represent the age-pattern of mortality in a given group (namely, a life insurance portfolio or a pension plan), we have to choose a life table or a mortality law. However, the mortality actually experienced by the group in future years may "systematically" differ from the one we have assumed. This may occur for various reasons. For example:

- because of poor past experience, we have chosen a life table relying on mortality experienced in other populations;
- the future trend in mortality differs from the forecasted one (expressed by a projected table).

So, whatever hypothesis has been assumed, the future level and trend in mortality are random. Then, an *uncertainty risk* arises, namely a risk due to the uncertainty in the representation of the mortality scenario. Hence, systematic deviations from the expected values can occur, which combine with ordinary random fluctuations. See Sect. 2.3.10 for an introduction to this topic.

Example 3.10.3. We refer to the cohort already considered in Example 3.10.2. First, we assume the age-pattern of mortality described by the Heligman-Pollard law with the parameters adopted in Example 3.10.2. Then, we suppose that the future mortality follows the Heligman-Pollard law in which parameters G and H are replaced by

$$\bar{G} = 0.000022875; \quad \bar{H} = 1.0878$$

We denote by $\mathbb{E}[N_t|G,H]$ and $\mathbb{E}[N_t|\bar{G},\bar{H}]$, $t = 1,2,\ldots$, the expected values based on the first and the second assumption respectively.

Figure 3.10.6 illustrates the behavior of 50 paths of the random sequence

$$N_1, N_2, \ldots, N_{10}$$

(i.e. limited to the first 10 years), simulated according to the new assumption about the mortality. The dashed line represents the expected values

$$\mathbb{E}[N_1|G,H], \mathbb{E}[N_2|G,H], \ldots, \mathbb{E}[N_{10}|G,H]$$

whereas the dotted line represents the expected values

$$\mathbb{E}[N_1|\bar{G},\bar{H}], \mathbb{E}[N_2|\bar{G},\bar{H}], \ldots, \mathbb{E}[N_{10}|\bar{G},\bar{H}]$$

around which the simulated paths develop.

The process risk causes the random fluctuations around the $\mathbb{E}[N_t|\bar{G},\bar{H}]$'s, whilst the uncertainty risk originates the systematic deviations from the $\mathbb{E}[N_t|G,H]$'s.

□

3.10.5 The impact of mortality / longevity risk on life insurance

The impact of mortality / longevity risk on the results of a life insurance portfolio depends on the features of the insurance products involved.

For example, an actual mortality lower than anticipated leads to insurer's profits when just benefits in the case of death are concerned. This can be originated either by random fluctuations in mortality, or by an overestimation of the probabilities of death. On the contrary, a mortality higher than expected may cause insurer's losses. Thus, the term mortality risk (see Sect. 3.10.2) expresses a downside risk for the insurer providing death benefits.

Conversely, when a life annuity portfolio is involved, an actual mortality lower than anticipated causes losses as only benefits in the case of survival are concerned.

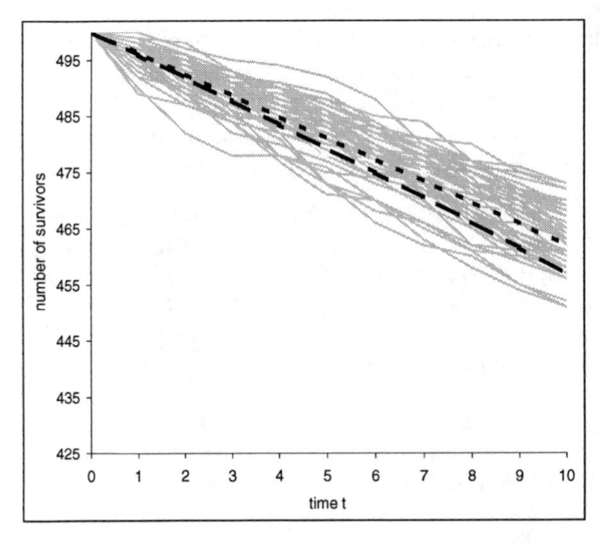

Fig. 3.10.6 Simulated number of survivors: systematic deviations (and random fluctuations)

Heavy negative results can be the consequence of an overestimation of the probabilities of death, and hence an underestimation of the probabilities of survival; in particular, this can be caused by a mortality trend leading to unanticipated mortality improvements. The term longevity risk is then used to express a downside risk for the insurer providing benefits in case of survival, e.g. life annuities.

3.11 References and suggestions for further reading

A number of textbooks of actuarial mathematics deal with life tables and mortality models, in both an age-discrete and an age-continuous context. The reader can refer for example to [10], [16], [20], [25], [26], [47], and [49].

The textbook [5] is particularly devoted to mortality analysis, graduation methods and mortality laws. For risk classification in life insurance and the numerical rating system in particular, the reader should refer to [15].

The reader interested in various perspectives on forecasting mortality can consult [52]. The textbook [46] is specifically devoted to mortality projections and the impact of future mortality trends on the costs of life annuities.

Historical aspects, also concerning the construction of life tables and mortality modeling in general, are dealt with in [27].

Mortality data can be found on a number of websites. As regards population mortality, we cite the following databases.

- The *Human Mortality Database* (HMD) is maintained by the Department of Demography at the University of California, Berkeley (USA), and the Max Planck

Institute for Demographic Research, Rostock (Germany). It provides information about mortality for 37 countries, based on official observations. HMD is available at www.mortality.org or www.humanmortality.de.

- The *Human Life-Table Database* (HLD) provides national life tables published officially, as well as non-official tables produced by researchers. HLD includes countries not included in HMD, because of lack of official sources. Three scientific institutions are jointly developing the HLD: the Max Planck Institute for Demographic Research, Rostock (Germany), the Department of Demography at the University of California, Berkeley (USA), and the Institut National d'Études Démographiques, Paris (France). HLD is available at www.lifetable.de.
- The database maintained by the *World Health Organization* (WHO) provides data for 130 countries, based on civil registration systems. Mortality data are subdivided by causes of death. WHO is available at www.who.int/en.

Mortality data related to insurance markets can be found on several websites. We cite the following ones.

- The American Academy of Actuaries provides mortality tables constructed by the *Commissioners Standard Ordinary (CSO) Task Force*. See www.actuary.org/life/cso_0702.asp.
- Mortality data related to the United Kingdom insurance and pension market are provided by *Continuous Mortality Investigation* (CMI) Library, available at www.actuaries.org.uk/knowledge/cmi. Data concern assured lives, annuitants, pensioners. Mortality projections are also available.

Chapter 4
Life insurance: pricing

4.1 Life insurance products

A short description of the main features of life insurance products is provided in this Section, which mainly aims at paving the way to premium calculation and other quantitative assessments.

4.1.1 General aspects

The object of a life insurance contract is to pay benefits depending on events concerning the lifetime of one or more individuals. The amount of benefits can be quantified in various ways. The following arrangements are of practical interest.

1. *Amount of benefits stated at policy issue* (namely, *fixed benefits*). In this case, we can have:

   a. benefit with a constant amount;
   b. benefit with an amount varying according to a stated rule (e.g. exponentially increasing, arithmetically decreasing, and so on).

2. *Initial amount of benefits stated at policy issue, then varying because of some linking mechanism*. Various linking models have been proposed and implemented; in particular we find:

   a. inflation-linked benefits;
   b. unit-linked benefits (namely, linked to the value of the unit of an investment fund);
   c. increasing benefits via profit participation, for example:
      i. bonus mechanisms (adopted in the UK);
      ii. revaluation mechanisms (adopted in continental Europe).

A. Olivieri, E. Pitacco, *Introduction to Insurance Mathematics*,
DOI 10.1007/978-3-642-16029-5_4,

Insurance products with benefits of type 1 are dealt with in this Chapter, whereas benefits of type 2 will be described in Chap. 7.

Besides the *insurer*, the parties involved in an insurance contract are:

- the *insured* (or the insureds), whose lifetime determines the payment of benefits;
- the *contractor* (or *policyholder*), who makes the contract and pays the premium(s);
- the *beneficiary*, who receives the benefits.

Two, or even three, of the parties above mentioned can coincide, depending on the type of benefits provided by the insurance contract. In what follows we will disregard insurance products involving more than one individual as the insured party.

The following categories of life insurance products can be singled out.

1. Insurance products providing *benefit in the case of survival.*
   - Their aim is to provide the beneficiary (who can coincide with the contractor and the insured) with deferred amounts;
   - the benefit is either a lump sum or an annuity;
   - typical products are the pure endowment and the life annuities.
2. Insurance products providing *benefit in the case of death.*
   - These products aim to cover the death risk and the related financial consequences;
   - the benefit is usually a lump sum (whereas annuities are less common) paid to the beneficiary (while the contractor and the insured can coincide);
   - the term insurance and the whole life insurance belong to this category.
3. Insurance products *combining death and survival benefits.*
   - In these products, usually the benefit is certain, although paid at a random time;
   - the benefit is a lump sum; two distinct beneficiaries are usually involved, one for the benefit in the case of death, and the other (who can coincide with the insured and the contractor) for the benefit in the case of survival;
   - a typical product is the endowment insurance.

Further categories could be added in order to enlarge the framework of life insurance (also in accordance to legislation and market practice in many countries). For example, products providing disability benefits constitute an important category, as well as products in which benefits are linked to the insured's health conditions. Moreover, disability or health-related benefits can be packaged in insurance products providing benefits related to the insured's lifetime, thus constituting *supplementary* (or *rider*) *benefits*. In the following, we will disregard these types of benefits.

Various premium arrangements, meeting the benefits, can be conceived. In particular, we can have:

1. a *single premium*, paid at the policy issue;

2. a sequence of *periodic premiums*, the first one paid at the policy issue, and the following ones paid, for example, at the policy anniversaries.

Whatever the premium arrangement, at any time the policyholder should be in a credit position (and, hence, the insurer in a debt position). We will focus on this feature in Sect. 4.4.1.

Premiums, benefits and expenses are the monetary ingredients of any life insurance product. The related cash-flow streams develop throughout the policy duration. In order to achieve an equilibrium situation, premiums must meet benefits and expenses. Hence, when benefits are stated and expenses assessed, premiums must be consequently determined. Conversely, if the amount of premiums is chosen by the policyholder, the benefits which can be financed by the premiums (net of expenses) have to be calculated.

The relation which links premiums, on the one hand, to benefits and expenses, on the other, must rely on a *premium calculation principle*. A principle commonly adopted in life insurance technique is the *equivalence principle* (see Sect. 1.7.4, and Cases 4a and 4b in particular), according to which the expected present value (shortly, the *actuarial value*) of benefits (and expenses) must be calculated.

In Sect. 4.2 we address the basics of expected present values in life insurance covers. For brevity, we will use the term "discounting" to denote the calculation of expected present values. In Sects. 4.3 to 4.4 we will focus on the application of this principle to net premium calculation. Finally, in Sect. 4.5 we will deal with the calculation of premiums also allowing for expenses.

4.1.2 Alterations of a life insurance contract

The "natural" conclusion of a life insurance contract occurs either at the maturity of the contract itself (and with the payment of the survival benefit, if any), or at the insured's death (and with the payment of the death benefit, if any).

Nevertheless, according to usual policy conditions, the policyholder has the right to alter some contract features. The following *alterations* are of practical interest:

- early termination;
- conversion.

In general, any alteration is determined by the exercise of an insured's option, and implies a change in future cash-flow streams (premiums, benefits, expenses).

The *early termination* of an insurance contract usually occurs because of cessation of the payment of periodic premiums. If the insurance contract provides a benefit certain, in the case of early termination a cash amount, called the *surrender value*, is paid to the policyholder. The surrender value is linked to the policyholder's credit, as we will see in Sect. 5.7.

As regards products only providing a benefit in the case of survival at maturity, the cessation of premium payments leads to a reduction of the sum insured, rather than a cash payment. This restriction to the possibility of surrendering clearly aims at reducing the risk of adverse selection in policyholders' choices.

Further, no surrender (and hence no cash payment) takes place when the policyholder's credit is very small. This happens in the first policy years (the first and the second, in particular) of a large range of insurance products with periodic premiums, as well as throughout the whole duration of short-term insurances just providing a death benefit. In all these cases the cessation of premium payment simply leads to the *lapse* of the contract.

The policyholder's credit can be used to help finance a *conversion* of an insurance policy, that is, a change in some elements of the policy itself. For example, an increase in the sum insured can be financed by both the policyholder's credit and future increased premiums.

Another example of conversion is given by the transformation of an insurance contract into a *paid-up* insurance contract, namely one for which no further premium payments are required. The sum insured is of course reduced, and its amount is determined accounting for the policyholder's credit, as we will see in Sect. 5.7.

4.2 Discounting cash-flows

Each insurance contract originates *cash-flows*. In particular, we will denote as *cash-inflow* stream any sequence of amounts cashed by the insurer, and *cash-outflow* stream any sequence of amounts paid by the insurer.

In what follows, we focus on discounting cash-inflows and outflows.

4.2.1 Premiums, benefits, expenses

The cash-inflow originated by an insurance contract consists of a sequence of *premiums*. In particular, it can reduce to a *single premium*, cashed at the policy issue.

The cash-outflows, namely the amounts paid by the insurer, consist of:

1. the *benefits*,
2. the *expenses*.

Most of the items of the cash-flow streams are deferred (that is, cashed or paid by the insurer after the policy issue), and random, as they depend on the random lifetime of the insured. Thus, the amounts which will be actually cashed and paid depend on the outcome of the lifetime.

Example 4.2.1. Figure 4.2.1 provides an example of cash-flow streams in a life insurance contract.

- Benefits are defined as follows:

  - the amount C_h, $h = 1, 2, \ldots, 5$ is paid at time h if the insured dies between time $h-1$ and h; thus, the C_h's are *death benefits*;
  - the amount S is paid at time 5 (that is, at maturity), if the insured is alive at that time; thus, S is a *survival benefit*.

- The expense EX_0 is paid at the policy issue, i.e. at time $t = 0$.
- Premiums P_0, P_1, P_2 are cashed at time $t = 0, 1, 2$ respectively; however, the second and the third one are cashed provided that the insured is alive at time $t = 1$ and $t = 2$ respectively.

Note that only the premium P_0 and the expense EX_0 are immediate and hence certain.

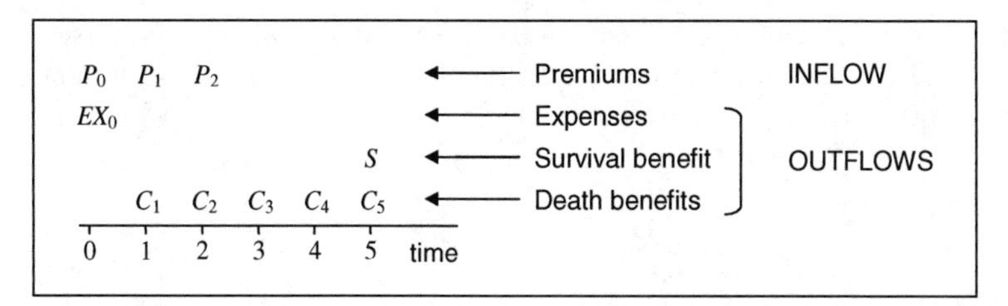

Fig. 4.2.1 Cash-inflows and outflows in a life insurance contract

Figure 4.2.2 illustrates the actual outcomes of the cash-flows, in the case of death in the second year (panel (a)), and in the case of survival at maturity (panel (b)). Clearly, all missing items (compared to those in Fig. 4.2.1) do not belong to the actual cash-flows.

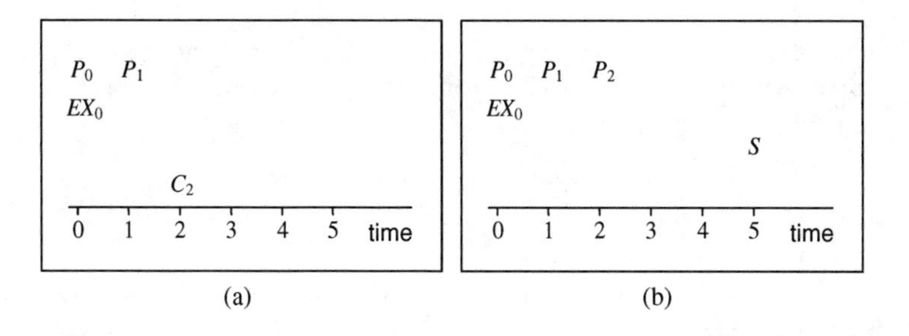

Fig. 4.2.2 Actual outcomes of cash-inflows and outflows

□

In what follows, we first focus on outflows originated by benefits, namely disregarding expenses. We denote the time of policy issue as time 0, and assume the year as the time unit.

4.2.2 A lump sum benefit in the case of death

Assume that the amount C will be paid at time h to the beneficiaries if the insured dies between time $h-1$ and h, i.e. during the h-th policy year (where h is given).

We denote with Y the random present value (at time 0) of this benefit. We assume that x is the insured's age at policy issue, and denote with K_x her curtate remaining lifetime (see Sect. 3.2.7). Hence

$$Y = \begin{cases} C(1+i)^{-h} & \text{if } K_x = h-1 \\ 0 & \text{otherwise} \end{cases} \tag{4.2.1}$$

where i is the interest rate used for discounting.

Further, we denote with ${}_{h-1|1}q_x$ the probability of dying in year h-th (see Sects. 3.2.5 and 3.2.6, and formula (3.2.16) in particular). Then, the expected present value is given by

$$\mathbb{E}[Y] = C(1+i)^{-h}\,{}_{h-1|1}q_x \tag{4.2.2}$$

4.2.3 A lump sum benefit in the case of survival

Assume that the amount S will be paid at time m (a stated time) to the beneficiaries (the insured in particular) if the insured is alive at that time.

We still denote with Y the random present value (at time 0) of this benefit. Hence

$$Y = \begin{cases} S(1+i)^{-m} & \text{if } K_x \geq m \\ 0 & \text{otherwise} \end{cases} \tag{4.2.3}$$

Further, we denote with ${}_m p_x$ the probability of being alive at time m (see Sect. 3.2.5, and formulae (3.2.9) and (3.2.10) in particular). Then, the expected present value of the benefit is given by

$$\mathbb{E}[Y] = S(1+i)^{-m}\,{}_m p_x \tag{4.2.4}$$

4.2.4 Combining benefits

Formulae (4.2.1) and (4.2.2), as regards benefits in the case of death, and formulae (4.2.3) and (4.2.4), as regards benefits in the case of survival, constitute the building blocks for more complex structures expressing random present values and expected present values. We note that expected present values can be easily derived, thanks to the additivity of the expectation.

For example, assume that:

- the amount C_h will be paid at time h to the beneficiaries if the insured dies between time $h-1$ and h, $h \leq m$, where m denotes the policy term;
- the amount S will be paid at time m to the beneficiaries (the insured in particular) if the insured is alive at that time.

Note that this benefit structure has been considered in Example 4.2.1, with $m = 5$. Further, setting $S = 0$ and $C_h = C$ for $h = 1, 2, \ldots, m$, we find the insurance product illustrated as Case 4b in Sect. 1.7.4.

The random present value of the benefits is given by:

$$Y = \begin{cases} C_1 (1+i)^{-1} & \text{if } K_x = 0 \\ C_2 (1+i)^{-2} & \text{if } K_x = 1 \\ \ldots & \ldots \\ C_m (1+i)^{-m} & \text{if } K_x = m-1 \\ S (1+i)^{-m} & \text{if } K_x \geq m \end{cases} \tag{4.2.5}$$

Then, the expected present value is as follows:

$$\mathbb{E}[Y] = \sum_{h=1}^{m} C_h (1+i)^{-h} {}_{h-1|1}q_x + S (1+i)^{-m} {}_m p_x \tag{4.2.6}$$

By combining survival benefits, we can define the cash-flow streams originated by *life annuities*. Assume, in particular, that the amount S is paid to the insured (namely, the "annuitant"), at times $1, 2, \ldots$, while she is alive. Then, we have:

$$Y = \begin{cases} 0 & \text{if } K_x = 0 \\ S(1+i)^{-1} & \text{if } K_x = 1 \\ S(1+i)^{-1} + S(1+i)^{-2} & \text{if } K_x = 2 \\ S(1+i)^{-1} + S(1+i)^{-2} + S(1+i)^{-3} & \text{if } K_x = 3 \\ \ldots & \ldots \end{cases} \tag{4.2.7}$$

or, according to the notation commonly adopted in financial mathematics:

$$Y = \begin{cases} 0 & \text{if } K_x = 0 \\ S a_{\overline{1}|} & \text{if } K_x = 1 \\ S a_{\overline{2}|} & \text{if } K_x = 2 \\ S a_{\overline{3}|} & \text{if } K_x = 3 \\ \ldots & \ldots \end{cases} \tag{4.2.8}$$

In compact terms, (4.2.8) can be written as follows:

$$Y = S a_{\overline{K_x}|} \tag{4.2.9}$$

The expected present value is then given by

$$\mathbb{E}[Y] = \mathbb{E}[S\, a_{\overline{K_x}|}] = S \sum_{h=1}^{\omega-x} a_{\overline{h}|}\; {}_{h|1}q_x \tag{4.2.10}$$

where ω denotes the maximum attainable age.

An alternative expression for the expected present value of a life annuity can be found thanks to the identity

$$ {}_k p_x = \sum_{h=k}^{\omega-x} {}_{h|1}q_x \tag{4.2.11}$$

that holds for any integer k. After a little algebra, we obtain the following expression

$$\mathbb{E}[Y] = S\,(1+i)^{-1}\,{}_1p_x + S\,(1+i)^{-2}\,{}_2p_x + S\,(1+i)^{-3}\,{}_3p_x + \dots \tag{4.2.12}$$

which has an easy direct interpretation.

Remark Formulae (4.2.10) and (4.2.12) reflect, in a modern form, calculation procedures proposed in the second half of the 17th century. Indeed, formula (4.2.10) generalizes the calculation procedure proposed in 1671 by the Dutch prime minister Jan de Witt, while formula (4.2.12) was proposed in 1693 by Edmond Halley, the famous astronomer. It is worth noting that formula (4.2.12) is computationally more straightforward, whereas formula (4.2.10) is much more interesting for further developments. In fact, as (4.2.10) directly refers to the random number K_x, de Witt's method can be easily adopted to calculate higher moments, e.g. the variance of the random present value, $a_{\overline{K_x}|}$, of a life annuity.

4.2.5 Actuarial values: terminology and notation

In the previous sections, we have shown that, for any given set of benefits, a random present value, Y, can be defined, and this value is a function of the remaining random lifetime of the insured, K_x. The expected value of Y is the actuarial value of the benefits. As seen above, its calculation consists in discounting the benefits, and relies on a life table and the interest rate i, which constitute the *technical basis*.

We now define the terminology used to denote cash-flows related to life insurance contracts, and the notation for the relevant actuarial values. Although reference is mainly to benefits, some of the following actuarial values are of interest also when evaluating inflows arising from periodic premiums, as well as outflows related to expenses.

The actuarial value of 1 monetary unit payable at time m if the insured (currently age x) is alive at that time, is given by:

$$ {}_mE_x = (1+i)^{-m}\;{}_mp_x \tag{4.2.13}$$

(see (4.2.4)). This benefit is provided by the *pure endowment* insurance.

Consider a sequence of unitary amounts, payable at the beginning of each year as long as the insured is alive. The benefit is provided by the *whole life annuity (paid in advance)*. Its actuarial value is given by:

$$\ddot{a}_x = \sum_{h=0}^{\omega-x} {}_hE_x \tag{4.2.14}$$

If the annual amounts are payable for at most m years, we have the *temporary life annuity (paid in advance)*, whose actuarial value is:

$$\ddot{a}_{x:m\rceil} = \sum_{h=0}^{m-1} {}_hE_x \tag{4.2.15}$$

Conversely, if the annual amounts are payable as long as the insured is alive, but starting from time r, we have the *deferred life annuity (paid in advance)*. The actuarial value is given by:

$$_{r|}\ddot{a}_x = \sum_{h=r}^{\omega-x} {}_hE_x = \ddot{a}_x - \ddot{a}_{x:r\rceil} \tag{4.2.16}$$

Combining the restrictions defined above, we have:

$$_{r|}\ddot{a}_{x:m\rceil} = \ddot{a}_{x:r+m\rceil} - {}_{r|}\ddot{a}_x \tag{4.2.17}$$

Formulae similar to the previous ones express the actuarial values of sequences of unitary amounts, payable at the end of each year, namely the values of *life annuities paid in arrears*. We have:

$$a_x = \sum_{h=1}^{\omega-x} {}_hE_x = \ddot{a}_x - 1 \tag{4.2.18}$$

$$a_{x:m\rceil} = \sum_{h=1}^{m} {}_hE_x \tag{4.2.19}$$

$$_{r|}a_x = \sum_{h=r+1}^{\omega-x} {}_hE_x = a_x - a_{x:r\rceil} \tag{4.2.20}$$

$$_{r|}a_{x:m\rceil} = a_{x:r+m\rceil} - {}_{r|}a_x \tag{4.2.21}$$

The actuarial value of a unitary amount payable at time h if the insured dies between time h and $h+1$ is given by:

$$_{h|1}A_x = (1+i)^{-(h+1)}\,{}_{h|1}q_x \tag{4.2.22}$$

(see (4.2.2)).

The actuarial value of a unitary amount payable at the end of the year of death, if this occurs within m years, is as follows:

$$_mA_x = \sum_{h=0}^{m-1} {}_{h|1}A_x \tag{4.2.23}$$

This benefit is provided by the *term insurance* (or *temporary insurance*).

Conversely, if the amount is payable whenever the death occurs, we have the *whole life insurance*; its actuarial value is given by:

$$A_x = \sum_{h=0}^{\omega-x} {}_{h|1}A_x \tag{4.2.24}$$

Death benefit can be restricted to time intervals which start after a given period r (the *deferred period*) has been elapsed since policy issue. Then, we have the following actuarial values:

$$ {}_{r|m}A_x = \sum_{h=r}^{r+m-1} {}_{h|1}A_x \tag{4.2.25}$$

$$ {}_{r|}A_x = \sum_{h=r}^{\omega-x} {}_{h|1}A_x \tag{4.2.26}$$

Of course:

$$A_x = {}_rA_x + {}_{r|}A_x \tag{4.2.27}$$

Combining the benefits provided by the pure endowment and the term insurance (whose actuarial values are given by formulae (4.2.13) and (4.2.23) respectively), we obtain the benefit provided by the *endowment insurance*. Its actuarial value is then:

$$A_{x,m\rceil} = {}_mE_x + {}_mA_x \tag{4.2.28}$$

We note that the resulting benefit consists in paying the unitary amount at the end of the year of death, if this occurs before m, or at time m at the latest.

Actuarial values (4.2.22) to (4.2.28) can be "adjusted" to allow (approximately) for benefit payment at the time of death, instead of the end of the year of death. Assuming that the probability distribution of the time of death is uniform over each year, we first define:

$$ {}_{h|1}\bar{A}_x = (1+i)^{-\left(h+\frac{1}{2}\right)} {}_{h|1}q_x = {}_{h|1}A_x\,(1+i)^{\frac{1}{2}} \tag{4.2.29}$$

Then, we find:

$$ {}_m\bar{A}_x = {}_mA_x\,(1+i)^{\frac{1}{2}} \tag{4.2.30}$$

$$\bar{A}_x = A_x\,(1+i)^{\frac{1}{2}} \tag{4.2.31}$$

$$\bar{A}_{x,m\rceil} = {}_mE_x + {}_m\bar{A}_x \tag{4.2.32}$$

Remark The notation defined in this Section is commonly used in the actuarial practice, especially in continental Europe. Nonetheless, it differs, to some extent, from the "standard" notation proposed at an international level. The interested reader can refer to some textbooks quoted in Sect. 4.6.

4.2.6 Actuarial values: inequalities

For any pair x, m, and any technical basis, the following *inequalities* hold:

$$ {}_{m}A_x \le A_x \le A_{x,m\rceil} \tag{4.2.33} $$

$$ {}_{m}E_x \le (1+i)^{-m} \le A_{x,m\rceil} \tag{4.2.34} $$

In particular, we note what follows.

- The inequality ${}_{m}A_x \le A_x$ holds because the benefit provided by the term insurance is just a "part" of that provided by the whole life insurance (see (4.2.27)). In practice, we have ${}_{m}A_x < A_x$, if x and m are not huge numbers.
- The inequality $A_x \le A_{x,m\rceil}$ holds because the endowment insurance pays the benefit not later than the whole life insurance (see the comment after formula (4.2.28)). In particular, we have $A_x = A_{x,m\rceil} = 1$ if $i = 0$; in fact, the benefit is paid certainly in both the insurance products, only the time of payment being random.
- The inequality ${}_{m}E_x \le (1+i)^{-m}$ is obvious, as the pure endowment benefit is paid only if the insured is alive at time m (and, in practice, we have ${}_{m}E_x < (1+i)^{-m}$).
- As the endowment insurance pays the benefit at time m at the latest, the inequality $(1+i)^{-m} \le A_{x,m\rceil}$ follows.

4.2.7 Actuarial values with zero interest rate

When a zero interest is assumed, namely no time-value of the money is accounted for, actuarial values only depend on probabilities assigned to the possible outcomes of the insured's lifetime. For example:

$$ {}_{m}A_x = \sum_{h=0}^{m-1} {}_{h|1}q_x = {}_{m}q_x \tag{4.2.35} $$

$$ {}_{m}E_x = {}_{m}p_x \tag{4.2.36} $$

On the contrary, when the payment of the benefit is certain, whatever the insured's lifetime, probabilities do not affect the actuarial value. This is the case, for example, of the whole life insurance and the endowment insurance. Indeed, we have:

$$ A_x = \sum_{h=0}^{\omega-x} {}_{h|1}q_x \tag{4.2.37} $$

$$ A_{x,m\rceil} = \sum_{h=0}^{m-1} {}_{h|1}q_x + {}_{m}p_x \tag{4.2.38} $$

Then, as seen in Sect. 4.2.6, we find:

$$A_x = A_{x,m\rceil} = 1 \tag{4.2.39}$$

A further interesting result concerns the relation between the life expectancy at age x (see Sect. 3.3.1) and the actuarial value of a life annuity. From (4.2.18) and (4.2.13), if $i = 0$ we have:

$$a_x = \sum_{h=1}^{\omega-x} {}_hp_x \tag{4.2.40}$$

and hence, via (3.3.4), we obtain:

$$\overset{\circ}{e}_x = a_x + \tfrac{1}{2} \tag{4.2.41}$$

4.2.8 The actuarial discount factor

Consider, for example, the actuarial value defined by (4.2.26), namely

$${}_{r|}A_x = \sum_{h=r}^{\omega-x} {}_{h|1}A_x = \sum_{h=r}^{\omega-x} (1+i)^{-(h+1)} {}_{h|1}q_x \tag{4.2.42}$$

Using relation (3.2.16), we can write

$${}_{r|}A_x = (1+i)^{-r} {}_rp_x \sum_{h=0}^{\omega-x-r} (1+i)^{-(h+1)} {}_{h|1}q_{x+r} \tag{4.2.43}$$

and finally:

$${}_{r|}A_x = {}_rE_x \, A_{x+r} \tag{4.2.44}$$

In Eq. (4.2.44) the actuarial value ${}_rE_x$ plays the role of r-year *actuarial discount factor* (at age x), thanks to which an actuarial value of deferred benefits (that is, ${}_{r|}A_x$) can be expressed in terms of an actuarial value of immediate benefits (A_{x+r}).

Note that, while ${}_rE_x$ is the actuarial value at time 0 (age x) of 1 unit payable at time r (age $x+r$) in the case of survival, $\frac{1}{{}_rE_x}$ is the amount payable at time r whose actuarial value at time 0 is 1 unit. Thus, $\frac{1}{{}_rE_x}$ can be meant as the r-year *actuarial accumulation factor* (at age x).

Relations similar to (4.2.44) are based on formula (3.2.11). In particular, we have:

$${}_{r+m}E_x = {}_rE_x \, {}_mE_{x+r} \tag{4.2.45}$$

$${}_{r|}\ddot{a}_x = {}_rE_x \, \ddot{a}_{x+r} \tag{4.2.46}$$

Remark Relations like (4.2.44), (4.2.45), and (4.2.46) require, of course, that the same technical basis is adopted in both the terms of the right-hand side. In particular, relation (4.2.46) relies on mortality (and interest) assumption at age x, e.g. $x = 40$, while the annuity duration can be greater than 50 years, say. Given the uncertainty in future mortality trends, such an assumption is rather unrealistic.

4.2.9 Actuarial values: further relations

Interesting relations can be found by comparing cash-flow streams and the related actuarial values.

A *perpetuity* is a stream of perpetual payments. We now refer to a perpetuity of annual payments of 1 monetary unit in advance. The present value, $\ddot{a}_{\overline{\infty}|}$, of a perpetuity in advance is given, for $i > 0$, by

$$\ddot{a}_{\overline{\infty}|} = 1 + (1+i)^{-1} + (1+i)^{-2} + \cdots = \frac{1+i}{i} \tag{4.2.47}$$

Denoting with d the discount rate (or *rate of interest-in-advance*), namely $d = \dfrac{i}{1+i}$, we have:

$$\ddot{a}_{\overline{\infty}|} = \frac{1}{d} \tag{4.2.48}$$

Of course, we have $\ddot{a}_x < \ddot{a}_{\overline{\infty}|}$. In particular, to obtain a perpetuity, we have to add to the whole life annuity a deferred perpetuity, whose first payment is placed at the end of the year of death of the annuitant. Recalling that the actuarial value (at time 0) of a given amount C payable at the end of the year of death is $C A_x$, we find that the actuarial value of the deferred perpetuity is $\ddot{a}_{\overline{\infty}|} A_x$. Hence:

$$\ddot{a}_{\overline{\infty}|} = \ddot{a}_x + \ddot{a}_{\overline{\infty}|} A_x \tag{4.2.49}$$

Using (4.2.48), we obtain

$$1 = d\,\ddot{a}_x + A_x \tag{4.2.50}$$

Relation (4.2.50) can be interpreted as follows. A debt of 1 monetary unit (contracted at time 0) is repaid with annual interests d in advance as long as the debtor is alive and the final payment of 1 at the end of the year of her death. We note that relation (4.2.50), which expresses a lifetime repayment, generalizes in term of expected values the well known relation

$$1 = d\,\ddot{a}_{\overline{m}|} + (1+i)^{-m} \tag{4.2.51}$$

which expresses an m-year repayment.

Interesting relations between actuarial values can be expresses as *recursive formulae*, which can be useful in calculation procedure, but also suggest instructive interpretations.

First, we consider the whole life insurance. From Eq. (4.2.27) with $r = 1$, we have

$$A_x = {}_1A_x + {}_{1|}A_x \tag{4.2.52}$$

and, thanks to (4.2.44):

$$A_x = {}_1A_x + {}_1E_x\, A_{x+1} \tag{4.2.53}$$

In explicit terms:

$$A_x = (1+i)^{-1}\, {}_1q_x + (1+i)^{-1}\, {}_1p_x\, A_{x+1} \tag{4.2.54}$$

Interpretation is as follows: the actuarial value at age x, A_x, is the expected value of a random variable whose outcomes are the discounted unitary sum insured in the case of death (hence, with probability ${}_1q_x$), and the discounted actuarial value of a whole life insurance from age $x+1$ onwards, in the case of survival at that age (probability ${}_1p_x$).

As regards life annuities, we consider the following example. From (4.2.16) and using (4.2.46), with $r = 1$, we obtain:

$$\ddot{a}_x = 1 + (1+i)^{-1}\, {}_1p_x\, \ddot{a}_{x+1} \tag{4.2.55}$$

Interpretation is as follows: the actuarial value at age x, $\ddot{a}_x$, includes the payment of 1 immediately due, and the discounted actuarial value of a life annuity from age $x+1$ onwards, in the case of survival at that age, thus with probability ${}_1p_x$.

4.2.10 Actuarial values at times following the policy issue

Actuarial values have been sofar referred to time 0, i.e. the policy issue. Clearly, these valuations are required for premium calculation, as we will see in the next sections. However, also valuations at (integer) times t following the policy issue, namely at *policy anniversaries*, are of practical interest, for example when calculating reserves (for basic ideas on reserving in life insurance, refer to Sect. 1.7.6). The time interval between policy issue and a generic time t is called *duration since initiation*.

Actuarial values at time t rely on the assumption that the insured is alive at that time. Hence, if the insured is age x at policy issue, her curtate remaining lifetime is K_{x+t}, and probabilities referred to age $x+t$ must be used.

For example, the actuarial value at time t of a m-year term insurance, with a unitary sum insured, is given by:

$$ {}_{m-t}A_{x+t} = \sum_{h=0}^{m-t-1} {}_{h|1}A_{x+t} = \sum_{h=0}^{m-t-1} (1+i)^{-(h+1)}\, {}_{h|1}q_{x+t} \tag{4.2.56}$$

The actuarial value at time t of a pure endowment with maturity at time m is given by:

$$ {}_{m-t}E_{x+t} = (1+i)^{-(m-t)}\, {}_{m-t}p_{x+t} \tag{4.2.57}$$

As final example, the actuarial value at time t of a whole life annuity in advance is given by:

$$\ddot{a}_{x+t} = \sum_{h=0}^{\omega-x-t} {}_hE_{x+t} \tag{4.2.58}$$

4.3 Single premiums

As noted in Sect. 4.2.1, the cash-inflow originated by an insurance contract can, in particular, reduce to a single premium, cashed by the insurer at the policy issue. We start the discussion on premium calculation focussing on this particular case, yet of practical interest.

After recalling the equivalence principle (see Sect. 1.7.4), we deal with single premiums of insurance products providing benefits in the case of survival (Sects. 4.3.2 and 4.3.3), in the case of death (Sects. 4.3.4 and 4.3.5), and in both the cases (Sect. 4.3.6).

4.3.1 The equivalence principle

Refer to a generic life insurance contract, and denote with

- Y the random present value of the benefits;
- Π the single premium;
- Z the random present value of the insurer's result (profit, or loss).

Of course, we have:

$$Z = \Pi - Y \tag{4.3.1}$$

Assume the equivalence principle for the premium calculation. Hence, Π must be such that

$$\mathbb{E}[Z] = 0 \tag{4.3.2}$$

and then

$$\Pi = \mathbb{E}[Y] \tag{4.3.3}$$

As noted in Sect. 1.7.4, a safety loading should be added to the premium, in order to

1. provide the insurer with a positive expected result, namely a profit;
2. face possible adverse experience as regards

    a. yield from investments,
    b. insureds' mortality.

In particular, the safety loading can be directly included into the premium, as is common in life insurance. In this case it is referred to as an *implicit safety loading*. To this purpose, cash-flows are discounted adopting an appropriate interest rate i', and an appropriate life table, namely probabilities of death q' or survival p', which constitute the *pricing basis*, often denoted also as the *first-order basis*.

Remark We note that, unlike in Sect. 1.7.4, the expected present value $\mathbb{E}[Y]$ must be meant as calculated according to the first-order basis. More precisely, the random present value Y relies on the interest rate i', and the expected value is quantified by adopting probabilities q' (or p').

Clearly, the interest rate i' should be lower than that expected as the yield from investment, whereas the life table must be chosen looking at the type of benefit, as we will see in the following sections.

In particular, the life tables we will adopt in the numerical examples are constructed by assuming that the age pattern of mortality follows the first Heligman-Pollard law (see Sect. 3.4.2). Various alternative parameters are shown in Table 4.3.1, while some corresponding markers can be found in Table 4.3.2.

Table 4.3.1 Life tables derived from the first Heligman-Pollard law: parameters

	A	B	C	D	E	F	G	H
LT1	0.00054	0.01700	0.10100	0.00016	10.72	18.67	$1.83000E-05$	1.11000
LT2	0.00054	0.01700	0.10100	0.00014	10.72	18.67	$1.64700E-05$	1.11000
LT3	0.00054	0.01700	0.10100	0.00013	10.72	18.67	$1.46400E-05$	1.11000
LT4	0.00054	0.01700	0.10100	0.00014	10.72	18.67	$2.00532E-06$	1.13025
LT5	0.00054	0.01700	0.10100	0.00014	10.72	18.67	$1.06038E-06$	1.13705

Table 4.3.2 Life tables derived from the first Heligman-Pollard law: some markers

	$\overset{\circ}{e}_0$	$\overset{\circ}{e}_{40}$	$\overset{\circ}{e}_{65}$	Lexis	q_0	q_{40}	q_{80}
LT1	77.282	38.601	16.725	83	0.00684	0.00121	0.07178
LT2	78.288	39.568	17.485	84	0.00684	0.00109	0.06507
LT3	79.412	40.653	18.352	85	0.00684	0.00097	0.05826
LT4	85.128	46.133	22.350	90	0.00682	0.00029	0.03475
LT5	86.464	47.446	23.389	91	0.00682	0.00020	0.02984

The five life tables can be interpreted as follows. Table LT1 could be a population table (see Sect. 3.2.4), e.g. representing the mortality in a whole national population, constructed as a period table. Mortality in LT1 could be also meant as slightly reduced with respect to the observed population mortality, in order to allow for a (generic) selection or self-selection in the insured populations. Hence LT1, if adopted as the pricing basis, can constitute a prudential choice in particular for insurance products providing death benefits. Tables LT2 and LT3 could be market tables, constructed as period tables, representing the mortality among insureds; the selection process underlying LT3 is likely to be more rigorous than that underlying table LT2 (if any). Finally, tables LT4 and LT5 could be cohort tables extracted from projected tables (see Sects. 3.8.2 to 3.8.4), representing a weaker and, respectively, a stronger mortality improvement. These tables should be adopted in relation to life annuities.

In the following sections, actuarial values calculated for pricing purposes will be denoted with ${}_mE'_x$, ${}_mA'_x$, $\ddot{a}'_x$, and so on, to recall the underlying use of a pricing basis, namely a first-order basis.

Further, we will denote with TB a generic technical basis, when a compact notation can be useful. In particular, the first-order basis will be denoted with TB1, and the second-order basis, namely the scenario or realistic basis, with TB2. For example, the notation $\mathrm{TB1} = (0.02, \mathrm{LT4})$ will denote that, in the pricing basis, the interest rate $i' = 0.02$ and the life table LT4 have been assumed.

4.3.2 The pure endowment

The pure endowment insurance provides the beneficiary (who often coincides with the insured) with a lump sum benefit, S, at time m (namely, at maturity), if the insured is alive at that time. The time from policy issue to maturity is also called the duration of the contract.

The single premium is then given by:

$$\Pi = S\,{}_mE'_x = S\,(1+i')^{-m}\,{}_mp'_x \tag{4.3.4}$$

Note that ${}_mE'_x$ is the single premium for 1 monetary unit. It is often denoted as the "premium rate" (of the pure endowment).

It is worth noting that this insurance product is not very common, because the benefit is only paid in the case of survival (and yet the premium is high because, for usual values of x and m, the probability of being alive at maturity is high). A more common product is the pure endowment combined with the *counterinsurance* benefit, which provides the beneficiary with the premium reimbursement (the so-called *return of premium*) in the case the insured dies before the maturity. Clearly, the premium is raised in order to account for this supplementary (or rider) benefit.

Example 4.3.1. Table 4.3.3 shows the effect of age and duration on the single premium of a pure endowment, with $S = 1\,000$. For any given policy duration m, the premium decreases as the age x at the policy issue increases, because the probability of being alive at maturity decreases. For the same reason, the premium decreases as the duration increases, for any given age x.

Of course, the single premium depends on the technical basis underlying its calculation, namely the pricing basis, as illustrated in Table 4.3.3. We recall that, moving from table LT1 to LT5, a higher life expectancy (and, in general, an improved mortality) is allowed for.

□

Table 4.3.3 The single premium of a pure endowment; $S = 1000$, TB1 $= (0.02, \mathrm{LT1})$

x	$m = 5$	$m = 10$	$m = 15$
40	898.97	804.08	713.10
45	894.44	793.24	693.49
50	886.86	775.33	661.73
55	874.25	746.15	611.70
60	853.48	699.69	536.39

Table 4.3.4 The single premium of a pure endowment; $S = 1000$, $x = 45$, $m = 10$

Life table	$i' = 0$	$i' = 0.01$	$i' = 0.02$	$i' = 0.03$
LT1	966.96	875.37	793.24	719.51
LT2	970.19	878.30	795.90	721.91
LT3	973.44	881.24	798.56	724.33
LT4	990.76	896.93	812.77	737.22
LT5	993.34	899.26	814.88	739.14

Remark It is worth noting that the assumption of a zero interest rate (in Example 4.3.1, and in the following examples) does not necessarily imply that no time-value of the money is ultimately accounted for. Indeed, in most insurance products the contract is yearly credited with an interest which depends on the return of insurer's investments, through various *participation mechanisms*, as we will see in Chap. 7. Thus, $i' = 0$ simply means that no interest is allowed for in advance, namely when pricing the insurance product.

Example 4.3.2. It is interesting to compare the single premium of a pure endowment with the present value of a lump sum certainly paid at maturity. Table 4.3.5 allows us to compare these values. In particular:

- the rows corresponding to $i = 0, 0.01, 0.02, 0.03$ allow us to perceive the effect of the "mortality discounting", if compared to the corresponding columns in Table 4.3.3;
- if we discount adopting the interest rate $i = 0.02343$, we obtain a present value equal to the premium of a pure endowment with $x = 45$, $m = 10$, according to the pricing basis TB1 $= (0.02, \mathrm{LT1})$. In what follows, we turn back on this issue.

□

The calculation of the single premium of a pure endowment relies on a two-fold discounting, namely a "financial" discounting (via the interest rate i'), and a "mortality" discounting (via the probabilities p'). This feature is common to all insurance products. However, for some insurance products, among which the pure endowment, it is interesting to express the joint effect through an "equivalent" discount rate. As the effect of the mortality discounting depends on the insured's age throughout the policy duration, we denote this rate with $g_{x,m}$. It is defined by the following equation:

$$(1 + g_{x,m})^{-m} = (1 + i')^{-m} \, {}_m p'_x \tag{4.3.5}$$

Table 4.3.5 Present value of a lump sum certain; $S = 1000$, $m = 10$

i	$1000\,(1+i)^{-10}$
0	1 000.00
0.01	905.29
0.02	820.35
0.02343	793.24
0.03	744.09
0.04	675.56
0.05	613.91

The spread $g_{x,m} - i'$ can be interpreted as follows. For a given sum insured S and a given interest rate i', the actuarial value of a pure endowment accounts for mortality, and hence for the mutuality effect inside the pure endowment portfolio (see Sect. 1.7.4). Consequently, the premium Π turns out to be lower than the present value (at the same rate) of the lump sum S paid certainly at maturity. Hence, if a person is willing to invest the amount Π in a purely financial transaction providing her (or some beneficiary) with the amount S at maturity, a yield higher than i' is needed, in order to recover the mutuality effect (which clearly is not involved in a purely financial transaction). Thus, the extra-yield $g_{x,m} - i'$, often called the *mortality drag*, "covers" the mutuality effect.

Example 4.3.3. Table 4.3.6 shows the equivalent rates for various ages x at the commencement of the financial transaction, and various durations m. Of course, the rate (and hence the extra-yield) increases as the age or the duration increase.

Table 4.3.6 Equivalent discount rate $g_{x,m}$; TB1 = (0.02, LT1)

x	$m = 5$	$m = 10$	$m = 15$
40	0.02153	0.02205	0.02280
45	0.02256	0.02343	0.02470
50	0.02430	0.02577	0.02791
55	0.02724	0.02972	0.03331
60	0.03219	0.03636	0.04240

□

4.3.3 Life annuities

A life annuity provides the annuitant with a sequence of periodic amounts, while she is alive. The payment frequency may be monthly, quarterly, semi-annual, or annual. In the following, for the sake of brevity we only focus on annual payments, even though annuities payable more frequently than once a year can be of practical interest.

A number of types of life annuities are sold on insurance markets, and paid by pension plans as well. The following terminology is usual:

- a *voluntary life annuity* (or *purchased life annuity*) is a life annuity bought as a consequence of individual choice, that is, exercised on a voluntary basis;
- a *pension annuity* is a life annuity paid to an individual as a direct consequence of her membership of an occupational pension plan, or a life annuity bought because a compulsory purchase mechanism works.

Although the two kinds of life annuity share the same technical structure, the adverse selection effect is clearly higher in the voluntary annuities, and this should be accounted for when choosing the pricing basis.

As seen in Sect. 4.2.5, annual amounts can be paid either in advance or in arrears. Further, life annuities can be either immediate or deferred.

As regards the *payment profile*, the following types of life annuities can be singled out.

- *Level annuities* (sometimes called *standard annuities*) provide the annuitant with an annual income, b, which is constant in nominal terms. Thus, the payment profile is flat.
- In the *fixed-rate escalating annuity* (or *constant-growth annuity*) the annual benefit increases at a fixed annual rate, α, so that the sequence of payments is

$$b_1,\ b_2 = (1+\alpha)\,b_1,\ b_3 = (1+\alpha)^2\,b_1,\ \ldots$$

- Various types of *index-linked escalating annuities* are available in insurance and pension markets. In these annuities, annual benefits vary in line with an inflation index, or a stock index, or according to some profit participation mechanism, and so on. Some types of index-linking will be discussed in Chap. 7.

We now focus on an *immediate life annuity in arrears*, with flat payment profile. The single premium, Π, is given, according to the equivalence principle, by

$$\Pi = b\,a'_x = b\sum_{h=1}^{\omega-x} {}_hE'_x = b\sum_{h=1}^{\omega-x}(1+i')^{-h}\,{}_hp'_x \tag{4.3.6}$$

(see (4.2.18)) where b denotes the annual benefit.

Example 4.3.4. Table 4.3.7 shows the single premium of an immediate life annuity, given by formula (4.3.6), as a function of the pricing basis. Clearly, table LT4 or LT5 should be used for pricing, as they embed a forecast of the future mortality trend. The other tables are referred to only to show the dramatic differences in the resulting actuarial values.

□

The *complete life annuity* (or *apportionable annuity*) is a life annuity in arrears which provides a pro-rata adjustment on the death of the annuitant, consisting in a

Table 4.3.7 The single premium of an immediate life annuity; $b = 100, x = 65$

Life table	$i' = 0$	$i' = 0.01$	$i' = 0.02$	$i' = 0.03$
LT1	1622.55	1462.05	1325.15	1207.62
LT2	1698.55	1524.98	1377.64	1251.72
LT3	1785.24	1596.23	1436.66	1300.97
LT4	2185.04	1923.61	1706.88	1525.74
LT5	2288.92	2007.36	1774.94	1581.51

final payment proportional to the time elapsed since the last payment date. Assuming that the probability distribution of the time of death is uniform over each year, the single premium is approximately given by:

$$\Pi = b a'_x + \frac{b}{2} \bar{A}'_x \tag{4.3.7}$$

(see also (4.2.31)).

From (4.2.18), it follows that the single premium for a *life annuity in advance* is given by

$$\Pi = b \ddot{a}'_x = b (a'_x + 1) \tag{4.3.8}$$

When dealing with the life annuities we have just described, it is natural to look at the single premium as the result of an accumulation process, in particular carried out during (a part of) the working life of the annuitant (see Case 4a in Sect. 1.2.5). It is worth noting that, insurance products which extend over the whole accumulation period can be conceived. This is, typically, the case of the *deferred life annuity* whose deferred period coincides with the accumulation period. A reasonable premium arrangement should then consist of a sequence of periodic premiums. However, we stress that, the longer is the deferred period the higher is the risk borne by the insurer, provided that the pricing basis is stated at the policy commencement, or, at least, when each periodic premium is determined and paid. In particular, an unanticipated improvement in mortality can cause serious technical problems.

4.3.4 The term insurance

The term insurance (or *temporary insurance*) pays the sum insured C at the end of the year of death, if the insured dies prior to the term m.

This product faces the risk of a financial distress caused to a family by the early death of a member who provides the family with an income. As noted in Sect. 1.2.5, it is almost impossible to quantify in monetary terms the impact of an early death, in particular because of the unknown value of the loss of income. Thus, the sum insured should represent a tentative estimation of the random impact, and hence it should be chosen in relation to the insured's age, present income, presence of dependants, and so on.

This insurance product is very common in all the insurance markets. Given the purpose, the age at entry is usually not old (say, in the range 30 - 50). Various risk factors can be accounted for in determining the premium rates (see Sects. 3.6.1 and 3.6.2). Their assessment is performed through appropriate questions in the application form and, as to health conditions, possibly through a medical examination. In the presence of poor health conditions, or a risky occupation, the applicant can be accepted as a substandard risk (or impaired life; see Sect. 3.6.3); in this case, a higher premium rate is adopted.

The single premium of a term insurance is given by

$$\Pi = C\,{}_mA'_x \tag{4.3.9}$$

Example 4.3.5. Table 4.3.8 shows the single premium of a term insurance as a function of the interest rate i' and the life table. We note that, for this insurance product, the life table LT1 can constitute a prudential choice of the pricing basis, whereas LT2 or LT3 can represent the expected mortality among the insureds.

Table 4.3.8 The single premium of a term insurance; $C = 1\,000$, $x = 40$, $m = 10$

Life table	$i' = 0$	$i' = 0.01$	$i' = 0.02$	$i' = 0.03$
LT1	19.83	18.63	17.53	16.51
LT2	17.89	16.80	15.81	14.89
LT3	15.93	14.97	14.08	13.26

Table 4.3.9 shows the effect of the age at policy issue and the duration on the single premium of a term insurance. For any given duration m, the premium increases as the age increases, because of an increasing probability of dying before the policy term. For the same reason, the premium increases as the duration increases, for any given age x.

Table 4.3.9 The single premium of a term insurance; $C = 1\,000$, TB1 = (0.02,LT1)

x	$m = 5$	$m = 10$	$m = 15$
40	7.01	17.53	33.26
45	11.70	29.20	55.10
50	19.57	48.52	90.53
55	32.64	80.01	146.52
60	54.19	130.26	231.30

□

In formula (4.3.9) it has been assumed that the sum insured is constant over the whole policy duration. If, on the contrary, the benefit changes moving from year to year, we have the *term insurance with varying benefit*. Let C_{h+1} denote the sum paid at time $h+1$ if the insured dies between times h and $h+1$; then:

$$\Pi = \sum_{h=0}^{m-1} C_{h+1\; h|1}A'_x \tag{4.3.10}$$

In particular, the *decreasing term insurance* provides a decreasing benefit defined as follows:

$$C_1 = C;\; C_2 = \frac{m-1}{m}C;\; C_3 = \frac{m-2}{m}C;\; \ldots;\; C_m = \frac{1}{m}C \tag{4.3.11}$$

Hence, from (4.3.10), we obtain the single premium

$$\Pi = \frac{C}{m}\sum_{h=0}^{m-1}(m-h)\,{}_{h|1}A'_x \tag{4.3.12}$$

which, after a little algebra, can also be expressed as follows:

$$\Pi = \frac{C}{m}\sum_{h=0}^{m-1}{}_{m-h}A'_x \tag{4.3.13}$$

The decreasing term insurance is usually sold to guarantee a loan repayment carried out via amortization; indeed, the decreasing benefit is approximately in line with the outstanding debt.

4.3.5 The whole life insurance

The whole life insurance pays the sum insured C at the end of the year of death, whenever the death occurs.

$$\Pi = CA'_x \tag{4.3.14}$$

The main historical purpose of the whole life insurance was the financing of inheritance taxes. Currently, a typical aim of this insurance product is as follows:

- to cover the risk of death up to a certain age (60 or 65, say);
- to provide the insured with a lump sum at a certain age, by surrendering the policy (see Sect. 4.1.2).

Example 4.3.6. Table 4.3.10 shows the single premium of a whole life insurance, for various life tables and interest rates. It is interesting to note that, whatever the life table and the interest rate, the premium is much higher than that of the term insurance; this is obviously due to the fact that the benefit of the whole life insurance is certainly paid, the only random item being the time of payment.

Table 4.3.10 The single premium of a whole life insurance; $C = 1\,000$, $x = 40$

Life table	$i' = 0$	$i' = 0.01$	$i' = 0.02$	$i' = 0.03$
LT1	1000.00	682.24	473.72	334.94
LT2	1000.00	675.76	464.90	325.80
LT3	1000.00	668.57	455.20	315.82
LT4	1000.00	632.24	406.23	265.44
LT5	1000.00	623.78	395.14	254.36

Premiums for various ages at entry (and a fixed technical basis) can be found in Table 4.3.11. We note that the single premium increases as the age at entry increases; this is clearly due to the shortening of the residual lifetime.

Table 4.3.11 The single premium of a whole life insurance; $C = 1\,000$, TB1 $= (0.02, \text{LT1})$

x	$1000\,A'_x$
40	473.72
45	519.16
50	567.35
55	617.66
60	669.17

□

4.3.6 Combining survival and death benefits

Survival benefits and death benefits can be packaged in several ways. Figure 4.3.1 shows three combinations of two lump sum benefits, in the case of survival and death respectively.

We note that, depending on the insured's lifetime:

- according to arrangement A, there will be either no payment or one payment;
- arrangement B will generate either one payment or two payments;
- according to arrangement C, there will be one and only one payment.

Usually, combining survival and death benefits aims at achieving the certainty of some payments. Assuming this target, we note what follows:

- arrangement A fails the target, if the insured dies between time m and n;

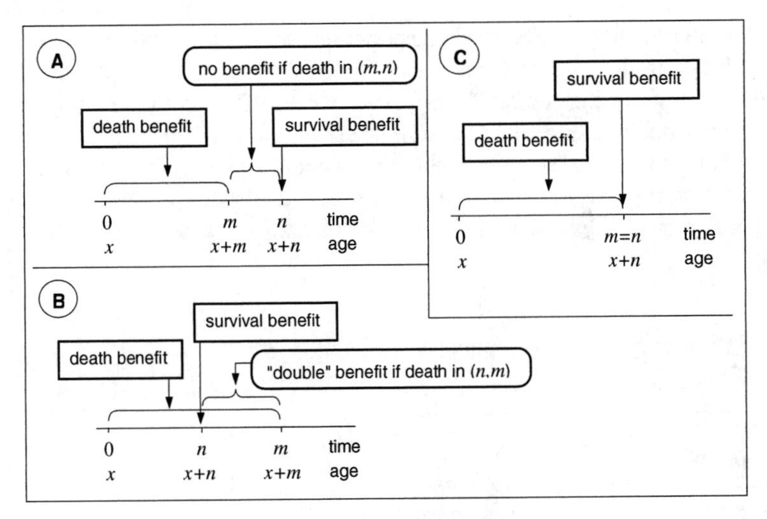

Fig. 4.3.1 Combining benefits in the case of death and survival

- arrangement B can lead to "over-insurance", because of the double payment in the case the insured dies between time n and m;
- arrangement C looks like the most appropriate; of course, the amounts of the death benefit and the survival benefit should be stated to comply with the specific policyholder's needs.

In the most common insurance products, in the framework of endowment insurance, benefits are arranged as in structure C; nonetheless, some insurance products follow the arrangement B.

4.3.7 Endowment insurance products

The *standard endowment insurance* (shortly, the endowment insurance), is defined as the combination of a pure endowment and a term insurance, with a common sum insured C and a common duration m. Hence, this type of endowment constitute an example of arrangement C defined in the previous Section.

We note that the purpose of this insurance product is twofold:

- savings, as the sum insured will be available (in the case of survival) at maturity;
- protection, as the sum insured will be paid to the beneficiaries in the case of the insured's death prior to maturity.

The single premium is given by

$$\Pi = C\left({}_{m}E'_{x} + {}_{m}A'_{x}\right) = C A'_{x,m\rceil} \tag{4.3.15}$$

Example 4.3.7. Table 4.3.12 displays the single premium as a function of the pricing basis. It is worth noting that, although the premium obviously depends on the life table adopted, a change in the life table has a very weak effect on the premium itself. This is clearly due to the fact that the sum insured is certainly paid, and only the time of payment is affected by the mortality assumption. Anyway, the premium is higher when the mortality is assumed to be higher (as expressed, for example, by table LT1). Hence, a mortality higher than that expected among insureds represents a prudential choice.

Table 4.3.12 The single premium of an endowment insurance; $C = 1\,000$, $x = 50$, $m = 15$

Life table	$i' = 0$	$i' = 0.01$	$i' = 0.02$	$i' = 0.03$
LT1	1000.00	866.51	752.26	654.32
LT2	1000.00	866.01	751.37	653.11
LT3	1000.00	865.51	750.47	651.90

□

Example 4.3.8. Table 4.3.13 shows, in particular, the splitting of the single premium of an endowment insurance into the single premiums of the two components, namely the pure endowment ($C\,{}_mE'_x$) and the term insurance ($C\,{}_mA'_x$), for various ages at entry. We note that the single premium of the endowment insurance has a small increase in spite of a significant increase in the age at entry, whereas the premiums of the two components strongly depend on the age itself, decreasing and increasing respectively, as the age increases. We also note that, for all the ages addressed, the actuarial value of the survival benefit is much higher than that of the death benefit, because the probability of being alive at maturity is higher than the probability of dying prior to maturity (at least in the range of ages we have considered). This result seems to be in contrast with the effect of the life table on the single premium, which has been stressed in Example 4.3.7; then, a deeper analysis of the nature of the endowment insurance is needed.

Table 4.3.13 Components of the single premium of an endowment insurance; $C = 1\,000$, $m = 15$, TB1 = (0.02, LT1)

		pure endowment + term insurance		benefit-certain + acceleration benefit	
x	$1\,000A'_{x,15\rceil}$	$1\,000\,{}_{15}E'_x$	$1\,000\,{}_{15}A'_x$	$1\,000(1+i')^{-15}$	$1\,000\left(A'_{x,15\rceil} - (1+i')^{-15}\right)$
40	746.36	713.10	33.26	743.01	3.35
45	748.59	693.49	55.10	743.01	5.57
50	752.26	661.73	90.53	743.01	9.25
55	758.23	611.70	146.52	743.01	15.21
60	767.69	536.39	231.30	743.01	24.67

□

The (usual) definition of the endowment insurance, as the combination of a pure endowment and a term insurance, does not allow us to correctly capture the technical contents of this insurance product, and consequently the risk borne by the insurer (whereas it does allow to understand the twofold purpose of the endowment insurance). We now address this point.

From the identity

$$A'_{x,m\rceil} = (1+i')^{-m} + \left(A'_{x,m\rceil} - (1+i')^{-m}\right) \tag{4.3.16}$$

namely

$$\begin{aligned} A'_{x,m\rceil} = {} & (1+i')^{-m} \\ & + \underbrace{\sum_{h=0}^{m-1} (1+i')^{-(h+1)} {}_{h|1}q'_x + (1+i')^{-m} {}_m p'_x}_{A'_{x,m\rceil}} \\ & - (1+i')^{-m} \underbrace{\left(\sum_{h=0}^{m-1} {}_{h|1}q'_x + {}_m p'_x\right)}_{1} \end{aligned} \tag{4.3.17}$$

we obtain:

$$\begin{aligned} \Pi = C A'_{x,m\rceil} & = C(1+i')^{-m} + C\left(A'_{x,m\rceil} - (1+i')^{-m}\right) \\ & = C(1+i')^{-m} + \sum_{h=0}^{m-2} \underbrace{C\left((1+i')^{-(h+1)} - (1+i')^{-m}\right)}_{\Gamma_h} {}_{h|1}q'_x \end{aligned} \tag{4.3.18}$$

We note that each Γ_h is the present value of the "acceleration" in the benefit payment due to the insured's death before maturity (and, anyhow, before the last year of the insurance cover). Thus, formula (4.3.18) suggests the following interpretation: the single premium of an endowment insurance can be obtained as the present value of an amount certain at maturity plus the actuarial value of the *acceleration benefit*.

Hence, the endowment insurance can be seen as a product providing the beneficiary with a payment acceleration in the case of death. It follows that the risk borne by the insurer is the risk of insured's death prior to maturity.

If $i' = 0$, trivially we have $\Gamma_h = 0$ for all h, and, of course $\Pi = C$.

Example 4.3.9. The last column in Table 4.3.13 shows the actuarial value of the acceleration benefit. Of course, the value increases as the age increases, because of the higher probability of dying prior to maturity. The actuarial value of the acceleration benefit turns out to be the only term depending on the insured's lifetime (see the last line of formula (4.3.18)), and this explains how a mortality higher than

that expected inside the endowment portfolio is a prudential choice (as remarked in Example 4.3.7).
□

The survival benefit, S, in an endowment insurance can be different from the death benefit, C. The single premium is then given by

$$\Pi = S\,{}_mE'_x + C\,{}_mA'_x \tag{4.3.19}$$

In particular, when $S > C$ we have the product sometimes called the *endowment insurance with additional survival benefit*. Setting $S = (1+\alpha)C$, we obtain:

$$\Pi = \alpha C\,{}_mE'_x + CA'_{x,\overline{m|}} \tag{4.3.20}$$

Note that also this product implements the arrangement C (see Sect. 4.3.6).

The insurance product built up combining a pure endowment with a whole life insurance is denoted as the *double endowment insurance*. Its single premium is given by:

$$\Pi = C\left({}_mE'_x + A'_x\right) = CA'_{x,\overline{m|}} + C\,{}_{m|}A'_x \tag{4.3.21}$$

This product constitutes an example of arrangement B, with $m = \omega - x$.

Remark We note that, whenever a death benefit is included in the insurance contract, this can be assumed to be payable at the time of death rather than at the end of the year death. Then, the approximations expressed by formulae (4.2.30) to (4.2.32) can be used for the single premiums of the term insurance, the whole life insurance, and the endowment insurance products.

4.3.8 The expected profit: a first insight

The assessment of expected profits is one of the main topics in actuarial mathematics. Section 5.5 and Sects. 6.2 to 6.4 are specifically devoted to this important aspect of life insurance. Nevertheless, some basic ideas already emerge from single premium calculation models.

As stated in Sect. 4.3.1, we assume that, for all the insurance products, the single premium Π is calculated as the actuarial value of the benefits provided by the insurance policy, and that, in the relevant calculations, the first-order technical basis, denoted by TB1, is assumed.

For example, referring to a whole life insurance providing the benefit C, we have $\Pi = CA'_x$ (see (4.3.14)). Let CA''_x denote the actuarial value of the benefit, calculated by adopting the second-order basis, TB2. Thus, CA''_x provides us with a "realistic" evaluation of the benefit, in terms of

- the interest rate, i'', which should represent the estimated investment yield;
- the probabilities of death, q'', which should represent the portfolio mortality actually expected.

Let $\overline{PL}$ denote the expected present value of the profit / loss, also called the *profit margin*; we have:

$$\overline{PL} = \Pi - CA''_x = C(A'_x - A''_x) \tag{4.3.22}$$

Example 4.3.10. Table 4.3.14 shows the profit margin, in absolute terms and relative terms (that is, referred to the single premium), for the term insurance, the whole life insurance, and the endowment insurance. A' and A'' generically denote the expected present value of benefits, according to TB1 and TB2 respectively. As regards the interest rate, it should be noted that we have assumed $i' = i'' = 0.03$; it follows that the profit margins originate from the spread between the first-order mortality and the second-order mortality. Further, it is interesting to note that, in relative terms, the effect of the mortality spread is very high in the term insurance, in which the benefit is only paid in the case of death before the policy term, while it is extremely low in the endowment insurance, in which the benefit is certainly paid within the policy term.

Table 4.3.14 Profit margins. TB1 = (0.03, LT1); TB2 = (0.03, LT3)

Insurance product	$\Pi = CA'$	CA''	$\overline{PL}$	$\overline{PL}/\Pi$
Term insurance; $C = 1000$; $x = 40$, $m = 10$	16.51	13.26	3.25	19.69%
Whole life insurance; $C = 1000$; $x = 40$	334.94	315.82	19.12	5.71%
Endowment insurance; $S = C = 1000$; $x = 50$, $m = 15$	654.32	651.90	2.42	0.37%

Table 4.3.15 shows the profit margins originated by the spread between interest rates. Indeed, it has been assumed that the mortality adopted in the first-order basis coincides with the mortality actually expected in the portfolio. It is worth noting that the effect is much higher in the whole life insurance and the endowment insurance than in the term insurance; this happens because the former insurance products have more important financial contents, as we will see in Chap. 5.

Table 4.3.15 Profit margins. TB1 = (0.02, LT3); TB2 = (0.03, LT3)

Insurance product	$\Pi = CA'$	CA''	$\overline{PL}$	$\overline{PL}/\Pi$
Term insurance; $C = 1000$; $x = 40$, $m = 10$	14.08	13.26	0.82	5.82%
Whole life insurance; $C = 1000$; $x = 40$	455.20	315.82	139.38	30.62%
Endowment insurance; $S = C = 1000$; $x = 50$, $m = 15$	750.47	651.90	98.57	13.13%

□

4.4 Periodic premiums

The expression *periodic premiums* denotes a wide range of premium arrangements, which share the common target of meeting reasonable policyholders' wishes. Different premium arrangements originate different technical and financial problems. We start dealing with this issue by focussing on a simple preliminary example.

4.4.1 An example

We refer to a term insurance, with $m = 5$ and sum assured $C = 1$. The single premium is then given by $\Pi = {}_5A'_x$. Assume that, instead of the single premium Π, a sequence of annual premiums

$$P_0,\ P_1,\ P_2,\ P_3,\ P_4 \tag{4.4.1}$$

payable at times $t = 0, 1, \ldots, 4$ respectively, is agreed. Each premium will be paid at the beginning of a policy year, provided that the insured will be alive at that time. Hence, the stream of annual premiums is a random inflow, which constitutes a temporary life annuity.

We start considering the case of two premiums only, and we set:

$$P_2 = P_3 = P_4 = 0$$

For the premium calculation, we adopt the equivalence principle (see Sect. 4.3.1). First, we note that, as the premium P_1 will be paid provided that the insured will be alive at time $t = 1$, the sequence of premiums has a random present value, X, given by

$$X = P_0 + \begin{cases} P_1\,(1+i')^{-1} & \text{if } K_x \geq 1 \\ 0 & \text{otherwise} \end{cases} \tag{4.4.2}$$

The related actuarial value is then

$$\mathbb{E}[X] = P_0 + P_1\,(1+i')^{-1}\,{}_1p'_x = P_0 + P_1\,{}_1E'_x \tag{4.4.3}$$

(assuming that the pricing basis is also adopted for discounting premiums).

According to the equivalence principle, as the expected value of benefits is given by ${}_5A'_x$, the premiums must be solutions of the following equation:

$$P_0 + P_1\,{}_1E'_x = {}_5A'_x \tag{4.4.4}$$

We note that Eq. (4.4.4) has two unknowns, and thus ∞ solutions. We start considering two particular solutions:

$$P_0 = {}_5A'_x,\ P_1 = 0 \tag{4.4.5a}$$

$$P_0 = 0,\ P_1 = \frac{{}_5A'_x}{{}_1E'_x} \tag{4.4.5b}$$

As regards these solutions, the following features should be stressed.

a. Solution (4.4.5a) trivially yields the single premiums.
b. Solution (4.4.5b) is not feasible. To this regard, we note what follows.

- Solution (4.4.5b) complies with the equivalence principle; thus, its unfeasibility concerns another criterion (not yet declared).
- If the policyholder "lapses" (i.e. abandons) the contract at time 1, before paying premium P_1, she has obtained a one-year insurance cover free of any charge, namely without contributing to mutuality.
- In the case of lapse, for the insurer a practical difficulty arises in obtaining the payment.

Remarks under point b above leads to the conclusion that the insurer should never be in a credit position. This requirement is denoted as the *financing condition*.

The expected present value of the first year cover is given by ${}_1A'_x$. Hence, feasible solutions, namely solutions fulfilling the financing condition, are the pairs P_0, P_1 such that

$$P_0 \geq {}_1A'_x \tag{4.4.6}$$

In particular, it can be proved that

$$P_0 = {}_1A'_x \Rightarrow P_1 = {}_4A'_{x+1} \tag{4.4.7}$$

Thus:

- premium P_0 exactly meets the expected cost of the insurance cover in the first year, so that at time 1 the insurer is neither in a credit nor in a debt position;
- premium P_1 can be interpreted as the single premium for the residual duration of the contract, provided that the insured is alive at time 1.

Conversely,

$$P_0 > {}_1A'_x \Rightarrow P_1 < {}_4A'_{x+1} \tag{4.4.8}$$

Hence:

- the insurer cashes at policy issue more than what is needed to meet the expected cost in the first year, whereas at time 1 cashes less than what is needed to cover the residual duration;
- at time 1, an amount is available (namely, an *asset*) to face the insufficiency of P_1; a corresponding debt (a *liability*) arises at the end of the first year.

The assets and liability originated by the premium arrangement defined by inequalities in (4.4.8) are the two aspects of the *mathematical reserve* of the insurance contract.

We note that, in the case of a single premium, the only cash-inflow (at the policy issue) originates a similar situation, in terms of assets and liability.

We now move, still referring to the 5-year term insurance, to premiums arrangements in which all the annual premiums are positive.

First, assume that the annual premiums are defined as follows:

$$P_h = {}_1A'_{x+h}; \quad h = 0, 1, \dots, 4 \tag{4.4.9}$$

We note that each premium (which will be paid provided that the insurer will be alive at the time of payment) fulfills the equivalence principle on a one-year basis. A similar premium arrangement is very common in non-life insurance, as we will see in Chap. 9. The annual premiums defined by (4.4.9) are commonly called the *natural premiums*, and are denoted with $P_h^{[N]}$.

Of course, the sequence of natural premiums also fulfills the equivalence principle over the whole policy duration. In fact, we have:

$$\sum_{h=0}^{4} {}_hE'_x P_h^{[N]} = \sum_{h=0}^{4} {}_hE'_x \, {}_1A'_{x+h} = \sum_{h=0}^{4} {}_{h|1}A'_x = {}_5A'_x \tag{4.4.10}$$

The financing condition is also fulfilled. In particular, at each policy anniversary there is neither credit nor debt.

We recall that ${}_1A'_{x+h} = (1+i')^{-1}\, q'_{x+h}$. Hence, the natural premiums of the term insurance are increasing if the annual probabilities of death are increasing, and this usually happens for ages and durations involved in this type of cover.

To avoid a sequence of increasing premiums, an arrangement based on *level premiums* can be applied. According to the equivalence principle, the annual premium P must be the solution of the following equation:

$$P \sum_{h=0}^{4} {}_hE'_x = {}_5A'_x \tag{4.4.11}$$

(which generalizes Eq. (4.4.4)), and then we obtain

$$P = \frac{{}_5A'_x}{\ddot{a}'_{x:\overline{5}|}} \tag{4.4.12}$$

Example 4.4.1. In Fig. 4.4.1 the natural premiums and the level premiums of a term insurance are plotted. The pricing basis is TB1 $= (0.02, \text{LT1})$; we have assumed $x = 40$, $m = 5$. Then, the annual level premium is $P = 1.46$. We note that, in the first policy years the insurer cashes more than what is needed to meet the annual expected costs (expressed by the natural premiums), whereas in the last years the amounts cashed are lower than the expected costs. Thus, a reserving process is required, aiming to "transfer" money from the initial years to the final ones.

□

Equation (4.4.12) can be written as follows:

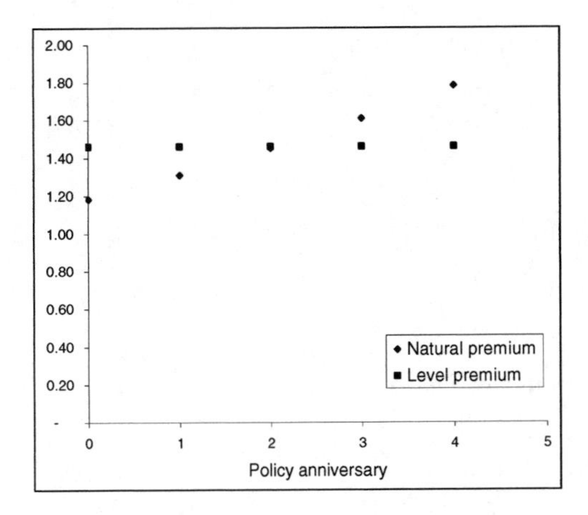

Fig. 4.4.1 Term insurance: natural premiums and annual level premiums

$$P = \frac{\sum_{h=0}^{4} {}_hE'_x \, {}_1A'_{x+h}}{\sum_{h=0}^{4} {}_hE'_x} \tag{4.4.13}$$

Hence, the annual level premium P turns out to be the weighted arithmetic mean of the natural premiums $P_h^{[N]} = {}_1A'_{x+h}$; the weights are given by the ${}_hE'_x$, and thus are proportional to the probability of being alive and paying the premium, and the time from policy issue to time of payment.

4.4.2 Level premiums

We now refer to a generic life insurance contract, and start discussing level premium arrangements.

According to the equivalence principle, we must have:

$$\text{actuarial value of premiums} = \text{actuarial value of benefits} \tag{4.4.14}$$

Provided that the single premium Π is calculated according to the same principle, we must then have:

$$\text{actuarial value of premiums} = \Pi \tag{4.4.15}$$

Let s denote the duration of the premium payment. In a number of insurance products, level premiums can be paid over the whole policy duration m, namely $s = m$. On the contrary, in some products the payment period must be shortened, so that $s < m$. As we will see, one reason for shortening the payment period is the fulfilling of the financing condition. When $s < m$, the level premium will be denoted by $P(s)$, whereas $s = m$ should be understood when just the symbol P is used. Some examples follow. In all the examples, we assume a unitary amount insured.

In a pure endowment insurance with duration m, we have

$$P(s) = \frac{{}_mE'_x}{\ddot{a}'_{x:\overline{s}|}} \tag{4.4.16}$$

In a term insurance with duration m, we have

$$P(s) = \frac{{}_mA'_x}{\ddot{a}'_{x:\overline{s}|}} \tag{4.4.17}$$

In an endowment insurance with duration m, we have

$$P(s) = \frac{A'_{x,\overline{m}|}}{\ddot{a}'_{x:\overline{s}|}} \tag{4.4.18}$$

Usually, in all these products we have $s = m$.

In a whole life insurance, the premium payment can, in principle, extend over the whole policy duration, that is

$$P(\omega - x) = \frac{A'_x}{\ddot{a}'_x} \tag{4.4.19}$$

In practice, however, the premium payment is restricted to s years, so that $x+s = 70$ or 75, say. Then, we have:

$$P(s) = \frac{A'_x}{\ddot{a}'_{x:\overline{s}|}} \tag{4.4.20}$$

Example 4.4.2. Table 4.4.1 shows single premiums and level premiums with various payment durations, for some insurance products; $\mathrm{TB1} = (0.02, \mathrm{LT1})$, and the sum insured is $C = 1\,000$ (or $S = 1\,000$) in all the cases.

□

4.4.3 Natural premiums

Consider a life insurance product, and

1. refer to the $(h+1)$-th year of contract, $h = 0, 1, \ldots$, namely the interval between time h and $h+1$;
2. assume that the insured is alive at time h;
3. single-out the benefits that fall due in the $(h+1)$-th year;
4. calculate the actuarial value at time h of the benefits referred to in step 3; this actuarial value is also called the *expected annual cost* (of the benefits).

The *natural premiums* of the contract are, by definition, the expected annual costs. The technical basis adopted in step 4 is usually the pricing basis.

Table 4.4.1 Single premium and annual level premiums for some insurance products

Insurance product	x	m	Π	s	$P(s)$
Pure endowment	45	10	793.24		
				5	165.72
				10	87.60
Term insurance	40	10	17.53		
				5	3.66
				10	1.93
Endowment insurance	50	15	752.26		
				5	157.63
				10	83.74
				15	59.54
Whole life insurance	40		473.72		
				10	52.07
				20	29.02
				30	21.80
				$\omega - x$	17.65

Natural premiums provide, on the one hand, important technical information about the time profile of the insurer's expected costs. On the other, natural premiums not necessarily constitute a practicable arrangement for the premium payment, as we will see in what follows.

For example, the natural premiums of a m-year term insurance with $C = 1$ are defined as follows:

$$P_h^{[N]} = {}_1A'_{x+h} = (1+i')^{-1}\,{}_1q'_{x+h};\ h = 0, 1, \dots, m-1 \tag{4.4.21}$$

Hence, for ages and durations usually involved in this insurance product, the natural premiums are increasing throughout the policy duration.

In a term insurance with sum assured C_{h+1} in the case of death in year $h+1$, we have:

$$P_h^{[N]} = C_{h+1}\,{}_1A'_{x+h} = C_{h+1}\,(1+i')^{-1}\,{}_1q'_{x+h};\ h = 0, 1, \dots, m-1 \tag{4.4.22}$$

We note that, if the C_{h+1}'s decrease as h increases (see, for example, the decreasing term insurance defined in Sect. 4.3.4), the natural premiums may decrease.

In both the types of term insurance, natural premiums constitute a possible arrangement for premium payment.

Example 4.4.3. Figures 4.4.2, 4.4.3 and 4.4.4 show the time profile of natural premiums of term insurances. The pricing basis is TB1 $= (0.02, \text{LT1})$. In Fig. 4.4.2, natural premiums and level premiums of a term insurance with $C = 1\,000$, $x = 40$, $m = 10$ are compared. Fig. 4.4.3 shows the behavior of the natural premiums for various ages at policy issue. We note that, the higher is the age the higher is the increase in natural premiums; this is due to the fact that the annual probabilities of death increase at an increasing rate. Finally, Fig. 4.4.4 illustrates the behavior of natural premiums in a decreasing term insurance, with $C_h = \frac{m-h+1}{m} C$ (see (4.3.11)). In panel (a), we see that, if the payment of level premiums is stated over the whole policy duration, the financing condition is not fulfilled, and the insurer immediately enters a credit position. Conversely, the condition is fulfilled if the premium payment period is properly shortened ($s = 7$), as displayed in panel (b).

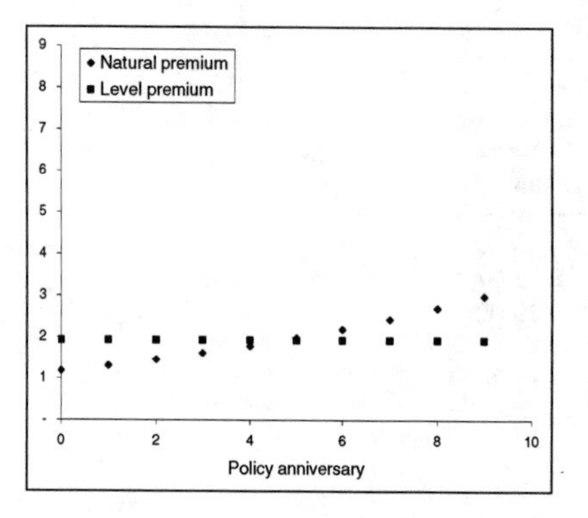

Fig. 4.4.2 Natural premiums and level premiums of a term insurance

Fig. 4.4.3 Natural premiums of a term insurance for various ages at the policy issue

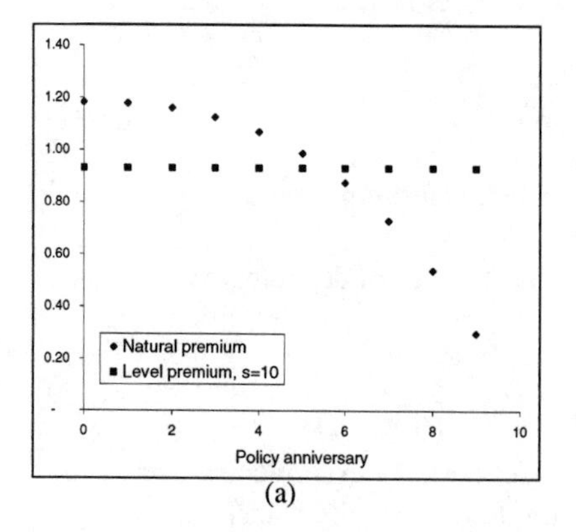

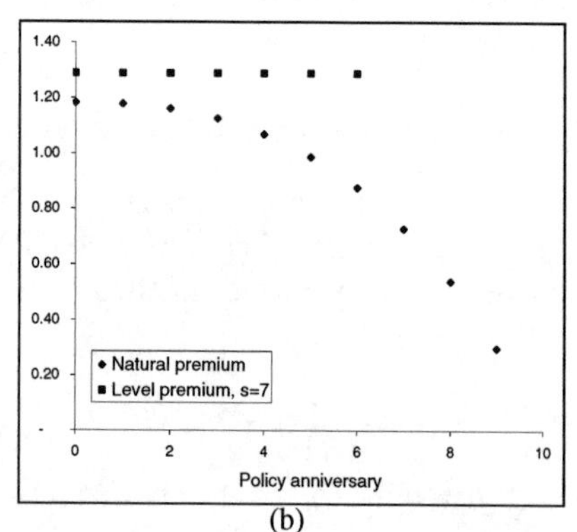

Fig. 4.4.4 Natural premiums, level premiums, and shortened level premiums of a decreasing term insurance

□

We now consider a pure endowment, with a unitary sum insured. As this insurance products only provides a benefit in the case of survival at maturity (time m), we have:

$$P_h^{[N]} = 0 \tag{4.4.23a}$$

$$P_{m-1}^{[N]} = {}_1E'_{x+m-1} = (1+i')^{-1}\,{}_1p'_{x+m-1} \tag{4.4.23b}$$

Obviously, natural premiums do not represent a reasonable arrangement of premium payment for this insurance product.

A (standard) endowment insurance is built-up combining a pure endowment with a term insurance, both with duration m. Then, referring to a unitary sum insured, we find:

$$P_h^{[N]} = {}_1A'_{x+h} = (1+i')^{-1}\,{}_1q'_{x+h}; \quad h = 0,1,\ldots,m-2 \tag{4.4.24a}$$

$$\begin{aligned} P_{m-1}^{[N]} &= {}_1A'_{x+m-1} + {}_1E'_{x+m-1} \\ &= (1+i')^{-1}\,({}_1q'_{x+m-1} + {}_1p'_{x+m-1}) = (1+i')^{-1} \end{aligned} \tag{4.4.24b}$$

Clearly, also in this case natural premiums do not represent a reasonable arrangement of premium payment, because of the presence of the pure endowment component.

4.4.4 *Single premium, natural premiums and level premiums: some relations*

The single premium Π, according to the equivalence principle, is the actuarial value (at the policy issue) of the benefits provided by an insurance contract (see (4.3.3)). Assume a policy duration of m years (it is understood that what follows can be applied to lifelong contracts also, by setting $m = \omega - x$). Intuitively, the actuarial value of the benefits can be expressed as the sum of the actuarial values (at the policy issue) of the benefits falling due in the various policy years. Each of these actuarial values can be expressed as ${}_hE'_x\,P_h^{[N]}$, $h = 0,1,\ldots,m-1$, since $P_h^{[N]}$ quantifies the expected value at time h of the benefits pertaining to year $h+1$, provided that the insured is alive at that time.

Hence, we have:

$$\Pi = \sum_{h=0}^{m-1} {}_hE'_x\,P_h^{[N]} \tag{4.4.25}$$

An example is provided by formula (4.4.10).

Assume that annual level premiums, P, are paid throughout the whole policy duration. The equivalence principle requires that relation (4.4.15) is fulfilled. From (4.4.25) we then obtain:

$$P\,\ddot{a}_{x:\overline{m}|} = \Pi \tag{4.4.26}$$

and hence

$$P = \frac{\sum_{h=0}^{m-1} {}_hE'_x P_h^{[N]}}{\sum_{h=0}^{m-1} {}_hE'_x} \tag{4.4.27}$$

It turns out that the level premium P is a weighted arithmetic mean of the natural premiums $P_h^{[N]}$'s, with the ${}_hE'_x$'s as the weights. An example is provided by the annual level premium of a term insurance, as expressed by (4.4.13).

We stress that expression (4.4.27), and the related interpretation hold with the proviso that the level premiums are paid as long as the policy is in-force. Thus, they do not hold in the case of shortened level premiums, as, for example, in formula (4.4.20).

The *actuarial saving premium*, or *reserve premium*, denoted as $P_h^{[AS]}$, is defined as follows:

$$P_h^{[AS]} = P - P_h^{[N]} \tag{4.4.28}$$

(in the case of annual level premiums). Clearly, when $P_h^{[AS]} > 0$ a share of the premium P is accumulated ("reserved") to meet future benefits, whereas, when $P_h^{[AS]} < 0$ an amount higher than the premium P is needed to meet benefits falling due in the current year, and hence resources previously accumulated must be used.

We note that the pair $(P_h^{[N]}, P_h^{[AS]})$, for $h = 0, 1, \ldots$, constitutes a splitting of the annual premium. A more important splitting will be discussed in Sect. 5.4.3.

4.4.5 Single recurrent premiums

Premium arrangements other than those consisting of either level premiums or natural premiums can be conceived in order to gain flexibility in the time profile of premium payment. An interesting example is provided by *single recurrent premiums*. We describe ideas underlying single recurrent premiums and their implementation referring to two examples, namely the pure endowment and the whole life insurance.

Refer to a pure endowment insurance with maturity at time m, and assume that, in order to meet the benefit, a sequence of payments, i.e. the single recurrent premiums, $\Pi_0, \Pi_1, \ldots, \Pi_{m-1}$, is arranged. The premium Π_h, paid at time h, funds the benefit ΔS_h deferred $m - h$ years, which constitutes an increase in the "cumulated benefit".

In formal terms, the link among premiums, benefits and cumulated benefits is described by the following relations:

$$\begin{aligned} \Pi_0 &= \Delta S_0 \, {}_mE'_x \; ; \quad S_1 = \Delta S_0 \\ \Pi_1 &= \Delta S_1 \, {}_{m-1}E'_{x+1} \; ; \quad S_2 = S_1 + \Delta S_1 \\ \Pi_2 &= \Delta S_2 \, {}_{m-2}E'_{x+2} \; ; \quad S_3 = S_2 + \Delta S_2 \\ & \ldots \qquad \ldots \\ \Pi_{m-1} &= \Delta S_{m-1} \, {}_1E'_{x+m-1} \; ; \quad S_m = S_{m-1} + \Delta S_{m-1} \end{aligned} \tag{4.4.29}$$

The amount S_m turns out to be the (total) sum insured, progressively financed by the single recurrent premiums Π_h's. Each one of the premiums can be stated at the time of payment. It follows that S_m is ultimately known at time $m-1$ only, when the last premium is paid.

As the total amount S_m consists, from a technical point of view, of m pure endowments, the related accumulation process relies on both interest and mutuality, so that S_m turns out to be greater than the result of a purely financial accumulation.

Example 4.4.4. Table 4.4.2 provides an example of pure endowment insurance financed by a sequence of single recurrent premiums. Data are as follows: $x = 50$, $m = 10$, TB1 $= (0.02, \text{LT1})$. The resulting accumulation process is then compared to the financial accumulation of amounts equal to the single recurrent premiums. According to this process, we have, for $h = 0, 1, \ldots, m-1$:

$$\Delta M_h = \Pi_h \, (1 + i')^{m-h}; \;\; M_{h+1} = M_h + \Delta M_h$$

with $M_0 = 0$.

Table 4.4.2 Pure endowment insurance and financial accumulation

h	Π_h	ΔS_h	S_{h+1}	ΔM_h	M_{h+1}
0	100	128.98	128.98	121.90	121.90
1	100	126.02	255.00	119.51	241.41
2	100	123.09	378.08	117.17	358.57
3	100	120.17	498.25	114.87	473.44
4	100	117.27	615.53	112.62	586.06
5	120	137.26	752.79	132.49	718.55
6	120	133.81	886.59	129.89	848.44
7	120	130.36	1016.95	127.34	975.79
8	120	126.91	1143.86	124.85	1100.63
9	120	123.46	1267.32	122.40	1223.03

□

Refer to a whole life insurance, and assume that, in order to meet the benefit, a sequence of payments (the single recurrent premiums), $\Pi_0, \Pi_1, \ldots, \Pi_h, \ldots$, is arranged. The premium Π_h, paid at time h, funds the amount ΔC_h which, from time h onwards, constitutes a share of the cumulated sum assured. The following relations describe the link among premium Π_h, amount ΔC_h, and amount C_{h+1} which turns out to be the total sum assured at time h (payable in the case of death between h and $h+1$).

$$\begin{aligned} \Pi_0 &= \Delta C_0 A'_x \; ; \quad C_1 = \Delta C_0 \\ \Pi_1 &= \Delta C_1 A'_{x+1} \; ; \quad C_2 = C_1 + \Delta C_1 \\ &\dots \quad \dots \\ \Pi_h &= \Delta C_h A'_{x+h} \; ; \quad C_{h+1} = C_h + \Delta C_h \\ &\dots \quad \dots \end{aligned} \tag{4.4.30}$$

Example 4.4.5. Table 4.4.3 provides an example of whole life insurance financed by a sequence of s single recurrent premiums. Data are as follows: $x = 50$, $s = 25$, $\text{TB1} = (0.02, \text{LT1})$. Single recurrent premiums are assumed to be constant: $\Pi_h = 100$, for $h = 0, 1, \dots, 24$. It is interesting to compare the resulting time profile of the sum assured C_h to the (constant) amount C assured in a traditional whole life insurance financed via annual level premiums, $P(25) = 100$, payable for 25 years:

$$C A'_{50} = P(25)\, \ddot{a}'_{50:\overline{25}|}$$

We note that $C_{h+1} < C$ in the first 21 years, whereas later we have $C_{h+1} > C$. For example, if the insured dies in the 12-th year, according to the single recurrent premium arrangement, the benefit is $C_{12} = 1\,932.25$, whereas if she dies in the 23-rd year, the benefit is $C_{23} = 3\,546.20$; of course, if the traditional level premium arrangement has been adopted, in both the cases the benefit is $C = 3\,202.60$. We can conclude that, initially, in the level premium arrangement, the same cumulated amount of premiums has to meet a sum assured higher than that financed by single recurrent premiums. Hence, mutuality plays a more important role, and the insurer bears a higher mortality risk. Then, an inversion occurs, and the mortality risk progressively decreases.

□

In a whole life insurance financed via single recurrent premiums, if $i' = 0$ then we have $A'_{x+h} = 1$ for all h, and hence (see relations (4.4.30)):

$$\Delta C_h = \Pi_h; \;\; h = 0, 1, \dots \tag{4.4.31}$$

Thus, the whole life insurance degenerates in a pure accumulation at zero interest rate. In the case of death in the t-th year, the benefit paid is given by:

$$C_t = \sum_{h=0}^{t-1} \Pi_h \tag{4.4.32}$$

Hence, no mortality risk is borne by the insurer.

Table 4.4.3 Whole life insurance: single recurrent premiums and level premiums

h	ΔC_h	C_{h+1}	C
0	176.26	176.26	3 202.60
1	173.23	349.49	3 202.60
2	170.28	519.77	3 202.60
3	167.41	687.18	3 202.60
4	164.62	851.80	3 202.60
5	161.90	1 013.70	3 202.60
...	...	...	...
10	149.44	1 785.08	3 202.60
11	147.17	1 932.25	3 202.60
12	144.96	2 077.21	3 202.60
13	142.83	2 220.05	3 202.60
14	140.77	2 360.82	3 202.60
15	138.78	2 499.60	3 202.60
...	...	...	...
20	129.83	3 166.00	3 202.60
21	128.24	3 294.24	3 202.60
22	126.71	3 420.96	3 202.60
23	125.25	3 546.20	3 202.60
24	123.84	3 670.05	3 202.60
25	0.00	3 670.05	3 202.60
26	0.00	3 670.05	3 202.60
...	...	...	...

4.4.6 Some concluding remarks

Each premium arrangement determines a specific inflow stream (namely, the sequence of premiums cashed by the insurer), which offsets the expected outflow stream (the sequence of expected benefits). Specific implications, concerning the finance of an insurance contract, can be found by comparing the time profile of the two streams.

The following features of premium systems should be stressed.

1. In the case of a single premium, the inflow is clearly concentrated at the policy issue. Hence, an important share of the premium itself (possibly, the whole premium) is to be reserved, whatever the type of insurance product.
2. A premium arrangement based on single recurrent premiums clearly leads to the single premium situation, iterated as many times as many premiums are cashed by the insurer. "Scaled" reserving processes then originate.
3. A situation, which clearly appears as the opposite to the single premium regime, arises from the natural premium arrangement (when applicable). In this case, each premium exactly funds the benefits expected to fall due in the relevant year. No reserving process develops throughout the policy duration (but the need for a one-year based reserving, which reflects the expected progressive consumption of the premium in the mutuality mechanism).

4. Level premiums must be arranged (as regards the duration of the premium stream) so that the financing condition is fulfilled. In many insurance products, level premiums payment can extend over the whole policy duration. In any case, the premium stream must "anticipate" the expected benefits, and hence a reserving process follows.

To conclude, we stress that any reserving process implies an investment - disinvestment process, whose impact on the insurer's finance obviously depends on the magnitude of the amounts involved. Related consequences are:

- financial profit opportunities, thanks to a (positive) spread between the yield on investments and the interest rate credited to insurance contracts;
- disinvestment risk, due to a sudden need for liquidity because of benefits falling due at unexpected early dates, for example because of a mortality higher than expected, or an unanticipated number of surrenders.

4.5 Loading for expenses

The operations involved by an insurance contract, by the management of a portfolio, and by the management of the whole insurance company imply several types of expenses.

Expenses constitute one of the ingredients in the premium calculation, as already mentioned in Sect. 1.7.3 (see Fig. 1.7.4). To this purpose expense loadings must be determined, and then expense-loaded premiums.

4.5.1 Premium components

Figure 4.5.1 illustrates the shift from the equivalence premium to the amount actually paid by the contractor, singling out the main components of this amount.

We start from the equivalence premium, only allowing for the benefits. If this premium is calculated by adopting the first-order or pricing basis (as is common in life insurance), then it already includes an implicit safety loading. Conversely, when the equivalence premium relies on a second-order or scenario basis, a safety loading must be explicitly added. The premium including the safety loading (either implicit or explicit), but not allowing for expenses, is called the *net premium* (or *pure premium*).

Expenses can be accounted for, and then included into the premium via an appropriate loading, by adopting one of the two following approaches:

- a *global* (or *forfeiture*) *expense loading*, according to which the premium is simply increased by a percentage such that the resulting loading approximately meets all the expenses attributed to the contract;

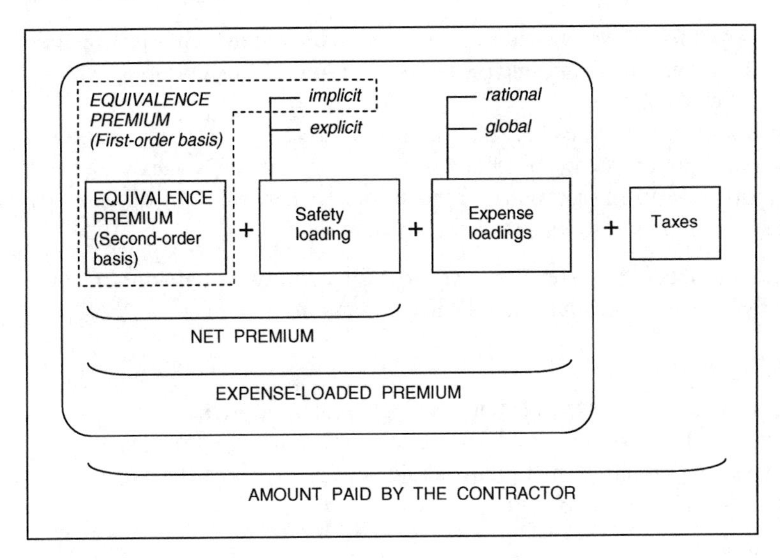

Fig. 4.5.1 Premium components

- a *rational expense loading*, based on recognizing several types of expenses and then determining the related loading components.

In what follows, we focus on the rational approach, whose greater complexity is compensated by the possibility of an appropriate quantification of the various expense components.

The premium including the expense loading is called the *expense-loaded premium* (or the *office premium*, or the *tariff premium*, or the *gross premium*).

The amount actually paid by the contractor may also include taxes which, however, do not involve technical aspects, as taxes are simply cashed by the insurer and then forwarded to tax authorities.

4.5.2 Expenses and loading for expenses

Expenses can be classified into three main groups.

1. *Acquisition expenses*. In this group all expenses connected with the issue of a new policy are included, in particular:

   a. agents' commission;
   b. medical examination (if any);
   c. policy writing.

   Acquisition expenses then constitute an initial cost to the insurer.

2. *Collection expenses*. These expenses are charged at the beginning of every period (every year, in particular) in which a periodic premium (for example, a level premium) is to be collected.
3. *General administration expenses*. All other insurer's expenses (not directly connected with the policy) are included in this group, for example: wages, data processing costs, investment costs, taxes, and so on. A share of these expenses is attributed to each policy, for the whole policy duration.

As regards the loading of premiums, we will focus on two premium arrangements, namely the single premium, and the level premiums.

- *Single premium*
  1. Acquisition expenses are loaded on the single premium itself.
  2. Of course, collection expenses do not concern this premium arrangement.
  3. The expected present value of future administration expenses attributed to the contract is loaded on the premium.
- *Level premiums*
  1. Although acquisition expenses constitute an initial cost to the insurer, the loading for these expenses is split into a sequence of annual amounts, each one loaded on the related premium. Hence, acquisition expenses are progressively recovered.
  2. Collection expenses are loaded year by year on the relevant annual premium.
  3. As administration expenses are attributed to the policy for the whole policy duration, if premiums are payable for the same duration then each premium is loaded with the annual share of expenses; conversely, if the premium payment period is shorter, then a higher share is loaded on each premium.

4.5.3 The expense-loaded premiums

Let $\Pi^{[T]}$ and $P^{[T]}$ denote the single premium and the annual level premium (omitting for brevity the duration of premium payment) respectively, loaded for expenses. Let [A], [C] and [G] denote the three groups of expenses, namely acquisition, collection and general administration expenses respectively.

For the single premium, we have:

$$\Pi^{[T]} = \Pi + \Theta^{[A]} + \Theta^{[G]} \tag{4.5.1}$$

where $\Theta^{[A]}$, $\Theta^{[G]}$ denote the two loading terms. The annual level premium is given by the following expression:

$$P^{[T]} = P + \Lambda^{[A]} + \Lambda^{[C]} + \Lambda^{[G]} \tag{4.5.2}$$

where $\Lambda^{[A]}$, $\Lambda^{[C]}$, and $\Lambda^{[G]}$ denote the three loading terms.

We now address an insurance contract with sum insured C, duration m, either single premium or annual level premiums payable for s years ($s \leq m$). Arguments similar to the following ones apply to life annuities.

The acquisition expenses are usually assumed proportional to the sum insured. Denoting by α the corresponding rate, we have for the related loadings:

$$\Theta^{[A]} = \alpha C \tag{4.5.3a}$$

$$\Lambda^{[A]} \ddot{a}'_{x:\overline{s}|} = \alpha C \tag{4.5.3b}$$

in the case of single premium and level premiums, respectively.

As an alternative, the acquisition expenses can be assumed proportional to the amount of the expense-loaded premium. The loading rate then depends on the number of level premiums. Denoting by $\delta(s)$ the corresponding rate, usually increasing as s increases (for a given policy duration m), we have:

$$\Theta^{[A]} = \delta(1)\,\Pi^{[T]} \tag{4.5.4a}$$

$$\Lambda^{[A]} \ddot{a}'_{x:\overline{s}|} = \delta(s)\,P^{[T]} \tag{4.5.4b}$$

The collection expenses are usually assumed to be proportional to the expense-loaded premium, at a rate we denote by β. Hence, the related loading is given by:

$$\Lambda^{[C]} = \beta\,P^{[T]} \tag{4.5.5}$$

The annual general administration expenses are commonly expressed as a proportion of the sum insured. Denoting by γ the corresponding rate, we have for the related loadings:

$$\Theta^{[G]} = \gamma C \ddot{a}'_{x:\overline{m}|} \tag{4.5.6a}$$

$$\Lambda^{[G]} \ddot{a}_{x:\overline{s}|} = \gamma C \ddot{a}'_{x:\overline{m}|} \tag{4.5.6b}$$

To assess the incidence of costs other than the expected present value of benefits, it is interesting to determine the total expense loading rate. In the case of a single premium, the total loading rate, θ, is given by:

$$\theta = \frac{\Pi^{[T]} - \Pi}{\Pi^{[T]}} = \frac{\Theta^{[A]} + \Theta^{[G]}}{\Pi^{[T]}} \tag{4.5.7}$$

Conversely, in the case of annual level premiums, the total loading rate, λ, is given by:

$$\lambda = \frac{P^{[T]} - P}{P^{[T]}} = \frac{\Lambda^{[A]} + \Lambda^{[C]} + \Lambda^{[G]}}{P^{[T]}} \tag{4.5.8}$$

To illustrate the calculation of an expense-loaded annual level premium, we first refer to a whole life insurance, with premiums payable for s years, sum assured C, age at policy issue x, and loading rate α for acquisition expenses. From Eq. (4.4.20) as regards the net premium, and Eqs. (4.5.3b), (4.5.5), and (4.5.6b) as regards the loadings, we obtain:

$$P^{[T]} = \frac{\frac{C}{\ddot{a}'_{x:s\rceil}}\left(A'_x + \alpha + \gamma \ddot{a}'_x\right)}{1-\beta} \tag{4.5.9}$$

We now refer to an endowment insurance, with duration m years, and level premiums payable for the whole policy duration. We assume the loading rate $\delta(s)$ for acquisition expenses. From Eq. (4.4.18) as regards the net premium, and Eqs. (4.5.4b), (4.5.5), and (4.5.6b) as regards the loadings (with $s = m$ in all the equations), we find:

$$P^{[T]} = \frac{C\left(\frac{A'_{x,m\rceil}}{\ddot{a}'_{x:m\rceil}} + \gamma\right)}{1-\beta-\frac{\delta(m)}{\ddot{a}'_{x:m\rceil}}} \tag{4.5.10}$$

Example 4.5.1. Consider a whole life insurance. Data are as follows: $C = 1\,000$, $x = 50$, $s = 15$. The pricing basis is $\mathrm{TB1} = (0.02, \mathrm{LT1})$. Assume, as the loading parameters: $\alpha = 0.02$, $\beta = 0.04$, $\gamma = 0.001$. We find

$$P = 44.90$$
$$P^{[T]} = 49.47$$

as the net premium and the expense-loaded premium, respectively. The total loading rate is then

$$\lambda = 0.0922$$

Refer to an endowment insurance. Data are as follows: $C = 1\,000$, $x = 50$, $m = s = 15$. The pricing basis is $\mathrm{TB1} = (0.02, \mathrm{LT1})$. Assume, as the loading parameters: $\delta(15) = 0.55$, $\beta = 0.04$, $\gamma = 0.0015$. We find

$$P = 59.54$$
$$P^{[T]} = 66.60$$

as the net premium and the expense-loaded premium, respectively. The total loading rate is then

$$\lambda = 0.1061$$

□

4.6 References and suggestions for further reading

A number of actuarial textbooks deal with technical and financial aspects of life insurance, and with pricing problems in particular. We quote the following ones: [10], [20], [25], [26], [47], and [49].

The book [40] is specifically devoted to life annuities and pensions. Conversely, a wide range of insurance products in the field of life (and health) insurance are described in [7].

The framework of life insurance can be enlarged to include products in which benefits depend not only on the lifetime but also, for example, on disability. For technical aspects of disability annuities and related products (in the area of health insurance) the reader can refer to [28].

Finally, we recall that the principles of Financial Mathematics (which underpin the calculation of actuarial values) are presented in [11], [54], and in the first five chapters of [37].

Chapter 5
Life insurance: reserving

5.1 Introduction

The insurer's debt position, which is an obvious implication of the single premium arrangement, must be realized also when other premium arrangements are adopted. This need clearly emerged in Sect. 4.4.1. We recall that an asset accumulation - decumulation process develops, throughout the policy duration, against the insurer's debt position. A technical tool for assessing the insurer's debt is provided by the so-called mathematical reserve.

The need for assessing the insurer's position with respect to an insurance policy emerges at any time during the policy duration. In particular, we can recognize:

- "ordinary" needs which emerge, for example, in relation to:
  - the balance-sheet, which must display the total insurer's debt towards the policyholders;
  - the sharing of profits with the policyholders, which, in particular, can be related to the proportion of assets contributed by each policy;
- "extraordinary" needs, related for example to the interruption of periodic premium payment, and hence the need for assessing the policyholder's credit and then
  - converting the policy into a "paid-up" one, namely a policy for which no further premium payment is required;
  - determining the amount to be paid by the insurer in the case of "surrender".

5.2 General aspects

We refer to a generic insurance policy, and focus on benefits and net premiums only. That is, we start disregarding expenses and related expense loadings. We assume that

A. Olivieri, E. Pitacco, *Introduction to Insurance Mathematics*,
DOI 10.1007/978-3-642-16029-5_5, © Springer-Verlag Berlin Heidelberg 2011

the policy term is m, but a generalization to lifelong policies is straightforward, by setting $m = \omega - x$ (where ω denotes as usual the maximum attainable age).

Let t_1, t_2 denote two integer times (policy anniversaries), with $0 \leq t_1 < t_2 \leq m$. We define the following notation, which proves to be useful when dealing with the definition of the mathematical reserve:

- $Y(t_1,t_2)$ denotes the random present value at time t_1 of the benefits which fall due in the time-interval (t_1,t_2);
- $X(t_1,t_2)$ denotes the random present value at time t_1 of the premiums to be cashed in the time-interval (t_1,t_2).

Remark The notation just defined generalizes the one we used in Chap. 4 to denote the random present values of benefits and premiums. Indeed, $Y = Y(0,m)$, $X = X(0,m)$.

Then, we define:

- $\mathrm{Ben}(t_1,t_2) = \mathbb{E}[Y(t_1,t_2)]$, i.e. the expected present value (or actuarial value) at time t_1 of the benefits which fall due in the time-interval (t_1,t_2);
- $\mathrm{Prem}(t_1,t_2) = \mathbb{E}[X(t_1,t_2)]$, i.e. the expected present value (or actuarial value) at time t_1 of the premiums to be cashed in the time-interval (t_1,t_2).

Remark It is worth commenting in some detail which of the benefits and premiums paid at the extremes of the time-interval (t_1,t_2), i.e. at times t_1 and t_2, are included in the quantities $Y(t_1,t_2)$ and $\mathrm{Ben}(t_1,t_2)$ (for benefits), $X(t_1,t_2)$ and $\mathrm{Prem}(t_1,t_2)$ (for premiums).
In general terms, if an amount is paid at a given time t because it is due at the beginning of year $(t,t+1)$, we say that it is paid at time t in advance. Conversely, if it is paid at time t because due at the end of year $(t-1,t)$, then we say that it is paid at time t in arrears. The rule we adopt when defining the flows included in the quantities $Y(t_1,t_2)$, $\mathrm{Ben}(t_1,t_2)$, $X(t_1,t_2)$ and $\mathrm{Prem}(t_1,t_2)$ is the following. Premiums and benefits paid at time t_1 in advance are included, while benefits paid at time t_1 in arrears are excluded. Benefits paid at time t_2 in arrears are included, while premiums and benefits paid at time t_2 in advance are excluded. Actually, the time-interval addressed by the quantities $Y(t_1,t_2)$, $\mathrm{Ben}(t_1,t_2)$, $X(t_1,t_2)$ and $\mathrm{Prem}(t_1,t_2)$ runs from the beginning of year (t_1,t_1+1) to the end of year (t_2-1,t_2). Of course, all the flows falling due at a time t, $t_1 < t < t_2$, are included in such quantities. The rule will clearly emerge in Example 5.2.1, as well as in the following Sections.

We now assume that the actuarial values rely on the first-order basis, i.e. the pricing basis TB1. The notations Ben′ and Prem′ reflect this hypothesis.

It is well known that the equivalence principle requires

$$\mathrm{Prem}'(0,m) = \mathrm{Ben}'(0,m) \tag{5.2.1}$$

On the contrary, all the following situations may occur, at least in principle, when intervals shorter than the whole policy duration are referred to:

$$\mathrm{Prem}'(0,t) \lesseqgtr \mathrm{Ben}'(0,t) \tag{5.2.2a}$$

$$\mathrm{Prem}'(t,m) \lesseqgtr \mathrm{Ben}'(t,m) \tag{5.2.2b}$$

(we recall that t is an integer time).

Further, it is usual to find that

$$\mathrm{Prem}'(t,t+1) \neq \mathrm{Ben}'(t,t+1) \tag{5.2.3}$$

where the term on the left-hand side denotes, for example, the annual level premium, whereas the term on the right-hand side denotes the natural premium.

Example 5.2.1. Consider a m-year term insurance, providing a unitary benefit (that is, $C = 1$), with single premium Π, or annual level premiums P payable for the whole policy duration. We have:

$$\mathrm{Ben}'(0,m) = {}_mA'_x$$

$$\mathrm{Prem}'(0,m) = \begin{cases} \Pi & \text{in the case of single premium} \\ P\ddot{a}'_{x:m\rceil} & \text{in the case of annual level premiums} \end{cases}$$

Further, for $t = 1,2,\dots,m-1$, we have:

$$\mathrm{Ben}'(t,m) = {}_{m-t}A'_{x+t}$$

$$\mathrm{Prem}'(t,m) = \begin{cases} 0 & \text{in the case of single premium} \\ P\ddot{a}'_{x+t:m-t\rceil} & \text{in the case of annual level premiums} \end{cases}$$

$$\mathrm{Ben}'(t,t+1) = {}_1A'_{x+t} = P_t^{[\mathrm{N}]}$$

$$\mathrm{Prem}'(t,t+1) = \begin{cases} 0 & \text{in the case of single premium} \\ P & \text{in the case of annual level premiums} \end{cases}$$

□

5.3 The policy reserve

5.3.1 *Definition*

Refer to the time-interval (t,m), with $t = 0,1,\dots,m$; let V_t denote the quantity such that:

$$\mathrm{Prem}'(t,m) + V_t = \mathrm{Ben}'(t,m) \tag{5.3.1}$$

Clearly, from Eq. (5.2.1) we obtain

$$V_0 = 0 \tag{5.3.2}$$

Conversely, for $t > 0$ the amount V_t fulfils the equivalence principle given that only "residual" benefits and premiums are referred to.

We note that if $\text{Ben}'(t,m) > \text{Prem}'(t,m)$, then the insurer is in a debt position. Hence, the financing condition can be simply expressed by the inequality $V_t \geq 0$, which means no credit position. From Eq. (5.3.1) we also note that, if $\text{Ben}'(t,m) > \text{Prem}'(t,m)$, the amount V_t together with the future premiums exactly meets the future benefits.

The quantity

$$V_t = \text{Ben}'(t,m) - \text{Prem}'(t,m) \tag{5.3.3}$$

is called the *prospective net reserve*. The adjective "prospective" denotes that the reserve refers to the "future" time interval, namely from time t onwards (the retrospective reserve will be shortly addressed in Sect. 5.3.6), whereas "net" recalls that we are not allowing for expenses and related loadings. Of course, the reserve we have defined is a "policy" reserve, as it refers to an insurance contract (the portfolio reserve will be dealt with in Sect. 6.1). The expression *mathematical reserve* is also used.

As already mentioned, the reserve, defined by (5.3.3), is assessed adopting the pricing basis TB1. Hence, it can be considered a prudential valuation of the insurer's debt. However, as the pricing basis leads to an implicit safety loading, the "degree" of prudence cannot be easily determined. An explicit approach to a safe-side assessment of the reserve will be presented in Sects. 6.1.2 and 6.1.3.

5.3.2 The policy reserve for some insurance products

The following examples are straightforward applications of formula (5.3.3), which defines the reserve. If not otherwise stated, we assume unitary benefits. We first consider insurance products financed by annual level premiums. It is understood that, for each product, the premium P must rely on the appropriate formula (see Sect. 4.4.2).

For a whole life insurance, with lifelong premiums, we find:

$$V_t = A'_{x+t} - P\,\ddot{a}'_{x+t} \tag{5.3.4}$$

In the case of s-year temporary premiums, we have:

$$V_t = \begin{cases} A'_{x+t} - P(s)\,\ddot{a}'_{x+t:s-t\rceil} & \text{if } t < s \\ A'_{x+t} & \text{if } t \geq s \end{cases} \tag{5.3.5}$$

The reserve of a term insurance, with premiums payable for the whole policy duration, is given by:

$$V_t = {}_{m-t}A'_{x+t} - P\,\ddot{a}'_{x+t:m-t\rceil} \tag{5.3.6}$$

For a pure endowment insurance, we have:

$$V_t = {}_{m-t}E'_{x+t} - P\,\ddot{a}'_{x+t:m-t\rceil} \tag{5.3.7}$$

and for an endowment insurance:

$$V_t = A'_{x+t,m-t\rceil} - P\ddot{a}'_{x+t:m-t\rceil} \tag{5.3.8}$$

We now address, for $t > 0$, insurance products financed by a single premium. For a pure endowment insurance, we have:

$$V_t = {}_{m-t}E'_{x+t} \tag{5.3.9}$$

whereas, for an immediate life annuity in advance, we find:

$$V_t = \ddot{a}'_{x+t} \tag{5.3.10}$$

When a premium arrangement based on single recurrent premiums is adopted, the reserve can be easily determined via iterated application of the single premium reserve formula. For example, consider a pure endowment insurance, and assume that, at time t, the amounts $\Delta S_0, \Delta S_1, \ldots \Delta S_{t-1}$ turn out to be financed according to the scheme presented in Sect. 4.4.5 (see relations (4.4.29)). The sum insured cumulated up to time t is then S_t. Hence, the reserve is given by

$$V_t = {}_{m-t}E'_{x+t} \sum_{h=0}^{t-1} \Delta S_h = S_t \, {}_{m-t}E'_{x+t} \tag{5.3.11}$$

In a whole life insurance, the sum assured cumulated up to time t is C_t. Then, we find:

$$V_t = A'_{x+t} \sum_{h=0}^{t-1} \Delta C_h = C_t A'_{x+t} \tag{5.3.12}$$

In particular, if $i' = 0$ we have (see (4.4.32)):

$$V_t = \sum_{h=0}^{t-1} \Pi_h \tag{5.3.13}$$

5.3.3 The time profile of the policy reserve

The policy reserve, V_t, is a function of time t. When analyzing its behavior against time, we assume that the insured is alive at time t.

As we have so far assumed that the reserve is calculated by adopting the pricing basis, the reserve itself at policy issue, namely at time $t = 0$, is equal to zero, whatever the premium arrangement (see (5.2.1) and (5.3.2)). However, in the case of a single premium, Π, it is usual to focus on the reserve immediately after cashing the premium itself, V_{0+}, hence setting:

$$V_{0+} = V_0 + \Pi = \Pi \tag{5.3.14}$$

As regards the value of the reserve at maturity, i.e. at time m, for a term insurance we clearly have:

$$V_m = 0 \tag{5.3.15}$$

Conversely, for a pure endowment and an endowment insurance with a unitary amount as the benefit in case of survival, we find:

$$V_m = 1 \tag{5.3.16}$$

We now move to the time profile for $t = 1, 2, \ldots$ (thus, restricting the analysis at the policy anniversaries). Since we have chosen as mortality assumptions numerical life tables (the input of the calculation procedures), although derived from an analytical model (the Heligman-Pollard law), the time profile of the reserve (the output) can be analyzed only in numerical terms. Notwithstanding, some arguments emerging from the numerical inspection have a wide range of application. A number of examples follow.

Example 5.3.1. The reserve of a single premium term insurance is plotted in Fig. 5.3.1, whereas the case of annual level premium is referred to in Fig. 5.3.2. In both the cases data are as follows: sum assured $C = 1\,000$, $x = 40$, $m = 10$; the pricing basis is TB1 $= (0.02, \text{LT1})$.

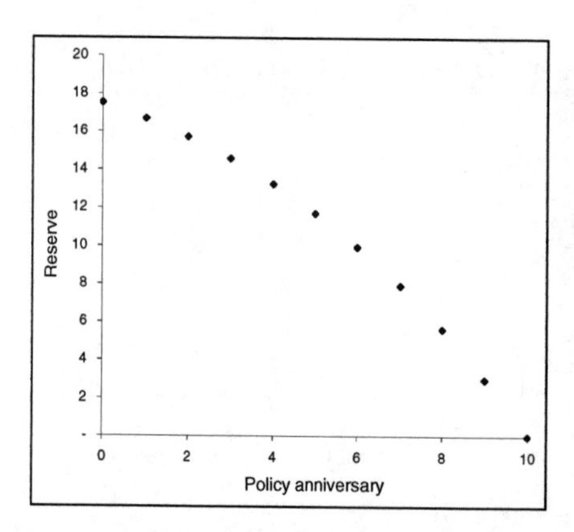

Fig. 5.3.1 Term insurance; single premium

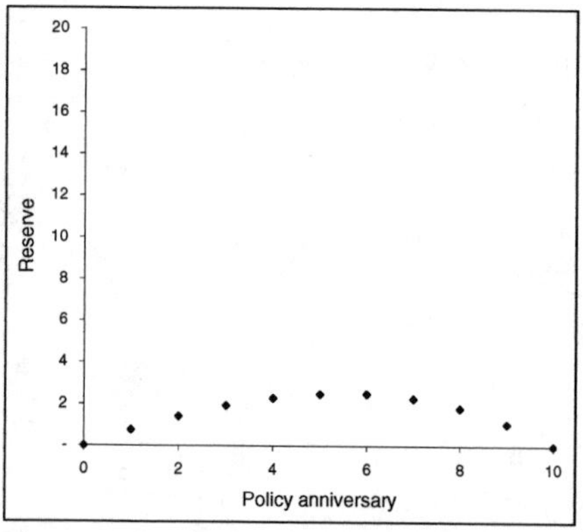

Fig. 5.3.2 Term insurance; annual level premiums

In Fig. 5.3.3, the reserves corresponding to various ages at entry are plotted. The other data are unchanged. Conversely, Fig. 5.3.4 displays the reserves related to various policy durations, age $x = 40$.

□

The following features of the reserve of the term insurance should be pointed out:

- the reserve is, in any case, very small if compared to the sum assured;

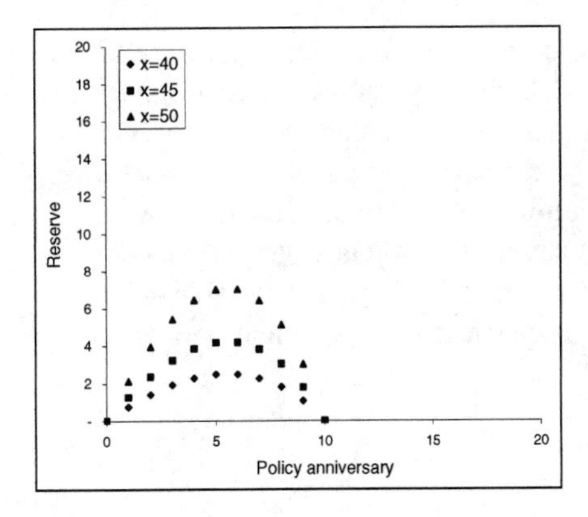

Fig. 5.3.3 Term insurances, with various ages at entry; annual level premiums

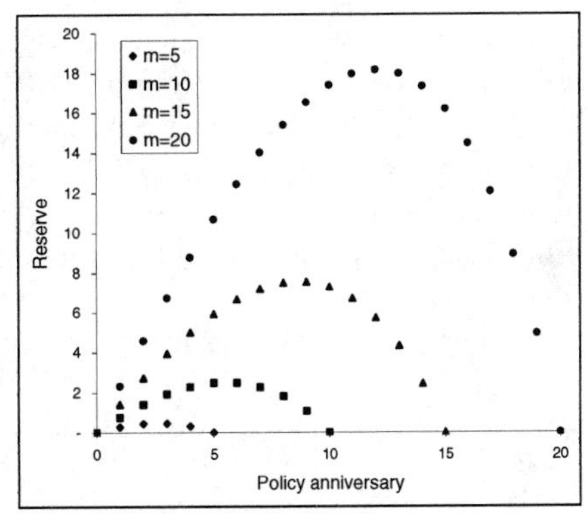

Fig. 5.3.4 Term insurances, with various durations; annual level premiums

- in the case of a single premium, the premium itself is progressively used according to the mutuality mechanism, and hence the reserve decreases throughout the policy duration;
- in the case of annual level premiums, the reserve initially grows, since the level premium slightly exceeds the corresponding natural premium (see Sect. 4.4.3, and Example 4.4.3 in particular), then it decreases, and is equal to zero at the end, because the insurer has no obligation if the insured is alive at maturity;
- still in the case of annual level premiums, the reserve profile is higher when the age at entry is higher; this can be explained in terms of variation of the natural premiums throughout the policy duration (again, see Sect. 4.4.3, and Example 4.4.3).

Figure 5.3.5 explains the variation (either positive or negative) in the reserve value, in the case of annual premiums.

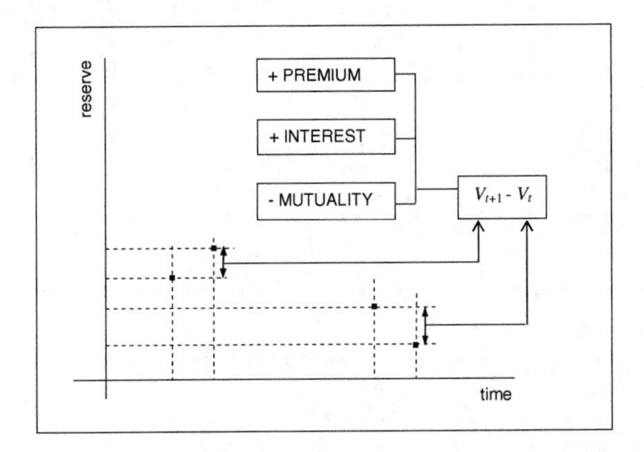

Fig. 5.3.5 Annual variations in the reserve of a term insurance (annual level premiums)

Example 5.3.2. We refer to a decreasing term insurance (see Sect. 4.3.4). Data are as follows: $x = 40$, $m = 10$, $\text{TB1} = (0.02, \text{LT1})$. The sum assured is given by $C_{h+1} = \frac{10-h}{10} 1\,000$. The reserve profile in the case of a single premium is plotted in Fig. 5.3.6. Conversely, Fig. 5.3.7 displays the reserve in the case of annual level premiums payable for the whole policy duration. The violation of the financing condition is apparent. Shortening the premium payment period leads to the reserve profiles plotted in Figs. 5.3.8 and 5.3.9. In particular, the former shows an insufficient shortening ($s = 8$), whereas the latter displays a feasible arrangement ($s = 7$).

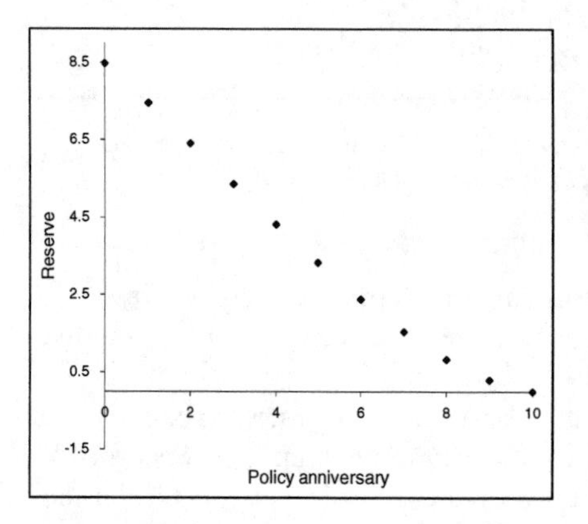

Fig. 5.3.6 Decreasing term insurance; single premium

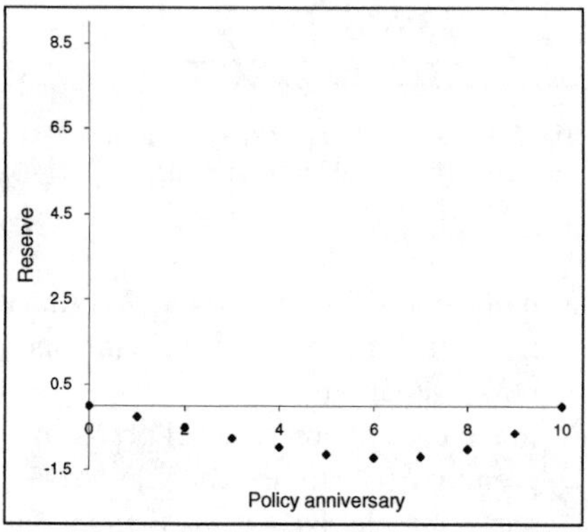

Fig. 5.3.7 Decreasing term insurance; annual level premiums

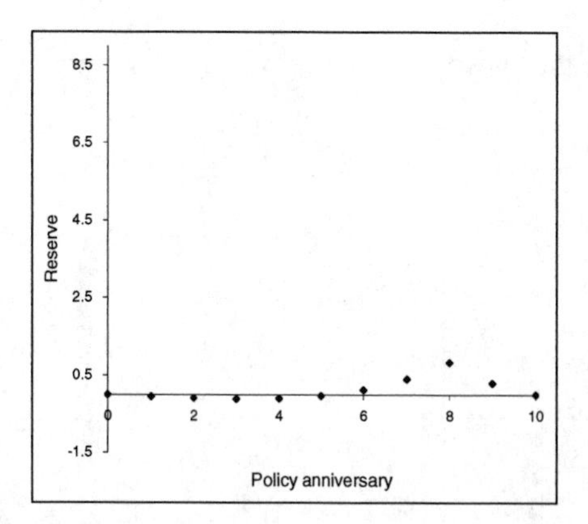

Fig. 5.3.8 Decreasing term insurance; shortened annual level premiums ($s = 8$)

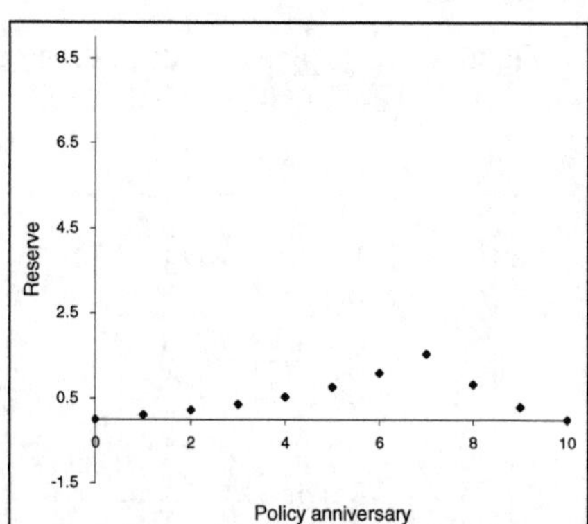

Fig. 5.3.9 Decreasing term insurance; shortened annual level premiums ($s = 7$)

□

Example 5.3.3. The reserve of a single premium pure endowment is plotted in Fig. 5.3.10, whereas the case of annual level premium is referred to in Fig. 5.3.11. In both the cases data are as follows: sum assured $C = 1\,000$, $x = 40$, $m = 10$, $\mathrm{TB1} = (0.02, \mathrm{LT1})$.

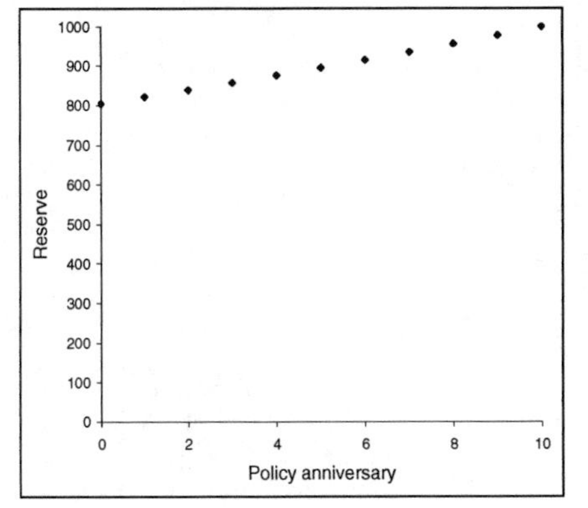

Fig. 5.3.10 Pure endowment; single premium

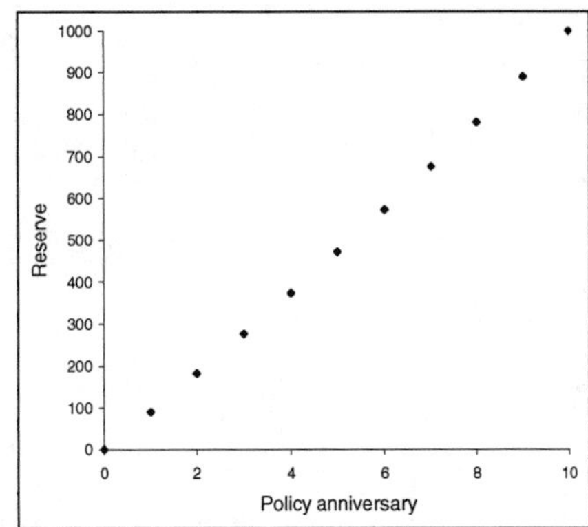

Fig. 5.3.11 Pure endowment; level premiums

□

The reserve of a pure endowment is increasing throughout the whole policy duration. Figure 5.3.12 shows the causes of annual increments in the reserve, in the case of annual premiums. In particular, we recall that each individual reserve is annually credited with a share of reserves released by the insureds who died in that year (see also Case 4a in Sect. 1.7.4, and Fig. 1.7.7 in particular).

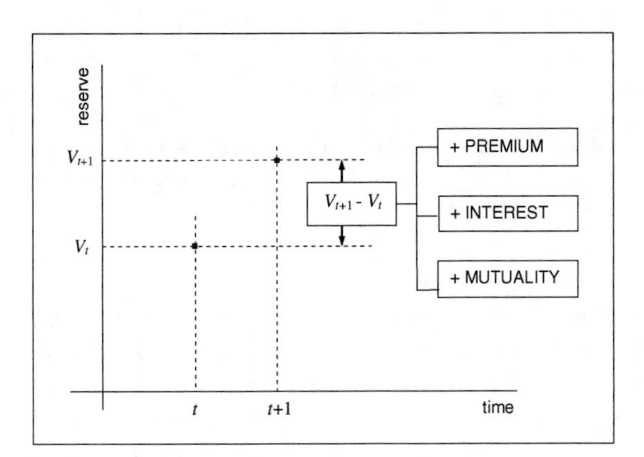

Fig. 5.3.12 Annual variation in the reserve of a pure endowment (annual level premiums)

Example 5.3.4. Figures 5.3.13 and 5.3.14 refer to an endowment insurance, with single premium and annual level premiums, respectively. Data are as for the pure endowment.

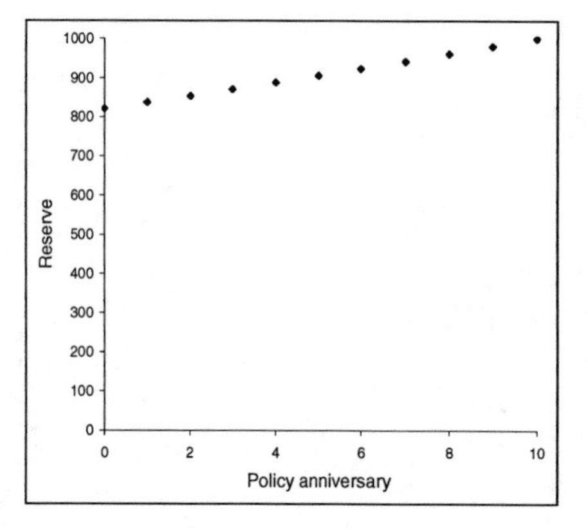

Fig. 5.3.13 Endowment insurance; single premium

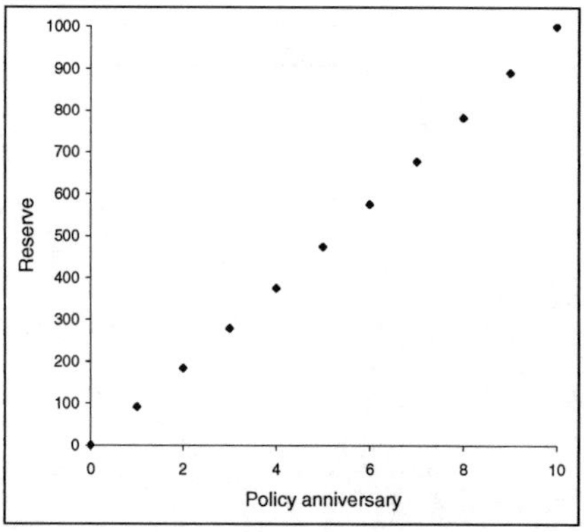

Fig. 5.3.14 Endowment insurance; annual level premiums

□

The time profile of the reserve of an endowment insurance almost coincides with that of a pure endowment. In fact, the difference between the two reserves is the reserve of a term insurance (assuming that the same technical basis is adopted in the three insurance products), and hence it is very small, as already noted. It is worth noting, however, that the rationale underlying the annual variations in the reserve of an endowment insurance is quite different. Indeed, the payment of death benefits to insureds who die implies that shares of each individual reserve are annually subtracted from the reserve itself. See Fig. 5.3.15, in which the mutuality effect works in a negative sense with respect to insureds who are still alive.

Example 5.3.5. Figures 5.3.16 and 5.3.17 refer to a whole life insurance, with single premium and annual level premiums payable for $s = 20$ years, respectively. Data are as follows: $C = 1000$, $x = 50$, $\text{TB1} = (0.02, \text{LT1})$.
□

The time profile of the reserve of a whole life insurance is increasing, in both the case of single premium and annual level premiums, and tends to the sum assured C. In the case of annual premiums, we note that, when all the premiums have been paid, the behavior of the reserve coincides with that of the single premium reserve.

Example 5.3.6. The reserve of a single-premium immediate life annuity (in arrears) is plotted in Fig. 5.3.18. Data are as follows: $b = 100$, $x = 65$, $\text{TB1} = (0.02, \text{LT4})$.
□

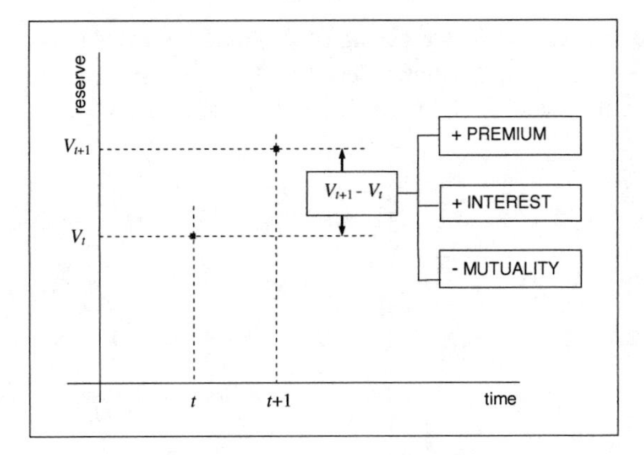

Fig. 5.3.15 Annual variation in the reserve of an endowment insurance (annual level premiums)

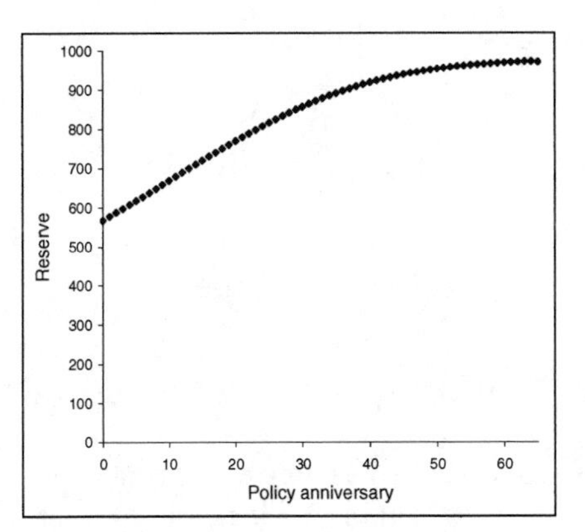

Fig. 5.3.16 Whole life insurance; single premium

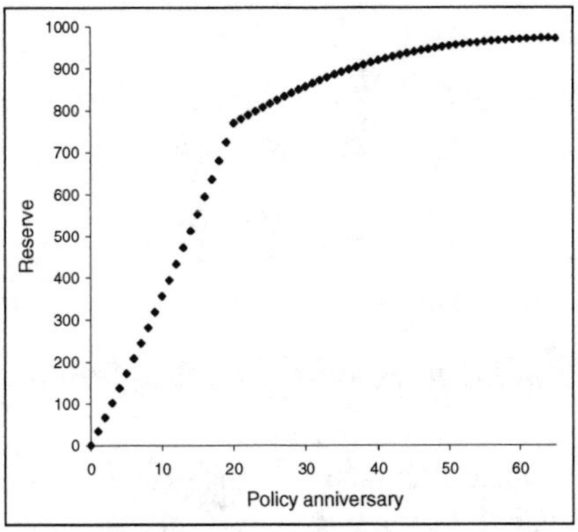

Fig. 5.3.17 Whole life insurance; temporary annual level premiums

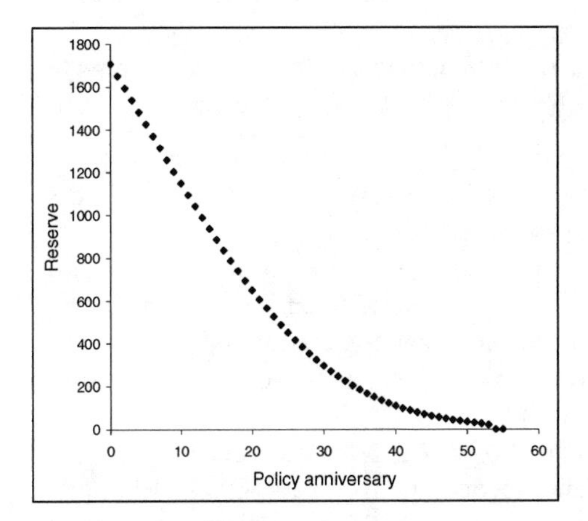

Fig. 5.3.18 Single-premium immediate life annuity

The reserve of an immediate life annuity is decreasing throughout the whole policy duration. Figure 5.3.19 shows the causes of annual decrements in the reserve. We note, in particular, that the mutuality mechanism works as in the pure endowment.

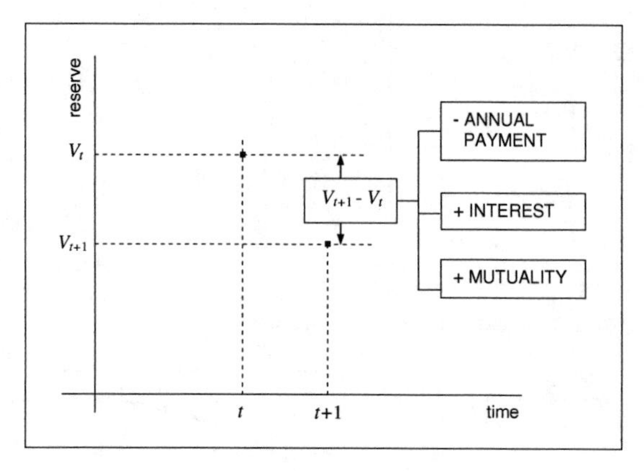

Fig. 5.3.19 Annual variation in the reserve of an immediate life annuity (single premium)

5.3.4 Change in the technical basis

In some circumstances the reserve must be calculated by adopting a technical basis (called the *reserving basis*, or *valuation basis*) other than the pricing basis used for determining the premiums. Such a need can arise, for example, because:

- a "realistic" assessment of the insurer's debt is required, in order to single out the safety component included in the reserve;
- an important change in the financial or demographic scenario makes the reserve (assessed according to the pricing basis) either no longer prudential, or conversely too high.

The former issue will be addressed in Sect. 6.1.3; how to allow for the consequences of a change in the scenario is the topic of the present Section.

Assume that a significant change in the scenario is accounted for when assessing the reserves. This change can be due, for example, to an important variation observed in the mortality, or to different forecasts about the return on investments. The consequent variation in the reserve (when positive) can constitute a compulsory action, imposed by the supervisory authority.

Figure 5.3.20 sketches the consequences of a change in the scenario. First, the new scenario is expressed by an updated second-order basis, TB2*, which, in its turn, suggests the adoption of a new first-order basis, TB1*. This basis will be used as a pricing basis, and hence adopted in pricing as well as reserving, for policies written after the scenario change. Conversely, premiums of in-force policies cannot be changed, since policy conditions are guaranteed at the policy issue. Thus, for these policies the basis TB1* is only used to update the reserves.

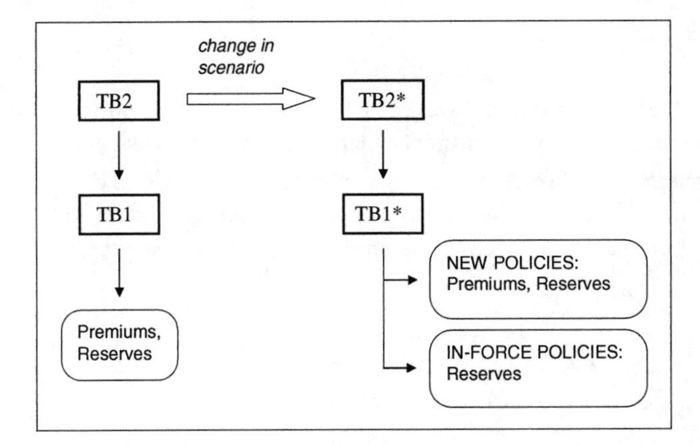

Fig. 5.3.20 Shift to new technical bases because of a change in scenario

Several approaches to the reserve updating are available, at least in principle. We focus on some approaches, referring to an endowment insurance with annual level premiums payable for the whole policy duration. As usual, x denotes the insured's age at policy issue, m the duration, C the sum insured in both the cases of death and survival. We assume that the shift in the technical basis occurs at time τ; the updated reserve will be denoted with $V_t^{[u]}$, for $t \geq \tau$. Further, we assume that the shift implies an increase in the reserve; hence, $V_\tau^{[u]} > V_\tau$.

The updated reserve is defined as the amount that, at time τ, together with the actuarial value of the future premiums, meets (according to the equivalence principle) the actuarial value of the future benefits; both the actuarial values rely on the new basis TB1*. In formal terms:

$$V_\tau^{[u]} + P\ddot{a}^*_{x+\tau:m-\tau\rceil} = CA^*_{x+\tau,m-\tau\rceil} \tag{5.3.17}$$

In more general terms, the equivalence principle requires that the following condition is fulfilled:

$$(V_\tau + \Delta V_\tau) + (P + \Delta P)\,\ddot{a}^*_{x+\tau:m-\tau\rceil} = CA^*_{x+\tau,m-\tau\rceil} \tag{5.3.18}$$

Condition (5.3.18) is an equation in the two unknowns ΔV_τ and ΔP. Particular solutions of (5.3.18) suggest practicable approaches to the problem of the reserve updating. It is understood that, whatever is the particular solution chosen, the insurer is charged with the payment of the amounts ΔV_τ and ΔP.

1. Set

$$\Delta V_\tau = V_\tau^{[u]} - V_\tau \tag{5.3.19}$$

and hence $\Delta P = 0$; Eq. (5.3.18) reduces to (5.3.17). This approach implies an immediate rise in the reserve (at time τ), and hence turns out to be the most prudential. For all integer t, $t \geq \tau$, we then have:

$$V_t^{[u]} = C A^*_{x+t,m-t\rceil} - P \ddot{a}^*_{x+t:m-t\rceil} \tag{5.3.20}$$

2. Less prudential approaches consist in a lower rise, ΔV_τ, in the reserve, followed by premium integrations ("paid" by the insurer) which amortize the missing share of the required increment in the reserve, namely the amount $V_\tau^{[u]} - V_\tau - \Delta V_\tau$. A particular approach in this category can be of prominent practical interest. Let P^* denote the annual premium according to the pricing basis TB1*, namely the premium such that $P^* \ddot{a}^*_{x:m\rceil} = C A^*_{x,m\rceil}$. Then, set

$$\Delta P = P^* - P \tag{5.3.21}$$

From (5.3.18), it follows:

$$\Delta V_\tau = C A^*_{x+\tau,m-\tau\rceil} - P^* \ddot{a}^*_{x+\tau:m-\tau\rceil} - V_\tau \tag{5.3.22}$$

It is worth noting that the resulting reserve, $V_\tau + \Delta V_\tau$, coincides with the reserve, $V_\tau^* = C A^*_{x+\tau,m-\tau\rceil} - P^* \ddot{a}^*_{x+\tau:m-\tau\rceil}$, which will pertain to new policies issued according to the basis TB1*. Hence, the advantage of this particular approach consists in a reserve accumulation process coinciding with that for the new policies. For all integer t, $t \geq \tau$, we then have:

$$V_t^* = C A^*_{x+t,m-t\rceil} - P^* \ddot{a}^*_{x+t:m-t\rceil} \tag{5.3.23}$$

3. Set $\Delta V_\tau = 0$; hence, from (5.3.18) we obtain:

$$\Delta P = \frac{C A^*_{x+\tau,m-\tau\rceil} - P \ddot{a}^*_{x+\tau:m-\tau\rceil} - V_\tau}{\ddot{a}^*_{x+\tau:m-\tau\rceil}} = \frac{V_\tau^{[u]} - V_\tau}{\ddot{a}^*_{x+\tau:m-\tau\rceil}} \tag{5.3.24}$$

Thus, the whole required update in the reserve is amortized in $m - \tau$ years. For all integer t, $t \geq \tau$, denoting with $\tilde{V}_t$ the resulting reserve, we then have:

$$\tilde{V}_t = C A^*_{x+t,m-t\rceil} - (P + \Delta P)\, \ddot{a}^*_{x+t:m-t\rceil} \tag{5.3.25}$$

with ΔP given by (5.3.24). Clearly, this approach does not provide a prudential solution.

The solutions we have described lead to the reserve profiles sketched in Fig. 5.3.21(a) (for simplicity, the reserve profile is represented by a solid line, i.e. disregarding the jumps corresponding to annual premiums).

Of course, other technical solutions are available, even outside the framework designed by condition (5.3.18). We just mention the following one.

4. Set $\Delta V_\tau = 0$; then, set

$$Q = \frac{V_\tau^{[u]} - V_\tau}{\ddot{a}^*_{x+\tau:r-\tau\rceil}} \tag{5.3.26}$$

with $r < m$. Hence, the premium integration, Q, amortizes the required increase in the reserve in a period shorter than the residual policy duration (see Fig. 5.3.21(b)).

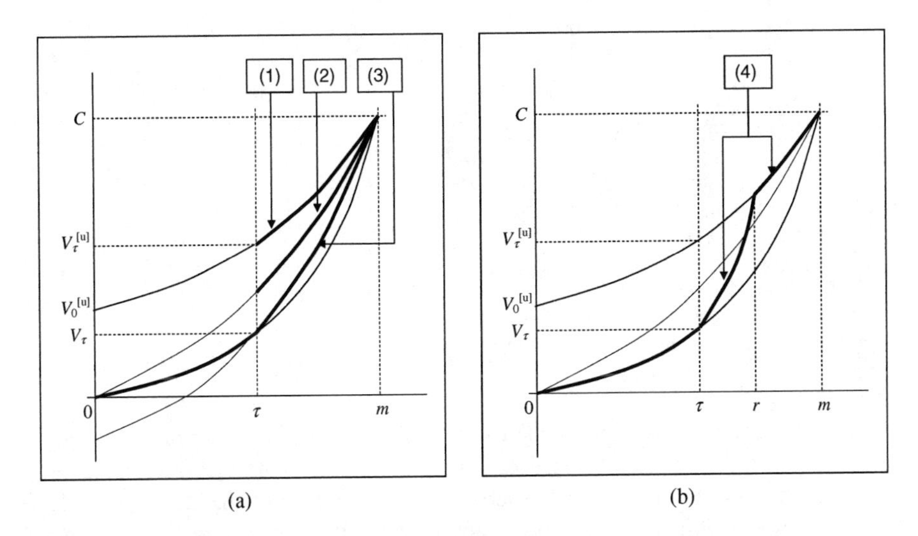

Fig. 5.3.21 Updating the reserve because of a shift in the technical basis

Example 5.3.7. Refer to an endowment insurance with annual level premiums payable for the whole policy duration. Data are as follows: $C = 1000$, $x = 50$, $m = 15$, $\mathrm{TB1} = (0.03, \mathrm{LT1})$. The resulting annual premium is $P = 55.13$. At time $\tau = 8$, because of a decrease in interest rates, the technical basis shifts to $\mathrm{TB1}^* = (0.01, \mathrm{LT1})$. The resulting annual premium is $P^* = 64.27$. Table 5.3.1 displays the reserve V_t which relies on the basis TB1, and the reserves $V_t^{[u]}$, V_t^*, and $\tilde{V}_t$, calculated according to formulae (5.3.20), (5.3.23), and (5.3.25) respectively.
□

5.3.5 The reserve at fractional durations

The analysis of the time profile of the reserve has been so far restricted to the policy anniversaries, namely integer durations since the policy issue. The extension to fractional durations is, however, of practical interest. For example, the need for calculating the policy reserve (and the portfolio reserve, as well) at times other than the policy anniversaries arises when assessing the items of the balance-sheet.

Table 5.3.1 Updating the reserve because of a shift in the technical basis

t	V_t	$V_t^{[u]}$ (1)	V_t^* (2)	$\tilde{V}_t$ (3)
0	0.00			
1	53.59			
2	108.64			
3	165.21			
4	223.37			
5	283.19			
6	344.75			
7	408.16			
8	473.51	570.03	509.62	473.51
9	540.95	628.54	576.35	545.16
10	610.63	687.83	643.97	617.76
11	682.71	747.99	712.59	691.43
12	757.42	809.14	782.33	766.30
13	835.00	871.41	853.35	842.55
14	915.74	934.97	925.83	920.37
15	1 000.00	1 000.00	1 000.00	1 000.00

The calculation of the exact value of the policy reserve at all durations can be carried out in a time-continuous setting. In such a setting, a mortality law must be available, instead of a numerical life table. In the actuarial practice, however, it is rather common to work in a time-discrete framework (as we are actually doing), and to obtain approximations to the exact value of the reserve via interpolation procedures, in particular by adopting linear interpolation formulae. Here we illustrate the interpolation approach, focussing on some examples.

Consider an insurance policy, for example a term insurance, with premium arrangement based on natural premiums. The reserve is, of course, equal to zero at all the policy anniversaries, before cashing the premium which falls due at that time; thus $V_t = 0$ for all integer t. Immediately after cashing the premium, the insurer's debt (and the corresponding asset) is clearly equal to the premium itself; hence, denoting with V_{t+} the reserve after cashing the premium, we have:

$$V_{t+} = P_t^{[N]}; \; t = 0, 1, \ldots \tag{5.3.27}$$

Then, the premium is used throughout the year according to the mutuality mechanism and, again, we have $V_{t+1} = 0$. At time $t+r$, with $0 < r < 1$, we let:

$$V_{t+r} = (1-r)\, V_{t+} = (1-r)\, P_t^{[N]} \tag{5.3.28}$$

The resulting time profile of the reserve is plotted in Fig. 5.3.22.

As the second example, we refer to an insurance product (e.g. an endowment insurance) with annual level premiums P. After cashing the premium which falls due at time t, the reserve increases from V_t to

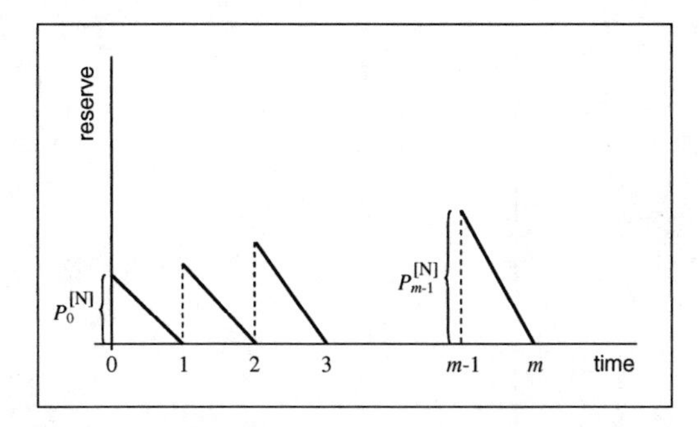

Fig. 5.3.22 Interpolated reserve profile in the case of natural premiums

$$V_{t^+} = V_t + P \tag{5.3.29}$$

Then, the linear interpolation yields:

$$V_{t+r} = (1-r)\,V_{t^+} + r\,V_{t+1} = [(1-r)\,V_t + r\,V_{t+1}] + (1-r)\,P \tag{5.3.30}$$

See Fig. 5.3.23. We note, in particular, the following aspects.

- Interpolating between V_t (instead of V_{t^+}) and V_{t+1} would cause an apparent underestimation of the reserve at all times between t and $t+1$ (again, see Fig. 5.3.23).
- The "use" of the premium P depends on the specific insurance product addressed. For example, if we consider an endowment insurance, the share of the premium used to cover death benefits according to the mutuality mechanism is decreasing throughout the policy duration (as we will see in Sect. 5.4.3); this fact determines a time profile of the reserve like that plotted in Fig. 5.3.24.

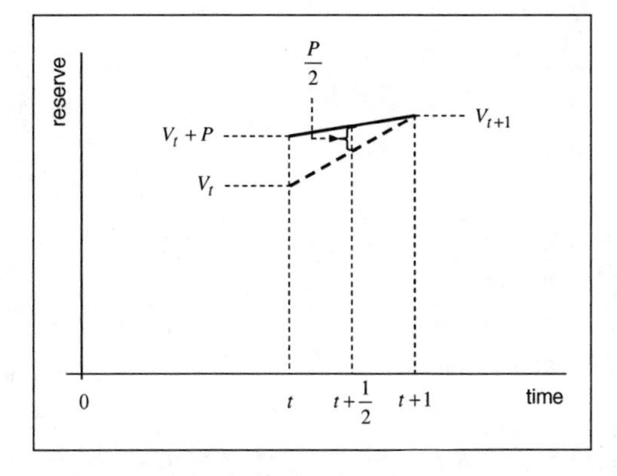

Fig. 5.3.23 Reserve interpolation in the case of annual level premiums

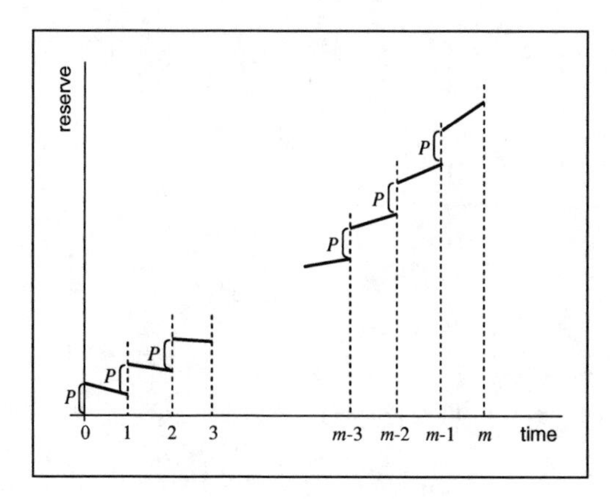

Fig. 5.3.24 Interpolated reserve profile in the case of annual level premiums: an example

As the third example, we consider an insurance product (for example, a term insurance, or a pure endowment, or an endowment insurance), with a single premium Π. In this case, there is no jump in the reserve profile, but at the payment of the single premium, when the reserve jumps from $V_0 = 0$ to $V_{0+} = \Pi$. Then, the interpolation procedure is as follows:

$$V_r = (1-r)\,V_{0+} + r\,V_1 \tag{5.3.31a}$$

$$V_{t+r} = (1-r)\,V_t + r\,V_{t+1} \quad \text{for } t = 1,2,\ldots \tag{5.3.31b}$$

See Fig. 5.3.25.

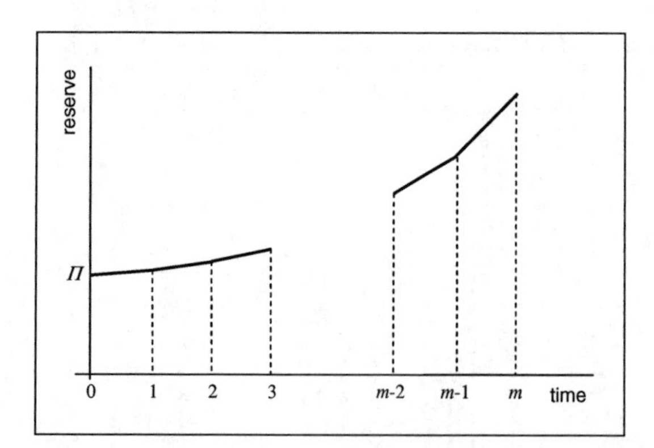

Fig. 5.3.25 Interpolated reserve profile in the case of single premium

Finally, we refer to single-premium life annuities, providing an annual benefit b. The jumps in the reserve profile correspond to the annual payments of the benefit, as illustrated in Fig. 5.3.26. For a life annuity in arrears (panel (a)), taking as usual $V_{0+} = \Pi$, the interpolation is as follows:

$$V_r = (1-r)\,V_{0+} + r\,(V_1 + b) \tag{5.3.32a}$$
$$V_{t+r} = (1-r)\,V_t + r\,(V_{t+1} + b) \;\; \text{for} \;\; t = 1,2,\ldots \tag{5.3.32b}$$

where $V_t = a'_{x+t}$. For a life annuity in advance (panel (b)), taking again $V_{0+} = \Pi$, the interpolation is as follows:

$$V_r = (1-r)\,(V_{0+} - b) + r\,V_1 \tag{5.3.33a}$$
$$V_{t+r} = (1-r)\,(V_t - b) + r\,V_{t+1} \;\; \text{for} \;\; t = 1,2,\ldots \tag{5.3.33b}$$

with $V_t = \ddot{a}'_{x+t}$.

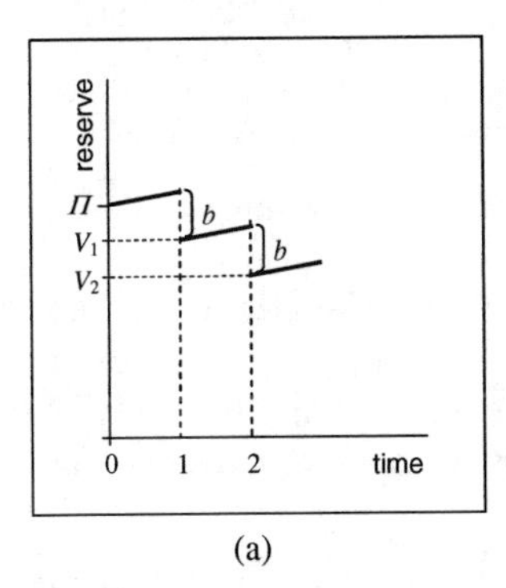

(a)

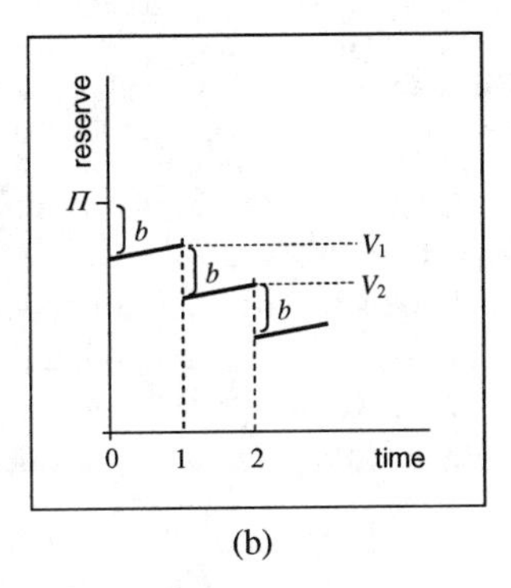

(b)

Fig. 5.3.26 Interpolated reserve profile for life annuities

5.3.6 The retrospective reserve

The (prospective) policy reserve has been defined as the balancing term, V_t, which transforms inequality (5.2.2b) into relation (5.3.1). Looking at inequality (5.2.2a), and hence referring to the time interval $(0,t)$, we can define the so-called retrospective reserve.

Let B_t denote the amount such that

$$\text{Prem}'(0,t) = B_t + \text{Ben}'(0,t) \tag{5.3.34}$$

The amount B_t can be interpreted as the actuarial value (at the policy issue) of the benefit that the insurer should pay at time t if the insured decides (at that time) to abandon the contract, stopping the premium payment and renouncing all the benefits which fall due after time t.

Clearly, this interpretation holds if $B_t > 0$, namely $\text{Prem}'(0,t) > \text{Ben}'(0,t)$. Actually, this inequality should be verified, as we will see in what follows.

The benefit, W_t, whose actuarial value at time $t = 0$ is given by B_t, is then defined by the following relation:

$$B_t = W_t \, {}_tE'_x \tag{5.3.35}$$

Hence, we find:

$$W_t = \frac{1}{{}_tE'_x}\left(\text{Prem}'(0,t) - \text{Ben}'(0,t)\right) \tag{5.3.36}$$

The quantity W_t is called the *retrospective reserve*. Note that the term $\frac{1}{{}_tE'_x}$, namely the actuarial accumulation factor (see Sect. 4.2.8), plays the role of referring the valuation at time t.

Remark The interpretation of W_t as the amount to be paid by the insurer in the case the policyholder abandons the contract, although interesting under a theoretical perspective, requires in practice various adjustments. For example, expenses should be accounted for, and penalties could be applied in determining the amount paid by the insurer. We will return on these issues in Sect. 5.7.

The following examples are straightforward applications of formula (5.3.36), which defines the retrospective reserve.

In insurance products which provide a death benefit (term insurance, whole life insurance, endowment insurance), the insurer's liability is given by the coverage of the death risk over the time interval $(0,t)$. Thus, assuming a unitary benefit, and annual level premiums payable throughout the whole policy duration, we have, for all these products:

$$W_t = \frac{1}{{}_tE'_x}\left(P\,\ddot{a}'_{x:t\rceil} - {}_tA'_x\right) \tag{5.3.37}$$

where P denotes the annual premium related to the specific product addressed.

In a pure endowment with annual level premiums, we have:

$$W_t = \frac{1}{{}_tE'_x} P\,\ddot{a}'_{x:t\rceil} \tag{5.3.38}$$

as this product does not provide any benefit in the time interval $(0,t)$ (of course, if $t < m$, where m denotes the policy term).

In the case of a single premium (given, according to the equivalence principle, by the actuarial value of the benefits), we have for an endowment insurance:

$$W_t = \frac{1}{{}_tE'_x}\left(A'_{x,m\rceil} - {}_tA'_x\right) \tag{5.3.39}$$

Replacing $A'_{x,m\rceil}$ with A'_x or ${}_mA'_x$, we have the retrospective reserve for the whole life insurance and the term insurance, respectively.

For a single-premium pure endowment, we have:

$$W_t = \frac{1}{{}_tE'_x}\,{}_mE'_x \tag{5.3.40}$$

Remark In spite of the adjective "retrospective", the reserve we are dealing with cannot be interpreted as an ex-post quantification of the "past" liabilities (namely, those preceding time t) of the insurer and the insured. From (5.3.36), it is apparent that the calculation of the retrospective reserve first relies on the valuation at time 0 of the benefits and premiums pertaining to the interval $(0,t)$ (and hence "future" with respect to time 0), and then on a valuation at time t via the actuarial accumulation factor $\frac{1}{{}_tE'_x}$.

Let us go back to the reserve of the single-premium pure endowment (see (5.3.40)). We note that, for this insurance product, the prospective reserve is given by $V_t = {}_{m-t}E'_{x+t}$. Further, we have ${}_mE'_x = {}_tE'_x\,{}_{m-t}E'_{x+t}$ (see (4.2.45)), and hence:

$$W_t = V_t \tag{5.3.41}$$

thus, the prospective and the retrospective reserve coincide. Result (5.3.41) holds under rather general conditions. This topic is beyond the scope of this Chapter. So, we will simple provide a further example, and some final remarks as well.

We refer to a whole life insurance, with annual level premium P payable for the whole policy duration. The single premium is, of course, given by A'_x. The following relations hold:

$$A'_x = {}_tA'_x + {}_tE'_x\,A'_{x+t} \tag{5.3.42a}$$

$$\ddot{a}'_x = \ddot{a}'_{x:t\rceil} + {}_tE'_x\,\ddot{a}'_{x+t} \tag{5.3.42b}$$

$$P\ddot{a}'_x = A'_x \tag{5.3.42c}$$

The prospective reserve for this insurance product is given by (5.3.4). By using relations (5.3.42), we find:

$$V_t = \frac{A'_x - {}_tA'_x}{{}_tE'_x} - P\frac{\ddot{a}'_x - \ddot{a}'_{x:t\rceil}}{{}_tE'_x} = \frac{1}{{}_tE'_x}\left(P\ddot{a}'_{x:t\rceil} - {}_tA'_x\right) = W_t \tag{5.3.43}$$

that is, the coincidence between the prospective and the retrospective reserve.

Whenever relations similar to those expressed by formulae (5.3.42) hold, we have the coincidence between the two reserves, provided that the same technical basis is used for both the reserves. However, relations of this type do not hold, for example, in relation to some insurance products which provide benefits depending on the lifetimes of more than one individual. In those products, the reserve at time t depends on which insureds are alive at that time, i.e. on the "status" of the insured group, whereas the retrospective reserve, which first requires a valuation at time 0, can only represent a weighted average of the "possible" prospective reserves at time t.

5.3.7 The actuarial accumulation process

To introduce some interesting relations between the reserving process and the premium flows, we will just refer to an example, provided by an m-year term insurance with annual level premiums payable for the whole policy duration. We assume a unitary sum insured.

The natural premiums of the term insurance are expressed by (4.4.21), namely $P_h^{[N]} = {}_1A'_{x+h} = (1+i')^{-1}\,q'_{x+h}$, for $h = 0, 1, \ldots, m-1$. The reserve premiums, $P_h^{[AS]}$, are defined by (4.4.28).

Consider the actuarial value at the policy issue of the reserve premiums pertaining to the first t policy years. This value is given by

$$\sum_{h=0}^{t-1} P_h^{[AS]}\,{}_hE'_x = P \sum_{h=0}^{t-1} {}_hE'_x - \sum_{h=0}^{t-1} {}_1A'_{x+h}\,{}_hE'_x \tag{5.3.44}$$

From the following relations

$$\sum_{h=0}^{t-1} {}_hE'_x = \ddot{a}'_{x:t\rceil} \tag{5.3.45a}$$

$$\sum_{h=0}^{t-1} {}_1A'_{x+h}\,{}_hE'_x = {}_tA'_x \tag{5.3.45b}$$

we then find that the actuarial value of the reserve premiums can be expressed as follows:

$$P\,\ddot{a}'_{x:t\rceil} - {}_tA'_x = {}_tE'_x\,W_t \tag{5.3.46}$$

(see also (5.3.37)). Finally, we obtain:

$$W_t = \frac{1}{{}_tE'_x} \sum_{h=0}^{t-1} P_h^{[AS]}\,{}_hE'_x = \sum_{h=0}^{t-1} P_h^{[AS]} \frac{1}{{}_{t-h}E'_{x+h}} \tag{5.3.47}$$

Thus, the retrospective reserve is the result of the actuarial accumulation of the reserve premiums pertaining to the policy years preceding the time of valuation of the reserve itself. On the one hand, this interpretation can be useful in interpreting the time profile of the reserve (see Sect. 5.3.3). On the other, a different splitting of the annual premium allow us to interpret the policy reserve as the result of a purely financial accumulation process. As we will see in Sect. 5.4, this alternative splitting of the annual premiums is of paramount importance in interpreting the intermediation role of a life insurer.

We just mention that, conversely, the prospective reserve at time t can be expressed as minus the actuarial value (at that time) of the future reserve premiums, namely:

$$V_t = -\sum_{h=0}^{m-t-1} P_{t+h}^{[AS]}\,{}_hE'_{x+t} \tag{5.3.48}$$

5.4 Risk and savings

The first topic addressed in this Section relates to recursive procedures for the calculation of the policy reserve. Nonetheless, the practical interest of the topic goes well beyond computational aspects. In fact, the topic itself constitutes the starting point for an in depth analysis of the role of a life insurance company. In particular, technical aspects will emerge, concerning the life insurer as a player in both the financial intermediation and the risk pooling process.

5.4.1 A (rather) general insurance product

We refer to an insurance product, with the following characteristics: term m, age at policy issue x, sum insured in the case of death C, sum insured in the case of survival at maturity S, annual level premiums, P, payable for the whole policy duration, and hence given by:

$$P = \frac{C\, {}_mA'_x + S\, {}_mE'_x}{\ddot{a}'_{x:m\rceil}} \tag{5.4.1}$$

For example,

- setting $S = 0$, $C > 0$, we have the term insurance, with constant sum assured;
- setting $S > 0$, $C = 0$, we find the pure endowment;
- setting $S = C > 0$, we have the (standard) endowment insurance;
- setting $S > C > 0$, we have the endowment insurance with additional survival benefit.

A number of possible generalizations allow us to recognize other insurance products. For example,

- setting $S = 0$, $C > 0$, $m = \omega - x$, we find the whole life insurance;
- setting $S = 0$, and replacing C with a sequence $C_1, C_2, \ldots, C_m$, we have the term insurance with varying benefit, and, in particular, the decreasing term insurance;
- replacing P with a sequence $P_0, P_1, \ldots, P_{m-1}$, we can represent arrangements based on variable premiums; in particular:
  - with $P_0 = P_1 = \cdots = P_{s-1}$, $P_s = P_{s+1} = \cdots = P_{m-1} = 0$, we have arrangements based on level premiums payable over a shortened period ($s < m$);
  - setting $P_0 > 0$, $P_1 = P_2 = \cdots = P_{m-1} = 0$, we represent the single premium arrangement;
  - the natural premium arrangement is obviously represented by setting $P_h = P_h^{[N]}$, for $h = 0, 1, \ldots, m-1$.

Other generalizations allow us to represent various types of life annuities. Notwithstanding, in what follows we refer to the insurance product defined at the beginning of this Section.

5.4.2 Recursive equations

The policy reserve, at time t, of the insurance product defined above is given by:

$$V_t = \text{Ben}'(t,m) - \text{Prem}'(t,m) = C\,{}_{m-t}A'_{x+t} + S\,{}_{m-t}E'_{x+t} - P\,\ddot{a}'_{x+t:m-t\rceil} \quad (5.4.2)$$

We can also write:

$$V_t = C\,{}_1A'_{x+t} - P + C\,{}_{1|m-t-1}A'_{x+t} + S\,{}_{m-t}E'_{x+t} - P\,{}_{1|}\ddot{a}'_{x+t:m-t-1\rceil} \quad (5.4.3)$$

and, after a little algebra, we get to the following expression:

$$V_t + P = C\,{}_1A'_{x+t} + V_{t+1}\,{}_1E'_{x+t} \quad (5.4.4)$$

or, in more explicit terms:

$$V_t + P = C\,(1+i')^{-1}\,q'_{x+t} + V_{t+1}\,(1+i')^{-1}\,p'_{x+t} \quad (5.4.5)$$

Recursive Eq. (5.4.5) is called the *Fouret equation* (1891). We note the following features.

- Actuarial values in (5.4.5) are referred at time t, as both the financial and the probabilistic evaluation are referred at that time (that is, the insured is assumed to be alive at time t).
- Equation (5.4.5) describes an "equilibrium" situation in the time interval $(t, t+1)$: the assets available at time t (the reserve V_t, and the premium P just cashed) exactly meet the liabilities which fall due at time $t+1$, namely:
  - the sum assured C, in the case of death;
  - the reserve V_{t+1}, which is needed either to continue the policy in the case of survival (if $t+1 < m$), or to be paid as sum S at maturity (if $t+1 = m$)

  (see Fig. 5.4.1, upper panel).
- The policy reserve can be calculated by an iterative application of (5.4.5): starting from $V_0 = 0$, the equation allows us to calculate $V_1, V_2, \ldots, V_m$ (with the "final" check $V_m = S$); conversely, starting from $V_m = S$, we can calculate $V_{m-1}, V_{m-2}, \ldots, V_0$ (with $V_0 = 0$).

  Alternative expressions of (5.4.5) are the following ones:

$$(V_t + P)\,(1+i') = C\,q'_{x+t} + V_{t+1}\,p'_{x+t} \quad (5.4.6)$$

$$V_t + P = (C - V_{t+1})\,(1+i')^{-1}\,q'_{x+t} + V_{t+1}\,(1+i')^{-1} \quad (5.4.7)$$

$$(V_t + P)\,(1+i') = (C - V_{t+1})\,q'_{x+t} + V_{t+1} \quad (5.4.8)$$

We note the following aspects.

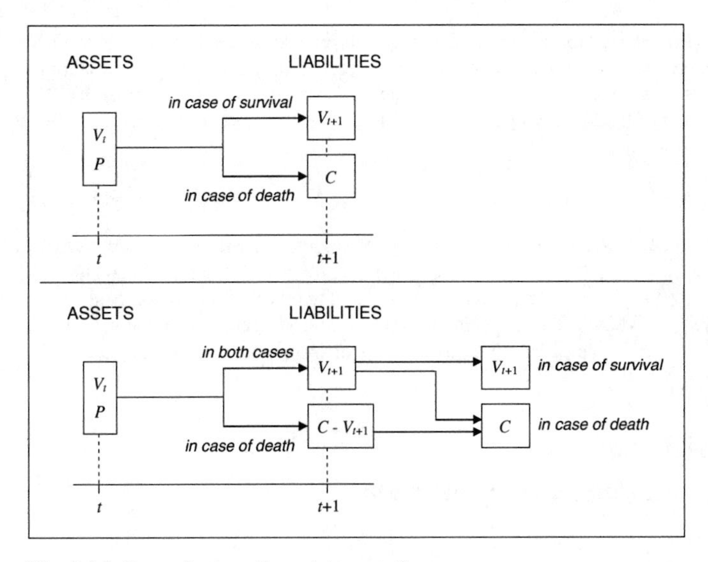

Fig. 5.4.1 Recursive equations: interpretations

- In Eq. (5.4.6) and Eq. (5.4.8) (called the *Kanner equation*, 1869), the financial evaluation is referred to time $t+1$ (whereas the probabilistic evaluation is still referred to time t).
- In Eqs. (5.4.7) and (5.4.8), the reserve V_{t+1} appears as a liability certain at time $t+1$ (that is, in both the cases of death and survival), whereas the death benefit (if any) is split into two shares,

$$C = (C - V_{t+1}) + V_{t+1} \tag{5.4.9}$$

  namely:

  – the amount $C - V_{t+1}$, which is called the *sum at risk* (or the *net amount at risk*), to stress that it is not yet available but funded (year by year) via the mutuality mechanism;
  – the amount V_{t+1}, which is not "at risk", as the reserve has to be used anyhow (sooner or later)

  (see Fig. 5.4.1, lower panel).
- In the case of no death benefit ($C = 0$), or a death benefit smaller than the reserve ($C < V_t$), the amount at risk is negative; in these cases, if the insured dies in the year, the sum at risk (the whole reserve, in the case $C = 0$) is released for mutuality, and thus contributes to financing the benefits pertaining to the policies still in-force.

Remark 1 It is worth stressing that the term "risk" is used, in this context, according to its traditional actuarial meaning, that is "risk of death". Other risk causes (e.g. the investment risk) are not involved.

Remark 2 Recursive Eqs. (5.4.5) to (5.4.8) can be easily interpreted also referring to a portfolio of policies. Let N_t denote the (given) number of policies in-force at time t, and N_{t+1} the random number of policies in-force at time $t+1$, namely the number of insureds still alive. Further, let D_t denote the random number of insureds dying in the year; thus, $D_t = N_t - N_{t+1}$. Refer, for example, to Eq. (5.4.6). We can write:

$$(V_t N_t + P N_t)(1+i') = C N_t \, q'_{x+t} + V_{t+1} N_t \, p'_{x+t} \tag{5.4.10}$$

On the left-hand side of Eq. (5.4.10) we find the amount of resources (reserves and premiums) pertaining to policies in-force at time t, cumulated to time $t+1$. As regards the right-hand side of the equation, we first note that $N_t \, p'_{x+t} = \mathbb{E}[N_{t+1}]$ is the expected number of policyholders alive at time $t+1$, whereas $N_t \, q'_{x+t} = \mathbb{E}[D_t]$ is the expected number of policyholders dying in the year. The interpretation of the right-hand side of Eq. (5.4.10) in terms of insurer's expected obligations is then straightforward.

5.4.3 Risk premium and savings premium

From Eq. (5.4.7), we obtain:

$$P = [(C - V_{t+1})\,(1+i')^{-1}\, q'_{x+t}] + [V_{t+1}\,(1+i')^{-1} - V_t] \tag{5.4.11}$$

so that the two following components of the annual premium can be recognized:

$$P_t^{[R]} = (C - V_{t+1})\,(1+i')^{-1}\, q'_{x+t} \tag{5.4.12a}$$

$$P_t^{[S]} = V_{t+1}\,(1+i')^{-1} - V_t \tag{5.4.12b}$$

The two components are called the *risk premium* and the *savings premium*, respectively.

The savings premiums maintain the reserving process. In fact, from (5.4.12b) we find:

$$V_{t+1} = (V_t + P_t^{[S]})\,(1+i') \tag{5.4.13}$$

and then:

$$V_{t+1} = P_0^{[S]}\,(1+i')^{t+1} + P_1^{[S]}\,(1+i')^{t} + \cdots + P_t^{[S]}\,(1+i') \tag{5.4.14}$$

It turns out that the policy reserve is the result of the financial accumulation of the savings premiums. Conversely, the risk premium is the premium of a one-year term insurance to cover the sum at risk.

We note that the two premium components are not necessarily both positive. In particular, if the sum at risk is negative, the risk premium is negative. See the following numerical examples for further details.

Example 5.4.1. Table 5.4.1 refers to a term insurance, with annual level premiums (denoted by P_t, as in following examples other premium arrangements will be addressed), payable for the whole policy duration. In particular, the decomposition of the annual premium into risk premium and savings premium is displayed. Further, the natural premiums and the time profiles of the reserve and the sum at risk are shown. Data are as follows: $C = 1\,000$, $x = 50$, $m = 10$, $\text{TB1} = (0.02, \text{LT1})$. It is interesting to note that the risk premiums are very close to the natural premiums, as the reserve is very small and hence the sum at risk almost coincides with the sum assured.

Table 5.4.1 Term insurance (annual level premiums)

t	P_t	$P_t^{[N]}$	$P_t^{[R]}$	$P_t^{[S]}$	V_t	$C - V_t$
0	5.40	3.31	3.31	2.09	0.00	—
1	5.40	3.68	3.66	1.74	2.14	997.86
2	5.40	4.08	4.05	1.35	3.95	996.05
3	5.40	4.52	4.49	0.91	5.40	994.60
4	5.40	5.01	4.98	0.42	6.44	993.56
5	5.40	5.56	5.52	−0.12	7.00	993.00
6	5.40	6.17	6.13	−0.73	7.01	992.99
7	5.40	6.84	6.80	−1.40	6.41	993.59
8	5.40	7.58	7.56	−2.16	5.11	994.89
9	5.40	8.41	8.41	−3.01	3.01	996.99
10	—	—	—	—	0	1 000.00

Table 5.4.2 refers to a single-premium term insurance. Clearly, $P_0 = \Pi = C\,{}_mA'_x$. Data are as above. Natural premiums coincide, of course, with those in Table 5.4.1; in fact, natural premiums only depend on the benefit structure, while they are independent of the specific premium arrangement. All the savings premiums, but the first one, are negative, and represent the "use" of the reserve in the mutuality process.

Table 5.4.2 Term insurance (single premium)

t	P_t	$P_t^{[N]}$	$P_t^{[R]}$	$P_t^{[S]}$	V_t	$C - V_t$
0	48.52	3.31	3.16	45.35	0.00	—
1	0.00	3.68	3.52	−3.52	46.26	953.74
2	0.00	4.08	3.91	−3.91	43.60	956.40
3	0.00	4.52	4.35	−4.35	40.48	959.52
4	0.00	5.01	4.85	−4.85	36.85	963.15
5	0.00	5.56	5.41	−5.41	32.64	967.36
6	0.00	6.17	6.03	−6.03	27.78	972.22
7	0.00	6.84	6.73	−6.73	22.19	977.81
8	0.00	7.58	7.52	−7.52	15.76	984.24
9	0.00	8.41	8.41	−8.41	8.41	991.59
10	—	—	—	—	0.00	1 000.00

□

Example 5.4.2. A pure endowment is referred to in Table 5.4.3. Data are as follows: $S = 1000$, $x = 50$, $m = 10$, $\text{TB1} = (0.02, \text{LT1})$. As $C = 0$, all the risk premiums are negative, and hence we find $P_t^{[S]} > P$ for all t. This means that the premium P is insufficient to maintain the reserving process, which in fact needs for the contributions provided by the reserves of the policies terminating because of the insureds' death.

Table 5.4.3 Pure endowment (annual level premiums)

t	P_t	$P_t^{[N]}$	$P_t^{[R]}$	$P_t^{[S]}$	V_t	$C - V_t$
0	86.30	0.00	−0.29	86.60	0.00	—
1	86.30	0.00	−0.66	86.96	88.33	−88.33
2	86.30	0.00	−1.11	87.41	178.80	−178.80
3	86.30	0.00	−1.66	87.96	271.53	−271.53
4	86.30	0.00	−2.33	88.63	366.68	−366.68
5	86.30	0.00	−3.14	89.45	464.42	−464.42
6	86.30	0.00	−4.12	90.43	564.95	−564.95
7	86.30	0.00	−5.30	91.61	668.48	−668.48
8	86.30	0.00	−6.72	93.02	775.29	−775.29
9	86.30	971.98	−8.41	94.71	885.68	−885.68
10	—	—	—	—	1 000.00	−1 000.00

□

Example 5.4.3. Table 5.4.4 refers to a (standard) endowment insurance. Data are as follows: $C = S = 1000$, $x = 50$, $m = 10$, $\text{TB1} = (0.02, \text{LT1})$. All the entries in the Table can be obtained as the sum of the corresponding entries in Tables 5.4.1 and 5.4.3. We note that all the risk premiums and the savings premiums are positive. This suggests to look at the endowment insurance as the combination of an m-year financial transaction and a sequence of one-year term insurances, as shown in Table 5.4.5. The interpretation is as follows. An individual, instead of purchasing a m-year endowment insurance with sum insured C, and hence paying the annual premiums P, could in each year:

- invest the amount $P_t^{[S]}$ in a fund, managed by a financial institution, and annually credited with the interest rate i';
- pay the amount $P_t^{[R]}$ to an insurer to buy a one-year term insurance for a sum assured such that the sum itself plus the balance of the fund is equal to C.

It is easy to check that, in both the case of survival and the case of death prior to maturity, the amounts paid by the individual and the benefits obtained by the beneficiaries coincide with the corresponding outflows and inflows of the endowment insurance. It is worth stressing, however, that the "equivalence" between the endowment insurance and the set of transactions described above relies on some important assumptions that, at least to some extent, are rather unrealistic. In particular, the financial transaction should guarantee a constant interest rate i', as a (traditional) endowment insurance does. As regards the one-year term insurances, the life table adopted for calculating the premiums could be changed throughout the m years, thanks to mortality improvements in the population, and hence with an advantage to the insured; on the contrary, if medical examinations are required, the death probabilities could be raised because of worsened health conditions. In conclusion, while the interpretation we have sketched is useful to understand the two-fold role of a life insurance company, it should not be meant as aiming to prove analogies among the results of different transactions.

Table 5.4.4 Endowment insurance (annual level premiums)

t	P_t	$P_t^{[N]}$	$P_t^{[R]}$	$P_t^{[S]}$	V_t	$C-V_t$
0	91.71	3.31	3.01	88.69	0.00	—
1	91.71	3.68	3.00	88.70	90.46	909.54
2	91.71	4.08	2.95	88.76	182.75	817.25
3	91.71	4.52	2.83	88.87	276.94	723.06
4	91.71	5.01	2.65	89.05	373.12	626.88
5	91.71	5.56	2.38	89.32	471.42	528.58
6	91.71	6.17	2.00	89.70	571.96	428.04
7	91.71	6.84	1.50	90.20	674.90	325.10
8	91.71	7.58	0.84	90.86	780.40	219.60
9	91.71	980.39	0.00	91.71	888.69	111.31
10	—	—	—	—	1000.00	0.00

Table 5.4.5 The endowment insurance as a combination of transactions

Year	A m-year financial transaction		A sequence of m one-year term insurances	
$(t,t+1)$	Payment (at time t)	Result (at time $t+1$)	Payment (at time t)	Result (at time $t+1$)
$(0,1)$	$P_0^{[S]}$	V_1	$P_0^{[R]}$	$C-V_1$
$(1,2)$	$P_1^{[S]}$	V_2	$P_1^{[R]}$	$C-V_2$
$(2,3)$	$P_2^{[S]}$	V_3	$P_2^{[R]}$	$C-V_3$
...	...	...	...	...
$(m-2,m-1)$	$P_{m-2}^{[S]}$	V_{m-1}	$P_{m-2}^{[R]}$	$C-V_{m-1}$
$(m-1,m)$	$P_{m-1}^{[S]}=P$	$V_m=C$	$P_{m-1}^{[R]}=0$	$C-V_m=0$

□

Example 5.4.4. Table 5.4.6 refers to a single-premium immediate life annuity (in arrears). Data are as follows: $b=100$, $x=65$, TB1 $=(0.02,\text{LT4})$. The technical structure of a life annuity requires a generalization of the recursive equations. We generalize Eq. (5.4.7) as follows:

$$V_t + P_t = -V_{t+1}\,(1+i')^{-1}\,q'_{x+t} + V_{t+1}\,(1+i')^{-1} + b\,(1+i')^{-1}\,p'_{x+t} \tag{5.4.15}$$

where $V_0 = 0$, $P_0 = \Pi = a'_x$, and $P_t = 0$, for $t = 1,2,\ldots$. From (5.4.15), after a little algebra we obtain:

$$P_t = [(V_{t+1}+b)\,(1+i')^{-1} - V_t] + [(-V_{t+1}-b)\,(1+i')^{-1}\,q'_{x+t}] \tag{5.4.16}$$

and then

$$P_t^{[R]} = (-V_{t+1}-b)\,(1+i')^{-1}\,q'_{x+t} \tag{5.4.17a}$$

$$P_t^{[S]} = (V_{t+1}+b)\,(1+i')^{-1} - V_t \tag{5.4.17b}$$

Hence, $P_t^{[R]} < 0$ for all t, and, for $t = 1,2,\ldots$, as $P_t = 0$ then $P_t^{[S]} = -P_t^{[R]} > 0$. Thus, reserves released by the annuitants dying in the various years maintain the reserves of the surviving annuitants, according to the mutuality mechanism. As regards the natural premiums, we have, for all t:

$$P_t^{[N]} = b(1+i')^{-1}\,p'_{x+t} \tag{5.4.18}$$

From Eq. (5.4.15), after a little algebra, we also obtain:

$$b = \Big((V_t+P_t)(1+i') - V_{t+1}\Big) + \Big(-P_t^{[R]}\,(1+i')\Big) \tag{5.4.19}$$

Equation (5.4.19) shows that part of the annual benefit is financed by the policy reserve, whereas the remaining part is financed by a share of the reserves released, namely via the mutuality effect. Table 5.4.6 shows that the mutuality effect becomes more and more important as t increases, clearly because of an increasing mortality among annuitants.

□

5.4.4 Life insurance products versus financial accumulation

Consider the whole life insurance, financed via single recurrent premiums, and assume $i' = 0$ (see Sect. 4.4.5). As already noted, according to this arrangement no mortality risk is borne by the insurer. The formal proof is straightforward. In the case of death in year t, the sum paid to the beneficiary is $C_t = \sum_{h=0}^{t-1} \Pi_h$ (see Eq. (4.4.32)); the reserve at time t is $V_t = \sum_{h=0}^{t-1} \Pi_h$ (see Eq. (5.3.13)). Thus, $C_t = V_t$, and hence the sum at risk is equal to zero.

In general, any product in which the death benefit coincides with the policy reserve is just a financial accumulation product. In fact, from

Table 5.4.6 Life annuity in arrears (single premium)

t	P_t	$P_t^{[N]}$	$P_t^{[R]}$	$P_t^{[S]}$	V_t	$b = 100$	
						$(V_t + P_t)(1+i') - V_{t+1}$	$-P_t^{[R]}(1+i')$
0	1706.88	97.48	−9.81	1716.69	—	90.00	10.00
1	0.00	97.41	−10.72	10.72	1651.02	89.06	10.94
2	0.00	97.33	−11.70	11.70	1594.97	88.07	11.93
3	0.00	97.23	−12.76	12.76	1538.81	86.99	13.01
4	0.00	97.13	−13.89	13.89	1482.60	85.83	14.17
5	0.00	97.01	−15.11	15.11	1426.43	84.60	15.40
...	...	...	...	...	...	...	...
10	0.00	96.16	−22.43	22.43	1149.01	77.12	22.88
11	0.00	95.92	−24.16	24.16	1094.87	75.36	24.64
12	0.00	95.65	−25.97	25.97	1041.41	73.51	26.49
13	0.00	95.35	−27.87	27.87	988.73	71.57	28.43
14	0.00	95.01	−29.85	29.85	936.93	69.56	30.44
15	0.00	94.63	−31.90	31.90	886.11	67.45	32.55
...	...	...	...	...	...	...	...
20	0.00	91.93	−43.17	43.17	650.12	55.97	44.03
21	0.00	91.20	−45.57	45.57	607.16	53.52	46.48
22	0.00	90.37	−47.99	47.99	565.78	51.05	48.95
23	0.00	89.46	−50.43	50.43	526.04	48.56	51.44
24	0.00	88.45	−52.88	52.88	488.01	46.07	53.93
25	0.00	87.34	−55.32	55.32	451.70	43.57	56.43
...	...	...	...	...	...	...	...

$$C_t = V_t \tag{5.4.20}$$

it follows:

$$P_t^{[R]} = 0 \tag{5.4.21}$$

and hence

$$P_t^{[S]} = P \tag{5.4.22}$$

so that the reserve coincides with the accumulation of the premiums P (or P_t, or Π_t).

A financial accumulation product can be transformed into a "real" insurance product via redefinition of the death benefit C_t, which can be expressed as a function of the reserve V_t, such that the following inequality holds:

$$C_t > V_t \tag{5.4.23}$$

(instead of (5.4.20)). This transformation can be mandatory because of regulation, or can be useful for tax purposes, etc. Some examples follow; in the related figures, the dashed line represents the policy reserve.

1. Choose the amount K, and set:

$$C_t = V_t + K \tag{5.4.24}$$

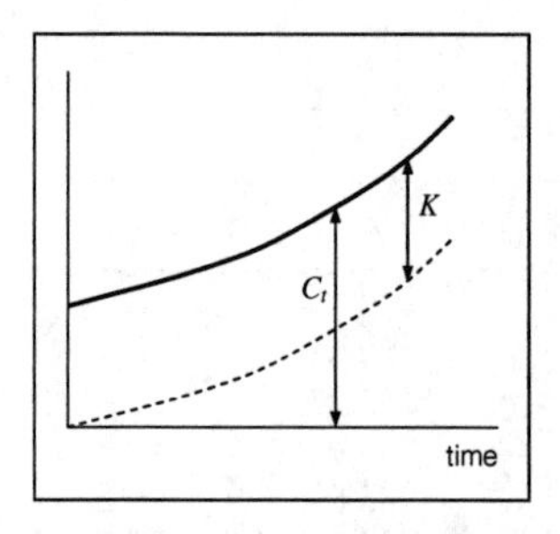

Fig. 5.4.2 Constant sum at risk

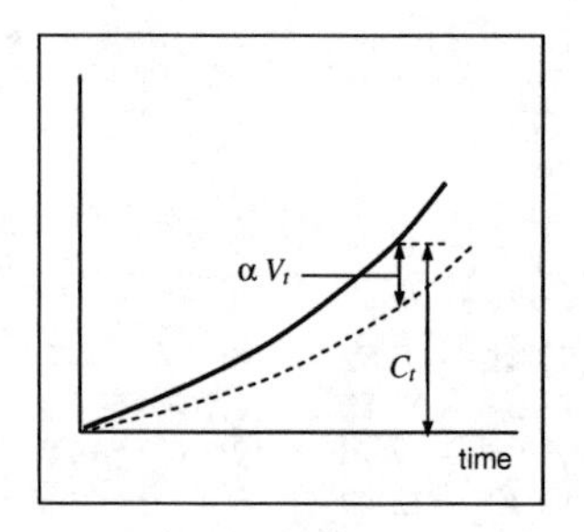

Fig. 5.4.3 Proportional sum at risk

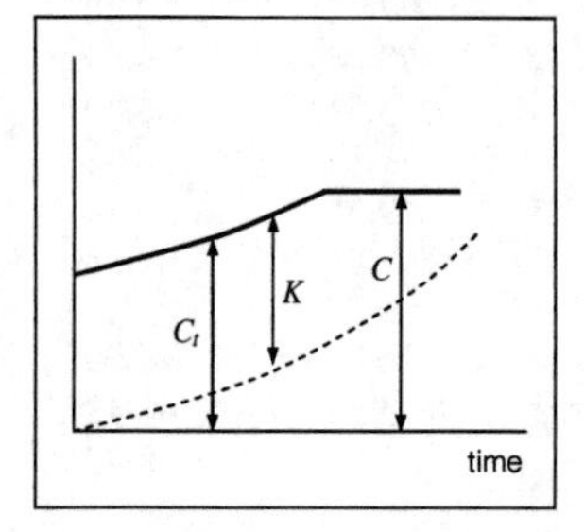

Fig. 5.4.4 Sum at risk with an upper bound

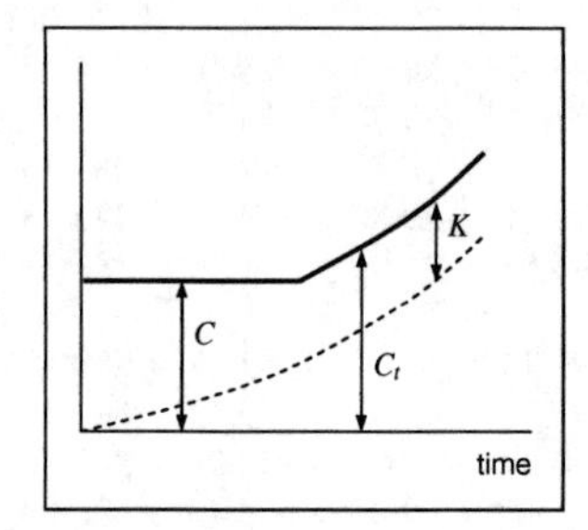

Fig. 5.4.5 Sum at risk with a lower bound

Thus, the sum at risk is K; see Fig. 5.4.2.

2. Choose the rate α, and set:

$$C_t = (1+\alpha)V_t \tag{5.4.25}$$

Thus, the sum at risk is αV_t; see Fig. 5.4.3.

3. Choose the amounts K and C, and set:

$$C_t = \min\{V_t + K, C\} \tag{5.4.26}$$

Thus, the sum at risk is $\min\{K, C - V_t\}$; see Fig. 5.4.4.

4. Choose the amounts K and C, and set:

$$C_t = \max\{V_t + K, C\} \tag{5.4.27}$$

Thus, the sum at risk is $\max\{K, C - V_t\}$; see Fig. 5.4.5.

Remark We note that in case 3 above it may turn out $C_t < V_t$, namely if V_t increases above C. Given the purposes of the policy design, it is not acceptable that $C_t < V_t$. Thus, the death benefit

$$C_t = \max\{\min\{V_t + K, C\}, V_t\} \tag{5.4.28}$$

should rather be considered instead of (5.4.26).

5.5 Expected profits: a further insight

The approach to the profit assessment we have described in Sect. 4.3.8 simply relies on a comparison between actuarial values of benefits, namely between the actuarial value calculated by adopting the scenario basis, i.e. TB2, and the actuarial value assumed as the single premium, hence calculated by adopting the prudential basis, i.e. TB1.

A deeper analysis of expected profits requires further steps. In particular:

1. premium arrangements other than that based on a single premium must be allowed for;
2. as life insurance contracts usually have a multi-year duration, it can be useful to attribute a share of the (total) expected profit to each policy year; hence, annual profits are defined, showing the profit emerging throughout the policy duration;
3. further elements, which can constitute sources of profit / loss should be taken into account, and typically
   - expenses and expense loadings;
   - lapses, surrenders, and policy alterations.

Issues 1 and 2 are dealt with in the present Section; indeed, the mathematical reserve provides a tool for a "natural" definition of expected annual profits. Conversely, topic 3 will be discussed in Chap. 6, in the framework of a life portfolio analysis.

5.5.1 Expected annual profits

We refer to Eq. (5.4.6), which can also be written as follows:

$$(V_t + P)\,(1+i') - C\,q'_{x+t} - V_{t+1}\,p'_{x+t} = 0 \tag{5.5.1}$$

Equation (5.5.1) relates to policy year $(t, t+1)$, and expresses a balance between resources (the reserve at the beginning of the year and the premium) and expected obligations (the sum in the case of death and the reserve at the end of the year). The balance relies on the adoption of the same technical basis, namely the first-order basis TB1, in all the elements of Eq. (5.5.1), and this, in its turns, follows from the assumptions adopted in defining the policy reserve (see Sect. 5.3.1).

Conversely, assume that:

- a realistic estimate of the yield from the investment of the amount $V_t + P$ is expressed by the interest rate i'';
- the mortality in the portfolio can be described in realistic terms by probabilities q''_{x+t}.

Thus, the scenario basis TB2 can be introduced into Eq. (5.5.1). The shift to TB2 clearly results in a different meaning of some quantities. Actually, we obtain:

$$(V_t + P)(1 + i'') - C q''_{x+t} - V_{t+1}\, p''_{x+t} = \overline{PL}_{t+1} \tag{5.5.2}$$

where $\overline{PL}_{t+1}$ ($\gtreqless 0$) denotes the *expected annual profit / loss* arising from the spread between TB1 and TB2. We note that $\overline{PL}_{t+1}$ is referred to time $t+1$, for a policy assumed to be in-force at time t.

Remark Equation (5.5.2) can be easily interpreted also referring to a portfolio of policies. A similar interpretation has been provided for Eq. (5.4.6) (see Remark 2 in Sect. 5.4.2). Let N_t denote the (given) number of policies in-force at time t, and N_{t+1} the random number of policies in-force at time $t+1$, namely the number of insureds still alive. Then, we can write:

$$N_t V_t + N_t P + (N_t V_t + N_t P)\, i'' - C N_t\, q''_{x+t} - V_{t+1} N_t\, p''_{x+t} = N_t\, \overline{PL}_{t+1} \tag{5.5.3}$$

All quantities can be interpreted as in Eq. (5.4.10). In particular: $N_t\, p''_{x+t} = \mathbb{E}[N_{t+1}]$, $N_t\, q''_{x+t} = \mathbb{E}[D_t]$. Note, however, that the expected numbers are now calculated according to TB2. In Eq. (5.5.3) we can recognize some of the (main) items of the *Profit & Loss Statement* (briefly *P & L*). In general, the P & L Statement refers to a specific period (say, a year), and indicates how the profit / loss originates from income net of expenditure. As we are only addressing one generation of policies, and we are disregarding expenses and related loadings as well as lapses and surrenders, the resulting representation is extremely simplified (see Table 5.5.1). Further, an obvious adjustment in the benefits is needed when referring to the last year of the policy duration. Note that the items classified as "Expenditure" in the P & L Statement enter Eq. (5.5.3) as negative terms.

Table 5.5.1 Actuarial values as items of a P & L statement

P & L STATEMENT	
Income	
Premiums	$P N_t$
Income from investments	$(V_t N_t + P N_t)\, i''$
Expenditure	
Benefits paid	$C\, \mathbb{E}[D_t]$
Change in liabilities	$V_{t+1}\, \mathbb{E}[N_{t+1}] - V_t N_t$
Profit	$\overline{PL}_{t+1}$

5.5.2 Splitting the annual profit

We now refer to Eq. (5.4.8), which can also be written as follows:

$$(V_t + P)(1 + i') - (C - V_{t+1})\, q'_{x+t} - V_{t+1} = 0 \tag{5.5.4}$$

Adopting the scenario basis TB2, as in Eq. (5.5.2), we have:

$$(V_t + P)(1 + i'') - (C - V_{t+1}) q''_{x+t} - V_{t+1} = \overline{PL}_{t+1} \tag{5.5.5}$$

Then, by subtracting (5.5.4) from (5.5.5), we obtain the so-called *contribution formula* (proposed by S. Homans, 1863):

$$(V_t + P)(i'' - i') + (C - V_{t+1})(q'_{x+t} - q''_{x+t}) = \overline{PL}_{t+1} \tag{5.5.6}$$

which suggests the splitting of the expected annual profit into two terms:

$$\overline{PL}_{t+1}^{[\text{fin}]} = (V_t + P)(i'' - i') \tag{5.5.7a}$$

$$\overline{PL}_{t+1}^{[\text{mort}]} = (C - V_{t+1})(q'_{x+t} - q''_{x+t}) \tag{5.5.7b}$$

The quantity $\overline{PL}_{t+1}^{[\text{fin}]}$ is the *financial margin*, namely the component of the expected annual profit originated by the spread between the interest rates, $i'' - i'$. Clearly, as $V_t + P > 0$, the financial margin is positive if and only if $i'' > i'$.

The component $\overline{PL}_{t+1}^{[\text{mort}]}$ is the *mortality / longevity margin*, which arises from the spread between the mortality rates. We note that

- if $C - V_{t+1} > 0$, the mortality / longevity margin is positive if and only if $q'_{x+t} > q''_{x+t}$;
- if $C - V_{t+1} < 0$, the mortality / longevity margin is positive if and only if $q'_{x+t} < q''_{x+t}$.

Thus, the sign of the sum at risk is the driving factor in the choice of the life table to be adopted in the first-order basis, TB1, in order to obtain implicit safety loadings, and hence positive expected profits. For pricing insurance products with a positive sum at risk (for example: the term insurance, the whole life insurance, the endowment insurance) a life table with a mortality higher than that actually expected in the portfolio should be chosen. On the contrary, products with a negative sum at risk (the pure endowment and the life annuities) require a mortality assumption lower than the mortality actually expected.

Example 5.5.1. Table 5.5.2 refers to a term insurance. Policy data are as follows: $C = 1\,000$, $x = 40$, $m = 10$; annual level premiums, P, are payable throughout the whole policy duration. The pricing basis is $\text{TB1} = (0.02, \text{LT1})$; we then find: $P = 1.93$. Expected profits are calculated by adopting the second-order basis $\text{TB2} = (0.03, \text{LT2})$. We note that the poor financial content of the term insurance implies very low financial profits, whereas more important contributions to the expected profits come from the mortality assumptions.

Table 5.5.3 refers to an endowment insurance. Policy data are as follows: $C = 1\,000$, $x = 50$, $m = 15$. Annual level premiums, P, are payable throughout the whole policy duration. The pricing basis is $\text{TB1} = (0.02, \text{LT1})$; we then find: $P = 59.54$. Expected profits are calculated by adopting the second-order basis $\text{TB2} = (0.03, \text{LT2})$. Unlike the term insurance, the endowment insurance has important financial contents, so that the spread between interest rates originates significant contributions to the expected profits. On the contrary, mortality profits are low, and definitely decreasing as the sum at risk decreases. As we will see in Sect. 7.3, the financial profit is shared with policyholders, through an adjustment of benefits.

Table 5.5.2 Term insurance: expected profits

t	V_t	$\overline{PL}_t$	$\overline{PL}_t^{[\text{fin}]}$	$\overline{PL}_t^{[\text{mort}]}$
0	0.00	–	–	–
1	0.76	0.14	0.02	0.12
2	1.40	0.16	0.03	0.13
3	1.92	0.18	0.03	0.15
4	2.29	0.20	0.04	0.16
5	2.48	0.22	0.04	0.18
6	2.49	0.24	0.04	0.20
7	2.27	0.27	0.04	0.22
8	1.81	0.29	0.04	0.25
9	1.06	0.31	0.04	0.27
10	0.00	0.33	0.03	0.30

Table 5.5.3 Endowment insurance: expected profits

t	V_t	$\overline{PL}_t$	$\overline{PL}_t^{[\text{fin}]}$	$\overline{PL}_t^{[\text{mort}]}$
0	0.00	–	–	–
1	57.54	0.91	0.60	0.32
2	116.11	1.50	1.17	0.33
3	175.74	2.10	1.76	0.34
4	236.46	2.70	2.35	0.35
5	298.33	3.32	2.96	0.36
6	361.40	3.94	3.58	0.36
7	425.75	4.57	4.21	0.36
8	491.45	5.20	4.85	0.35
9	558.59	5.85	5.51	0.34
10	627.30	6.50	6.18	0.32
11	697.70	7.15	6.87	0.28
12	769.96	7.81	7.57	0.24
13	844.26	8.47	8.29	0.18
14	920.85	9.14	9.04	0.10
15	1000.00	9.80	9.80	0.00

□

5.5.3 The expected total profit

The sequence of expected profits / losses $\overline{PL}_1, \overline{PL}_2, \ldots, \overline{PL}_m$, which are originated yearly by the policy, can be interpreted as a temporary life annuity. The expected present value of this annuity, $\overline{PL}$, according to the scenario basis TB2, is given by:

$$\overline{PL} = \sum_{t=0}^{m-1} \overline{PL}_{t+1} \, (1+i'')^{-(t+1)} \, {}_t p''_x \tag{5.5.8}$$

which can be interpreted as the expected value of the total profit / loss, expressed as a present value at time 0.

It is possible to check that, assuming $V_0 = 0$ and $V_m = S$, and plugging Eq. (5.5.2) into (5.5.8), we find the following expression:

$$\overline{PL} = \sum_{t=0}^{m-1} P\,(1+i'')^{-t}\,{}_tp''_x - \sum_{t=0}^{m-1} C\,(1+i'')^{-(t+1)}\,{}_{t|1}q''_x - S\,(1+i'')^{-m}\,{}_mp''_x \qquad (5.5.9)$$

in which the policy reserve does not appear. We note that the result expressed by (5.5.9) holds thanks to the use of the TB2 for discounting the expected annual profits.

Equation (5.5.9) can also be written as follows:

$$\begin{aligned}\overline{PL} = & \sum_{t=0}^{m-2} {}_tp''_x\,(1+i'')^{-(t+1)}\left[P\,(1+i'') - C\,q''_{x+t}\right] + \\ & {}_{m-1}p''_x\,(1+i'')^{-m}\left[P\,(1+i'') - C\,q''_{x+m-1} - S\,p''_{x+m-1}\right]\end{aligned} \qquad (5.5.10)$$

The quantities in brackets, namely

$$\overline{CF}_{t+1} = P\,(1+i'') - C\,q''_{x+t}; \quad t = 0, 1, \ldots, m-2 \qquad (5.5.11a)$$

$$\overline{CF}_m = P\,(1+i'') - C\,q''_{x+m-1} - S\,p''_{x+m-1} \qquad (5.5.11b)$$

represent the *expected annual cash-flows*, referred to a policy in-force at time t or $m-1$ respectively, each cash-flow being cumulated at the end of the relevant year.

Thus, the expected total profit is the expected present value of the life annuity which consists of the expected annual cash-flows. In formal terms:

$$\overline{PL} = \sum_{t=0}^{m-1} \overline{CF}_{t+1}\,(1+i'')^{-(t+1)}\,{}_tp''_x \qquad (5.5.12)$$

Hence, the reserve profile affects the expected annual profits and then the emergence of profit throughout time, i.e. the *timing of the profit*, while it does not affect the total amount of the expected profit.

Example 5.5.2. We refer to the insurance products addressed in Example 5.5.1. We find

- for the term insurance: $\overline{PL} = 1.93$;
- for the endowment insurance: $\overline{PL} = 55.90$.

□

As regards the effect of the reserve on the emerging of expected profits, the following example can help in understanding this aspect.

Example 5.5.3. Refer to an endowment insurance with annual premiums payable for the whole policy duration. Data are as follows: $C = 1000$, $x = 50$, $m = 15$;

$TB1 = (0.02, LT1)$, $TB2 = (0.03, LT2)$. Figure 5.5.1 displays the policy reserves calculated with the interest rates 0, 0.02 (namely i'), and 0.04; possible negative values have been replaced by 0.

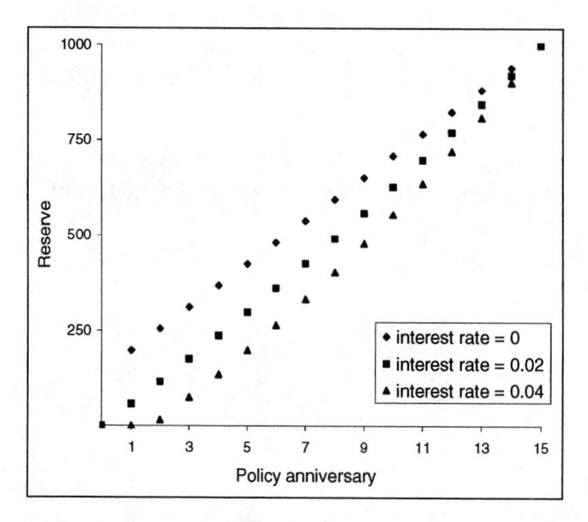

Fig. 5.5.1 Endowment insurance (annual level premiums)

Table 5.5.4 shows the annual profits corresponding to the three reserve profiles. It clearly emerges that high reserve values (compared to those obtained using the interest rate i') imply a heavy expected loss in the first year, which is recovered by positive expected profits in the following years. Conversely, low reserve values lead to an accelerated emerging of expected profits, compensated by expected losses in the following years.

□

Some results, which emerge from Example 5.5.3, can be generalized. In particular, it can be proved that

- a lower interest rate adopted in the reserve calculation implies higher reserve values, and hence a "delay" in profit emerging;
- a higher interest rate adopted in the reserve calculation implies lower reserve values, and hence an "acceleration" in profit emerging.

5.5.4 Cash-flows, profits, premium margins

By comparing Eqs. (5.5.11) to Eq. (5.5.2), we find (as $V_m = S$) the following relations:

Table 5.5.4 Endowment insurance (annual level premiums)

t	$V_t^{(0.00)}$	$\overline{PL}_t^{(0.00)}$	$V_t^{(0.02)}$	$\overline{PL}_t^{(0.02)}$	$V_t^{(0.04)}$	$\overline{PL}_t^{(0.04)}$
0	0.00	—	0.00	—	0.00	—
1	198.08	−139.20	57.54	0.91	0.00	58.28
2	254.82	8.01	116.11	1.50	16.10	41.90
3	311.50	9.72	175.74	2.10	74.82	−0.37
4	368.13	11.42	236.46	2.70	135.75	−0.95
5	424.72	13.12	298.33	3.32	199.00	−1.55
6	481.32	14.82	361.40	3.94	264.71	−2.17
7	537.95	16.51	425.75	4.57	333.02	−2.83
8	594.66	18.21	491.45	5.20	404.11	−3.51
9	651.51	19.89	558.59	5.85	478.16	−4.24
10	708.55	21.58	627.30	6.50	555.39	−5.00
11	765.86	23.26	697.70	7.15	636.07	−5.81
12	823.54	24.95	769.96	7.81	720.48	−6.66
13	881.69	26.63	844.26	8.47	808.99	−7.58
14	940.46	28.31	920.85	9.14	902.00	−8.56
15	1000.00	30.00	1000.00	9.80	1000.00	−9.62

$$\overline{PL}_{t+1} = \overline{CF}_{t+1} + V_t\,(1+i'') - V_{t+1}\,p''_{x+t};\quad t = 0,1,\ldots,m-2 \tag{5.5.13a}$$

$$\overline{PL}_m = \overline{CF}_m + V_{m-1}\,(1+i'') \tag{5.5.13b}$$

In respect of the annual profit / loss, the role of the policy reserve, and its change in particular, then consists in attributing shares of premiums to policy years, so shifting from "cash-based" valuations (the $\overline{CF}$'s) to "pertinence-based" valuations (the $\overline{PL}$'s).

Example 5.5.4. Profit profile and cash-flow profile are compared in Tables 5.5.5 and 5.5.6, which refer to a term insurance and an endowment insurance respectively. Policy data and technical bases TB1 and TB2 are as in Example 5.5.1.

Table 5.5.5 Term insurance (annual level premiums)

t	$\overline{PL}_t$	$\overline{CF}_t$	$(P-P'')(1+i'')$
1	0.14	0.90	0.23
2	0.16	0.78	0.23
3	0.18	0.65	0.23
4	0.20	0.51	0.23
5	0.22	0.35	0.23
6	0.24	0.17	0.23
7	0.27	−0.03	0.23
8	0.29	−0.25	0.23
9	0.31	−0.49	0.23
10	0.33	−0.76	0.23

Table 5.5.6 Endowment insurance (annual level premiums)

t	$\overline{PL}_t$	$\overline{CF}_t$	$(P-P'')(1+i'')$
1	0.91	58.28	4.84
2	1.50	57.95	4.84
3	2.10	57.58	4.84
4	2.70	57.17	4.84
5	3.32	56.72	4.84
6	3.94	56.22	4.84
7	4.57	55.66	4.84
8	5.20	55.04	4.84
9	5.85	54.36	4.84
10	6.50	53.60	4.84
11	7.15	52.76	4.84
12	7.81	51.82	4.84
13	8.47	50.79	4.84
14	9.14	49.65	4.84
15	9.80	−938.67	4.84

□

Of course, different time profiles of the reserve lead to different premium attributions and hence, as shown in Example 5.5.3, to different profit profiles. A very particular reserve profile and the related profit profile will be presented in Sect. 5.5.5.

Moreover, specific profit profiles can be generated by adopting a different approach to profit assessment. An interesting approach is described in what follows.

Refer to Eq. (5.5.9), and set:

$$\ddot{a}''_{x:\overline{m}|} = \sum_{t=0}^{m-1} (1+i'')^{-t}\, {}_tp''_x \tag{5.5.14}$$

$$ {}_mA''_x = \sum_{t=0}^{m-1} (1+i'')^{-(t+1)}\, {}_{t|1}q''_x \tag{5.5.15}$$

$$ {}_mE''_x = (1+i'')^{-m}\, {}_mp''_x \tag{5.5.16}$$

The expected total profit can be expressed as follows:

$$\overline{PL} = P\,\ddot{a}''_{x:\overline{m}|} - C\,{}_mA''_x - S\,{}_mE''_x \tag{5.5.17}$$

Let P'' denote the "second-order premium", namely the annual level premium calculated by adopting the scenario basis TB2, such that:

$$P''\,\ddot{a}''_{x:\overline{m}|} = C\,{}_mA''_x + S\,{}_mE''_x \tag{5.5.18}$$

The expected total profit / loss can then be expressed as the actuarial value of the temporary life annuity whose items are the annual *premium margins*:

$$\overline{PL} = (P - P'')\,\ddot{a}''_{x:\overline{m}|} \tag{5.5.19}$$

The following aspects should be stressed.

- The result expressed by Eq. (5.5.19) is extremely intuitive: indeed, the expected total profit / loss is due to the spread between the premium charged to the policyholder (P) and the premium fulfilling the equivalence principle under realistic assumptions (P''), which clearly leads to a zero expected profit.
- Equation (5.5.19) generalizes to the case of annual premiums the result expressed by Eq. (4.3.22) for the single premium arrangement.
- According to Eq. (5.5.19), we could assume as the expected annual profit the amount

$$\overline{PL}_t = (P - P'')\,(1 + i''); \quad t = 1, 2, \dots, m \tag{5.5.20}$$

  so originating a flat profit profile. Note, however, that this can lead to a significant acceleration in the emerging of profits (see Example 5.5.5).

Example 5.5.5. From tables 5.5.5 and 5.5.6, which refer to a term insurance and an endowment insurance respectively, it clearly appears that, in both the insurance products, the assumption (5.5.20) leads to a significant acceleration in the profit profile.
□

5.5.5 Expected profits according to best-estimate reserving

Consider the expected present value of future benefits net of future premiums, according to the scenario basis TB2, that is, in formal terms:

$$V_t^{[\mathrm{BE}]} = C\,{}_{m-t}A''_{x+t} + S\,{}_{m-t}E''_{x+t} - P\,\ddot{a}''_{x+t:\overline{m-t}|} \tag{5.5.21}$$

The quantity $V_t^{[\mathrm{BE}]}$ is usually called the *best-estimate reserve*.

In particular, we have:

$$V_0^{[\mathrm{BE}]} = C\,{}_{m}A''_{x} + S\,{}_{m}E''_{x} - P\,\ddot{a}''_{x:\overline{m}|} \tag{5.5.22}$$

and hence (see Eqs. (5.5.18) and (5.5.19)):

$$V_0^{[\mathrm{BE}]} = (P'' - P)\,\ddot{a}''_{x:\overline{m}|} = -\overline{PL} \tag{5.5.23}$$

Thus, the quantity $-V_0^{[\mathrm{BE}]} = \overline{PL}$ represents the "value" of the policy (at the time of its issue), meant as the expected present value of profits / losses originated by the policy itself throughout its duration.

Assume now, for the policy reserve V_t, the following values:

$$V_0 = 0; \quad V_t = V_t^{[\mathrm{BE}]},\ \text{for } t = 1, 2, \dots, m-1; \quad V_m = S \tag{5.5.24}$$

By using Eq. (5.5.2), with the reserves as defined by (5.5.24), after a little algebra we obtain the following results:

$$\overline{PL}_1 = -V_0^{[BE]}(1+i'') \tag{5.5.25a}$$

$$\overline{PL}_t = 0; \quad t = 2,\ldots,m \tag{5.5.25b}$$

Thus, the expected total profit / loss completely emerges in the first policy year.

Remark The particular profit profile originated by the best-estimate reserve witnesses the existence of two basic approaches to profit emerging. The *Deferral & Matching* approach is a traditional feature of actuarial models. The basic idea underlying this approach is that the total profit arises progressively throughout time. The profit assessment procedure basically consists of two steps:

- assessment of annual results (typically: cash-flows and profits);
- calculation of the total profit as the expected present value of annual results.

The *Assets & Liabilities* approach is a feature of financial models. The profit assessment procedure basically consists of two steps:

- the total profit is given by the difference between the value of assets (e.g the single premium, or the credit for future periodic premiums) and the value of liabilities (the insurer's obligations);
- possible annual profits are only given by changes in the values of assets and liabilities.

5.6 Reserving for expenses

Equation (5.3.3) defines the "net reserve", in which benefits and net premiums are only involved. We can extend the definition, and thus define the "total reserve", in which expenses and loading for expenses are also included:

$$V_t^{[\text{tot}]} = \text{Ben}'(t,m) + \text{Exp}'(t,m) - \text{Prem}'(t,m) - \text{Load}'(t,m) \tag{5.6.1}$$

where $\text{Exp}'(t,m)$ and $\text{Load}'(t,m)$ represent the actuarial values at time t of future expenses and expense loadings, respectively, calculated according to the first-order basis. It turns out that $V_t^{[\text{tot}]}$ can be determined including the future expenses in the insurer's liabilities and allowing for the expense-loaded premiums instead of the net premiums.

Of course, we also have

$$V_t^{[\text{tot}]} = V_t + V_t^{[E]} \tag{5.6.2}$$

where $V_t^{[E]} = \text{Exp}'(t,m) - \text{Load}'(t,m)$ just allows for expenses and expense loadings.

Notwithstanding, it is much more useful to deal separately with the various expense components and the related loadings. First, we note that the need for reserving arises because of a time-mismatching between the insurer's inflow and outflow streams. So, as regards expenses and related loadings we can exclude premium collection expenses, as these are supposed to occur at the same time the relevant loading is cashed.

Acquisition costs can also be excluded from further analysis in the case of a single premium. Conversely, in the case of periodic premiums payable for s years, we can define the (negative) *acquisition cost reserve*, which in fact represents the insurer's credit for the related loadings to be cashed in future years:

$$V_t^{[A]} = \begin{cases} -\Lambda^{[A]} \ddot{a}'_{x+t:s-t\rceil} & \text{for } t \le s-1 \\ 0 & \text{for } t \ge s \end{cases} \tag{5.6.3}$$

The quantity

$$V_t^{[Z]} = V_t + V_t^{[A]} \tag{5.6.4}$$

is called the *Zillmer reserve*. In general, we have $V_t^{[Z]} \le V_t$, and in particular, in the first policy years, we may find $V_t^{[Z]} < 0$. We note that the Zillmer reserve implies a "clearing" between insurer's credit and debt, and, (also) for this reason, in many countries zillmerization is not allowed when assessing the balance-sheet portfolio reserve.

General administration expenses do not originate any reserve if the premiums are payable for the whole policy duration, that is if $s = m$. On the contrary, if $s < m$ the *reserve for general administration expenses* is defined as follows:

$$V_t^{[G]} = \begin{cases} \gamma C \ddot{a}'_{x+t:m-t\rceil} - \Lambda^{[G]} \ddot{a}'_{x+t:s-t\rceil} & \text{for } t \le s-1 \\ \gamma C \ddot{a}'_{x+t:m-t\rceil} & \text{for } t \ge s \end{cases} \tag{5.6.5}$$

In the case of a single premium we have:

$$V_t^{[G]} = \gamma C \ddot{a}'_{x+t:m-t\rceil} \tag{5.6.6}$$

In some countries (in particular in Continental Europe), it is usual to define the following reserve:

$$V_t^{[I]} = V_t + V_t^{[G]} \tag{5.6.7}$$

which is called in Germany the *Inventardeckungscapital*.

It is easy to prove that the reserve, $V_t^{[\mathrm{tot}]}$, allowing for all the expenses and the related loadings, as well as for benefits and net premiums, can be expressed as follows:

$$V_t^{[\mathrm{tot}]} = V_t + V_t^{[A]} + V_t^{[G]} \tag{5.6.8}$$

Example 5.6.1. We refer to the insurance products and the related data considered in Example 4.5.1. Tables 5.6.1 and 5.6.2 display the various reserves allowing for expenses and related loadings.

□

Table 5.6.1 Whole life insurance (level premiums; $s = 15$)

t	V_t	$V_t^{[A]}$	$V_t^{[G]}$	$V_t^{[Z]}$	$V_t^{[I]}$	$V_t^{[tot]}$
0	0.00	0.00	0.00	0.00	0.00	0.00
1	42.57	−18.85	0.76	23.72	43.33	24.48
2	85.79	−17.68	1.55	68.11	87.34	69.66
3	129.69	−16.49	2.35	113.20	132.04	115.55
4	174.28	−15.27	3.17	159.01	177.45	162.18
5	219.57	−14.03	4.02	205.54	223.59	209.56
...	...	...	...	...	...	...
12	559.31	−4.60	10.74	554.71	570.05	565.45
13	611.76	−3.11	11.86	608.64	623.62	620.50
14	665.46	−1.58	13.02	663.88	678.49	676.90
15	720.56	0.00	14.25	720.56	734.81	734.81
16	730.68	0.00	13.74	730.68	744.42	744.42
...	...	...	...	...	...	...
24	807.47	0.00	9.82	807.47	817.29	817.29
25	816.33	0.00	9.37	816.33	825.70	825.70
...	...	...	...	...	...	...

Table 5.6.2 Endowment insurance (level premiums; $s = m = 15$)

t	V_t	$V_t^{[A]}$	$V_t^{[G]}$	$V_t^{[Z]}$	$V_t^{[I]}$	$V_t^{[tot]}$
0	0.00	0.00	0	0.00	0.00	0.00
1	57.54	−34.52	0	23.02	57.54	23.02
2	116.11	−32.38	0	83.73	116.11	83.73
3	175.74	−30.19	0	145.54	175.74	145.54
4	236.46	−27.97	0	208.49	236.46	208.49
5	298.33	−25.70	0	272.63	298.33	272.63
...	...	...	...	...	...	...
12	769.96	−8.43	0	761.53	769.96	761.53
13	844.26	−5.70	0	838.56	844.26	838.56
14	920.85	−2.90	0	917.95	920.85	917.95
15	1000.00	0.00	0	1000.00	1000.00	1000.00

5.7 Surrender values and paid-up values

As mentioned in Sect. 4.1.2, the calculation of surrender values and paid-up values should account for the policyholder's credit at the time of the contract alteration. The net policyholder's credit (that is, the amount which allows for benefits, expenses and expense-loaded premiums) is given by the reserve $V_t^{[tot]}$, defined by (5.6.8). As this reserve coincides in many cases with the Zillmer reserve $V_t^{[Z]}$ (see, for instance, Table 5.6.2 in Example 5.6.1), we just focus on the Zillmer reserve.

The *surrender value*, denoted as R_t, can be determined as follows:

$$R_t = \varphi(t)\, V_t^{[Z]} \tag{5.7.1}$$

Note that the function $\varphi(t)$ ($0 \le \varphi(t) \le 1$, and usually equal to 0 for $t = 1,2$ only), aims at penalizing the surrendering policyholders. Commonly, the penalty decreases as t increases, and to this purpose the function should be increasing. The penalty can be justified as follows:

- from a legal point of view, the policyholder breaks the contract;
- from an economic point of view, the insurer can recover, via the penalty, future profits expected from the contract.

Other formulae are also commonly adopted in insurance practice. For endowment insurance products, with maturity at time m and annual level premiums for $s = m$ years, the so-called *proportional rule* is frequently adopted. If C denotes the sum insured, we have:

$$R_t = \frac{t}{m}\, C\,(1+i^*)^{-(m-t)} \tag{5.7.2}$$

Thus, a share of the sum insured, proportional to the number of annual premiums already paid, is discounted at a rate i^*, higher than the interest rate in the technical basis. Formula (5.7.2) can be justified looking at the time profile of the policy reserve in an endowment insurance, which is very close to a linear profile (see, for example, Fig. 5.3.14). The discounting rate i^* can be used as a parameter to allow for both zillmerisation and penalty.

To illustrate the *reduction* of the sum insured when converting an insurance contract into a paid-up one, we refer to a m-year pure endowment, with sum insured S and annual level premiums payable for the whole policy duration.

Assume that the policyholder asks for the reduction at time t, namely after paying the annual premiums at times $0, 1, \dots, t-1$. A share of the Zillmer reserve, $V_t^{[Z]}$, is then used (as a "single" premium) to finance the paid-up contract, namely the reduced benefit at maturity, $S^{[\mathrm{red}]}$, and the general administration expenses (quantified as results from (4.5.6a)) for the residual duration.

In formal terms, $S^{[\mathrm{red}]}$ is the solution of the following equation:

$$\bar{\varphi}(t)\, V_t^{[Z]} = S^{[\mathrm{red}]} \left({}_{m-t}E'_{x+t} + \gamma\, \ddot{a}'_{x+t:\overline{m-t|}} \right) \tag{5.7.3}$$

The function $\bar{\varphi}(t)$ ($0 < \bar{\varphi}(t) \le 1$) determines a penalty charged to the policyholder when shifting to the paid-up contract, and can be justified similarly to the surrender penalty (see above). However, as the contract goes on, we usually have $\bar{\varphi}(t) \ge \varphi(t)$.

Formula (5.7.3) relies on the equivalence principle, and hence leads to a result consistent with this actuarial calculation principle. Nonetheless, other (approximate) formulae are often adopted in insurance practice. For example, according to the *proportional rule*, in an endowment insurance with maturity at time m and annual level premiums for $s = m$ years, the amount $S^{[\mathrm{red}]}$ can be determined as follows:

$$S^{[\mathrm{red}]} = \frac{t}{m}\, S \tag{5.7.4}$$

5.8 References and suggestions for further reading

All the actuarial textbooks on life insurance deal with the calculation of reserves. Hence, the reader can refer to [10], [20], [25], [26], [47], and [49].

The traditional approach to the profit assessment at the policy level is proposed by [47], whereas [26] places special emphasis on mortality profits.

Chapter 6
Reserves and profits in a life insurance portfolio

6.1 The portfolio reserve

When shifting from individual reserves to the portfolio reserve, various specific problems arise, although many basic ideas about the individual reserving process keep their validity.

In particular, as in the individual case, the portfolio reserve can be looked at under two different perspectives:

- an amount which quantifies the expected insurer's liability for future benefits, net of future premiums;
- assets, provided by the accumulation of (part of) the premiums, facing the liability mentioned above.

The current reserve of an in-force portfolio can be calculated as the sum of the individual policy reserves. In particular, when referring to a portfolio which consists of a generation of identical policies (as assumed for simplicity in the following), the portfolio reserve is determined by the individual reserve and the size of the portfolio itself. When focussing on the evolution of a portfolio, its estimated size must be taken into account.

Further, the portfolio riskiness (due to random fluctuations in mortality, in interest rates, and so on) can be of interest, and hence the portfolio reserve could be assessed explicitly allowing for risks (rather than simply relying on a generic safety loading).

In this Section we address the following topics:

- the evaluation of future portfolio reserves, starting from the individual (net premium) reserve, as defined in Sect. 5.3, thus using the same technical basis, that is, the first-order basis (Sect. 6.1.1);
- the definition of the portfolio reserve by adopting a different approach to the assessment of the insurer's obligations, namely allowing for the riskiness inherent in the liability, although, for simplicity, we will only focus on the mortality risk (Sects. 6.1.2 to 6.1.5).

A. Olivieri, E. Pitacco, *Introduction to Insurance Mathematics*,
DOI 10.1007/978-3-642-16029-5_6,

6.1.1 Future portfolio reserves

We refer to the insurance product we have addressed while describing the risk and savings components of the life insurance business (for the relevant notation, and the expression of the annual level premium, see Sect. 5.4.1). We disregard expenses and related loadings.

We focus on a portfolio initially consisting of N_0 "identical" policies (with N_0 a given number). At (future) time t, the individual reserve is equal to V_t for each policy still in-force. Assume that the portfolio is closed with respect to new policies (namely, it is a "generation" of policies), and let N_t denote the random number of policies in-force at time t. Hence, the *portfolio reserve* at time t is represented by the random amount $V_t^{[P]}$ defined as follows:

$$V_t^{[P]} = N_t V_t; \quad t = 1, 2, \dots \tag{6.1.1}$$

Future portfolio reserves can be assessed by assuming a sequence, $n_1, n_2, \dots$, of outcomes of the random numbers $N_1, N_2, \dots$. For any given sequence, the estimated portfolio reserve is given by:

$$\hat{V}_t^{[P]} = n_t V_t; \quad t = 1, 2, \dots \tag{6.1.2}$$

In particular, assume that the only cause of exit is the insured's death. Then, we can set, for example:

$$n_t = \mathbb{E}[N_t] = N_0 \, {}_t p_x''; \quad t = 1, 2, \dots \tag{6.1.3}$$

where ${}_t p_x''$ denotes the probability of an insured age x being alive at age $x+t$, according to a second-order basis, namely a realistic basis. In this case, we obtain:

$$\hat{V}_t^{[P]} = \mathbb{E}[V_t^{[P]}] = \mathbb{E}[N_t] \, V_t; \quad t = 1, 2, \dots \tag{6.1.4}$$

Thus, $\hat{V}_t^{[P]}$ is the *expected portfolio reserve* (according to information available at time 0).

Example 6.1.1. Refer to an endowment insurance. Data are as follows: $S = C = 1000$, $x = 50$, $m = 15$, $\text{TB1} = (0.02, \text{LT1})$. The initial portfolio size is $N_0 = 1\,000$. Table 6.1.1 shows the expected numbers of policies in-force and the expected portfolio reserve, according to probabilities ${}_t p_x''$ derived from life tables LT2 and LT3 respectively.

□

Table 6.1.1 The expected portfolio reserve

		LT2		LT3	
t	V_t	$\mathbb{E}[N_t]$	$\mathbb{E}[V_t^{[P]}]$	$\mathbb{E}[N_t]$	$\mathbb{E}[V_t^{[P]}]$
0	0.00	1000.00	0.00	1000.00	0.00
1	57.54	996.96	57368.75	997.29	57388.07
2	116.11	993.59	115366.58	994.30	115448.58
3	175.74	989.87	173955.34	990.98	174151.18
4	236.46	985.76	233091.59	987.32	233461.39
5	298.33	981.22	292726.25	983.27	293340.38
6	361.40	976.20	352804.19	978.81	353744.67
7	425.75	970.67	413263.88	973.87	414625.97
8	491.45	964.57	474036.92	968.43	475930.96
9	558.59	957.85	535047.83	962.42	537601.18
10	627.30	950.45	596213.64	955.80	599572.90
11	697.70	942.31	657443.73	948.52	661777.20
12	769.96	933.35	718639.69	940.50	724140.06
13	844.26	923.52	779695.29	931.68	786582.62
14	920.85	912.74	840496.65	921.99	849021.63
15	1000.00	900.92	900922.57	911.37	911370.10

6.1.2 Safe-side reserve versus best-estimate reserve

The traditional approach to reserving (in most countries of Continental Europe) relies on the adoption of the first-order basis in discounting future benefits and premiums (see Sect. 5.3.1). Hence, the (individual) reserve constitutes a prudential (or "safe-side") evaluation of the insurer's liability. However, the "degree" of prudence cannot be easily determined. We also recall that, in the case of significant changes in the scenario, a consequent shift to a new reserving basis is required, as described in Sect. 5.3.4.

A different approach to reserving, which explicitly allows for risks and for a chosen prudence target, can be defined. In what follows, we refer to a term insurance (namely, with $S = 0$), with annual level premiums P payable throughout the whole policy duration, and, in particular:

- we disregard expenses and expense loadings;
- we focus on the mortality risk only, thus disregarding investment risks, lapses, and so on.

Although these assumptions lead to a very simplified setting, many important features of the different approach to reserving can be captured.

The (traditional) prospective reserve at time t, for the insurance product we are dealing with, is given by:

$$V_t = C\,{}_{m-t}A'_{x+t} - P\,\ddot{a}'_{x+t:m-t\rceil} \tag{6.1.5}$$

If we assume, for discounting future benefits and future premiums, the second-order (or realistic) basis, we obtain the "best estimate" assessment of the policy reserve, shortly the (individual) *best-estimate reserve* (or *central-estimate reserve*):

$$V_t^{[\mathrm{BE}]} = C\,{}_{m-t}A''_{x+t} - P\,\ddot{a}''_{x+t:m-t\rceil} \tag{6.1.6}$$

(see also Sect. 5.5.5).

As the first-order basis relies on a mortality higher than that included in the second-order basis, we have:

$${}_{m-t}A''_{x+t} < {}_{m-t}A'_{x+t} \tag{6.1.7a}$$

$$a''_{x+t:m-t\rceil} > a'_{x+t:m-t\rceil} \tag{6.1.7b}$$

and hence

$$V_t^{[\mathrm{BE}]} < V_t \tag{6.1.8}$$

The difference $V_t - V_t^{[\mathrm{BE}]}$ represents the safety margin implied by the adoption of the first-order basis in the assessment of the policy reserve V_t. In particular, adverse fluctuations in mortality can be faced thanks to this margin.

6.1.3 The risk margin

Moving from a single policy to a portfolio, in particular a generation of "identical" policies, we assume that, at time t, the portfolio consists of N_t policies (with N_t a given number). The traditional reserve is then given by $N_t\,V_t$ (see Sect. 6.1.1), and the best-estimate reserve by $N_t\,V_t^{[\mathrm{BE}]}$.

As the safety margin aims at facing the portfolio riskiness, it can also be denoted as the *risk margin*. Thus, the risk margin at the portfolio level, RM_t, is given by:

$$RM_t = N_t\,(V_t - V_t^{[\mathrm{BE}]}) \tag{6.1.9}$$

However, a sound approach to the management of the insurer's risks requires an appropriate quantification of the relevant impact on portfolio results. This means that, rather than starting from a generic prudential assessment of the reserve and then finding the resulting risk margin according to (6.1.9), the risk margin should be determined depending on the insurer's risk profile, quantified by a convenient risk measure.

We refer to the following random quantities, all defined for the portfolio:

- N_{t+h}, namely the random number of policies still in-force at time $t+h$; then, for $h = 0, 1, \ldots, m-t-1$, $D_{t+h} = N_{t+h} - N_{t+h+1}$ denotes the random number of insureds dying between time $t+h$ and $t+h+1$;
- $Y^{[\mathrm{P}]}(t,m)$, defined as the random present value at time t of future benefits, that is

$$Y^{[P]}(t,m) = C \sum_{h=0}^{m-t-1} (1+i'')^{-(h+1)} D_{t+h} \tag{6.1.10}$$

- $X^{[P]}(t,m)$, defined as the random present value at time t of future premiums, that is

$$X^{[P]}(t,m) = P \sum_{h=0}^{m-t-1} (1+i'')^{-h} N_{t+h} \tag{6.1.11}$$

- $Z^{[P]}(t,m)$, defined as the random present value at time t of the portfolio result over the residual portfolio duration (see below for the formal definition).

To define $Z^{[P]}(t,m)$, we note what follows. It can be shown, with a little algebra, that, if we calculate the expected values of $Y^{[P]}(t,m)$ and $X^{[P]}(t,m)$ according to the second-order basis, we obtain:

$$\mathbb{E}[Y^{[P]}(t,m)] - \mathbb{E}[X^{[P]}(t,m)] = N_t \left[C_{m-t}A''_{x+t} - P \ddot{a}''_{x+t:m-t\rceil} \right] = N_t V_t^{[BE]} \tag{6.1.12}$$

namely, the best-estimate portfolio reserve. We assume that an amount of assets equal to $N_t V_t^{[BE]}$ is available at time t, so that the reserve plus the future premiums meet the future benefits. Hence, we define the random present value of the portfolio result as follows:

$$Z^{[P]}(t,m) = N_t V_t^{[BE]} + X^{[P]}(t,m) - Y^{[P]}(t,m) \tag{6.1.13}$$

From (6.1.12), we find that, according to the realistic basis, $\mathbb{E}[Z^{[P]}(t,m)] = 0$.

If only the amount $N_t V_t^{[BE]}$ is available to meet future benefits net of future premiums, the probability of a negative result is very high. Thus, a further amount should be available. Appropriate risk measures can help in determining this amount. Of course, risk measures should rely on the probability distribution of $Z^{[P]}(t,m)$, which can be estimated via stochastic simulation (see Sect. 3.10.3). Once the function $\Phi_{Z^{[P]}(t,m)}$, defined as

$$\Phi_{Z^{[P]}(t,m)}(z) = \mathbb{P}[Z^{[P]}(t,m) \le z] \tag{6.1.14}$$

has been constructed, we can assume, for example, the VaR at a given confidence level as the risk measure (see Sect. 1.5.4). Thus, if $1-\alpha$ denotes the confidence level, we can set:

$$RM_t = -VaR_\alpha \tag{6.1.15}$$

If we assume that the amount RM_t "belongs" to the portfolio (and, actually, it should be financed, at least to some extent, by the safety loadings embedded in the premiums already cashed), the random present value of the portfolio result can be redefined as follows:

$$Z^{[P][RM]}(t,m) = Z^{[P]}(t,m) + RM_t = N_t V_t^{[BE]} + RM_t + X^{[P]}(t,m) - Y^{[P]}(t,m) \tag{6.1.16}$$

Figure 6.1.1 shows a graph of the probability distribution of $Z^{[P]}(t,m)$ (assumed to be continuous, for simplicity), whereas Fig. 6.1.2 refers to the distribution of $Z^{[P][RM]}(t,m)$.

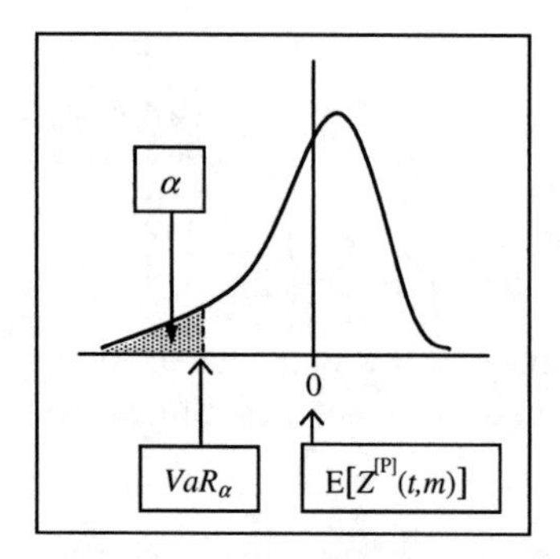

Fig. 6.1.1 The prob. distribution of $Z^{[P]}(t,m)$

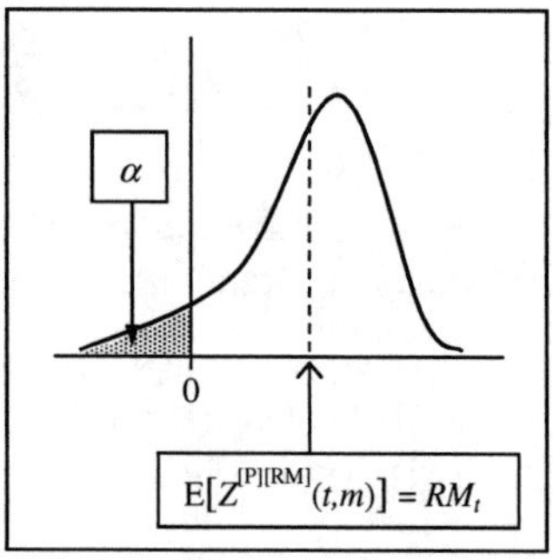

Fig. 6.1.2 The prob. distribution of $Z^{[P][RM]}(t,m)$

Example 6.1.2. We refer to a portfolio of 10-year term insurances, with $C = 1\,000$ and annual premium $P = 1.93$ payable for the whole policy duration. Age at entry is $x = 40$. The first-order basis is $\mathrm{TB1} = (0.02, \mathrm{LT1})$, whereas the realistic basis is $\mathrm{TB2} = (0.03, \mathrm{LT2})$. Table 6.1.2 shows the safe-side reserve, the best-estimate reserve, and the reserve including the risk margin at time $t = 3$, for three portfolio sizes. The following aspects should be stressed:

- the lower is α, the higher is the risk margin;
- for any given α, the risk margin depends on the portfolio size: small portfolios require a risk margin very high in relative terms, as can be seen comparing $N_3\,V_3^{[BE]} + RM_3$ to $N_3\,V_3^{[BE]}$ (which is obviously proportional to the portfolio size).

Table 6.1.2 Safe-side reserve, best-estimate reserve and reserve including risk margin

N_3	$N_3\,V_3$	$N_3\,V_3^{[BE]}$	$V_3^{[P]} = N_3\,V_3^{[BE]} + RM_3$		
			$\alpha = 0.25$	$\alpha = 0.10$	$\alpha = 0.05$
100	192.00	29.03	550.05	1418.41	2173.86
1000	1920.05	290.32	2282.10	4382.54	5835.17
10000	19200.45	2903.22	7295.59	13519.05	17494.89

□

Remark Various approaches to the assessment of the risk margin can be proposed, and have been actually adopted in the insurance technique. Although approaches based on appropriate risk measures (like the VaR and the TailVaR) are rigorous, other procedures can simplify the assessment of the risk margin. This is the case, for example, of the calculation procedure proposed within the project "Solvency 2". See also Sect. 6.1.5.

6.1.4 The portfolio liability and beyond

Assume that the portfolio reserve, $V_t^{[P]}$, is calculated as the best-estimate reserve plus the risk margin. Further, assume that the risk margin is given by the VaR at a stated confidence level, as described in Sect. 6.1.3; clearly, it depends on the portfolio size N_t. In formal terms, we have:

$$V_t^{[P]} = N_t V_t^{[BE]} + RM_t(N_t) \tag{6.1.17}$$

for $t = 0, 1, 2, \ldots$.

As in Sect. 6.1.1, refer to a portfolio initially consisting of N_0 "identical" policies (with N_0 a given number). At (future) time t, the portfolio reserve, defined by (6.1.17), is a random amount, as the portfolio size N_t is a random number. For any given sequence $n_1, n_2, \ldots$ of numbers of policies in-force, we obtain the estimated future portfolio reserve:

$$\hat{V}_t^{[P]} = n_t V_t^{[BE]} + RM_t(n_t) \tag{6.1.18}$$

Example 6.1.3. We refer to the portfolio described in Example 6.1.2. We assume that the portfolio initially consists of $N_0 = 1000$ policies; we set $n_t = \mathbb{E}[N_t]$, for $t = 1, 2, \ldots, 10$, according to the life table LT2. We assume $\alpha = 0.25$. Table 6.1.3 shows the best-estimate reserve and the portfolio reserve $\hat{V}_t^{[P]}$ including the risk margin. Note that the negative values of $V_1^{[BE]}$ and $V_2^{[BE]}$ have been replaced by 0.

Table 6.1.3 Best-estimate reserve and reserve including risk margin

t	n_t	$V_t^{[BE]}$	$n_t V_t^{[BE]}$	$\hat{V}_t^{[P]}$
0	1000.00	0.00	0.00	0.00
1	998.91	0.00	0.00	1235.15
2	997.71	0.00	0.00	1801.19
3	996.38	0.29	289.27	2269.33
4	994.90	0.81	801.55	2068.91
5	993.27	1.18	1169.16	1816.03
6	991.47	1.38	1370.71	1641.64
7	989.47	1.40	1382.41	1566.16
8	987.26	1.19	1177.86	1732.03
9	984.82	0.74	727.85	1258.37
10	982.11	0.00	0.00	0.00

□

Portfolio liabilities are counterbalanced by assets. If the assets have to be assessed at their market value, assumed as the "true" or "fair" value, also the related liabilities should be assessed, for consistency, at market value. Thus, the so-called *mark-to-market* approach to liability assessment should be adopted. However, a problem arises: is a (reliable) market value of liabilities available ?

As insurer's liabilities are only traded in markets which cannot provide a reliable fair value (for example, the reinsurance market), the application of the mark-to-market approach is restricted to liabilities which can be perfectly hedged by assets traded on appropriate markets. This is the case, in particular, of the liabilities related to unit-linked insurance products (see Chap. 7).

Conversely, Eqs. (6.1.17) and (6.1.18) implement the so-called *mark-to-model* approach to the assessment of the portfolio liabilities. This approach relies on an actuarial model whose output should provide a reasonably fair value of the liabilities.

More assets than those just backing the fair value of the liabilities are usually needed to face risks. To this purpose, shareholders' capital must be allocated and assigned to the portfolio. The amount to be allocated to a portfolio (and, more in general, to a life insurance business) is determined according to a stated solvency target. Thus, the total amount of assets backing the insurer's liabilities (assessed in terms of fair value) and the shareholders' capital must fulfill the *adequacy requirement*, as stated by the supervisory authorities, or by the company management, if the latter results in an amount higher than that required by the authorities.

The shareholders' capital needed to fulfill the adequacy requirement can be called the *required capital*, whereas the (possibly) remaining shareholders' capital constitutes the *excess capital* (see Fig. 6.1.3).

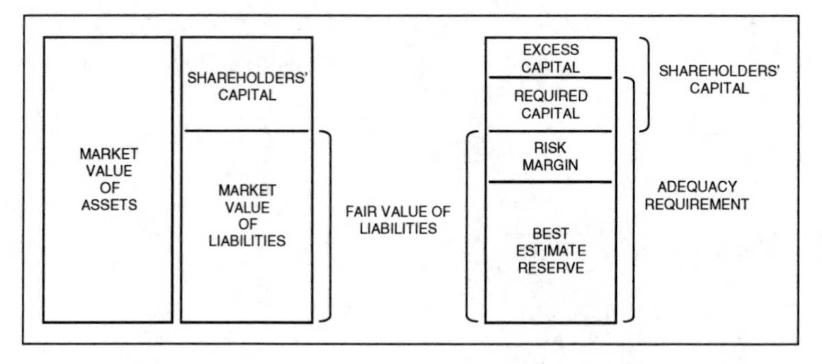

Fig. 6.1.3 Assets, liabilities and shareholders' capital (1)

6.1.5 Risk margin: the "Cost of Capital" approach

We describe an approach to the calculation of the risk margin, which constitutes a practicable alternative to the VaR approach.

We define the *target capital* at time t, TC_t, as the amount of assets net of liabilities, which is required (in particular, by the supervisory authority) for solvency purposes; assets are assessed at their market value, liabilities at their best-estimate value. We assume that the target capital consists of two components, namely the risk margin, RM_t, and the *solvency capital requirement* , SCR_t (see Fig. 6.1.4). Thus:

$$TC_t = RM_t + SCR_t \tag{6.1.19}$$

Hence, the adequacy requirement defined in Sect. 6.1.4 is fulfilled by: (1) the best-estimate reserve, (2) the risk margin, and (3) the solvency capital requirement.

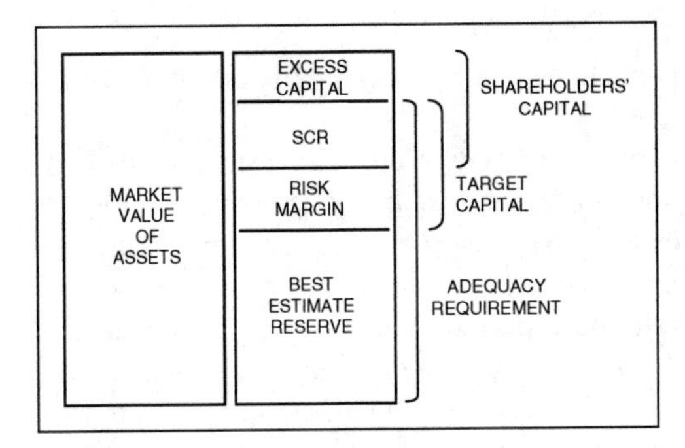

Fig. 6.1.4 Assets, liabilities and shareholders' capital (2)

Further, we assume that SCR_t can be determined (at least approximately) by adopting a given formula. The risk margin, RM_t, is then defined as the cost of the solvency capital which is required for the run-off of the portfolio in the case of insurer's default at the end of the current year. Hence:

- the risk margin makes possible the run-off of the portfolio after default;
- without risk margin, no other insurer would be available to be charged with the portfolio itself;
- the risk margin "belongs" to the policyholders, because in the case of default it must be transferred together with the portfolio; thus, it is not a part of the shareholders' capital.

To illustrate the procedure for calculating the target capital, we refer to a portfolio of identical policies, with total duration m years. We denote by ρ the risk discount rate; of course, $\rho > r_{\text{f}}$, where r_{f} is the risk-free rate. We assume that:

- ρ is the return required on the shareholders' capital;
- the capital allocated to the portfolio is invested at the risk-free rate.

Hence, the spread $\rho - r_{\text{f}}$ represents the cost of capital not covered by the investment yield.

The risk margin at time t is formally defined as follows:

$$RM_t = (\rho - r_f)\left(\widehat{SCR}_{t+1}(1+r_f)^{-1} + \widehat{SCR}_{t+2}(1+r_f)^{-2} + \cdots + \widehat{SCR}_{m-1}(1+r_f)^{-m+t+1}\right) \tag{6.1.20}$$

where $\widehat{SCR}_{t+h}$ denotes an estimate, at time t, of the solvency capital requirement at time $t+h$. Such an estimate should be based on the projection of the quantities involved in the calculation of the solvency capital requirement at time t (or in the approximation adopted).

Finally, the target capital at time t is given by Eq. (6.1.19).

6.2 The total profit

The assessment of expected profits constitutes one of the most important topics in life insurance mathematics.

A first insight into the assessment of expected profits has been provided in Sect. 4.3.8, just comparing actuarial values of benefits at the policy issue, calculated by adopting a pricing basis and a scenario basis, respectively. A further insight has been given in Sect. 5.5. By using recurrent equations of the policy reserve, expected annual profits have been defined. Then, the expected total profit has been defined, in terms of either the expected annual profits or the expected annual cash-flows.

Several items must still be added in order to get to a more complete setting. In particular, we have to account for the insurer's expenses and the related premium loadings, as well as policyholders' lapses and surrenders.

Moreover, by referring to a portfolio of policies, and allowing for the portfolio reserve, we can define a more natural framework, closer to the valuation needs which emerge in the insurance practice. Although this extended framework makes it possible to perform risk analysis, for brevity we will deal with expected values only.

We focus on the insurance product we have addressed while describing the risk and savings components of the life insurance business (for the relevant notation, and the expression of the net annual level premium, see Sect. 5.4.1). To start, we disregard expenses and related loadings, as well as lapses and surrenders. These items will be included into the model in Sect. 6.4.

6.2.1 The life fund

We refer to a portfolio of "identical" policies, issued at time $t = 0$. The only causes of exit are the death of the insured and the maturity. We assume that the portfolio consists of a generation of policies, hence closed to new entries.

We define the following quantities:

- N_0, the (given) initial number of policies in the portfolio;
- N_t, the random number of policies in the portfolio at time t, $t = 1, 2, \ldots, m$;

- $D_t = N_t - N_{t+1}$, the random number of insureds dying between time t and $t+1$;
- $F_t^{[P]}$, the amount of the *portfolio fund* (or *life fund*) at time t, $t = 0, 1, \ldots, m$.

We assume

$$F_0^{[P]} = 0 \tag{6.2.1}$$

Then, according to information available at time 0, the behavior of the portfolio fund is described by the following recursive relations:

$$(F_t^{[P]} + PN_t)(1+i'') - CD_t = F_{t+1}^{[P]}; \quad t = 0, 1, \ldots, m-2 \tag{6.2.2a}$$

$$(F_{m-1}^{[P]} + PN_{m-1})(1+i'') - CD_{m-1} - SN_m = F_m^{[P]} \tag{6.2.2b}$$

where i'' denotes the (realistic) estimated yield from the fund investment.

We note that:

- the only cause of randomness we are allowing for is the mortality in the portfolio; random yields could be introduced into the model, but this would result in a much higher complexity;
- the random amount $F_t^{[P]}$ (if positive) represents the portfolio assets cumulated up to time t, excluding allocations of shareholders' capital, and release of profits as well;
- the random amount $F_t^{[P]}$ might take negative values, for example because of an unexpected high mortality; in this case, money should be borrowed to reinstate the fund;
- the assumption $F_0^{[P]} = 0$ is rather unrealistic (although it makes some interpretations much easier); an initial allocation of assets to the portfolio lowers, of course, the probability of negative values for $F_t^{[P]}$ (see, for example, Sect. 2.7.3);
- the final value of the fund, $F_m^{[P]}$, is net of the benefits paid to the insureds alive at maturity (see Eq. (6.2.2b)); hence, it represents the *random total profit* cumulated at the end of the portfolio duration.

Remark The adjective "total" is herein used to denote the profit related to the whole duration of the portfolio.

6.2.2 The expected life fund and the expected total profit

To simplify the notation, we denote the expected values as follows: $\bar{N}_t = \mathbb{E}[N_t]$, $\bar{F}_t^{[P]} = \mathbb{E}[F_t^{[P]}]$, and so on.

The calculation of expected values of the life fund relies on mortality assumptions. To this purpose, we adopt a realistic life table, which constitutes, together with the interest rate i'', the scenario basis, namely TB2.

We note that (according to information available at time 0)

$$\bar{N}_t = N_0\,{}_t p''_x \tag{6.2.3}$$
$$\bar{D}_t = N_0\,{}_{t|1} q''_x \tag{6.2.4}$$

with ${}_t p''_x$ and ${}_{t|1} q''_x$ taken from the realistic life table. Given assumption (6.2.1) and recursive relations (6.2.2), we find:

$$(\bar{F}_t^{[P]} + P\,\bar{N}_t)(1+i'') - C\,\bar{D}_t = \bar{F}_{t+1}^{[P]}; \quad t = 0,1,\dots,m-2 \tag{6.2.5a}$$
$$(\bar{F}_{m-1}^{[P]} + P\,\bar{N}_{m-1})(1+i'') - C\,\bar{D}_{m-1} - S\,\bar{N}_m = \bar{F}_m^{[P]} \tag{6.2.5b}$$

The amount $\bar{F}_m^{[P]}$ is the *expected total profit*, cumulated at the end of the portfolio duration. Indeed, at time m the portfolio is no longer uncumbered with any obligation.

Remark It is worth noting that the relationships among the random quantities ($F_t^{[P]}$, N_t, D_t) are linear (see Eqs. (6.2.2)). Hence, the same linear relations link the relevant expected values (see Eqs. (6.2.5)). This feature also regards future developments, and allows us to express relationships directly in terms of expected values.

By solving Eqs. (6.2.5), we obtain the following explicit expression for the expected life fund:

$$\bar{F}_{t+1}^{[P]} = \sum_{h=0}^{t} \left(P\,\bar{N}_h(1+i'') - C\,\bar{D}_h\right)(1+i'')^{t-h}; \quad t = 0,1,\dots,m-2 \tag{6.2.6a}$$
$$\bar{F}_m^{[P]} = \sum_{h=0}^{m-1} \left(P\,\bar{N}_h(1+i'') - C\,\bar{D}_h\right)(1+i'')^{m-(h+1)} - S\,\bar{N}_m \tag{6.2.6b}$$

The following quantities

$$\overline{CF}_{h+1}^{[P]} = P\,\bar{N}_h(1+i'') - C\,\bar{D}_h; \quad h = 0,1,\dots,m-2 \tag{6.2.7a}$$
$$\overline{CF}_m^{[P]} = P\,\bar{N}_{m-1}(1+i'') - C\,\bar{D}_{m-1} - S\,\bar{N}_m \tag{6.2.7b}$$

are the values, at the end of the relevant year, of the expected annual cash-flows. Thus, the expected fund $\bar{F}_{t+1}^{[P]}$ is the accumulated value of all the expected annual cash-flows up to time $t+1$. In particular, $\bar{F}_m^{[P]}$, namely the expected value of the total profit cumulated at maturity, is the accumulated value of all the expected annual cash-flows.

Example 6.2.1. Table 6.2.1 refers to a portfolio of term insurance policies; according to the notation defined in Sect. 5.4.1, we then have $S = 0$, $C > 0$. Data are as follows: $N_0 = 10\,000$, $C = 1\,000$, $x = 40$, $m = 10$. Annual level premiums, P, are payable throughout the whole policy duration. The pricing basis is TB1 $= (0.02, \mathrm{LT1})$. We then find: $P = 1.93$. Expected values are calculated by adopting the scenario basis TB2 $= (0.03, \mathrm{LT2})$.

We note that the expected annual cash-flows cannot be interpreted as expected annual profits. Although the premium inflow is, in the first years, higher than the

Table 6.2.1 The life fund in a term insurance portfolio

t	$\bar{F}_t^{[P]}$	$\bar{N}_t$	$P\,\bar{N}_t$	$C\,\bar{D}_t$	$\overline{CF}_{t+1}^{[P]}$
0	0.00	10000.00	19265.25	10879.26	8963.95
1	8963.95	9989.12	19244.29	12035.23	7786.38
2	17019.25	9977.09	19221.10	13316.31	6481.43
3	24011.26	9963.77	19195.45	14735.32	5036.00
4	29767.59	9949.03	19167.06	16306.38	3435.69
5	34096.31	9932.73	19135.65	18045.01	1664.71
6	36783.91	9914.68	19100.88	19968.17	−294.26
7	37593.17	9894.71	19062.41	22094.41	−2460.13
8	36260.84	9872.62	19019.85	24443.96	−4853.51
9	32495.15	9848.18	18972.76	27038.79	−7496.85
10	25973.16	9821.14			

benefit outflow, a share of this difference must be reserved to meet the benefit outflow in the last years, when premiums are no longer sufficient.

Table 6.2.2 refers to a portfolio of (standard) endowment insurances; according to the notation defined in Sect. 5.4.1, we then have $S = C$. Data are as follows: $N_0 = 10000$, $C = 1000$, $x = 50$, $m = 15$. Annual level premiums, P, are payable throughout the whole policy duration. The pricing basis is $\text{TB1} = (0.02, \text{LT1})$. We then find: $P = 59.54$. Expected values are calculated by adopting the scenario basis $\text{TB2} = (0.03, \text{LT2})$.

Table 6.2.2 The life fund in an endowment insurance portfolio

t	$\bar{F}_t^{[P]}$	$\bar{N}_t$	$P\,\bar{N}_t$	$C\,\bar{D}_t$	$\overline{CF}_{t+1}^{[P]}$
0	0.00	10000.00	593576.97	30447.33	582804.13
1	582804.13	9969.55	593576.97	33663.70	577720.57
2	1178008.83	9935.89	591572.66	37208.60	572111.24
3	1785460.33	9898.68	589357.30	41112.36	565925.66
4	2404949.80	9857.57	586909.51	45407.36	559109.43
5	3036207.73	9812.16	584206.01	50128.03	551604.16
6	3678898.12	9762.03	581221.43	55310.67	543347.41
7	4332612.47	9706.72	577928.29	60993.35	534272.79
8	4996863.64	9645.73	574296.81	67215.59	524310.13
9	5671079.68	9578.51	570294.86	81441.57	501422.98
10	6354597.82	9504.50	561038.96	98315.09	474064.76
13	8452858.09	9235.21	549855.02	107842.36	458508.31
14	9164952.14	9127.37	543434.19	9127368.69	−8567631.47
15	872269.24	9009.23			

It is apparent that, also in this case, the expected cash-flows cannot be interpreted as expected annual profits. Indeed, a significant share of the difference between each annual premium inflow and death benefit outflow must be reserved to meet the survival benefit outflow at maturity.

□

6.2.3 The total profit: an alternative interpretation

Let $\overline{PL}^{[P]}$ denote the present value at time 0 of the expected total profit (or loss), that is

$$\overline{PL}^{[P]} = \bar{F}_m^{[P]} (1+i'')^{-m} \tag{6.2.8}$$

From (6.2.6b) we obtain:

$$\overline{PL}^{[P]} = \sum_{h=0}^{m-1} P \bar{N}_h (1+i'')^{-h} - \sum_{h=0}^{m-1} C \bar{D}_h (1+i'')^{-(h+1)} - S \bar{N}_m (1+i'')^{-m} \tag{6.2.9}$$

Let P'' denote the second-order premium, namely the premium such that, at the policy level, we have

$$P'' \ddot{a}''_{x:m\rceil} = C\, {}_mA''_x + S\, {}_mE''_x \tag{6.2.10}$$

that is:

$$\sum_{h=0}^{m-1} P''\, {}_hp''_x (1+i'')^{-h} = \sum_{h=0}^{m-1} C\, {}_{h|1}q''_x (1+i'')^{-(h+1)} + S\, {}_mp''_x (1+i'')^{-m} \tag{6.2.11}$$

At the portfolio level, Eq. (6.2.11) yields:

$$\sum_{h=0}^{m-1} P'' \bar{N}_h (1+i'')^{-h} = \sum_{h=0}^{m-1} C \bar{D}_h (1+i'')^{-(h+1)} + S \bar{N}_m (1+i'')^{-m} \tag{6.2.12}$$

Looking at Eq. (6.2.9), we finally obtain the following relations:

$$\overline{PL}^{[P]} = \sum_{h=0}^{m-1} (P-P'') \bar{N}_h (1+i'')^{-h} \tag{6.2.13}$$

$$\bar{F}_m^{[P]} = \sum_{h=0}^{m-1} (P-P'') \bar{N}_h (1+i'')^{m-h} \tag{6.2.14}$$

The quantity $(P-P'')\bar{N}_h$ represents the expected annual premium margin at the portfolio level. Thus, the expected total profit cumulated at maturity, which coincides with the expected life fund $\bar{F}_m^{[P]}$, originates from the expected premium margins, as is rather intuitive.

Remark We note that a result quite similar to that expressed by Eq. (6.2.13) has been obtained at the policy level; see Eq. (5.5.19).

Example 6.2.2. We refer to a term insurance portfolio and an endowment insurance portfolio. Data are as in Example 6.2.1. Table 6.2.3 shows level premiums, second-order level premiums, and premium margins for the two insurance products. Table 6.2.4 shows the consequent expected premium margins in the two portfolios.

Table 6.2.3 Individual premium margins

Insurance product	P	P''	$P-P''$
Term insurance	1.93	1.71	0.22
Endowment insurance	59.54	54.84	4.70

Table 6.2.4 Expected premium margins in the portfolio

t	$(P-P'')\bar{N}_t$	
	Term insurance	Endowment insurance
0	2213.33	47009.59
1	2210.92	46866.46
2	2208.26	46708.21
3	2205.31	46533.29
4	2202.05	46340.03
5	2198.44	46126.57
6	2194.45	45890.92
7	2190.03	45630.91
8	2185.14	45344.18
9	2179.73	45028.20
10		44680.25
11		44297.39
12		43876.53
13		43414.35
14		42907.39

□

6.3 Expected annual profits

The model so far developed provides a synthetic information on the portfolio expected profit, that is, the expected total profit cumulated at maturity, $\bar{F}_m^{[P]}$, and its present value, $\overline{PL}^{[P]}$. More detailed results can be achieved by introducing new elements into the model. In particular, the sequence of *annual profits*, namely the *timing of the profit*, is of great interest under both a theoretical perspective (as seen in Sect. 5.5) and a practical perspective as well. To this purpose, the portfolio reserve must be accounted for.

6.3.1 The expected surplus and the expected annual profits

The portfolio reserve at a future time t is the random amount $V_t^{[P]} = N_t V_t$ (see Sect. 6.1). Conversely, the estimated portfolio reserve is given by $\hat{V}_t^{[P]} = n_t V_t$, where n_t is the estimated number of policies in the portfolio at that time. In particular, the expected portfolio reserve is given by $\hat{V}_t^{[P]} = \mathbb{E}[V_t^{[P]}] = N_0\, {}_t p_x'' V_t$ (thus, in this case: $n_t = N_0\, {}_t p_x''$).

As seen in Sect. 6.2.1, the random amount $F_t^{[P]}$ represents (if positive) the portfolio assets at time t. Conversely, the portfolio liability is expressed by the reserve $V_t^{[P]}$. Hence, the difference $F_t^{[P]} - V_t^{[P]}$ represents the (random) *cumulated surplus* at time t, as we have excluded capital allocations to the portfolio. This difference also represents the *Net Asset Value* (*NAV*) pertaining to the portfolio itself.

To describe the evolution of the cumulated surplus, in terms of expected values, we refer to Eq. (6.2.5a). Then, we subtract $\hat{V}_{t+1}^{[P]}$ in both the left-hand side and the right-hand side of the equation, and add and subtract $\hat{V}_t^{[P]}(1+i'')$ in the left-hand side. Obvious adjustments are required for Eq. (6.2.5b). We obtain:

$$(\bar{F}_t^{[P]} - \hat{V}_t^{[P]})(1+i'') + (\hat{V}_t^{[P]} + P\bar{N}_t)(1+i'') - C\bar{D}_t - \hat{V}_{t+1}^{[P]} = \bar{F}_{t+1}^{[P]} - \hat{V}_{t+1}^{[P]}; \quad t = 0,1,\dots,m-2 \tag{6.3.1a}$$

$$(\bar{F}_{m-1}^{[P]} - \hat{V}_{m-1}^{[P]})(1+i'') + (\hat{V}_{m-1}^{[P]} + P\bar{N}_{m-1})(1+i'') - C\bar{D}_{m-1} - S\bar{N}_m = \bar{F}_m^{[P]} \tag{6.3.1b}$$

with $\bar{F}_0^{[P]} = 0$ (see assumption (6.2.1)), and, of course, $\hat{V}_0^{[P]} = 0$. Further, note that we set $\hat{V}_m^{[P]} = 0$, as we have assumed that the life fund at time m is net of the outflow for maturity benefits.

Let $\overline{PL}_{t+1}^{[P]}$ denote the expected annual variation in the cumulated surplus, namely the annual contribution to the cumulated surplus, which clearly represents the *expected annual profit* (or *loss*):

$$\overline{PL}_{t+1}^{[P]} = (\bar{F}_{t+1}^{[P]} - \hat{V}_{t+1}^{[P]}) - (\bar{F}_t^{[P]} - \hat{V}_t^{[P]}) \tag{6.3.2}$$

From Eqs. (6.3.1), we obtain the following expressions:

$$\overline{PL}_{t+1}^{[P]} = (\bar{F}_t^{[P]} - \hat{V}_t^{[P]})i'' + (\hat{V}_t^{[P]} + P\bar{N}_t)(1+i'') - C\bar{D}_t - \hat{V}_{t+1}^{[P]}; \quad t = 0,1,\dots,m-2 \tag{6.3.3a}$$

$$\overline{PL}_m^{[P]} = (\bar{F}_{m-1}^{[P]} - \hat{V}_{m-1}^{[P]})i'' + (\hat{V}_{m-1}^{[P]} + P\bar{N}_{m-1})(1+i'') - C\bar{D}_{m-1} - S\bar{N}_m \tag{6.3.3b}$$

Looking at Eqs. (6.3.3), two components of the annual profit can be singled out. The quantity

$$\overline{PL}_{t+1}^{[P][NAV]} = (\bar{F}_t^{[P]} - \hat{V}_t^{[P]})i'' \tag{6.3.4}$$

represents the interest on the NAV. Conversely, the component

$$\overline{PL}_{t+1}^{[P][I]} = (\hat{V}_t^{[P]} + P\bar{N}_t)(1+i'') - C\bar{D}_t - \hat{V}_{t+1}^{[P]}; \quad t = 0,1,\dots,m-2 \tag{6.3.5a}$$

$$\overline{PL}_{m}^{[P][I]} = (\hat{V}_{m-1}^{[P]} + P\bar{N}_{m-1})(1+i'') - C\bar{D}_{m-1} - S\bar{N}_m \tag{6.3.5b}$$

represents the *industrial profit*.

By using Eqs. (6.2.7), we obtain the following expression

$$\overline{PL}_{t+1}^{[P][I]} = \overline{CF}_{t+1}^{[P]} + \hat{V}_t^{[P]}(1+i'') - \hat{V}_{t+1}^{[P]}; \quad t = 0,1,\dots,m-1 \tag{6.3.6}$$

(with $\hat{V}_0^{[P]} = \hat{V}_m^{[P]} = 0$) which shows that the industrial profits arise from the portfolio cash-flows adjusted by accounting for the variation in the portfolio reserve, $\hat{V}_t^{[P]} - \hat{V}_{t+1}^{[P]}$, and the interest on the reserve at the beginning of the year, $\hat{V}_t^{[P]}\, i''$.

Remark We note that a relation between annual profits and annual cashflows, quite similar to that expressed by Eq. (6.3.6), holds at policy level as shown by Eqs. (5.5.13).

Example 6.3.1. Tables 6.3.1 and 6.3.2 show the expected portfolio reserve, the NAV, the expected annual profits, and the related components, in a term insurance and in an endowment insurance portfolio, respectively. Data are as in Example 6.2.1. It is interesting to compare the expected annual profits with the expected annual cash-flows (see Tables 6.2.1 and 6.2.2): the role of the portfolio reserve (and, in particular, of the variation in the portfolio reserve) clearly appears.

Table 6.3.1 Expected portfolio reserve and expected profits in a term insurance portfolio

t	$\hat{V}_t^{[P]}$	$\bar{F}_t^{[P]} - \hat{V}_t^{[P]}$	$\overline{PL}_t^{[P]}$	$\overline{PL}_t^{[P][NAV]}$	$\overline{PL}_t^{[P][I]}$
0	0.00	0.00	–	–	–
1	7585.09	1378.85	1378.85	0.00	1378.85
2	14016.80	3002.45	1623.59	41.37	1582.23
3	19130.89	4880.37	1877.92	90.07	1787.85
4	22745.28	7022.31	2141.94	146.41	1995.53
5	24658.24	9438.07	2415.77	210.67	2205.10
6	24646.35	12137.56	2699.48	283.14	2416.34
7	22462.43	15130.74	2993.18	364.13	2629.05
8	17833.19	18427.65	3296.92	453.92	2843.00
9	10456.74	22038.41	3610.76	552.83	3057.93
10	0.00	25973.16	3934.75	661.15	3273.59

□

Table 6.3.2 Expected portfolio reserve and expected profits in an endowment insurance portfolio

t	$\hat{V}_t^{[P]}$	$\bar{F}_t^{[P]} - \hat{V}_t^{[P]}$	$\overline{PL}_t^{[P]}$	$\overline{PL}_t^{[P][NAV]}$	$\overline{PL}_t^{[P][I]}$
0	0.00	0.00	–	–	–
1	573687.47	9116.66	9116.66	0.00	9116.66
2	1153665.85	24342.98	15226.32	273.50	14952.82
3	1739553.40	45906.93	21563.95	730.29	20833.66
4	2330915.86	74033.94	28127.01	1377.21	26749.80
5	2927262.45	108945.28	34911.34	2221.02	32690.32
6	3528041.94	150856.18	41910.90	3268.36	38642.54
7	4132638.76	199973.72	49117.54	4525.69	44591.85
8	4740369.25	256494.39	56520.68	5999.21	50521.46
9	5350478.29	320601.39	64106.99	7694.83	56412.16
10	5962136.36	392461.46	71860.07	9618.04	62242.03
11	6574437.28	472221.46	79760.00	11773.84	67986.15
12	7186396.87	560004.42	87782.96	14166.64	73616.32
13	7796952.88	655905.21	95900.79	16800.13	79100.66
15	0.00	872269.24	112283.58	22799.57	89484.01

Remark The rationale underlying the definition of the expected annual profits we have just proposed is quite different from that underpinning the profit assessment within the framework presented in Sect. 5.5. We note that, according to Eq. (5.5.2), interest originate from the reserve and the annual premium only, whereas, according to the life fund logic, the investment of both the cumulated surplus and the (portfolio) reserve plus the annual premiums is accounted for (see, for example, Eqs. (6.3.3), (6.3.4), and (6.3.5)). Indeed, the logic underlying Eq. (5.5.2) does not allow for any profit accumulation. The two approaches can be considered as particular implementations of a more general model for profit assessment. The approach leading to (5.5.2) is based on the so-called *profits-released* assumption, whereas the approach involving the life fund analysis is based on the *profits-retained* assumption. Nonetheless, thanks to the splitting of the annual profit into the industrial profit and the interest on the NAV, we can easily recognize in the industrial component (see Eqs. (6.3.5)) the logic underlying (5.5.2).

6.3.2 The role of the portfolio reserve

By solving Eqs. (6.3.1), we obtain the following explicit expressions for the expected cumulated surplus:

$$\bar{F}_{t+1}^{[P]} - \hat{V}_{t+1}^{[P]} = \sum_{h=0}^{t} \left((\hat{V}_h^{[P]} + P\,\bar{N}_h)(1+i'') - C\bar{D}_h - \hat{V}_{h+1}^{[P]} \right)(1+i'')^{t-h}; \quad t = 0,1,\ldots,m-2 \tag{6.3.7a}$$

$$\bar{F}_m^{[P]} = \sum_{h=0}^{m-1} \left((\hat{V}_h^{[P]} + P\,\bar{N}_h)(1+i'') - C\bar{D}_h - \hat{V}_{h+1}^{[P]} \right)(1+i'')^{m-(h+1)} - S\bar{N}_m \tag{6.3.7b}$$

From Eqs. (6.3.5) it follows that (6.3.7b) can also be written as follows:

$$\bar{F}_m^{[P]} = \sum_{h=0}^{m-1} \overline{PL}_{h+1}^{[P][I]} (1+i'')^{m-(h+1)} \tag{6.3.8}$$

which expresses the cumulated surplus as the accumulated value of the industrial profits. Further, we have:

$$\overline{PL}^{[P]} = \sum_{h=0}^{m-1} \overline{PL}_{h+1}^{[P][I]} (1+i'')^{-(h+1)} \tag{6.3.9}$$

It is possible to prove that, assuming $\hat{V}_0^{[P]} = \hat{V}_m^{[P]} = 0$, Eq. (6.3.7b) yields:

$$\bar{F}_m^{[P]} = \sum_{h=0}^{m-1} \left(P\,\bar{N}_h(1+i'') - C\bar{D}_h\right)(1+i'')^{m-(h+1)} - S\bar{N}_m \tag{6.3.10}$$

which coincides with (6.2.6b). Indeed, all the other reserve values annul each other.

According to (6.3.10), the expected profit cumulated at maturity coincides with the accumulated value of the expected annual cash-flows. Hence, the reserve profile does not affect the amount of the total profit, while it does affect the annual industrial profits (see Eqs. (6.3.5)), and then the *timing* of the total profit.

Example 6.3.2. We refer to an endowment insurance portfolio. Data are as in Example 6.2.1. Table 6.3.3 shows the timing of the industrial profits, originated by three reserve profiles. The profiles correspond to different interest rates in the reserving basis, namely 0.00, 0.02 (which coincides with the interest rate in the pricing basis TB1), and 0.04. Possible negative reserve values have been replaced with 0. We note that an interest rate lower than that in the pricing basis implies a profit delay, whereas a higher interest rate implies a profit acceleration. These results can be compared to those displayed in Table 5.5.4 of Example 5.5.3.

Table 6.3.4 shows the profit timing originated by two reserve profiles: the best-estimate reserve and the best-estimate reserve plus the risk margin. The risk margin has been calculated as described in Sect. 6.1.5; in particular:

- we have assumed $\rho = 0.08$, $r_f = 0.02$;
- SCR_t has been set equal to 10% of the best-estimate reserve;

Again, possible negative reserve values have been replaced with 0.
□

The reserve profile based on the best-estimate reserve implies a very particular timing of the industrial profits, as shown in Table 6.3.4 (and as already seen at policy level; see in particular Eqs. (5.5.25)). A formal proof of this specific profit profile, with reference to a portfolio, follows.

The policy best-estimate reserve of an endowment insurance is defined by Eq. (5.5.21). At the portfolio level, we assume:

$$\hat{V}_t^{[P][BE]} = \mathbb{E}[N_t]\, V_t^{[BE]} \tag{6.3.11}$$

Table 6.3.3 Expected profits according to various reserving profiles

t	$\hat{V}_t^{[P](0.00)}$	$\overline{PL}_t^{[P][I](0.00)}$	$\hat{V}_t^{[P](0.02)}$	$\overline{PL}_t^{[P][I](0.02)}$	$\hat{V}_t^{[P](0.04)}$	$\overline{PL}_t^{[P][I](0.04)}$
0	0.00	–	0.00	–	0.00	–
1	1974761.61	−1391957.48	573687.47	9116.66	0.00	582804.13
2	2531909.48	79815.55	1153665.85	14952.82	159977.92	417742.65
3	3083449.20	96528.81	1739553.40	20833.66	740608.75	−3720.25
4	3628830.21	113048.12	2330915.86	26749.80	1338135.16	−9382.49
5	4167452.93	129351.62	2927262.45	32690.32	1952629.86	−15241.21
6	4698665.52	145415.16	3528041.94	38642.54	2584119.24	−21306.33
7	5221760.78	161212.11	4132638.76	44591.85	3232579.39	−27589.16
8	5735973.20	176713.19	4740369.25	50521.46	3897932.22	−34102.66
9	6240476.31	191886.22	5350478.29	56412.16	4580042.06	−40861.75
10	6734380.40	206695.95	5962136.36	62242.03	5278712.75	−47883.67
11	7216730.95	221103.84	6574437.28	67986.15	5993685.54	−55188.44
12	7686507.78	235067.89	7186396.87	73616.32	6724638.22	−62799.32
13	8142625.36	248542.42	7796952.88	79100.66	7471185.57	−70743.43
14	8583934.49	261477.94	8404966.48	84403.30	8232881.85	−79052.41
15	0.00	273821.06	0.00	89484.01	0.00	−87763.16
	$\bar{F}_{15}^{[P](0.00)} = 872269.24$		$\bar{F}_{15}^{[P](0.02)} = 872269.24$		$\bar{F}_{15}^{[P](0.04)} = 872269.24$	
	$\overline{PL}^{[P](0.00)} = 559876.43$		$\overline{PL}^{[P](0.02)} = 559876.43$		$\overline{PL}^{[P](0.04)} = 559876.43$	

Table 6.3.4 Expected profits according to various reserving profiles

t	$\hat{V}_t^{[P][BE]}$	$\overline{PL}_t^{[P][I][BE]}$	$\hat{V}_t^{[P][BE]} + RM_t$	$\overline{PL}_t^{[P][I][BE+RM]}$
0	0.00	–	0.00	–
1	6131.41	576672.72	266864.82	315939.31
2	584035.92	0.00	852553.45	37.89
3	1173668.24	0.00	1446631.95	3609.34
4	1774803.95	0.00	2048703.30	7253.27
5	2387157.50	0.00	2658305.54	10968.29
6	3010376.38	0.00	3274906.24	14752.63
7	3644035.08	0.00	3897896.70	18604.13
8	4287628.93	0.00	4526586.26	22520.14
9	4940567.92	0.00	5160196.43	26497.55
10	5602170.71	0.00	5797855.36	30532.71
11	6271658.81	0.00	6438592.59	34621.41
12	6948151.37	0.00	7081334.30	38758.85
13	7630660.67	0.00	7724899.52	42939.58
14	8318088.80	0.00	8367997.34	47157.48
15	9009225.74	0.00	9009225.74	51405.79
	$\bar{F}_{15}^{[P][BE]} = 872269.24$		$\bar{F}_{15}^{[P][BE+RM]} = 872269.24$	
	$\overline{PL}^{[P][BE]} = 559876.43$		$\overline{PL}^{[P][BE+RM]} = 559876.43$	

with $\mathbb{E}[N_t] = N_0 \, {}_tp''_x$, that is, calculated by adopting the second-order basis TB2.

We then find:

$$\hat{V}_t^{[P][BE]} = N_0\, {}_tp''_x \left(C\, {}_{m-t}A''_{x+t} + S\, {}_{m-t}E''_{x+t} - P\ddot{a}''_{x+t:m-t\rceil}\right)$$

$$= N_0\, {}_tp''_x \sum_{h=0}^{m-t-1} \left(C\, {}_{h|1}q''_{x+t} - P\, {}_hp''_{x+t}(1+i'')\right)(1+i'')^{-(h+1)} + S\, {}_{m-t}p''_{x+t}(1+i'')^{-(m-t)}$$

$$= \sum_{h=0}^{m-t-1} \left(C\bar{D}_{t+h} - P\bar{N}_{t+h}(1+i'')\right)(1+i'')^{-(h+1)} + S\bar{N}_m(1+i'')^{-(m-t)} \tag{6.3.12}$$

and finally we have:

$$\hat{V}_t^{[P][BE]} = -\sum_{h=0}^{m-t-1} \overline{CF}_{t+h+1}^{[P]}(1+i'')^{-(h+1)} \tag{6.3.13}$$

From Eq. (6.3.13), we immediately obtain the following recursive expression:

$$\hat{V}_{t+1}^{[P][BE]} = \hat{V}_t^{[P][BE]}(1+i'') + \overline{CF}_{t+1}^{[P]} \tag{6.3.14}$$

In particular, Eq. (6.3.13) yields:

$$\hat{V}_0^{[P][BE]} = -\sum_{h=0}^{m-1} \overline{CF}_{h+1}^{[P]}(1+i'')^{-(h+1)} = -\bar{F}_m^{[P]}(1+i'')^{-m} = -\overline{PL}^{[P]} \tag{6.3.15}$$

Assume that, for the portfolio reserve, the following profile is chosen:

$$\hat{V}_0^{[P]} = 0; \tag{6.3.16a}$$

$$\hat{V}_t^{[P]} = \hat{V}_t^{[P][BE]}; \quad t = 1, 2, \ldots, m-1 \tag{6.3.16b}$$

$$\hat{V}_m^{[P]} = 0 \tag{6.3.16c}$$

It follows that the industrial profits (see Eq. (6.3.6)) are given by:

$$\overline{PL}_1^{[P][I][BE]} = -\hat{V}_1^{[P][BE]} + \overline{CF}_1^{[P]} = \sum_{h=0}^{m-1} \overline{CF}_{h+1}^{[P]}(1+i'')^{-h} = \overline{PL}^{[P]}(1+i'') \tag{6.3.17a}$$

$$\overline{PL}_{t+1}^{[P][I][BE]} = \hat{V}_t^{[P][BE]}(1+i'') - \hat{V}_{t+1}^{[P][BE]} + \overline{CF}_{t+1}^{[P]} = 0;\ t = 1, 2, \ldots, m-1 \tag{6.3.17b}$$

Thus, the expected profit entirely emerges in the first year, while the annual profits are identically equal to zero in all the following years.

Remark The result expressed by Eqs. (6.3.17) holds provided that the scenario assumed at time 0 (namely, the technical basis TB2) keeps its validity throughout the whole duration of the portfolio. Conversely, if changes in the scenario suggest, at some point, the shift to a different technical basis, the profit profile will change, and hence profit will emerge at the time of the shift.

6.4 Expected annual profits: a more general setting

Further elements must be added to our model, in order to build-up a more realistic framework for profit assessment. To this purpose, we allow for the following elements:

- expenses and related loadings;
- lapses and surrenders.

As regards the expense loadings, we refer to the loading structure described in Sect. 4.5.3. We assume that a realistic estimate of expenses can be expressed by the same structure, although expense parameters possibly differ from those used in calculating the premium loading. Let EX_t denote the expense at time t related to a generic policy. Referring to policies with annual premiums payable throughout the whole policy duration m, and assuming that expenses are charged at the beginning of each year, we have:

$$EX_0 = \alpha'' C + \beta'' P^{[T]} + \gamma'' C \tag{6.4.1a}$$

$$EX_t = \beta'' P^{[T]} + \gamma'' C; \quad t = 1, 2 \ldots, m-1 \tag{6.4.1b}$$

where α'', β'', and γ'' denote the realistic estimation of the expense parameters. Note that the term $\alpha'' C$ in Eq. (6.4.1a) should be replaced by $\delta''(m) P^{[T]}$ when the acquisition costs are expressed in terms of the annual premium.

The random number of in-force policies, N_t, must be redefined because of the presence of lapses / surrenders. We assume that lapses / surrenders occur at the end of the generic policy year, before paying the premium due for the following year. Let A_t denote the number of policyholders who abandon the contract at time t. Then, we have:

$$N_{t+1} = N_t - D_t - A_{t+1} \tag{6.4.2}$$

Let w_t denote the probability of abandoning the contract at time t, conditional on belonging to the portfolio at that time. Hence, the expected values can be calculated as follows:

$$\bar{N}_t = N_0 \, {}_t p''_x \prod_{h=1}^{t} (1 - w_h) \tag{6.4.3}$$

$$\bar{D}_t = \bar{N}_t \, {}_1 q''_{x+t} \tag{6.4.4}$$

$$\bar{A}_{t+1} = \bar{N}_t (1 - {}_1 q''_{x+t}) \, w_{t+1} \tag{6.4.5}$$

Let R_t denote the surrender value at time t (see Sect. 5.7). The behavior of the portfolio fund, $F_t^{[P]}$, in terms of expected values, is described by the following recursive equations:

$$(\bar{F}_t^{[P]} + P^{[T]} \bar{N}_t - EX_t \, \bar{N}_t)(1+i'') - C\bar{D}_t - R_{t+1} \bar{A}_{t+1} = \bar{F}_{t+1}^{[P]}; \quad t = 0,1,\ldots,m-2 \tag{6.4.6a}$$

$$(\bar{F}_{m-1}^{[P]} + P^{[T]} \bar{N}_{m-1} - EX_{m-1} \, \bar{N}_{m-1})(1+i'') - C\bar{D}_{m-1} - S\bar{N}_m = \bar{F}_m^{[P]} \tag{6.4.6b}$$

which generalize Eqs. (6.2.5). In Eq. (6.4.6b), it is assumed, as is reasonable, that $A_m = 0$.

The expected annual profits and the related components can be determined by generalizing Eqs. (6.3.3), (6.3.4), and (6.3.5). The estimated portfolio reserve, $\hat{V}_t^{[P]}$, can either allow for the expenses and the related premium loadings or not. As we have proved (see Sect. 6.3.2), the time profile of the reserve does not affect the total profit, whereas it does affect the annual profits (see Example 6.4.1).

The expected annual profits are then expressed by the following equations:

$$\overline{PL}_{t+1}^{[P]} = (\bar{F}_t^{[P]} - \hat{V}_t^{[P]})i'' + (\hat{V}_t^{[P]} + P^{[T]} \bar{N}_t - EX_t \, \bar{N}_t)(1+i'') - C\bar{D}_t - R_{t+1} \bar{A}_{t+1} - \hat{V}_{t+1}^{[P]};$$
$$t = 0,1,\ldots,m-2 \tag{6.4.7a}$$

$$\overline{PL}_m^{[P]} = (\bar{F}_{m-1}^{[P]} - \hat{V}_{m-1}^{[P]})i'' + (\hat{V}_{m-1}^{[P]} + P^{[T]} \bar{N}_{m-1} - EX_t \, \bar{N}_{m-1})(1+i'') - C\bar{D}_{m-1} - S\bar{N}_m \tag{6.4.7b}$$

The interest on the NAV is given by

$$\overline{PL}_{t+1}^{[P][NAV]} = (\bar{F}_t^{[P]} - \hat{V}_t^{[P]})i'' \tag{6.4.8}$$

whereas the industrial component of the annual profit is expressed as follows:

$$\overline{PL}_{t+1}^{[P][I]} = (\hat{V}_t^{[P]} + P^{[T]} \bar{N}_t - EX_t \, \bar{N}_t)(1+i'') - C\bar{D}_t - R_{t+1} \bar{A}_{t+1} - \hat{V}_{t+1}^{[P]};$$
$$t = 0,1,\ldots,m-2 \tag{6.4.9a}$$

$$\overline{PL}_m^{[P][I]} = (\hat{V}_{m-1}^{[P]} + P^{[T]} \bar{N}_{m-1} - EX_t \, \bar{N}_{m-1})(1+i'') - C\bar{D}_{m-1} - S\bar{N}_m \tag{6.4.9b}$$

In particular, as $\hat{V}_0^{[P]} = 0$, we have:

$$\overline{PL}_1^{[P][I]} = (P^{[T]} \bar{N}_0 - EX_0 \, \bar{N}_0)(1+i'') - C\bar{D}_0 - R_1 \bar{A}_1 - \hat{V}_1^{[P]} \tag{6.4.10}$$

We note what follows:

- the first year expenses, EX_0, include the acquisition costs (see Eq. (6.4.1a)), and hence are usually much higher than the expenses in the following years;
- the acquisition costs, in the case of periodic premiums, are progressively amortized throughout the premium payment period (see Eqs. (4.5.3b) and (4.5.4b));
- it follows that, the higher is the reserve $\hat{V}_1^{[P]}$, the lower (and possibly negative) is the first year industrial profit; in particular, this happens if $\hat{V}_1^{[P]}$ is set equal to the net premium reserve, rather than the Zillmer reserve (see Sect. 5.6).

Remark As already seen at policy level (see the Remark in Sect. 5.5.1), the equations which define the expected annual profits contain the (main) items of the industrial *Profit & Loss Statement* (briefly *P & L*). Table 6.4.1 displays the items of Eqs. (6.4.9) according to a P & L format. An obvious adjustment in the benefits is needed when referring to the last year of the policy duration.

Table 6.4.1 Actuarial values as items of a P & L statement

INDUSTRIAL P & L STATEMENT	
Income	
Premiums	$P^{[\mathrm{T}]}\,\bar{N}_t$
Income from investments	$(\hat{V}_t^{[\mathrm{P}]} + P^{[\mathrm{T}]}\,\bar{N}_t - EX_t\,\bar{N}_t)\; i''$
Expenditure	
Benefits paid	$C\bar{D}_t + R_{t+1}\bar{A}_{t+1}$
Portfolio expenses	$EX_t\,\bar{N}_t$
Change in liabilities	$\hat{V}_{t+1}^{[\mathrm{P}]} - \hat{V}_t^{[\mathrm{P}]}$
Profit (or *Loss*)	$\overline{PL}_{t+1}^{[\mathrm{P}][\mathrm{I}]}$

Example 6.4.1. We refer to an endowment insurance portfolio. Data are as in Example 6.2.1, namely: $N_0 = 10\,000$, $S = C = 1\,000$, $x = 50$, $m = 15$. Annual level premiums, $P^{[\mathrm{T}]}$, are payable throughout the whole policy duration. The pricing basis is $\mathrm{TB1} = (0.02, \mathrm{LT1})$. We then find: $P = 59.54$. We assume the following expense loading parameters (see Sect. 4.5.3): $\delta(15) = 0.55$, $\beta = 0.04$, $\gamma = 0.0015$. Hence, $P^{[\mathrm{T}]} = 66.60$. Surrender values are given by:

$$R_t = \begin{cases} 0; & t = 1,2 \\ 0.90\,V_t^{[\mathrm{Z}]}; & t \geq 3 \end{cases} \tag{6.4.11}$$

The scenario basis is $\mathrm{TB2} = (0.03, \mathrm{LT2})$. As regards surrenders and expenses, we consider the two following cases.

1. Expense realistic expectation: $\delta'' = \delta$, $\beta'' = \beta$, $\gamma'' = \gamma$;
   surrender probabilities: $w_t = 0$; $t = 1, 2, \ldots, 15$.
2. Expense realistic expectation: $\delta'' = 0.58$, $\beta'' = 0.04$, $\gamma'' = 0.0018$;
   surrender probabilities: see Table 6.4.2.

Tables 6.4.3 and 6.4.4 show the expected profits in case 1 and 2, respectively. It is worth noting that, in both cases, the Zillmer reserve leads to a "smoother" annual profit profile, whereas the net premium reserve causes a heavy loss in the first year (as we can argue from Eq. (6.4.10)). Indeed, according to the net premium reserve, the acquisition costs are charged to the first year result, disregarding the progressive amortization. Conversely, this fact does not affect the amount of the total profit, which is independent of the reserve profile.

□

Table 6.4.2 Probability of surrender

t	w_t	t	w_t	t	w_t	t	w_t	t	w_t
1	0.05	4	0.03	7	0.03	10	w_t	13	0.03
2	0.02	5	0.03	8	0.03	11	w_t	14	0.00
3	0.06	6	0.03	9	0.03	12	w_t	15	0.00

Table 6.4.3 Expected profits - Case 1

t	Net premium reserve				Zillmer reserve			
	$\hat{V}_t^{[P]}$	$\overline{PL}_t^{[P]}$	$\overline{PL}_t^{[P][NAV]}$	$\overline{PL}_t^{[P][I]}$	$\hat{V}_t^{[P]}$	$\overline{PL}_t^{[P]}$	$\overline{PL}_t^{[P][NAV]}$	$\overline{PL}_t^{[P][I]}$
0	0.00	–	–	–	0.00	–	–	–
1	573687.47	−338323.21	0.00	−338323.21	229504.95	5859.32	0.00	5859.32
2	1153665.85	34574.46	−10149.70	44724.16	831961.82	12095.97	175.78	11920.19
3	1739553.40	41392.01	−9112.46	50504.47	1440674.18	18567.20	538.66	18028.55
4	2330915.86	48438.80	−7870.70	56309.51	2055205.00	25270.44	1095.67	24174.77
5	2927262.45	55709.71	−6417.54	62127.25	2675059.85	32201.45	1853.79	30347.66
6	3528041.94	63197.63	−4746.25	67943.87	3299682.92	39354.06	2819.83	36534.22
7	4132638.76	70893.17	−2850.32	73743.49	3928452.98	46719.92	4000.45	42719.47
8	4740369.25	78784.41	−723.52	79507.93	4560679.68	54288.20	5402.05	48886.15
9	5350478.29	86856.50	1640.01	85216.49	5195599.99	62045.24	7030.70	55014.54
10	5962136.36	95091.34	4245.70	90845.64	5832375.25	69974.16	8892.05	61082.10
11	6574437.28	103467.17	7098.45	96368.73	6470088.83	78054.50	10991.28	67063.22
12	7186396.87	111958.15	10202.46	101755.69	7107744.78	86261.79	13332.91	72928.88
13	7796952.88	120533.88	13561.20	106972.68	7744267.65	94567.03	15920.77	78646.26
14	8404966.48	129158.95	17177.22	111981.73	8378503.97	102936.24	18757.78	84178.46
15	0.00	137792.39	21051.99	116740.40	0.00	111329.87	21845.87	89484.01
	$\bar{F}_{15}^{[P]} = 839525.38$				$\bar{F}_{15}^{[P]} = 839525.38$			
	$\overline{PL}^{[P]} = 538859.40$				$\overline{PL}^{[P]} = 538859.40$			

Table 6.4.4 Expected profits - Case 2

t	Net premium reserve				Zillmer reserve			
	$\hat{V}_t^{[P]}$	$\overline{PL}_t^{[P]}$	$\overline{PL}_t^{[P][NAV]}$	$\overline{PL}_t^{[P][I]}$	$\hat{V}_t^{[P]}$	$\overline{PL}_t^{[P]}$	$\overline{PL}_t^{[P][NAV]}$	$\overline{PL}_t^{[P][I]}$
0	0.00	–	–	–	0.00	–	–	–
1	545003.10	−333308.95	0.00	−333308.95	218029.70	−6335.55	0.00	6335.55
2	1074062.91	51481.77	−9999.27	61481.04	774556.45	24014.83	−190.07	24204.89
3	1522352.77	60449.50	−8454.82	68904.32	1260791.60	22504.21	530.38	21973.84
4	1978681.38	52594.93	−6641.33	59236.26	1744634.34	25080.81	1205.50	23875.30
5	2410364.18	58324.96	−5063.48	63388.44	2202695.71	31946.39	1957.93	29988.46
6	2817905.84	63928.11	−3313.73	67241.84	2635511.68	38653.80	2916.32	35737.48
7	3201783.16	69400.72	−1395.89	70796.61	3043589.18	45200.54	4075.93	41124.61
8	3562446.72	74738.22	686.13	74052.09	3427407.77	51583.20	5431.95	46151.25
9	3900322.75	79935.10	2928.28	77006.82	3787421.56	57797.33	6979.45	50817.88
10	4215815.33	84984.82	5326.33	79658.49	4124061.49	63837.47	8713.37	55124.10
11	4509308.99	89879.85	7875.88	82003.98	4437737.94	69697.06	10628.49	59068.57
12	4781171.70	94611.56	10572.27	84039.29	4728843.79	75368.42	12719.40	62649.02
13	5031758.48	99170.21	13410.62	85759.60	4997758.10	80842.69	14980.46	65862.23
14	5424139.66	86811.44	16385.72	70425.72	5407062.09	69888.63	17405.74	52482.89
15	0.00	92508.29	18990.07	73518.22	0.00	75430.72	19502.39	55928.33
	$\bar{F}_{15}^{[P]} = 725510.54$				$\bar{F}_{15}^{[P]} = 725510.54$			
	$\overline{PL}^{[P]} = 465677.61$				$\overline{PL}^{[P]} = 465677.61$			

6.5 References and suggestions for further reading

Best-estimate reserves and risk margins constitute recent issues in the actuarial framework. The interested reader should then refer to recent literature addressing risk management and solvency in the insurance business: for example, [21] and [50]; see also [42].

Expected profits in the life insurance business are dealt with, under different perspectives, in various textbooks. For example, [17] presents an interesting scheme for assessing profits arising from one generation of policies.

We recall that the traditional approach to the profit assessment at the policy level is described by [47], whereas [26] places special emphasis on mortality profits.

Assessment of profits in a profit-testing framework is dealt with by [20]. A modern approach to profits (not restricted to life insurance) is proposed in Chap. 16 of [4].

Effects of solvency regulation on the emergence of profit are discussed in Chap. 9 of [9]. A thorough analysis of expected profits in the context of evaluation procedures for life insurance portfolios is provided by [42].

Chapter 7
Finance in life insurance: linking benefits to the investment performance

7.1 Introduction

The life insurance products described in the previous chapters are characterized by fixed benefits (and premiums), i.e. the amount of benefits and premiums is stated at issue.

Remark It is worth stressing that the expression "fixed benefits" should not be meant as "constant benefits" but, as specified above, as benefits whose amount is definitively assigned at policy issue. If the amount of benefits varies in time but in a way specified at issue, we still refer to such an arrangement as with fixed benefits. The same terminology applies to premiums; a rating arrangement based on natural premiums, for example, is considered to be with fixed premiums, as their amount is univocally defined at policy issue.

In this Chapter we examine life insurance products whose benefit amount depends either on the return on investments, market interest rates, stock-market indexes, or other financial index. The purpose is to provide a return on the investment of the policyholder which is higher than the usual technical interest rate, and in any case in line with the prevailing market rates for the class of assets backing the reserve. We recall that since the technical interest rate is guaranteed for the whole (long) duration of the contract, it must be set at a low level, to avoid major risks for the insurer. Clearly, insurance products for which there is an interest in realizing a linking to a financial index are those with a large savings component, i.e. those with a large reserve (and then, a large investment) in respect of the insured amount, such as endowments, whole life assurances, life annuities.

As we are going to describe in this Chapter, the linking of benefits to the investment performance can be realized in different ways. Basically, what makes the difference is how financial risk is shared between the insurer and the policyholder. In this respect, the following classes of life insurance products can be identified:

- policies with embedded financial guarantees (participating, with-profit and universal life policies);
- policies without financial guarantees (unit-linked and universal life policies);

A. Olivieri, E. Pitacco, *Introduction to Insurance Mathematics*,
DOI 10.1007/978-3-642-16029-5_7, © Springer-Verlag Berlin Heidelberg 2011

- policies with explicit financial guarantees (unit-linked and universal life policies with minimum guarantees, index-linked policies, variable annuities).

To better understand the technical features of the life insurance policies with benefits linked to the investment performance, it is interesting to investigate how the various mechanisms that we are going to examine were introduced. As mentioned above, the products we refer to are those with a large reserve in respect of the insured amount, i.e. those whose underlying purpose is not only the insurance protection, but also savings.

The idea behind policy designs with a linking to the investment performance is to share investment profits, and possibly losses, between the insurer and the policyholder. The idea of sharing profits with the policyholders is not new in life insurance. There is a tradition in the UK and German markets which dates back to the Nineteen century, and which in earlier times concerned industrial or balance-sheet profits. In UK, in particular, the traditional product realizing the linking of benefits to profits is the *with-profit policy*, under which in each year the benefit amount is increased by the so-called *bonus*.

Apart from these examples, life insurance products have been characterized by fixed benefits until the Sixties of the last century. At that time, some important innovations were introduced. In UK, the early forms of *unit-linked policies* were designed, under which the return on investments is not guaranteed; the assets backing the reserve can then be selected out of classes of securities riskier (and, on average, more rewarding) than the traditional ones (typically, government or high-quality bonds). Conversely, in the US a form of flexibility was introduced into the premiums, whose amount could be chosen (possibly within a given band) year by year by the policyholder.

Many innovations were introduced in the Seventies, in contrast to the dramatic decrease of the volume of the lines of business with a large savings component, due to the high inflation rates. It is worth noting that the long duration of life policies and the nature of the insurer's liability (which concerns the amount of the benefit, and not its real value) expose the policyholder to inflation risk. When the inflation rate is low and interest rates are positive in real terms, the risk is not perceived; vice versa, when the inflation rate is very high (as it was during the Seventies) the depreciation of the insured amount is too strong to make the product attractive in respect of alternative investment solutions. The life insurance industry reacted to the negative underwriting trends by linking benefits to the inflation rate or to the return on investment. Due to the limited availability of inflation-linked securities, most of the insurers focused on the linking to the investment returns, developing the so-called *participating policies*.

Participating policies have gained large market shares during the Eighties, in particular in continental Europe. At the same time, new products were developed; in the UK, in particular, the business of unit-linked policies, with and without financial guarantees, reached a large importance. In the US market, *Universal Life policies* were introduced, which are characterized by many options available to the policyholder in regard of the amount of premiums, possible withdrawals, selection of the asset mix, as well as by the detailed information provided to the policyholder about the costs charged by the insurer. During the Nineties, the innovations produced by financial engineering have supported the design of *index-linked policies*, which are single premium contracts whose return is linked to stock-market indexes. Nowadays, insurers are increasingly addressing *variable annuity policies* which, first developed in the US, offer savings opportunities during the working life of an individual, and then private pension solutions after her retirement.

In the following sections, we describe the technical features of the policies mentioned above.

7.2 Adjusting benefits

In this Section we illustrate what requirement must be satisfied when the benefit amount, first defined at policy issue following the pricing model for fixed benefits (see Chap. 4), is adjusted during the policy duration. This description is useful to understand how we can realize a linking of benefits to some index, as we discuss in Sect. 7.3. The actuarial structure defined below (see, in particular, Sect. 7.2.1) extends the basic actuarial model defined in the previous chapters.

7.2.1 The general case

Refer to a life insurance policy whose current duration is t years. With reference to the policy anniversary t, we denote by t^- the time just before premium payment. At time t^-, the reserve V_{t^-} has to be available, so to realize the actuarial balance between future benefits and future premiums:

$$V_{t^-} = \mathrm{Ben}'(t^-,m) - \mathrm{Prem}'(t^-,m) \tag{7.2.1}$$

(see (5.3.3)). To avoid any misunderstanding, we point out that (similarly to assumptions in (5.3.3)), the actuarial value $\mathrm{Ben}'(t^-,m)$ does not include the benefit due at the end of year $(t-1,t)$, while the actuarial value $\mathrm{Prem}'(t^-,m)$ includes the premium due at time t for year $(t,t+1)$ (see Sect. 7.2.2 for examples). The notation t^- only recalls that the quantities referred to are considered prior to possible adjustments occurring at time t. From (7.2.1) we obtain the balance condition

$$V_{t^-} + \mathrm{Prem}'(t^-,m) = \mathrm{Ben}'(t^-,m) \tag{7.2.2}$$

(see also (5.3.1)), where $\mathrm{Ben}'(t^-,m)$ represents the gross liability of the insurer at time t^- (the net liability corresponding to the difference $\mathrm{Ben}'(t^-,m) - \mathrm{Prem}'(t^-,m)$). Equation (7.2.2) shows us that such a liability is funded by the current assets, whose value is V_{t^-}, joint to the assets to be purchased with future premiums, whose current actuarial value is $\mathrm{Prem}'(t^-,m)$.

Assume now that at time t^- the benefit amount is adjusted, so that the value of benefits increases at the rate $j_t^{[B]}$. The technical basis is not changed. To maintain the actuarial balance between assets (current and future) and (gross) liabilities, the quantity in the left-hand side of (7.2.2) must also be increased at the rate $j_t^{[B]}$. Thus, the new balance condition is expressed as follows

$$(V_{t^-} + \mathrm{Prem}'(t^-,m))\,(1 + j_t^{[B]}) = \mathrm{Ben}'(t^-,m)\,(1 + j_t^{[B]}) \tag{7.2.3}$$

Equation (7.2.3) does not require that both the reserve and the future premiums are increased at the rate $j_t^{[B]}$ (for example, this would not be possible if $\mathrm{Prem}'(t^-,m) = 0$, i.e. if the policy is paid-up); what is required is that their total value is increased at the rate $j_t^{[B]}$. Different rates of adjustment of the reserve and the future premiums, respectively denoted by $j_t^{[V]}$ and $j_t^{[\Pi]}$, can be adopted, provided that the following balance condition is satisfied

$$V_{t^-}\,(1 + j_t^{[V]}) + \mathrm{Prem}'(t^-,m)\,(1 + j_t^{[\Pi]}) = \mathrm{Ben}'(t^-,m)\,(1 + j_t^{[B]}) \tag{7.2.4}$$

Since (7.2.2) must be fulfilled, (7.2.4) requires

$$V_{t^-}\,j_t^{[V]} + \mathrm{Prem}'(t^-,m)\,j_t^{[\Pi]} = \mathrm{Ben}'(t^-,m)\,j_t^{[B]} \tag{7.2.5}$$

(as we obtain by subtracting (7.2.2) from (7.2.4)). Equation (7.2.5) expresses that the adjustments of the reserve and premiums must be on actuarial balance with the benefit adjustment.

Equation (7.2.5) admits an infinite number of solutions, as it has three unknowns (namely, the three rates of adjustment $j_t^{[V]}$, $j_t^{[\Pi]}$, $j_t^{[B]}$). The amount $V_{t^-}\,j_t^{[V]}$, the so-called *reserve jump*, or simply *reserve adjustment*, is funded by the insurer, so to share profits with the policyholder. To keep control of the relevant cost charged to the insurer, in common practice first the value for $j_t^{[V]}$ is selected, according to policy conditions (see Sect. 7.2.3 and 7.3 for some examples). Then, still according to policy conditions, a value for the rate $j_t^{[\Pi]}$ is chosen. Finally, the rate $j_t^{[B]}$ is calculated so that (7.2.5) is satisfied. It is then interesting to obtain an expression for $j_t^{[B]}$ from (7.2.5). We find

$$j_t^{[B]} = \frac{j_t^{[V]}\,V_{t^-} + j_t^{[\Pi]}\,\mathrm{Prem}'(t^-,m)}{\mathrm{Ben}'(t^-,m)} \tag{7.2.6}$$

and, replacing $\mathrm{Ben}'(t^-,m)$ according to (7.2.2),

$$j_t^{[B]} = \frac{j_t^{[V]}\,V_{t^-} + j_t^{[\Pi]}\,\mathrm{Prem}'(t^-,m)}{V_{t^-} + \mathrm{Prem}'(t^-,m)} \tag{7.2.7}$$

Equation (7.2.7) shows us that the rate of adjustment of the benefit, $j_t^{[B]}$, is a weighted average of the rate of adjustment of the reserve, $j_t^{[V]}$, and the rate of adjustment of premiums, $j_t^{[\Pi]}$. The weights, respectively V_{t^-} and $\mathrm{Prem}'(t^-,m)$, change in time; in particular, since we are addressing insurance covers characterized by a significant savings component (such as endowments or whole life assurances), we should expect that the weight of level premiums decreases in time, while the weight of the reserve increases. This means that when t is small (in particular, close to time 0), we should expect that $j_t^{[B]}$ is closer to $j_t^{[\Pi]}$ than to $j_t^{[V]}$; vice versa, when t is high, and in particular close to the maturity m, we should expect that $j_t^{[B]}$ is closer to $j_t^{[V]}$ than to $j_t^{[\Pi]}$. Of course, whenever $\text{Prem}'(t^-,m) = 0$, such as for example in the case of single premium, or paid-up policy, or if $t = m$, then $j_t^{[B]} = j_t^{[V]}$ (see also examples in Sect.7.2.2).

Some remarks are useful to conclude this Section. We first note that the reserve to be set up at time t, after the benefit adjustment but before the premium payment, is

$$V_t = V_{t^-}\,(1 + j_t^{[V]}) \tag{7.2.8}$$

From (7.2.4) we obtain the relevant expression in terms of actuarial value of future benefits and premiums, namely

$$V_t = \text{Ben}'(t^-,m)\,(1 + j_t^{[B]}) - \text{Prem}'(t^-,m)\,(1 + j_t^{[\Pi]}) \tag{7.2.9}$$

If we let

$$\text{Ben}'(t,m) = \text{Ben}'(t^-,m)\,(1 + j_t^{[B]}) \tag{7.2.10}$$

$$\text{Prem}'(t,m) = \text{Prem}'(t^-,m)\,(1 + j_t^{[\Pi]}) \tag{7.2.11}$$

we can also write

$$V_t = \text{Ben}'(t,m) - \text{Prem}'(t,m) \tag{7.2.12}$$

which shows us that V_t is a prospective reserve (see (5.3.3)), as it is desirable, given that the adjustment has not been motivated by the need of revising the logic for the calculation of the reserve. Note that the reserve V_t is assessed considering the updated benefit and premium amounts, while the technical basis is unchanged.

From the notation used above, it should be clear that the adjustment only involves benefits, net premiums and the net reserve. However, if $j_t^{[\Pi]} > 0$, what is actually adjusted is the expense-loaded premium; this way, also the expense loading would be adjusted, which is not strictly required by the model. However, the adjustment of the expense loading could compensate the insurer for inflationary effects affecting administration expenses. Anyhow, we point out that if the contract is designed so that $j_t^{[\Pi]} > 0$, then in principle the loading set at issue for administration expenses should be lower than what applied to contracts designed so that $j_t^{[\Pi]} = 0$.

In the presentation above, we have excluded from the reserve at time t^- the benefit due at the end of year $(t-1,t)$. In particular, this means that the reserve V_{t^-} is set up just for those policies whose insured is still alive at time t. From a technical point of view, it would be possible to include in V_{t^-} also the benefit due at time t, in which case V_{t^-} would be meant as the reserve set up at time t^- for all the policies in-force at the beginning of the year, i.e. at time $t-1$. From a formal point of view, we can still adopt the model presented above. The weights in (7.2.7) would be different, as the reserve would be higher than what considered above. In the following, we make reference only to the first interpretation (i.e., V_{t^-} does not include the benefit due at time t, unless it is the maturity benefit at time $t = m$), as this is the prevailing approach in actuarial practice.

7.2.2 Addressing specific insurance products

In this Section, we show how the model for the adjustment of benefits described in Sect. 7.2.1 applies to specific insurance covers.

We first refer to a standard endowment insurance, issued at time 0, with maturity m, benefit C and annual level premium P. We assume that P has been calculated according to (4.4.18) with $s = m$. The policy is designed so that at each policy anniversary it is possible to adjust the benefits, following (7.2.7). We then need to extend the notation for benefits and premiums.

The premium to be paid at time t, after the adjustment at that time, is denoted by P_t. We set $P_0 = P$, while

$$P_t = P_{t-1}\,(1 + j_t^{[\Pi]}) \tag{7.2.13}$$

The death benefit payable at time $t+1$ is denoted by C_{t+1}; in particular, $C_1 = C$, while

$$C_{t+1} = C_t\,(1 + j_t^{[B]}) \tag{7.2.14}$$

Note that, following the more common practice, we are assuming that the adjustment is applied at time t just to policies in-force at that time.

For the survival benefit (payable at maturity), we denote by S_t the amount defined at time t. In particular, $S_0 = C$ is the amount defined at issue, while

$$S_t = S_{t-1}\,(1 + j_t^{[B]}) \tag{7.2.15}$$

is the amount defined at time t. Then, we simply have $S_t = C_{t+1}$ for $t = 0, 1, \ldots, m-1$, as it is reasonable given that we are addressing a standard endowment insurance. At time m, it is possible to make a final adjustment for in-force policies, so that $S_m = S_{m-1}\,(1 + j_m^{[B]})$, i.e. $S_m = C_m\,(1 + j_m^{[B]})$.

Remark As we have commented after (7.2.14), we are assuming that the death benefit due in year $(t-1, t)$ is the benefit last adjusted at the beginning of the year, i.e. at time $t-1$. When assessing the premium according to (4.4.18), we assume that the death benefit is paid at the end of the year of death; thus, at time t it would be possible to adjust the benefit before payment, as it occurs at time m for the maturity benefit. However, as we have already pointed out, the rate $j_t^{[B]}$ in (7.2.7) refers to a policy in-force at time t^-, thus excluding an adjustment of the death benefit due at time t. This is annoying in particular at maturity, when the in-force policies receive a higher amount than the policies reporting a death in the last year. However, in practice the death benefit is paid upon death; in the last year, in particular, the death benefit is thus paid on average before maturity. We further note that in practice insurers are willing to adjust the death benefit at the time of payment, in proportion to the time spent by the policy in the portfolio in the year of death. For the sake of brevity, we disregard this possibility and we continue to refer to benefits assessed as in (7.2.14).

For the reserve, we have (see (7.2.1) and (7.2.12))

$$V_{t^-} = C_t A'_{x+t,m-t|} - P_{t-1}\ddot{a}'_{x+t:m-t|} \tag{7.2.16}$$

$$V_t = C_{t+1} A'_{x+t,m-t|} - P_t\ddot{a}'_{x+t:m-t|} \tag{7.2.17}$$

for $t = 1, 2, \ldots, m-1$, while for $t = m$ we have

$$V_{m^-} = S_{m-1} \tag{7.2.18}$$

$$V_m = S_m \tag{7.2.19}$$

(clearly, for $t = 0$ we have $V_0 = 0$, while V_{0^-} is not defined, given that at time 0^- the policy does not yet exist).

We note that, thanks to (7.2.7), the rate of the adjustment of the benefit $j_t^{[B]}$ is an intermediate value between $j_t^{[V]}$ and $j_t^{[\Pi]}$ for $t = 1, 2, \ldots, m-1$; at time m, since no premium remains to be paid, we simply have $j_m^{[B]} = j_m^{[V]}$.

Example 7.2.1. Refer to a standard endowment insurance, issued at age $x = 50$, with maturity $m = 15$ and benefit $C = 1000$. Adopting the technical basis $\mathrm{TB1} = (0.02, \mathrm{LT1})$, we find $P = 59.54$. Table 7.2.1 quotes the development in time of the benefits if in each year the reserve is adjusted at the rate $j_t^{[V]} = 0.03$, while the premium remains unchanged. Note that in each year $0 < j_t^{[B]} \le 0.03$, and in particular $j_m^{[B]} = 0.03$, given that $j_t^{[B]}$ is the weighted average of $j_t^{[V]}$ and $j_t^{[\Pi]}$; note also that $j_t^{[B]}$ is increasing in time, due to the increasing weight of the reserve.

Table 7.2.1 Adjusting the benefits of an endowment insurance; $j_t^{[\Pi]} = 0$

t	$j_t^{[V]}$	$j_t^{[\Pi]}$	$j_t^{[B]}$	P_t	C_t	S_t	V_{t^-}	V_t
0				59.54		1 000.00		0.00
1	3%	0%	0.225%	59.54	1 000.00	1 002.25	57.54	59.27
2	3%	0%	0.452%	59.54	1 002.25	1 006.78	117.87	121.41
3	3%	0%	0.678%	59.54	1 006.78	1 013.61	181.13	186.57
4	3%	0%	0.903%	59.54	1 013.61	1 022.76	247.49	254.92
5	3%	0%	1.126%	59.54	1 022.76	1 034.28	317.14	326.65
6	3%	0%	1.345%	59.54	1 034.28	1 048.19	390.26	401.97
7	3%	0%	1.559%	59.54	1 048.19	1 064.53	467.08	481.10
8	3%	0%	1.766%	59.54	1 064.53	1 083.33	547.84	564.28
9	3%	0%	1.968%	59.54	1 083.33	1 104.65	632.81	651.79
10	3%	0%	2.161%	59.54	1 104.65	1 128.52	722.28	743.95
11	3%	0%	2.347%	59.54	1 128.52	1 155.00	816.59	841.09
12	3%	0%	2.523%	59.54	1 155.00	1 184.14	916.12	943.60
13	3%	0%	2.691%	59.54	1 184.14	1 216.01	1 021.30	1 051.94
14	3%	0%	2.850%	59.54	1 216.01	1 250.67	1 132.63	1 166.61
15	3%	0%	3.000%		1 250.67	1 288.19	1 250.67	1 288.19

Table 7.2.2 quotes the rate of adjustment of premiums that would be required in each year to obtain the same benefit adjustment plotted in Table 7.2.1, but setting $j_t^{[V]} = 0$. Since $j_t^{[B]}$ is the weighted average of $j_t^{[V]}$ and $j_t^{[\Pi]}$, it turns out $0 \le j_t^{[B]} < j_t^{[\Pi]}$, with $j_t^{[\Pi]}$ increasing in time due to the decreasing weight of future premiums. Clearly, this arrangement cannot be applied recursively, but could be of some interest if applied once during the policy duration (see Sect. 7.2.3). Note that, contrarily to the example in Table 7.2.2, $j_m^{[B]} = 0$, as the policy is paid-up at time m and the reserve is not adjusted.

Table 7.2.2 Adjusting the benefits of an endowment insurance; $j_t^{[V]} = 0$

t	$j_t^{[V]}$	$j_t^{[\Pi]}$	$j_t^{[B]}$	P_t	C_t	S_t	V_{t^-}	V_t
0				59.54		1 000.00		0.00
1	0%	0.243%	0.225%	59.68	1 000.00	1 002.25	57.54	57.54
2	0%	0.531%	0.452%	60.00	1 002.25	1 006.78	116.25	116.25
3	0%	0.870%	0.678%	60.52	1 006.78	1 013.61	176.32	176.32
4	0%	1.272%	0.903%	61.29	1 013.61	1 022.76	238.01	238.01
5	0%	1.751%	1.126%	62.37	1 022.76	1 034.28	301.60	301.60
6	0%	2.327%	1.345%	63.82	1 034.28	1 048.19	367.46	367.46
7	0%	3.026%	1.559%	65.75	1 048.19	1 064.53	436.05	436.05
8	0%	3.890%	1.766%	68.31	1 064.53	1 083.33	507.95	507.95
9	0%	4.983%	1.968%	71.71	1 083.33	1 104.65	583.92	583.92
10	0%	6.417%	2.161%	76.31	1 104.65	1 128.52	664.97	664.97
11	0%	8.405%	2.347%	82.72	1 128.52	1 155.00	752.53	752.53
12	0%	11.430%	2.523%	92.18	1 155.00	1 184.14	848.73	848.73
13	0%	16.892%	2.691%	107.75	1 184.14	1 216.01	957.07	957.07
14	0%	31.535%	2.850%	141.73	1 216.01	1 250.67	1 084.42	1 084.42
15	0%		0.000%		1 250.67	1 250.67	1 250.67	1 250.67

Tables 7.2.3 and 7.2.4 assume, respectively, $j_t^{[\Pi]} = j_t^{[V]}$ and $j_t^{[\Pi]} = 0.5\, j_t^{[V]}$. In the former case, we find trivially $j_t^{[B]} = j_t^{[V]}$; in the latter, $j_t^{[\Pi]} < j_t^{[B]} \le j_t^{[V]}$, with $j_t^{[B]}$ increasing in time. In both cases, the amount of the benefit is at any time higher than in the example of Table 7.2.1, and this is due to the higher premiums. If the cost of the update of the reserve is charged to the insurer, as it is usually the case, and the policyholder refers to $j_t^{[B]}$ to get an idea of the size of the profit distributed by the insurer, then arrangements in Tables 7.2.3 and 7.2.4 seem more appealing than arrangement in Table 7.2.1; we stress, however, that $j_t^{[B]}$ is also the result of the adjustment of premiums. In order to understand the size of the profit distributed by the insurer, reference should be made to $j_t^{[V]}$ only; see also Sect. 7.2.4 for further analyses in this regard.

□

We now refer to a whole life insurance, issued at time 0, with benefit C and annual level premiums P payable for s years, assessed according to (4.4.20). Equations (7.2.13) and (7.2.14), describing respectively the adjusted premium and the

Table 7.2.3 Adjusting the benefits of an endowment insurance; $j_t^{[\Pi]} = j_t^{[V]}$

t	$j_t^{[V]}$	$j_t^{[\Pi]}$	$j_t^{[B]}$	P_t	C_t	S_t	V_{t^-}	V_t
0				59.54		1 000.00		0.00
1	3%	3%	3%	61.33	1 000.00	1 030.00	57.54	59.27
2	3%	3%	3%	63.16	1 030.00	1 060.90	119.59	123.18
3	3%	3%	3%	65.06	1 060.90	1 092.73	186.44	192.03
4	3%	3%	3%	67.01	1 092.73	1 125.51	258.39	266.14
5	3%	3%	3%	69.02	1 125.51	1 159.27	335.77	345.85
6	3%	3%	3%	71.09	1 159.27	1 194.05	418.97	431.54
7	3%	3%	3%	73.23	1 194.05	1 229.87	508.37	523.62
8	3%	3%	3%	75.42	1 229.87	1 266.77	604.42	622.55
9	3%	3%	3%	77.68	1 266.77	1 304.77	707.61	728.84
10	3%	3%	3%	80.02	1 304.77	1 343.92	818.48	843.03
11	3%	3%	3%	82.42	1 343.92	1 384.23	937.65	965.78
12	3%	3%	3%	84.89	1 384.23	1 425.76	1 065.80	1 097.77
13	3%	3%	3%	87.43	1 425.76	1 468.53	1 203.72	1 239.83
14	3%	3%	3%	90.06	1 468.53	1 512.59	1 352.30	1 392.87
15	3%		3%		1 512.59	1 557.97	1 512.59	1 557.97

Table 7.2.4 Adjusting the benefits of an endowment insurance; $j_t^{[\Pi]} = 0.5\, j_t^{[V]}$

t	$j_t^{[V]}$	$j_t^{[\Pi]}$	$j_t^{[B]}$	P_t	C_t	S_t	V_{t^-}	V_t
0				59.54		1 000.00		0.00
1	3%	1.5%	1.613%	60.43	1 000.00	1 016.13	57.54	59.27
2	3%	1.5%	1.724%	61.34	1 016.13	1 033.65	118.73	122.29
3	3%	1.5%	1.835%	62.26	1 033.65	1 052.62	183.77	189.28
4	3%	1.5%	1.944%	63.19	1 052.62	1 073.08	252.89	260.47
5	3%	1.5%	2.052%	64.14	1 073.08	1 095.11	326.32	336.11
6	3%	1.5%	2.158%	65.10	1 095.11	1 118.74	404.34	416.47
7	3%	1.5%	2.262%	66.08	1 118.74	1 144.04	487.23	501.85
8	3%	1.5%	2.363%	67.07	1 144.04	1 171.07	575.32	592.58
9	3%	1.5%	2.462%	68.08	1 171.07	1 199.91	668.95	689.02
10	3%	1.5%	2.558%	69.10	1 199.91	1 230.60	768.54	791.60
11	3%	1.5%	2.652%	70.13	1 230.60	1 263.24	874.52	900.76
12	3%	1.5%	2.743%	71.19	1 263.24	1 297.90	987.40	1 017.02
13	3%	1.5%	2.832%	72.25	1 297.90	1 334.65	1 107.75	1 140.98
14	3%	1.5%	2.917%	73.34	1 334.65	1 373.58	1 236.23	1 273.31
15	3%		3.000%		1 373.58	1 414.79	1 373.58	1 414.79

adjusted benefits, apply also to this case (the former, clearly, for $t = 1,2,\dots,s-1$, and with $P_0 = P(s)$). For the reserve, we have

$$V_{t^-} = C_t A'_{x+t} - P_{t-1}\, \ddot{a}_{x+t:s-t|} \tag{7.2.20}$$

$$V_t = C_{t+1} A'_{x+t} - P_t\, \ddot{a}_{x+t:s-t|} \tag{7.2.21}$$

for $t = 1,2,\dots,s-1$, and

$$V_{t^-} = C_t A'_{x+t} \tag{7.2.22}$$
$$V_t = C_{t+1} A'_{x+t} \tag{7.2.23}$$

for $t = s, s+1, \ldots$. The rate of adjustment of the benefit, $j_t^{[B]}$, is an average of the rates of adjustment of the reserve and premiums as long as there remain premiums to be paid, i.e. for $t = 1, 2, \ldots, s-1$; conversely, for $t = s, s+1, \ldots$, it turns out $j_t^{[B]} = j_t^{[V]}$, as the policy is paid-up.

Example 7.2.2. We consider a whole life assurance, issued for a person age $x = 50$, with annual level premiums payable for $s = 15$ years, benefit $C = 1000$. Adopting the technical basis TB1 $= (0.02, \text{LT1})$, we find $P = 44.90$. Table 7.2.5 quotes the development in time of the benefits if in each year the reserve is adjusted at the rate $j_t^{[V]} = 0.03$, while the premium remains unchanged. Note that for $t = 1, 2, \ldots, 14$ we find $0 < j_t^{[B]} < 0.03$, given that $j_t^{[B]}$ is the weighted average of $j_t^{[V]}$ and $j_t^{[\Pi]}$; note also that $j_t^{[B]}$ is increasing in time, due to the increasing weight of the reserve. For $t = 15, 16, \ldots$, we find $j_t^{[B]} = j_t^{[V]}$, as the policy is paid-up.

Table 7.2.5 Adjusting the benefits of a whole life insurance; $j_t^{[\Pi]} = 0$

t	$j_t^{[V]}$	$j_t^{[\Pi]}$	$j_t^{[B]}$	P_t	C_t	V_{t^-}	V_t
0				44.90			0.00
1	3%	0%	0.221%	44.90	1 000.00	42.57	43.84
2	3%	0%	0.444%	44.90	1 002.21	87.09	89.70
3	3%	0%	0.667%	44.90	1 006.66	133.67	137.68
4	3%	0%	0.889%	44.90	1 013.37	182.40	187.87
5	3%	0%	1.109%	44.90	1 022.38	233.40	240.40
6	3%	0%	1.325%	44.90	1 033.72	286.77	295.38
7	3%	0%	1.538%	44.90	1 047.42	342.65	352.93
8	3%	0%	1.745%	44.90	1 063.53	401.17	413.21
9	3%	0%	1.946%	44.90	1 082.09	462.48	476.35
10	3%	0%	2.141%	44.90	1 103.15	526.74	542.54
11	3%	0%	2.328%	44.90	1 126.76	594.13	611.95
12	3%	0%	2.508%	44.90	1 152.99	664.85	684.79
13	3%	0%	2.680%	44.90	1 181.90	739.11	761.29
14	3%	0%	2.844%	44.90	1 213.58	817.18	841.69
15	3%		3.000%		1 248.09	899.32	926.30
16	3%		3.000%		1 285.53	939.32	967.50
17	3%		3.000%		1 324.09	980.79	1 010.22
18	3%		3.000%		1 363.82	1 023.78	1 054.49
19	3%		3.000%		1 404.73	1 068.30	1 100.35
20	3%		3.000%		1 446.87	1 114.40	1 147.83
...	...		...		...	...	...

□

Refer now to an immediate life annuity in arrears, issued at age x, with initial amount for the annual benefit b. We still denote by $j_t^{[B]}$ the rate of adjustment of the benefit at time t. Since the policy is funded with a single premium, from (7.2.7) we find $j_t^{[B]} = j_t^{[V]}$ at any time t, $t = 1,2,\dots$ The benefit adjusted at time t is then

$$b_t = b_{t-1}\,(1 + j_t^{[V]}) \tag{7.2.24}$$

with $b_0 = b$. According to assumptions underlying (7.2.7), the benefit paid at time t has been last adjusted at time $t-1$; thus, for the reserve we find

$$V_{t^-} = b_{t-1}\,a'_{x+t} \tag{7.2.25}$$
$$V_t = b_t\,a'_{x+t} \tag{7.2.26}$$

We note that since the benefit at time t is paid to in-force policies, it would be reasonable to adjust the benefit right before payment (so that b_t would be the benefit paid at time t), similarly to what happens for the maturity benefit of endowment policies. The choice of one solution or the other also depends on the frequency of payment of the benefit, which is here assumed annual, but can be monthly or other (in this latter case, it is reasonable to apply the adjustment at time t to benefits which will be paid after that time).

We do not give a numerical example on the adjustment of the annuity benefits, as we would simply find $j_t^{[B]} = j_t^{[V]}$ at any time t.

7.2.3 Implementing solutions

As noted in Sect. 7.2.1, Eq. (7.2.5) admits an infinite number of solutions (unless $\mathrm{Prem}'(t^-,m) = 0$, in which case we simply find $j_t^{[B]} = j_t^{[V]}$). Table 7.2.6 summarizes the logic commonly followed to select particular solutions; some of these solutions have already been considered in the numerical examples of Sect. 7.2.2, as we recall below.

Table 7.2.6 Solutions for the adjustment of benefits

Solution	$j_t^{[V]}$	$j_t^{[\Pi]}$	$j_t^{[B]}$
I	>0	0	>0
II	0	>0	>0
III	>0	>0	>0
IV	>0	<0	0

Solution I (considered in Table 7.2.1 for the endowment insurance and in Table 7.2.5 for the whole life assurance), is usually adopted when the insurer wants to share financial profits with the policyholder; actually, the cost of updating the reserve is charged to the insurer. As already pointed out, first $j_t^{[V]}$ is chosen, according to the financial profit gained by the insurer; then $j_t^{[B]}$ is calculated through (7.2.7). It turns out: $0 < j_t^{[B]} \leq j_t^{[V]}$ (see also remarks in this respect in Sects. 7.2.1 and 7.2.2). From an actuarial point of view, we note that (7.2.5) reduces to:

$$V_{t^-}\, j_t^{[V]} = \mathrm{Ben}'(t^-,m)\, j_t^{[B]} \tag{7.2.27}$$

which shows us that the increase of the cost of benefits, namely the right-hand side of (7.2.27), is immediately funded with an increase of the reserve (see left-hand side of (7.2.27)). The arrangement is also referred to as an *adjustment of benefits with constant premiums*, given that premiums remain unchanged.

Solution II (considered in Table 7.2.2 for the endowment insurance) is never applied recursively. Actually, the cost of adjusting the benefits is fully charged to the policyholder and it turns out $0 \leq j_t^{[B]} < j_t^{[\Pi]}$, with $j_t^{[\Pi]}$ much larger than $j_t^{[B]}$ when t is close to the maximum duration of premium payment. Such a solution finds application when an *insurability guarantee* (or *benefit increase option*) has been underwritten within policy conditions; according to this guarantee, the policyholder may apply for an increase of the benefit amount in face of some specific events (typically concerning her household, such as the birth of a child), without being adopted a reinforced technical basis. Without the guarantee, should a benefit increase be required by the policyholder, a revision of the premium rate could be applied, to prevent adverse selection. Equation (7.2.5) shows us that

$$\mathrm{Prem}'(t^-,m)\, j_t^{[\Pi]} = \mathrm{Ben}'(t^-,m)\, j_t^{[B]} \tag{7.2.28}$$

which means that the increase of the cost of benefits (quantity in the right-hand side of (7.2.28)) is amortized over the residual duration of premium payment (see left-hand side of (7.2.28)).

In solution III (considered in Tables 7.2.3 and 7.2.4 for the endowment insurance), the increase of the cost of benefits (right-hand side of (7.2.5)) is partially funded immediately, through the adjustment of the reserve, and partially amortized over the residual duration of premium payment, through the adjustment of premiums (see left-hand side of (7.2.5)). The adjustment of the reserve is charged to the insurer, which this way shares profits with the policyholder, while the adjustment of premiums is charged to the policyholder. First, the insurer selects $j_t^{[V]}$ consistently with the realized financial profit; then, $j_t^{[\Pi]}$ is set according to policy conditions. Usually, $j_t^{[\Pi]}$ is a proportion of $j_t^{[V]}$, namely $j_t^{[\Pi]} = \gamma_t\, j_t^{[V]}$ with $0 \leq \gamma_t \leq 1$. If $j_t^{[\Pi]} = j_t^{[V]}$ (i.e., $\gamma_t = 1$) for all t, then $j_t^{[B]} = j_t^{[\Pi]} = j_t^{[V]}$ for all t and the solution is referred to as the *adjustment scheme with three identical rates*. If $j_t^{[\Pi]} < j_t^{[V]}$ (i.e., $\gamma_t < 1$), then $j_t^{[\Pi]} < j_t^{[B]} \leq j_t^{[V]}$. Finally, if $j_t^{[\Pi]} = 0$ (i.e., $\gamma_t = 0$) for all t, we find again the case of constant premiums (i.e., solution I). When $\gamma_t > 0$, usually the policyholder may apply for setting $j_t^{[\Pi]} = 0$ from a given time t' onwards (so that $\gamma_t > 0$

for $t = 1,2,\ldots,t'-1$, while $\gamma_t = 0$ for $t = t', t'+1, \ldots, s-1$); this policy condition is called *stabilization of premiums*.

In solution IV, the adjustment of the reserve results in a premium reduction; actually, (7.2.5) reduces to

$$V_{t^-}\, j_t^{[V]} = \text{Prem}'(t^-, m)\,(-j_t^{[\Pi]}) \qquad (7.2.29)$$

The arrangement, which is not very common, can be of interest when the policyholder is not the beneficiary, e.g. in group insurance (in which the policyholder is the employer and the beneficiaries are the employees or their estate); in this case, the only way to let the policyholder participate directly to the profit of the insurer is through a premium reduction. Some triggers are required to avoid that the premium becomes too low.

As we have mentioned several times, the cost of the reserve adjustment is charged to the insurer, and is covered through financial profits. To avoid major risks, it is then important that the insurer plays a direct control on $j_t^{[V]}$, i.e. that such a rate is chosen depending on the investment yield g_t gained by the insurer in year $(t-1,t)$. We recall that when calculating the premium and the reserve, some interest (namely, those based on the technical interest rate i') is computed in advance, as all future flows are discounted with the rate i'. Then, the rate $j_t^{[V]}$ should depend on the difference between the yield on investments in year $(t-1,t)$ and the technical interest rate, i.e.

$$j_t^{[V]} = F(g_t - i') \qquad (7.2.30)$$

where $F(\cdot)$ is a given function (some examples are examined in detail in Sect. 7.3). We stress that only an appropriate choice of (7.2.30) avoids that the cost charged to the insurer is higher than the financial profit actually gained.

On the other hand, the policyholder may have some specific expectations in regard of the benefit adjustment. For example, the policyholder may be interested into recovering the depreciation of the benefit amount due to inflation. If we denote by s_t the observed inflation rate in year $(t-1,t)$, the target of the policyholder is then

$$j_t^{[B]} \geq s_t \qquad (7.2.31)$$

Given the relation linking the rates $j_t^{[B]}$, $j_t^{[V]}$ and $j_t^{[\Pi]}$, to reach target (7.2.31), while keeping $j_t^{[\Pi]} \leq j_t^{[V]}$ (as is reasonable, for commercial reasons), we need

1. $j_t^{[V]} \geq s_t$;
2. possibly, $j_t^{[\Pi]} > 0$ (but $j_t^{[\Pi]} \leq j_t^{[V]}$).

If the inflation rate is low, constraints (7.2.30) and (7.2.31) can be easily fulfilled simultaneously, as it is likely that condition 1 above is realized. Conversely, if the inflation rate is high, it can be difficult to meet (7.2.31), because in this case it is hard to realize condition 1. We note that when the inflation rate is low, constraint (7.2.31) loses importance. Indeed, in this case the policyholder's expectation is addressed to the return on investment, which is expected to be as high as possible, and in any case in line with the prevailing market rates.

Table 7.2.7 summarizes several arrangements which have been adopted in the market, or which have been investigated by insurers; some of these arrangements are no longer applied, or are not feasible in practice, as we comment below in more detail. To simplify the presentation, in Table 7.2.7 we assume that the technical interest rate is $i' = 0$; the case of a positive technical interest rate, i.e. $i' > 0$, is discussed in Sect. 7.3. If $i' = 0$, no interest is computed in advance, so that the difference $g_t - i'$ in (7.2.30) simply reduces to g_t, i.e. the realized investment yield. The notation $j_t^{[B]} = \varphi(j_t^{[V]}, j_t^{[\Pi]})$, which we use in Table 7.2.7, expresses that $j_t^{[B]}$ is the weighted average of $j_t^{[V]}$ and $j_t^{[\Pi]}$, as defined by (7.2.7); when it is clear, we write explicitly the value of such an average. Finally, the notation $j_t^{[\Pi]} = \psi(j_t^{[V]}, j_t^{[B]})$ is used to express that, once $j_t^{[V]}$ and $j_t^{[B]}$ have been chosen, $j_t^{[\Pi]}$ must be set so to fulfill the balance condition (7.2.5).

Table 7.2.7 Particular solutions for the adjustment of benefits

Model	$j_t^{[V]}$	$j_t^{[\Pi]}$	$j_t^{[B]}$	Remarks
1a	s_t	s_t	s_t	
1b	$h(s_t)$	$h(s_t)$	$h(s_t)$	$0 \leq h(s_t) < s_t$
1c	$h(s_t)$	$\phi(s_t)$	$\varphi(h(s_t), \phi(s_t))$	$0 \leq \phi(s_t) < h(s_t) < s_t$
2a	$\eta_t\, g_t$	$\eta_t\, g_t$	$\eta_t\, g_t$	$0 < \eta_t < 1$
2b	$\eta_t\, g_t$	$\gamma_t\, \eta_t\, g_t$	$\varphi(\eta_t\, g_t, \gamma_t\, \eta_t\, g_t)$	$0 < \eta_t < 1,\ 0 \leq \gamma_t \leq 1$
3a	$\eta_t\, g_t$	$\psi(\eta_t\, g_t, s_t)$	s_t	$0 < \eta_t < 1$
3b	$\eta_t\, g_t$	$\psi(\eta_t\, g_t, \alpha\, s_t)$	$\alpha\, s_t$	$0 < \eta_t < 1,\ 0 < \alpha < 1$
4a	$\eta_t\, g_t$	s_t	$\varphi(\eta_t\, g_t, s_t)$	$0 < \eta_t < 1$
4b	$\eta_t\, g_t$	$\min\{\eta_t\, g_t, s_t\}$	$\varphi(\eta_t\, g_t, \min\{\eta_t\, g_t, s_t\})$	$0 < \eta_t < 1$
4c	$\eta_t\, g_t$	$\max\{\eta_t\, g_t, s_t\}$	$\varphi(\eta_t\, g_t, \max\{\eta_t\, g_t, s_t\})$	$0 < \eta_t < 1$

Models 1a–1c realize a linking to inflation; we refer to the relevant policy design as *inflation-linked policies*. Model 1a, in particular, aims at recovering completely the depreciation of the benefit amount. For what stated above, the insurer could be willing to offer this arrangement if appropriate assets are available, i.e. inflation-linked securities whose return with certainty is not smaller than the inflation rate (as, in this case, the insurer would report $g_t = s_t$). Inflation-linked securities may be available, but usually just offering a partial indexation. For this reason, arrangement 1a has not been introduced in the market. Conversely, models 1b and 1c have been adopted. The function $h(s_t)$ is first chosen, according to the indexation of the available assets. For example,

$$h(s_t) = \begin{cases} s' & \text{if } \alpha\, s_t < s' \\ \alpha\, s_t & \text{if } s' \leq \alpha\, s_t < s'' \\ s'' & \text{if } \alpha\, s_t \geq s'' \end{cases} \tag{7.2.32}$$

where s', s'' and α are given values (clearly, $0 \le s' < s''$, while $0 < \alpha \le 1$). In model 1b, the same value is chosen for the rate of the adjustment of the reserve and premiums, and then of the benefit. In model 1c, premiums are increased at a lower rate than the reserve (so to reduce the direct cost to the policyholder), but this results in a lower increase of the benefit, given that because of (7.2.7) we find $\phi(s_t) < j_t^{[B]} \le h(s_t)$.

Arrangements 2a and 2b are typical of the *participating policies*. The aim, in this case, is to acknowledge to the policyholder a return on the reserve in line with the investment yield realized by the insurer. The assets are selected by the insurer, and they typically consist of bonds. The quantity η_t represents a *participating proportion*: the yield $\eta_t g_t$ is assigned to the policyholder, while $(1-\eta_t) g_t$ represents the financial profit of the insurer. In model 2a, the same value is set for $j_t^{[V]}$ and $j_t^{[\Pi]}$, and then for $j_t^{[B]}$. If the investment performance of the insurer is very good, this can result in a strong premium increase. In model 2b, then, premiums are allowed to increase at a lower rate than the reserve, clearly with $j_t^{[\Pi]} < j_t^{[B]} \le j_t^{[V]}$. If $\gamma_t = 0$ for all t, we find the case of the *participating policies with constant premiums*; as already mentioned, in some policy designs $\gamma_t > 0$ up to some time (chosen by the policyholder), after which $\gamma_t = 0$ (stabilization of premiums).

Arrangements 3a and 3b are fitted simultaneously to constraint (7.2.30) and (7.2.31): the reserve is increased depending on the realized investment yield, while the benefit is indexed to inflation. The rate of adjustment of the premium is assessed consistently, so to realize the actuarial balance (7.2.5). The resulting premium increase may be too high; thus, in model 3b just a partial indexation of the benefit is realized. We note that if $\alpha s_t < r_t$, then $j_t^{[\Pi]} < j_t^{[B]} \le j_t^{[V]}$; vice versa, if $\alpha s_t > r_t$, then $j_t^{[\Pi]} > j_t^{[B]} \ge j_t^{[V]}$

Similarly to arrangements 3a and 3b, arrangements 4a–4c represent mixed solutions in respect of inflation-linked and participating policies. In particular, model 4a assumes that the policyholder may afford an indexation of premiums (thanks to a presumable revaluation of her normal income based on the inflation rate). The main aim of model 4b is to prevent major increases of the premiums, while not disregarding the need for an indexation of the benefit amount. In solution 4c, the aim is to get the maximum possible increase of the benefit, taking as benchmarks the investment yield paid by the insurer and the inflation rate.

Inflation-linked policies were designed during the Seventies, in a period of high inflation. They have not gained importance in the market, in particular because of the unavailability of appropriate securities in many markets. Since the Eighties, a major role has been played by participating policies, to which we devote Sect. 7.3.

7.2.4 The yield to maturity for the policyholder

As we have stated previously, the purpose of the adjustment model examined so far is to provide a return on the investment of the policyholder which is higher than the technical interest rate. This is why the cost of updating the reserve is paid by the insurer. In this Section, we take the point of view of the policyholder and we discuss how the yield received from a life insurance policy can be measured.

Before going into details, it is worth making some remarks. Products for which it is reasonable to measure the yield received by the policyholder are those in which there is an accumulation process, i.e. endowments, whole life assurances, or other similar arrangements. The result of the accumulation is the benefit at maturity (for endowments) or the surrender value (for endowments and whole life assurances). In the following, for brevity we just refer to the benefit at maturity of an endowment insurance. We also recall that the premium paid by the policyholder is not the amount invested by the insurer on her behalf. As was discussed in Sect. 5.4.3, just a part of the premium, namely the savings premium, is reserved for savings purposes; however, the policyholder usually does not hold the information about the splitting of the premium, and thus quite naturally compares the benefit received with the expense-loaded premiums paid year by year.

Let us refer to a standard endowment, subject to adjustments as described in Sect. 7.2.2. As noted in Sect. 7.2.1, what is updated at the rate $j_t^{[\Pi]}$ is the expense-loaded premium, and not just the net premium (as would be required by the actuarial balance (7.2.5)). The expense-loaded premium paid at time t is then calculated as follows:

$$P_t^{[\mathrm{T}]} = P_{t-1}^{[\mathrm{T}]}\,(1 + j_t^{[\Pi]}) \tag{7.2.33}$$

with $P_0^{[\mathrm{T}]} = P^{[\mathrm{T}]}$ (where $P^{[\mathrm{T}]}$ is the initial expense-loaded premium, assessed at issue according to the selected technical basis and expense-loading parameters).

We define *yield to maturity on the expense-loaded premiums* the annual interest rate $i^{[\mathrm{T}]}$ satisfying the following equation

$$S_m = \sum_{t=0}^{m-1} P_t^{[\mathrm{T}]}\,(1 + i^{[\mathrm{T}]})^{m-t} \tag{7.2.34}$$

Clearly, $i^{[\mathrm{T}]}$ is the internal rate of return of the cash-flows paid and received by the policyholder, having assumed that the policy stays in-force until maturity.

In place of the expense-loaded premiums, it could be interesting to consider other quantities in (7.2.34), always referring to the payments by the policyholder. In particular, we can replace the expense-loaded premiums $P_t^{[\mathrm{T}]}$ with the savings premiums, $P_t^{[\mathrm{S}]}$, or with the net premiums, P_t. In some countries, a tax discount is applied to the taxpayer who has underwritten a life insurance contract; the size of such a discount usually depends on the amount of the premium paid. It could be interesting to calculate the yield to maturity on the expense-loaded premiums net of the tax discount, namely on $P_t^{[\mathrm{T}]}\,(1-\varepsilon)$ (where ε is the proportion of the tax discount). We denote by $i^{[\mathrm{S}]}$, $i^{[\Pi]}$, $i^{[\mathrm{TD}]}$ the yield to maturity obtained solving (7.2.34) after having replaced the expense-loaded premiums respectively with the savings premiums, the net premiums or the expense-loaded premiums net of the tax discount.

Example 7.2.3. We refer to the standard endowment insurance considered in Example 7.2.1. Take the expense-loading parameters of Example 4.5.1 (in Sect. 4.5.3); the initial expense-loaded premium then is $P^{[T]} = 66.60$. Just to provide an example, we assume that the tax discount proportion is $\varepsilon = 20\%$. Table 7.2.8 quotes the yield to maturity for the arrangement examined in Table 7.2.1 of Example 7.2.1; to facilitate the interpretation of the results, we quote the whole sequence of net premiums, savings premiums, office premiums gross and net of the tax discount. In each case, the premium sequence is compared to the benefit at maturity, which amounts to 1 288.19.

Table 7.2.8 Yield to maturity for an endowment insurance; $j_t^{[V]} = 3\%$, $j_t^{[\Pi]} = 0\%$, $S_m = 1\,288.19$

t	P_t	$P_t^{[S]}$	$P_t^{[T]}$	$0.80\,P_t^{[T]}$
0	59.54	56.42	66.60	53.28
1	59.54	56.29	66.60	53.28
2	59.54	56.17	66.60	53.28
3	59.54	56.08	66.60	53.28
4	59.54	56.00	66.60	53.28
5	59.54	55.96	66.60	53.28
6	59.54	55.96	66.60	53.28
7	59.54	56.01	66.60	53.28
8	59.54	56.12	66.60	53.28
9	59.54	56.32	66.60	53.28
10	59.54	56.63	66.60	53.28
11	59.54	57.07	66.60	53.28
12	59.54	57.67	66.60	53.28
13	59.54	58.48	66.60	53.28
14	59.54	59.54	66.60	53.28
yield to maturity	$i^{[\Pi]}$ =4.454%	$i^{[S]}$ =5.060%	$i^{[T]}$ =3.115%	$i^{[TD]}$ =5.763%

We note that whatever is the type of premium addressed, the yield to maturity is higher than the technical interest rate ($i' = 0.02$), thanks to the adjustment of the reserve paid by the insurer. Trivially, the yield to maturity is higher the lower is the premium amount considered. The difference between $i^{[TD]}$ and $i^{[T]}$ is due to the tax discount. The difference between $i^{[T]}$ and $i^{[\Pi]}$ is due to the expense-loading; the lower is the expense-loading applied by the insurer, the lower is this difference. Finally, the difference between $i^{[\Pi]}$ and $i^{[S]}$ is due to the cost of the sum at risk, i.e. to risk premiums. We recall that the risk premium is used for mutuality purposes, and hence does not contribute to the accumulation of the benefit at maturity.

Tables 7.2.9 and 7.2.10 refer, respectively, to the arrangements in Tables 7.2.3 and 7.2.4. In the former case, the benefit at maturity amounts to 1 557.97, in the latter to 1 414.79. Within each table, comparisons similar to those commented for Table 7.2.8 can be performed. It is more interesting to compare the yields to maturity in the different arrangements. We note that the size of the yield to maturity is more or less the same when the same type of premium is considered under the different arrangements; typically, the higher is the premium paid by the policyholder, the slightly lower is the yield to maturity. When referring to savings premiums, the yield to maturity is not affected by the specific arrangement; in all the three tables, actually we find $i^{[S]} = 5.060\%$. This can be justified noting that in all the three arrangements, the yield paid year by year by the insurer is 2% through the technical interest rate and 3% through the adjustment of the reserve. In Sect. 7.3 we will explain how 5.060% comes out (here we just note that $0.05060 = 1.02 \times 1.03 - 1$); see, in particular, Example 7.3.2.

Table 7.2.9 Yield to maturity for an endowment insurance; $j_t^{[V]} = 3\%$, $j_t^{[\Pi]} = 3\%$, $S_m = 1\,557.97$

t	P_t	$P_t^{[S]}$	$P_t^{[T]}$	$0.80\,P_t^{[T]}$
0	59.54	56.42	66.60	53.28
1	61.33	57.98	68.60	54.88
2	63.16	59.60	70.66	56.53
3	65.06	61.29	72.78	58.22
4	67.01	63.05	74.96	59.97
5	69.02	64.91	77.21	61.77
6	71.09	66.86	79.53	63.62
7	73.23	68.95	81.91	65.53
8	75.42	71.18	84.37	67.50
9	77.68	73.60	86.90	69.52
10	80.02	76.23	89.51	71.61
11	82.42	79.12	92.19	73.75
12	84.89	82.34	94.96	75.97
13	87.43	85.96	97.81	78.25
14	90.06	90.06	100.74	80.59
yield to maturity	$i^{[\Pi]} = 4.443\%$	$i^{[S]} = 5.060\%$	$i^{[T]} = 3.012\%$	$i^{[TD]} = 5.835\%$

□

From a financial point of view, the most appropriate measure for the yield to the policyholder is $i^{[S]}$. As we have noted in Example 7.2.3, only the savings premium contributes to the accumulation of the benefit at maturity; the risk premium and the expense loading are used to cover annual costs (namely, mutuality and expenses). As it has emerged in Example 7.2.3, $i^{[S]}$ corresponds to the return totally assigned to the contract by the insurer. However, unless the insurer provides the policyholder with detailed information about the costs of the contract, the policyholder can just refer to the expense-loaded premium, possibly net of the tax discount.

Table 7.2.10 Yield to maturity for an endowment insurance; $j_t^{[V]} = 3\%$, $j_t^{[\Pi]} = 1.5\%$, $S_m = 1414.79$

t	P_t	$P_t^{[S]}$	$P_t^{[T]}$	$0.80\,P_t^{[T]}$
0	59.54	56.42	66.60	53.28
1	60.43	57.13	67.60	54.08
2	61.34	57.87	68.62	54.89
3	62.26	58.64	69.64	55.72
4	63.19	59.45	70.69	56.55
5	64.14	60.30	71.75	57.40
6	65.10	61.21	72.83	58.26
7	66.08	62.19	73.92	59.13
8	67.07	63.26	75.03	60.02
9	68.08	64.45	76.15	60.92
10	69.10	65.78	77.29	61.84
11	70.13	67.28	78.45	62.76
12	71.19	69.01	79.63	63.70
13	72.25	71.00	80.83	64.66
14	73.34	73.34	82.04	65.63
yield to maturity	$i^{[\Pi]} = 4.449\%$	$i^{[S]} = 5.060\%$	$i^{[T]} = 3.065\%$	$i^{[TD]} = 5.798\%$

7.3 Participating policies

Participating policies are designed along the model examined in Sect. 7.2. As already noted in Sect. 7.2.3, the feature of the participating design stands in the way the rate of update of the reserve, $j_t^{[V]}$, is selected. Thus, in this Section we discuss about the choice of $j_t^{[V]}$. For participating policies, the rate $j_t^{[V]}$ is usually denoted as r_t, the so-called *revaluation rate*; in the following, we adopt this notation. We recall that premiums may either be updated, typically in proportion to r_t (see Table 7.2.7), or remain unchanged. Nowadays, the latter solution (namely, constant premiums) is the most common choice.

Participating policies usually provide some (implicit) guarantee for the return on the investment of the policyholder; consistently, the investment realized by the insurer is not too risky. Assets typically consist of bonds. The investment fund is internal, i.e. directly managed by the insurer; hence, the yield gained in a year not only reflects market conditions, but also the investment ability of the insurer. In some countries, the realized yield must be certified by an independent auditor; in this case, the assets backing the reserve must be objectively identifiable in respect of the overall assets of the insurer. A specific reporting is performed, and the fund is referred to as a *special (managed) fund* or *segregated fund*.

7.3.1 Participating policies with a guaranteed annual return

We let g_t denote the investment yield gained in year $(t-1,t)$ by the insurer on the assets backing the reserve of the participating business. When a certification is required, the rate g_t is the latest certified yield.

In the traditional participating arrangements, policy conditions define the total return on the investment of the policyholder in year $(t-1,t)$ as follows

$$\max\{i', \eta_t\, g_t\} \tag{7.3.1}$$

where η_t is a given proportion, the so-called *participating proportion*, and i' (as usual) the technical interest rate. Note that in (7.3.1) the technical interest rate is guaranteed in each year (given that (7.3.1) must be fulfilled for any t). The meaning of the participating proportion was already illustrated in Sect. 7.2.3. The quantity η_t can be chosen year by year by the insurer, not below a minimum value η' (stated in policy conditions, and often mandated by the supervisor) and below 1 (hence: $0 \leq \eta' \leq \eta_t < 1$; for example, $\eta' = 0.75$). In some arrangements, a waiting period is given (of 1 or 2 years), during which $\eta_t = 0$; this is justified by the fact that in the early years of the contract the credit of the insurer in respect of initial expenses is still too high (see Sect. 5.6). In general, an appropriate choice of η_t could allow to realize some smoothing of the total yield paid on the investment of the policyholder (for example, choosing a lower value for η_t than usual when the interest rate g_t is high, and vice versa a higher value η_t when g_t is low). However, in some markets it is usual to set η_t more or less constant in time. The fact that the supervisor requires $\eta_t < 1$ expresses that some profit must be retained by the insurer to face future adverse fluctuations.

Remark Of course, it could be possible, for the insurer, to retain profit also setting $\eta_t = 1$. Indeed, the application of appropriate asset management fees could replace the profit otherwise gained setting $\eta_t < 1$, and this is an approach that some insurers prefer. However, this is not a natural choice within the participating business, as participating policies embed financial guarantees, as we discuss in detail below. A financial guarantee implies some risk for the insurer, for which it must be rewarded, either charging a fee expressing the cost of the guarantee (but this is not usual for participating policies, as we comment below) or retaining profit. Vice versa, if no financial risk is charged to the insurer, than the insurer just needs to be paid for managing the policyholder's assets, and this is appropriately obtained through asset management fees.

Turning to the calculation of r_t, we recall that the total annual return on the investment of the policyholder must fulfil definition (7.3.1). First, we need to state what is the amount invested for the policyholder. Refer to a policy for which premiums are being paid. At the beginning of the year, i.e. at time $t-1$, after premium payment, the investment of the policyholder consists of the reserve V_{t-1} and the savings premium $P_{t-1}^{[S]}$ (we recall that the risk premium and the expense loading are used to fund annual costs, and hence do not contribute to the investment of the policyholder). At the end of the year, i.e. at time t, the value of the investment belonging to an in-force policy is V_t. According to (7.3.1), it must turn out

$$V_t = (V_{t-1} + P_{t-1}^{[S]})\,(1 + \max\{i', \eta_t\, g_t\}) \tag{7.3.2}$$

As described in Sect. 7.2.1, the rate r_t is used to update the reserve V_{t^-}, so to fulfil (7.3.2). According to (5.4.13), the reserve V_{t^-} can be expressed as follows

$$V_{t^-} = (V_{t-1} + P_{t-1}^{[S]})\,(1 + i') \tag{7.3.3}$$

We note that (7.3.3) holds also when a premium is not being paid, as the savings premium is defined also in this case (see Sect. 5.4.3).

The link between V_t and V_{t^-} is described by (7.2.8), with $j_t^{[V]} = r_t$. Replacing (7.3.3) into (7.2.8), it turns out

$$V_t = (V_{t-1} + P_{t-1}^{[S]})\,(1 + i')\,(1 + r_t) \tag{7.3.4}$$

Equating (7.3.2) to (7.3.4), we finally find

$$(1 + i')\,(1 + r_t) = 1 + \max\{i', \eta_t\, g_t\} \tag{7.3.5}$$

and hence

$$r_t = \max\left\{\frac{\eta_t\, g_t - i'}{1 + i'}, 0\right\} \tag{7.3.6}$$

The impossibility for r_t to fall below 0 is a consequence of the fact that the technical interest rate i' is guaranteed annually; actually, the ratio $\frac{\eta_t g_t - i'}{1+i'}$ can become negative only if $\eta_t\, g_t < i'$, namely if the realized investment yield is lower than the technical interest rate. Conversely, the investment yield above i', i.e. the difference $\eta_t\, g_t - i'$, must be divided by $1 + i'$ because interest based on the technical rate i' are computed in advance, as it is made explicit by (7.3.3).

Expression (7.3.6) corresponds to the pay-off of a financial option; indeed, the technical interest rate i' is the minimum annual return guaranteed on the investment of the policyholder.

Remark The financial option whose pay-off is described by (7.3.6) is considered to be an *embedded financial option*. While the insurer's liability is affected by such an option, the relevant cost is not explicitly charged to the policyholder. The reason can be found in the origins of participating policies. As noted in Sect. 7.1, participating policies were first designed during the Eighties of the last century. At that time, the spread between market rates and the technical rate was very high. The model described in Sect. 7.2 suggested how to pay to policyholders a return on investment in line with the yield realized by the insurer, while keeping the technical interest rate at the usual low levels. The adjustment rate in (7.3.6) represented a nice commercial solution; the risk originated by it was assumed to be negligible, given that the relevant financial option was deeply out-of-the-money. Indeed, a charge for the guarantee was not considered to be necessary. Conversely, nowadays the spread $\eta_t\, g_t - i'$ has reduced a lot, and the cost of the guarantee is no longer negligible. New definitions of r_t have been introduced, as we describe in Sect. 7.3.2. We will come back to the valuation of the guarantee in Sect. 7.5.

Since (7.3.1) guarantees in each year a return not lower than i', the yield realized above i' in a year, namely $\eta_t\, g_t - i'$, is *locked-in*, and this is shown by (7.3.6) (the option embedded in a traditional participating policy is then like a *cliquet option*).

Therefore, when assessing the rate g_t, the insurer must be sure that the return obtained in a year cannot be lost in subsequent years; in other words, the rate g_t must be permanently gained, i.e. it must have been cashed. A return based on the current value of assets is not appropriate, as the market value of assets is subject to depreciation (if market conditions change unfavorably). In principle, assets backing the reserve of a participating business are reported at historical cost (but clearly, several accounting rules apply, which we do not discuss).

Example 7.3.1. In order to show the effect of locking-in the return realized above an annual guaranteed level, we compare the following two arrangements: a participating policy without any guarantee for the annual return and a participating policy with guarantee (7.3.1). We stress that the former arrangement is purely notional, as no rational policyholder would accept it, and is quoted here only for comparison with the latter (which describes a real arrangement). If there is no guarantee, the annual return is simply $\eta_t g_t$. Assume that in both cases the technical interest rate is $i' = 0.02$. We stress that only under (7.3.1) the rate i' is the guaranteed annual return; otherwise, it is simply a computation rate (i.e. a rate used to calculate premiums and reserves; see Sect. 7.3.2 for further remarks and examples in this regard). Table 7.3.1 quotes a possible path for the annual return (with no guarantee) $\eta_t g_t$ and the corresponding annual return with guarantee (7.3.1), i.e. $\max\{\eta_t g_t, i'\}$. The rates $i_t^{[\text{ave}]}$ and $i_t^{[\text{ave,guar}]}$ are defined so that the following equations are, respectively, fulfilled:

$$(1+i_t^{[\text{ave}]})^t = \prod_{s=1}^{t}(1+\eta_s g_s) \tag{7.3.7}$$

$$(1+i_t^{[\text{ave,guar}]})^t = \prod_{s=1}^{t}(1+\max\{\eta_s g_s, i'\}) \tag{7.3.8}$$

Thus, the quantities $i_t^{[\text{ave,guar}]}$ and $i_t^{[\text{ave}]}$ are the average interest rates obtained in the time-interval $(0,t)$, either providing or not an annual guarantee.

The comparison between $\eta_t g_t$ and $\max\{\eta_t g_t, i'\}$ is straightforward. What is more interesting is to compare $i_t^{[\text{ave}]}$ to $i_t^{[\text{ave,guar}]}$. First note that $i_t^{[\text{ave}]}$ is at any time higher than 2%; so, if the target of the policyholder is to get an annual return which is on average at least 2%, for the particular path of $\eta_t g_t$ quoted in Table 7.3.1, there is no need of a guarantee. However, (7.3.1) requires that in each year the return is at least 2%, so that after having observed $\eta_t g_t < 2\%$, we find $i_t^{[\text{ave,guar}]} > i^{[\text{ave}]}$. Note that starting from the year in which $\eta_t g_t$ first falls below 2%, it turns out $i_s^{[\text{ave,guar}]} > i_s^{[\text{ave}]}$ in all the future years s, even if $\eta_s g_s \geq 2\%$; this is the effect of locking-in the extra-yield realized before it first occurs $\eta_t g_t < 2\%$.
□

In Sect. 7.2.4, we have stated that $i^{[S]}$ is an appropriate measure of the yield to maturity to the policyholder. In Example 7.2.3, we have noted in particular that, given the path of the return paid year by year by the insurer, the rate $i^{[S]}$ remains

Table 7.3.1 Annual return and annual average return, with and without guarantee (7.3.1)

t	$\eta_t\, g_t$	$i_t^{[\text{ave}]}$	$\max\{\eta_t\, g_t, i'\}$	$i_t^{[\text{ave,guar}]}$
1	5.000%	5.000%	5.000%	5.000%
2	4.500%	4.750%	4.500%	4.750%
3	4.000%	4.499%	4.000%	4.499%
4	3.000%	4.122%	3.000%	4.122%
5	2.000%	3.694%	2.000%	3.694%
6	1.500%	3.325%	2.000%	3.410%
7	2.500%	3.207%	2.500%	3.280%
8	2.000%	3.055%	2.000%	3.119%
9	3.000%	3.049%	3.000%	3.106%
10	1.000%	2.842%	2.000%	2.994%
11	2.000%	2.766%	2.000%	2.904%
12	2.500%	2.743%	2.500%	2.870%
13	1.500%	2.647%	2.000%	2.803%
14	5.000%	2.814%	5.000%	2.958%
15	5.000%	2.958%	5.000%	3.093%

the same whatever is the choice concerning the update of the premium. We can now comment more in detail. First, we note that $i^{[\text{S}]} = i_m^{[\text{ave,guar}]}$ (or $i^{[\text{S}]} = i_m^{[\text{ave}]}$, if no guarantee applies; we have already noticed that this is not a realistic situation). Thanks to (7.3.2), solving

$$S_m = \sum_{t=0}^{m-1} P_t^{[\text{S}]}\,(1+i^{[\text{S}]})^{m-t} \tag{7.3.9}$$

or solving (7.3.8) for $t = m$ is the same. Indeed, (7.3.1) expresses the annual return on the investment of the policyholder, and such an investment is formed through the savings premium. This justifies the findings of Example 7.2.3.

Example 7.3.2. In Example 7.2.3, we assumed $i' = 0.02$ and $j_t^{[\text{V}]} = 0.03$ in each year (i.e., $r_t = 0.03$ with the notation adopted in this Section). We are now able to say that the annual return paid in each year by the insurer is: $1.02 \times 1.03 - 1 = 0.05060$. Such a return is obtained on the investment of the policyholder, i.e. on the accumulation of savings premium. From this, it turns out $i^{[\text{S}]} = 0.05060$, whatever is the arrangement for the update of the annual premiums.
□

An alternative definition for the revaluation rate r_t is the following:

$$r_t = \max\left\{\frac{\eta_t\, g_t - i'}{1+i'}, r_{\min}\right\} \tag{7.3.10}$$

where $r_{\min}$ is a *minimum guaranteed annual revaluation rate* (in practice, it is usually referred to as the guaranteed rate; we prefer to avoid this terminology, as this can create some misunderstanding in respect of i', which is also guaranteed). If we take

$r_{\min} = 0$, we find (7.3.6) as a particular case. Commonly, (7.3.10) is adopted with a technical interest rate i' lower than what is otherwise usual; possibly, $i' = 0$. Then, $r_{\min}$ is set so that the usual guaranteed rate is provided; for example, if the insurer is willing to guarantee an annual return equal to 2%, it can either select $i' = 0.02$ and $r_{\min} = 0$, or $i' = 0$ and $r_{\min} = 0.02$, or $i' = 0.01$ and $r_{\min} = 0.01$, and so on. The effect of reducing i' in respect of the usual levels is to avoid computing in advance (some of) the interest which is guaranteed. For a policy reaching maturity, either computing in advance or not the guaranteed interest is to some extent the same (see Example 7.3.3); conversely, for a policy getting closed before maturity, the benefit may turn out to be lower if interest has not been computed in advance. We note that the type of financial option embedded in (7.3.10) is the same as in (7.3.6), clearly with different parameters; thus, similarly to (7.3.6), (7.3.10) implies the locking-in of extra-yields on investment.

Example 7.3.3. Refer to a participating standard endowment insurance, issued at age $x = 50$, with maturity $m = 15$ and net premium $P = 59.54$. We adopt the technical basis $\text{TB1} = (0, \text{LT1})$; thus, $i' = 0$. We set the minimum guaranteed revaluation rate $r_{\min} = 0.02$. Solving (4.4.18), we find $C = 858.75$. Table 7.3.2 quotes the development in time of the benefits, if in each year the reserve is adjusted at the rate $r_t = 0.05060$, while the premium remains unchanged.

Table 7.3.2 Participating endowment insurance; $i' = 0$, $r_{\min} = 0.02$, $j_t^{[\Pi]} = 0$

t	r_t	$j_t^{[\Pi]}$	$j_t^{[B]}$	P_t	C_t	S_t	V_{t^-}	V_t
0				59.54		858.75		0.00
1	5.060%	0%	0.335%	59.54	858.75	861.62	56.83	59.70
2	5.060%	0%	0.684%	59.54	861.62	867.52	116.45	122.34
3	5.060%	0%	1.044%	59.54	867.52	876.58	179.02	188.08
4	5.060%	0%	1.413%	59.54	876.58	888.96	244.70	257.08
5	5.060%	0%	1.785%	59.54	888.96	904.83	313.68	329.55
6	5.060%	0%	2.159%	59.54	904.83	924.37	386.15	405.69
7	5.060%	0%	2.531%	59.54	924.37	947.76	462.32	485.71
8	5.060%	0%	2.896%	59.54	947.76	975.21	542.43	569.87
9	5.060%	0%	3.252%	59.54	975.21	1 006.92	626.72	658.43
10	5.060%	0%	3.595%	59.54	1 006.92	1 043.12	715.47	751.67
11	5.060%	0%	3.924%	59.54	1 043.12	1 084.06	808.98	849.92
12	5.060%	0%	4.236%	59.54	1 084.06	1 129.98	907.60	953.52
13	5.060%	0%	4.530%	59.54	1 129.98	1 181.17	1 011.68	1 062.87
14	5.060%	0%	4.805%	59.54	1 181.17	1 237.93	1 121.63	1 178.39
15	5.060%	0%	5.060%		1 237.93	1 300.57	1 237.93	1 300.57

The example in Table 7.3.2 can be compared with the example in Table 7.2.1; the two examples differ for the technical interest rate, which is $i' = 0.02$ in Table 7.2.1 (see data in Example 7.2.1). Given the same premium amount, when $i' = 0$ (Table 7.3.2) the initial amount of the benefit is lower. Further, r_t is higher and, given that the observed yield is always higher than the minimum guaranteed level, we simply find $r_t = \eta_t g_t$. As a result, $j_t^{[B]}$ is higher in Table 7.3.2 than in Table 7.2.1. The benefit at maturity is almost the same in the two cases, as well as the reserve at any time; however, the death benefit in Table 7.3.2 is quite always lower than in Table 7.2.1. Thus, in case of early death, if some interest has not been computed in advance, it is likely that the benefit is lower. Table 7.3.3 summarizes the comparison between the arrangements in Tables 7.2.1 and 7.3.2.

Table 7.3.3 Participating endowment insurance with different technical interest rates

	arrangement in Table 7.2.1	arrangement in Table 7.3.2
initial premium, P	59.54	59.54
technical interest rate, i'	0.02	0
minimum guaranteed revaluation rate, $r_{\min}$	0	0.02
initial benefit, $C_1 = S_0$	1000	858.75
benefit at maturity, S_m	1288.19	1300.57
revaluation rate	$r_t = \max\{\frac{\eta_t g_t - 0.02}{1.02}, 0\}$	$r_t = \max\{\eta_t g_t, 0.02\}$
annual total return	$(1+i')(1+r_t) - 1$ $= 1.02 \times 1.03 - 1 = 5.06\%$	$r_t = 5.06\%$
yield to maturity, $i^{[S]}$	4.454%	4.568%

In Table 7.3.4 a further comparison is performed between arrangements considered in Tables 7.2.1 and 7.3.2, but with an alternative path for the observed investment yield (we consider the path of Example 7.3.1). Comments are straightforward. □

7.3.2 Participating policies with a guaranteed average return

As we have noted in Sect. 7.3.1, the insurer does not apply a fee for the financial options embedded in (7.3.6) and (7.3.10); indeed, the pricing of participating life insurance covers is the same as for covers with fixed-benefits. The technical justification stands in the fact that the model described in Sect. 7.2, for the adjustment of benefits, guarantees that the contract is always on actuarial balance; the economic justification, as we have mentioned in Sect. 7.3.1, stands in the fact that for many years the value of the financial options embedded in (7.3.6) and (7.3.10) has been negligible.

If we wonder about what benefits are affected by the guarantee, we first note that, because of (7.3.6) or (7.3.10), it is guaranteed that $j_t^{[B]} \geq 0$; thus, as it is natural, the benefit at maturity and the death benefit are affected by the guarantee. Also the reserve is affected by the guarantee, given that under (7.3.6) we have $r_t \geq 0$, while

Table 7.3.4 Participating endowment insurance with different technical interest rates

t	$\eta_t\, g_t$	$r_t = \max\{\frac{\eta_t\, g_t - 0.02}{1.02}, 0\}$				$r_t = \max\{\eta_t\, g_t, 0.02\}$			
		r_t	$j_t^{[B]}$	C_t	S_t	r_t	$j_t^{[B]}$	C_t	S_t
0					1 000.00				858.75
1	5.000%	2.941%	0.221%	1 000.00	1 002.21	5.000%	0.331%	858.75	861.59
2	4.500%	2.451%	0.369%	1 002.21	1 005.91	4.500%	0.608%	861.59	866.83
3	4.000%	1.961%	0.442%	1 005.91	1 010.35	4.000%	0.823%	866.83	873.96
4	3.000%	0.980%	0.293%	1 010.35	1 013.31	3.000%	0.831%	873.96	881.22
5	2.000%	0.000%	0.000%	1 013.31	1 013.31	2.000%	0.694%	881.22	887.34
6	1.500%	0.000%	0.000%	1 013.31	1 013.31	2.000%	0.831%	887.34	894.72
7	2.500%	0.490%	0.247%	1 013.31	1 015.81	2.500%	1.209%	894.72	905.53
8	2.000%	0.000%	0.000%	1 015.81	1 015.81	2.000%	1.105%	905.53	915.54
9	3.000%	0.980%	0.621%	1 015.81	1 022.11	3.000%	1.858%	915.54	932.55
10	1.000%	0.000%	0.000%	1 022.11	1 022.11	2.000%	1.375%	932.55	945.37
11	2.000%	0.000%	0.000%	1 022.11	1 022.11	2.000%	1.505%	945.37	959.60
12	2.500%	0.490%	0.402%	1 022.11	1 026.23	2.500%	2.040%	959.60	979.17
13	1.500%	0.000%	0.000%	1 026.23	1 026.23	2.000%	1.758%	979.17	996.39
14	5.000%	2.941%	2.767%	1 026.23	1 054.62	5.000%	4.701%	996.39	1 043.23
15	5.000%	2.941%	2.941%	1 054.62	1 085.64	5.000%	5.000%	1 043.23	1 095.40

under (7.3.10) we have $r_t \geq r_{\min}$. We recall that the surrender value, if any, is a part of the reserve (see Sect. 5.7); hence, also the surrender value is affected by the guarantee.

Considering that, because of the design, the financial guarantees in participating policies are embedded, which in particular means that no fee is applied, in recent times insurers have introduced new rules for the calculation of the revaluation rate r_t, aiming at reducing the cost of the guarantee. In particular, what has been weakened is the locking-in of realized extra-yields. In the following, we examine these modern designs of participating policies, assuming that $j_t^{[\Pi]} = 0$, as is common nowadays.

With reference to a participating policy, let us take the perspective of the accumulation of savings premium; in other words, we look only at the accumulation of the investment of the policyholder. For simplicity, we understand reference to an endowment policy (but what we illustrate can be also referred to other products, such as whole life assurances or life annuities).

Just for comparison, we first address a traditional policy with fixed-benefits. Such a product guarantees the investment yield i'.

Remark We stress that in this case i' is the return obtained by the policyholder on her investment, and not a minimum guaranteed rate as in participating policies. The guarantee stands in the fact that the yield paid by the insurer cannot be lower than i', but it will not be higher than i'.

The development in time of the investment of the policyholder can be described as follows (see Sect. 5.4.3)

$$V_t = (V_{t-1} + P_{t-1}^{[S]})\,(1+i') \tag{7.3.11}$$

and then

$$V_t = \sum_{s=0}^{t-1} P_s^{[S]} (1+i')^{t-s} \tag{7.3.12}$$

We denote as

$$f(s,t) = (1+i')^{t-s} \tag{7.3.13}$$

the accumulation factor applied to savings premiums.

We now refer to a participating policy with revaluation rate r_t as in (7.3.6). For making easier the comparison among different cases, in the following we denote such a revaluation rate as $r_t^{[1]}$. The development in time of the investment of the policyholder can be described as follows (see (7.3.2) and (7.3.4))

$$\begin{aligned} V_t &= (V_{t-1} + P_t^{[S]})(1+i')\left(1+\max\left\{\tfrac{\eta_t g_t - i'}{1+i'}, 0\right\}\right) \\ &= (V_{t-1} + P_t^{[S]})(1+\max\{\eta_t g_t, i'\}) \end{aligned} \tag{7.3.14}$$

Thus we can define the accumulation factor

$$f^{[1]}(t-1,t) = (1+\max\{\eta_t g_t, i'\}) \tag{7.3.15}$$

for year $(t-1,t)$ and, more in general for the time-interval (s,t)

$$f^{[1]}(s,t) = \prod_{h=s+1}^{t} (1+\max\{\eta_h g_h, i'\}) \tag{7.3.16}$$

It turns out

$$f^{[1]}(s,t) \geq f(s,t) \tag{7.3.17}$$

due to the locking-in effect.

If the revaluation rate r_t is defined as in (7.3.10), after examining the development in time of the investment of the policyholder, we find that the accumulation factor is

$$\begin{aligned} f^{[2]}(t-1,t) &= (1+i')\left(1+\max\left\{\tfrac{\eta_t g_t - i'}{1+i'}, r_{\min}\right\}\right) \\ &= \max\{(1+i')(1+r_{\min}), (1+\eta_t g_t)\} \end{aligned} \tag{7.3.18}$$

and more in general

$$f^{[2]}(s,t) = \prod_{h=s+1}^{t} \max\{(1+i')(1+r_{\min}), (1+\eta_h g_h)\} \tag{7.3.19}$$

Clearly, it turns out

$$f^{[2]}(s,t) \geq f(s,t) \tag{7.3.20}$$

due to the locking-in effect. Further, parameters i' and $r_{\min}$ in (7.3.19) are commonly chosen so that

$$f^{[2]}(s,t) = f^{[1]}(s,t) \tag{7.3.21}$$

for any time-interval (s,t). In the following, for comparison we refer to the revaluation rate defined by (7.3.10) as to the rate $r_t^{[2]}$.

Consider now the revaluation rate $r_t^{[3]}$ defined as follows

$$r_t^{[3]} = \frac{\eta_t\, g_t - i'}{1+i'} \tag{7.3.22}$$

Depending on the the difference $\eta_t\, g_t - i'$, the rate $r_t^{[3]}$ can take negative values; thus, no guarantee is embedded. The accumulation factor based on $r_t^{[3]}$ can be defined as follows

$$f^{[3]}(s,t) = \prod_{h=s+1}^{t} (1+\eta_h\, g_h) \tag{7.3.23}$$

as one can easily check writing the equations expressing the development in time of the investment of the policyholder. Under (7.3.23), we have

$$f^{[3]}(s,t) \gtreqless f(s,t) \tag{7.3.24}$$

given that no guarantee is applied. Further

$$f^{[3]}(s,t) \leq f^{[1]}(s,t) \tag{7.3.25}$$

$$f^{[3]}(s,t) \leq f^{[2]}(s,t) \tag{7.3.26}$$

We note that the accumulation factor $f^{[3]}(s,t)$ was considered in Example 7.3.1, for comparison with $f^{[1]}(s,t)$ (even if this notation was not yet introduced); we mentioned there that solution (7.3.23) is unworkable, as no rational policyholder would accept to receive no guarantee under a participating policy. However, the only way to avoid to lock-in extra-yields on investment is to let the revaluation rate r_t become negative, if necessary. Indeed, if $r_t < 0$, the reserve is reduced, so to offset (at least partially) the positive adjustments applied in previous years.

If the revaluation rate r_t is allowed to take negative values, as in (7.3.22), we may experience $j_t^{[B]} < 0$ (we recall that we are assuming $j_t^{[\Pi]} = 0$). Hence, the death and the benefit at maturity are no longer guaranteed; as already noted, this is not acceptable. If (7.3.22) is adopted, we can introduce an explicit guarantee on the death benefit, for example defining the death benefit at time t as follows:

$$C_t = C_1 \times \max\left\{ \prod_{s=1}^{t-1} (1+j_s^{[B]}), (1+j^{[B,guar]})^{t-1} \right\} \tag{7.3.27}$$

where: C_1 is the initial death benefit (i.e., the amount referred to for the calculation of premiums); $\prod_{s=1}^{t-1}(1+j_s^{[B]})$ is the revaluation obtained in the time-interval $(1,t-1)$ based on the observed investment yields; $j^{[B,guar]}$ is a minimum guaranteed revaluation rate of the death benefit. Note that, to avoid to lock-in past revaluations, the rate $j^{[B,guar]}$ is guaranteed just to the time of payment of the death benefit, i.e. it expresses the annual average minimum revaluation rate of the death benefit which is guaranteed.

A similar guarantee can be introduced for the benefit at maturity. Given that the benefit at maturity is the result of the accumulation of savings premiums, it is more natural to express the guarantee in terms of accumulation factor. For example, the following definition could be adopted

$$f^{[4]}(s,m) = \max\left\{ \prod_{h=s+1}^{m} (1+\eta_h g_h), (1+i')^{m-s} \right\} \tag{7.3.28}$$

where the return guaranteed to maturity is i'; the factor $f^{[4]}(s,m)$ is meant here to be applied just for the time-intervals (s,m), $s = 0,1,\dots,m-1$. Should we be interested in assessing the value of the investment at time t, $t < m$, for example because we need to define the surrender value, reference should be made to the accumulation factor $f^{[3]}(s,t)$. Note that

$$f^{[4]}(s,m) \geq (1+i')^{m-s} \tag{7.3.29}$$

so that, with reference to the benefit at maturity, each savings premium will be accumulated at an annual rate that on average is not lower than i'; indeed, i' in (7.3.28) represents the annual average minimum return guaranteed on the investment of the policyholder. Also in this case, past extra-yields are not locked-in. Other solutions are clearly possible; in particular, insurers have designed solutions which do not avoid to lock-in extra-yields, but the locking-in does not occur in each year. In particular, the rate $r_t^{[3]}$ has been adopted, but requiring that every k years (since time 0) the average return on the investment of the policyholder must be at least i'. In this case, the accumulation factor could be defined as follows

$$f^{[5]}(s,t) = f^{[5]}(s,z) \times \begin{cases} \prod_{h=z+1}^{t} (1+\eta_h g_h) & \text{if } z < t < z+k \\ \max\left\{\prod_{h=z+1}^{k} (1+\eta_h g_h), (1+i')^k\right\} & \text{if } t = z+k \end{cases} \tag{7.3.30}$$

with $z = 0,k,2k,\dots$ and $s \leq z$. Solution (7.3.30) implies a *partial lock-in* of extra-yields on investment. The period k is commonly set to 3 or 5 years; if $k = m$, we find (7.3.28) as a particular case, i.e. the yield would be guaranteed to maturity. We note that under (7.3.30) at some policy anniversaries, namely at time $k,2k,\dots$, the reserve cannot reduce (indeed, at time $k,2k,\dots$ the revaluation rate cannot be negative); as a consequence, the surrender value at such policy anniversaries receives some guarantee.

Example 7.3.4. Table 7.3.5 lists the accumulation factors experienced in face of a given path of the yield on investment $\eta_t g_t$. The several definitions introduced above for the accumulation factor have been considered. For the sake of brevity, only the time-interval $(0,t)$ has been addressed. The path for $\eta_t g_t$ is the same of Example 7.3.1 and Table 7.3.4.

Table 7.3.5 Alternative accumulation factors for participating policies

t	$\eta_t\, g_t$	$f(0,t)$	$f^{[1]}(0,t)$	$f^{[2]}(0,t)$	$f^{[3]}(0,t)$	$f^{[4]}(0,t)$	$f^{[5]}(0,t)$
		$i'=0.02$	$i'=0.02$	$i'=0$ $r_{\min}=0.02$		$i'=0.02$	$i'=0.02$ $k=3$
1	5.000%	1.02000	1.05000	1.05000	1.05000	1.05000	1.05000
2	4.500%	1.04040	1.09725	1.09725	1.09725	1.09725	1.09725
3	4.000%	1.06121	1.14114	1.14114	1.14114	1.14114	1.14114
4	3.000%	1.08243	1.17537	1.17537	1.17537	1.17537	1.17537
5	2.000%	1.10408	1.19888	1.19888	1.19888	1.19888	1.19888
6	1.500%	1.12616	1.22286	1.22286	1.21686	1.21686	1.21686
7	2.500%	1.14869	1.25343	1.25343	1.24729	1.24729	1.24729
8	2.000%	1.17166	1.27850	1.27850	1.27223	1.27223	1.27223
9	3.000%	1.19509	1.31685	1.31685	1.31040	1.31040	1.31040
10	1.000%	1.21899	1.34319	1.34319	1.32350	1.32350	1.32350
11	2.000%	1.24337	1.37006	1.37006	1.34997	1.34997	1.34997
12	2.500%	1.26824	1.40431	1.40431	1.38372	1.38372	1.39061
13	1.500%	1.29361	1.43239	1.43239	1.40448	1.40448	1.41147
14	5.000%	1.31948	1.50401	1.50401	1.47470	1.47470	1.48204
15	5.000%	1.34587	1.57921	1.57921	1.54844	1.54844	1.55614

The annual yield experienced on average starting from time 0 is at any time higher than 2% (see also Table 7.3.1); thus we have $f(0,t) \leq f^{[j]}(0,t)$, for $j = 1,2,\ldots,5$. For the same reason, $f^{[4]}(0,m) = f^{[3]}(0,m)$, i.e. the guarantee in (7.3.28) is not active at maturity. Due to the parameters, it always turns out $f^{[1]}(0,t) = f^{[2]}(0,t)$. Due to the lock-in, we find $f^{[1]}(0,t) \geq f^{[3]}(0,t)$, given that $f^{[3]}(0,t)$ is simply based on the experienced yield (with no guarantee). Finally, we note that $f^{[5]}(0,t) > f^{[3]}(0,t)$ from time $t = 12$, where a partial lock-in occurs (given that in the latest 3-years prior to time 12 the observed yield is on average lower than 2%). Due to the fact that the lock-in in (7.3.30) is just partial, it turns out $f^{[5]}(0,t) \leq f^{[1]}(0,t)$.
□

7.4 Unit-linked policies

The main feature of unit-linked policies is that the financial risk is borne by the policyholder. The underlying insurance cover is usually an endowment; the premium is invested into a reference fund, selected by the policyholder out of a basket designed by the insurer. Commonly, the lines of investment which are made available by the insurer implies different risk-return profiles; thus, the policyholder can opt for more conservative or more dynamic asset combinations. The line of investment selected by the policyholder may be changed later on, possibly against the payment of a fee (the so-called *switching fee*). If a *switching option* has been underwritten in policy conditions, the policyholder has the opportunity to change the investment line at some dates at no cost.

The asset management of unit-linked policies plays a primary role in this business; the discussion of the several issues involved, however, is outside the scope of this book. In the following we address only the actuarial issues which are involved in the management of unit-linked policies.

7.4.1 Definition of unit-linked benefits

The fund accumulated with premiums is called *policy fund* or *policy account*. Benefits are defined in terms of the policy fund available at the time of payment. More precisely:

- the survival benefit at maturity, the so-called *maturity benefit* is the current value at maturity of the policy fund;
- the death benefit is the current value of the policy fund at the time of death, to which a sum at risk is added, which is defined so that it is positive (or at least non-negative);
- the surrender value is the current value of the policy fund at the time of surrender, possibly net of a (small) surrender fee.

Given that benefits depend on the current value of the policy fund, a risk emerges for the policyholder, as such a value is unknown before payment. Guarantees may be provided; the sum at risk, for example, can be defined so that there is some embedded guarantee on the death benefit. However, it is more usual to define guarantees explicitly; a fee is then applied to meet the relevant cost.

Unit-linked policies are given this name because the reference fund is split into a notional number of units. Benefits could then be though of as the current value of the number of units which have been credited to the policy; such a number can be assessed adopting the actuarial model used for fixed benefits, as we describe in Sect. 7.4.2. What remains unknown before payment is the current value of a unit. In this perspective, the benefit can be considered to be expressed in account units other than the usual currency, whence the term unit-linked. However, as we will see in Sect.7.4.2, the number of units which define the benefit is usually known just at the beginning of the year of payment, and not earlier.

As account units, in principle reference can be made to any quantity whose value is likely to increase in time, such as gold, some foreign currency, real estate, securities, and so on. In practice, there is an Asset-Liability constraint: the insurer must be able to buy or replicate the reference units, to meet its liability without taking a (too strong) basis risk. Account units which have been adopted by insurers are foreign currency and investment funds. Nowadays, the standard choice is reference to investment funds.

As far as the underlying insurance cover is concerned, we mentioned above that the usual form is the endowment insurance. However, also whole life assurances can be realized with a unit-linked arrangement; in this case, there would be no maturity benefit. Also life annuities can be realized as unit-linked; however, the annual amount would fluctuate depending on the current value of the reference fund, thus originating a severe risk for the policyholder. In the following, we refer to an endowment insurance.

7.4.2 Unit-linked policies without guarantees

Consider a unit-linked endowment insurance, including no financial guarantee. At time t, a premium $P_t^{[T]}$ is paid, inclusive of expense loadings; the premium can be either constant in time or not, depending on policy conditions. We denote by Λ_t the total expense loading at time t; the initial commission is usually charged to the first premium. After issue, the loading includes collection and general administrative expenses, as well as management fees. The expense loading is typically proportional to the size of the premium and the policy fund.

The net premium P_t is invested into the reference fund. If we let w_t denote the current value of a unit, then

$$n_t = \frac{P_t}{w_t} \tag{7.4.1}$$

represents the number of units purchased by the insurer at time t with the net premium. Information about the current value w_t is made available to the policyholder; since the policyholder is bearing the financial risk, she must receive full information about the performance of the investment fund. The sum at risk originates a mutuality cost, which needs to be funded. In principle, the net premium must be split into risk and savings premium; in other words, the number n_t is split into two components: one (which we denote by $n_t^{[S]}$) is credited to the policy to contribute to savings, the other (which we denote by $n_t^{[R]}$) is used to meet mutuality costs. Trivially,

$$n_t = n_t^{[S]} + n_t^{[R]} \tag{7.4.2}$$

We note that if $n_t = 0$ (due to having set $P_t = 0$), then (7.4.2) would imply $n_t^{[S]} = -n_t^{[R]}$; if no premium is paid, annual costs (namely, the risk premium, as well as expenses) are met by taking money from the policy account. This shows that unit-linked arrangements allow quite easily for some flexibility in the choice of the annual premium. Actually, from a technical point of view, it is not necessary that a premium is paid in each year; as noted above, what is required is that the current policy account is large enough to meet annual costs. However, appropriate policy conditions must be designed in this case, such as those adopted in Universal Life policies (see Sect. 7.8). In the following, we mean that $P_t > 0$ and such that $n_t^{[S]} > 0$, as is more usual for the policy design that we are addressing.

We let N_t denote the number of units totally credited to the policy at time t, before premium payment. It is easy to understand that

$$N_t = \sum_{s=0}^{t-1} n_s^{[S]} \tag{7.4.3}$$

In order to perform the splitting (7.4.2), and then to calculate N_t, we need to assess the amount of benefits, the sum at risk in particular.

First, we define the policy fund at time t, as follows

$$F_t = N_t \, w_t \tag{7.4.4}$$

Note that F_t is assessed at current value, given that the financial risk is borne by the policyholder. As usual, assets back liabilities. The reserve at time t, representing the liability of the insurer, is simply defined as follows

$$V_t = F_t \tag{7.4.5}$$

The maturity benefit, which is the current value of the policy fund, is given by

$$S_m = F_m \tag{7.4.6}$$

at time m. Earlier to time m, we can assess the amount which is funded by current assets, as follows

$$S_t = F_t \tag{7.4.7}$$

It is worth noting that, contrarily to fixed-benefits and participating policies, the maturity benefit gradually accumulates in time (similarly to what happens in the case of single recurrent premiums; see Sect. 4.4.5); indeed, F_t is clearly the result of the payments which have been made to date.

The death benefit payable at time t is

$$C_t = F_t + K_t \tag{7.4.8}$$

where K_t is the sum at risk, defined so that $K_t \geq 0$. For example

$$C_t = F_t \, (1 + \alpha) \tag{7.4.9}$$

i.e. $K_t = \alpha \, F_t$, with $\alpha > 0$, or

$$C_t = F_t + G \tag{7.4.10}$$

i.e. $K_t = G$, with $G > 0$. We note that (7.4.10) embeds a financial guarantee, as (excluding the case $F_t < 0$) we always have $C_t \geq G > 0$. Conversely, no guarantee is embedded in (7.4.9), as $C_t = 0$ if $F_t = 0$. The quantity α in (7.4.9) is simply a proportion expressing the sum at risk; the quantity G in (7.4.10) can instead be meant as a minimum death benefit guaranteed.

The surrender value at time t is usually defined as follows

$$R_t = \varphi(t) \, F_t \tag{7.4.11}$$

where $1 - \varphi(t)$ represents the surrender fee at time t (commonly very close to 0).

We can now calculate the number of units, $n_t^{[S]}$, which are credited to the policy fund at time t, after receiving the premium; we note that

$$n_t^{[S]} = N_{t+1} - N_t \tag{7.4.12}$$

The number N_{t+1} must be assessed so that assets and liabilities of the contract are on actuarial balance in year $(t, t+1)$. Extending the recursive Eq. (5.4.8) for the reserve of insurance covers with fixed benefit, we can write the following equation relating to year $(t, t+1)$

$$(F_t + P_t)\frac{w_{t+1}}{w_t} = (C_{t+1} - F_{t+1})\, q'_{x+t} + F_{t+1} \tag{7.4.13}$$

Equation (7.4.13) can be easily understood if compared to (5.4.8):

- reference is to a policy in-force at time t;
- F_t represents the amount of assets available at time t, while P_t is the net premium cashed at that time;
- assets are invested into the reference fund, whose yield in year $(t, t+1)$ is

$$z_{t+1} = \frac{w_{t+1}}{w_t} - 1 \tag{7.4.14}$$

  We note that z_{t+1} is unknown at time t;
- whatever happens, namely if the insured is still alive at the end of the year or not, the policy fund F_{t+1} will be available;
- in case of death during the year, the sum at risk $C_{t+1} - F_{t+1}$ must be added to the policy fund, so to pay the death benefit C_{t+1} to beneficiaries.

Contrarily to (5.4.8), the equality in (7.4.13) is just notional, as not all the quantities involved are known for sure. Let assume death benefit (7.4.9); if we replace into (7.4.13) the definitions introduced above for the several quantities involved, we find

$$(N_t + n_t)\, w_{t+1} = \alpha N_{t+1}\, w_{t+1}\, q'_{x+t} + N_{t+1}\, w_{t+1} \tag{7.4.15}$$

Each term in (7.4.15) is proportional to w_{t+1}; this allows us to change the account unit, from the monetary unit to the reference fund unit. If we assume that w_{t+1} is strictly positive, after dividing (7.4.15) by w_{t+1} we obtain

$$N_t + n_t = \alpha N_{t+1}\, q'_{x+t} + N_{t+1} \tag{7.4.16}$$

which is a balance condition expressed in terms of investment units, in which all the quantities involved are deterministic at time t. Thus, (7.4.16) can be used to calculate N_{t+1} (i.e., $n_t^{[S]}$). It is worth noting that quantities in (7.4.16) are deterministic only at time t, i.e. after premium payment. Before that time, the number n_t is random, given that it depends on the current value of a unit (see (7.4.1)).

Solving (7.4.16), we find

$$N_{t+1} = \frac{N_t + n_t}{\alpha q'_{x+t} + 1} \tag{7.4.17}$$

and then

$$n_t^{[S]} = \frac{n_t - \alpha q'_{x+t}}{\alpha q'_{x+t} + 1} \tag{7.4.18}$$

$$n_t^{[R]} = (n_t + 1)\frac{\alpha q'_{x+t}}{\alpha q'_{x+t} + 1} \tag{7.4.19}$$

Of course, it turns out $n_t^{[S]} < n_t$.

The definition of the risk and the savings premium is now straightforward. We have

$$P_t^{[R]} = n_t^{[R]} w_t \tag{7.4.20}$$

$$P_t^{[S]} = n_t^{[S]} w_t \tag{7.4.21}$$

The development in time of the risk premium depends on several factors, namely the mortality rate (which is increasing throughout the policy duration), the size of the sum at risk (which is proportional to the size of the policy fund), and the current value of a unit. What is important to note is that, after premium payment, the balance condition (7.4.16) implies no financial risk for the insurer. This is a consequence of defining the benefits so that they are proportional to the current value of one unit. However, Eq. (7.4.16), if considered before time t, reveals a financial risk for the insurer, as the number of units which are purchased year by year is unknown, and then the number of units credited to the contract is also unknown, because these numbers depend on the value of one unit at the time of premium payment. Thus, the number of units which define a benefit (the maturity, the death or the surrender benefit) are known for certain just at the beginning of the year of possible payment.

Example 7.4.1. Consider a unit-linked endowment, with no financial guarantee, issued for a person age $x = 50$, maturity $m = 15$, death benefit $C_{t+1} = 1.10 F_{t+1}$ (then, $\alpha = 0.10$). The life table is LT1 (while no technical interest rate needs to be assigned, as the financial risk is not transferred to the insurer). Table 7.4.1 provides an example of development of benefits throughout time. Information about the number of units purchased and credited at any policy anniversary is also included, as well as information about the risk premium and the savings premium.

Note the decreasing behavior of n_t, due to the increasing value of a unit (while the net premium remains constant). The risk premium is increasing in time, and this is due to the death probabilities and to the fact that the sum at risk increases in time. The magnitude of the risk premium is anyhow very small, given that the sum at risk is not very large.

□

We have used above the terms risk premium and savings premium. This terminology is not the usual one for unit-linked policies. A unit-linked policy is mainly

Table 7.4.1 Unit-linked endowment insurance; $C_{t+1} = 1.10 F_{t+1}$

t	P_t	w_t	z_t	n_t	N_t	$n_t^{[S]}$	$P_t^{[R]}$	$P_t^{[S]}$	F_t	C_t	$C_t - F_t$
0	100	1.00		100.00	0.00	99.97	0.03	99.97	0.00		
1	100	1.04	4%	96.15	99.97	96.08	0.08	99.92	103.96	114.36	10.40
2	100	1.08	4%	92.46	196.05	92.34	0.13	99.87	212.04	233.25	21.20
3	100	1.12	4%	88.90	288.38	88.73	0.20	99.80	324.39	356.83	32.44
4	100	1.17	4%	85.48	377.11	85.24	0.28	99.72	441.16	485.28	44.12
5	100	1.22	4%	82.19	462.35	81.88	0.38	99.62	562.52	618.77	56.25
6	100	1.27	4%	79.03	544.24	78.64	0.50	99.50	688.63	757.50	68.86
7	100	1.32	4%	75.99	622.88	75.50	0.64	99.36	819.66	901.63	81.97
8	100	1.37	4%	73.07	698.38	72.47	0.82	99.18	955.78	1 051.36	95.58
9	100	1.42	4%	70.26	770.85	69.54	1.03	98.97	1 097.16	1 206.88	109.72
10	100	1.48	4%	67.56	840.39	66.69	1.28	98.72	1 243.98	1 368.38	124.40
11	100	1.54	4%	64.96	907.08	63.93	1.58	98.42	1 396.42	1 536.06	139.64
12	100	1.60	4%	62.46	971.02	61.25	1.93	98.07	1 554.63	1 710.10	155.46
13	100	1.67	4%	60.06	1 032.27	58.64	2.35	97.65	1 718.81	1 890.69	171.88
14	100	1.73	4%	57.75	1 090.92	56.10	2.85	97.15	1 889.11	2 078.02	188.91
15		1.80	4%		1 147.02				2 065.71	2 272.28	206.57

designed to provide an appropriate investment opportunity, joint to some capital protection in case of early death. The benefit corresponding to the policy fund is then simply addressed as the savings, or investment, of the policyholder, and the quantity $P_t^{[S]}$ is named the *invested premium* (or *invested amount*). In this perspective, the sum at risk is a *supplementary* (or *rider*) *benefit*, and then the quantity $P_t^{[R]}$ is dealt with as a *fee for supplementary* (or *rider*) *benefits*. Under (7.4.9), it is likely that the risk premium is increasing, as plotted in the example of Table 7.4.1. Insurers often prefer to apply a constant fee for the rider benefits, similarly to any other fee. Some approximations then result in respect of the example provided in Table 7.4.1, as a sort of level risk premium must be assessed. The magnitude of the risk premium is usually so low that such approximations are negligible in this case.

Let us now consider the death benefit (7.4.10). If we replace the various quantities in (7.4.13), we find

$$(N_t + n_t)\, w_{t+1} = G q'_{x+t} + N_{t+1}\, w_{t+1} \tag{7.4.22}$$

from which we obtain

$$N_{t+1} = N_t + n_t - \frac{G q'_{x+t}}{w_{t+1}} \tag{7.4.23}$$

The quantity w_{t+1} in (7.4.23) (and in (7.4.22)) is unknown; in order to calculate N_{t+1}, an estimate of w_{t+1} is required. This implies some financial risk for the insurer; such a risk is originated by the fact that the death benefit embeds a fixed benefit (which, as we have mentioned above, represents a guaranteed minimum benefit). This is a reason why the definition of the death benefit highly preferred by insurers in unit-linked policies with no guarantees is given by Eq. (7.4.9).

7.4.3 Unit-linked policies with financial guarantees

As described at the beginning of Sect. 7.4, in unit-linked policies the financial risk is borne by the policyholder. Partially, the risk can be transferred to the insurer, through the underwriting of appropriate guarantees.

Guarantees may relate to any of the benefits provided by the insurance cover:

- the *maturity guarantee* concerns the maturity benefit;
- the *death benefit guarantee* is given on the death benefit;
- the *surrender guarantee* concerns the surrender value.

In the following, we disregard the surrender guarantee, and we assume that the same type of guarantee is provided for the maturity and the death benefit. This way, we shorten a little bit the notation; anyhow, it is not difficult to address more general cases, in which the maturity and the death benefit guarantees are different.

The guarantee is defined specifying the minimum benefit amount. If, because of adverse financial trends, the policy fund at the time of payment is not high enough, the minimum amount will be paid. Generically, we let B_t be the benefit due at time t; if $t = 1,2,\dots,m-1$ it is a death benefit, while if $t = m$ it is the benefit paid at maturity, either in case of death or survival (given that we are assuming that the same guarantee is provided for the maturity and the death benefit).

The guaranteed amount can be stated according to different targets. The simplest case is to set a fixed guaranteed amount G. The benefit at time $t+1$ is then defined as follows

$$B_{t+1} = \max\{F_{t+1}, G\} \tag{7.4.24}$$

For an example, see Fig. 7.4.1.

The quantity

$$K_{t+1} = B_{t+1} - F_{t+1} = \max\{G - F_{t+1}, 0\} \tag{7.4.25}$$

represents the sum at risk, and corresponds to the pay-off of a put option (see also Sect. 7.5).

Alternative definitions of the guaranteed amount are chosen so that the difference $B_{t+1} - F_{t+1}$ corresponds to the pay-off of a given financial option.

Remark When a financial guarantee is underwritten, a financial risk emerges for the insurer. Such a risk needs to be hedged appropriately, through a suitable asset management strategy. Similarly to what noted for participating policies, before underwriting a guarantee, the insurer must investigate if it is possible to hedge it. Therefore, usually the insurer investigates the hedging strategies available on the market, and then selects the guarantee offered to the policyholder.

For example, under the benefit

$$B_{t+1} = \max\{F_{t+1}, \max\{F_s\}_{s=0,1,\dots,t}\} \tag{7.4.26}$$

it is guaranteed that the minimum amount paid at time $t+1$ is the highest value of the policy fund experienced at the previous policy anniversaries; see Fig.7.4.2. The guaranteed amount in this case is defined as follows

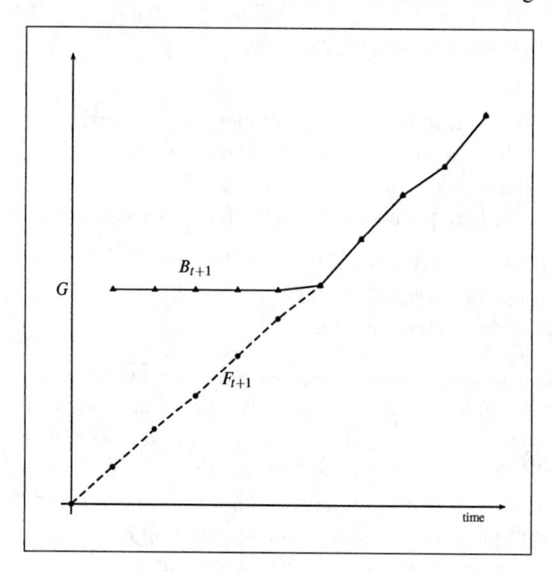

Fig. 7.4.1 Guaranteed benefit $B_{t+1} = \max\{F_{t+1}, G\}$; annual constant premiums

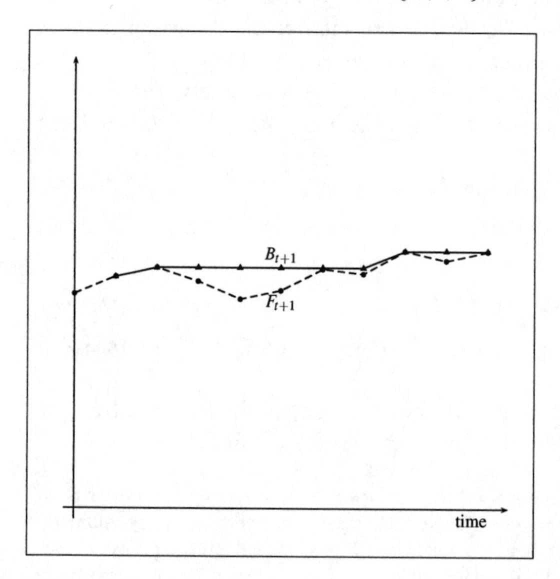

Fig. 7.4.2 Guaranteed benefit $B_{t+1} = \max\{F_{t+1}, \max\{F_s\}_{s=0,1,\ldots,t}\}$; single premium

$$G_t = \max\{F_s\}_{s=0,1,\dots,t} \tag{7.4.27}$$

(where the suffix t denotes that such an amount is known at the beginning of year $(t, t+1)$). The sum at risk is

$$K_{t+1} = B_{t+1} - F_{t+1} = \max\{\max\{F_s\}_{s=0,1,\dots,t} - F_{t+1}, 0\} \tag{7.4.28}$$

and this corresponds to the pay-off of a *ratchet option*.

With reference to the maturity benefit, the following guarantee

$$S_m = \max\{F_m, G_{m-1}\} \tag{7.4.29}$$

where

$$G_{m-1} = \sum_{t=0}^{m-1} P_t^{[S]} (1+i')^{m-t} \tag{7.4.30}$$

is similar to the guarantee embedded in the accumulation factor $f^{[4]}(s,m)$ for participating policies (see (7.3.28)).

Some remarks on the valuation of guarantees are presented in Sect. 7.5.

Remark It should be clear that in unit-linked policies the asset perspective is prevailing. It is enough to look at the way the reserve, i.e. the value of the insurer's liability, is defined when no guarantee is provided (see (7.4.5)). Indeed, unit-linked policies are considered to be an *asset-driven* business. In contrast, fixed-benefit policies, for which the definition of liabilities comes before the selection of assets, are considered to be a *liability-driven* business. The distinction mainly relies on the party bearing the financial risk, namely the insurer for liability-driven arrangements, the policyholder for asset-driven solutions. Typical of a liability-driven business is a conservative assessment of the liabilities, and assets as well; for an asset-driven business, a market-consistent valuation is instead the natural choice.
Participating policies, as well as unit-linked policies with financial guarantees are somewhat at an intermediate step between a liability-driven and an asset-driven business. Basically, participating policies are liability-driven, as is suggested by the approach adopted for the calculation of premiums and reserves. However, the benefit amount, and then the insurer's liability, is affected by the investment performance. Similarly, unit-linked policies with financial guarantees are asset-driven; however, since the guarantees transfer risk to the insurer, conservative valuation assumptions are required in this regard. In particular, an *additional reserve* in respect of the reserve (7.4.5) may be necessary, which should be assessed consistently with the cost of the guarantee.
Figure 7.4.3 provides a graphical representation of the comments developed in this Remark. The large arrows, in particular, show which is the starting point for the assessment of the value of assets and liabilities, or for their management: the liabilities for fixed-benefits and participating policies, the assets for unit-linked policies (with or without guarantees). In the case of participating policies, the small arrow expresses that the value of the liability must be updated according to the investment performance, while the small arrow in the case of unit-linked policies with guarantees recalls that the liability originated by the guarantee requires an appropriate hedging, and then an appropriate selection of assets.

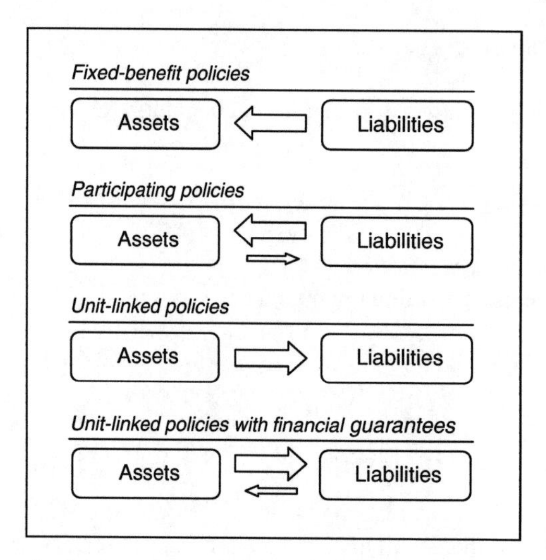

Fig. 7.4.3 Interaction between assets and liabilities

7.5 Financial options in unit-linked and participating policies

As we have already noted, it goes beyond the scope of this book to deal with issues related to asset management and asset valuation. In this Section we just provide a description of the structure of the financial options included in life insurance policies. In Sect. 7.5.2, in particular, we address the valuation of such options in one example, just aiming at outlining the main issues of such a valuation.

7.5.1 The structure of minimum guarantees

We first refer to a unit-linked endowment insurance, including a guarantee on the death and the maturity benefit. The guarantee is defined so that the benefit payable at maturity in case of death is the same as the maturity benefit. We disregard guarantees concerning the surrender value.

The benefit payable at time $t+1$ (in case of death or, if $t+1=m$, survival on death) is defined as follows

$$B_{t+1} = \max\{F_{t+1}, G_t\} \tag{7.5.1}$$

where G_t is the guaranteed amount, known at time t at the latest. Rearranging (7.5.1), the benefit can be expressed as follows

$$B_{t+1} = F_{t+1} + \max\{G_t - F_{t+1}, 0\} \tag{7.5.2}$$

or as follows

$$B_{t+1} = G_t + \max\{F_{t+1} - G_t, 0\} \tag{7.5.3}$$

According to (7.5.2), the benefit consists of the policy fund (whose value is unknown) and the pay-off of a *put option*, with strike G_t and underlying the reference fund. The maturity of the option is time $t+1$, the time of possible payment of the benefit. Conversely, according to (7.5.3) the benefit consists of a fixed benefit G_t, like a traditional policy, to which the pay-off of a call option is added. The strike, the underlying and the maturity of the call option are clearly the same as those of the put option (given that we are describing the same benefit). For unit-linked policies, the description provided by (7.5.2) is more natural than (7.5.3), as the main feature of the arrangement is to realize an investment in the reference fund. However, when addressing the calculation of the cost of the guarantee, sometimes it is easier to assess the cost of a call option, and then reference would be made to (7.5.3).

According to the way the strike is defined, we can investigate further the structure of the option. If $G_t = G$, constant, the option is European-like, while if G_t depends on the past performances of the policy fund, such as in (7.4.27), the option is path-dependent. If G_t is a function of the premiums paid, such as in (7.4.30), the guarantee is endogenous. If guarantee (7.4.27) is chosen and a single premium was paid, the value of the option just depends on the investment performance (whilst when it is endogenous its value also depends on choices concerning the invested amount).

If a guarantee is underwritten in a unit-linked policy, a fee is required to the policyholder. In general, it is easier to assess the cost of a European-like option than of a path dependent option, as well as it is easier to evaluate an option with exogenous guarantees than with endogenous guarantees. A market-consistent assessment is required, following the common practice for the pricing of financial derivatives. This requires a calibration to market data, even if the option is not traded directly on the market. Reference should be made to similar options. However, options traded in financial markets have many differences in respect of those included in life policies, such as the maturity (which is typically shorter for traded options), a different underlying, a different strike. Further, it must be noted that the exercise of options in (7.5.2) and (7.5.3) is subject not only to economic events (namely: it is convenient to exercise the call option if $F_{t+1} > G_t$, while it is convenient to exercise the put option if $F_{t+1} < G_t$), but also to the lifetime of the insured. Indeed, the benefit at time $t+1$ is payable in case of death (or survival); this aspect adds complexity to the valuation of the insurer's liability. Some details in this regard are provided in Sect. 7.5.2.

Addressing now surrender guarantees, we note that they may be expressed similarly to (7.5.1); clearly, the benefit would be the surrender value R_{t+1} instead of B_{t+1}. The guaranteed amount G_t is usually defined so to provide a financial protection to the investment of the policyholder; therefore, the amount G_t typically implies a minimum (annual or average) return on the amount invested. The exercise of the surrender guarantee depends on economic events (the exercise is convenient if $G_t > F_{t+1}$), but also on preferences of the policyholder (whether to maintain or not the policy). This latter aspect is very hard to model; surrender guarantees represent important costs for insurers, but their assessment is still an open problem, due to the difficulty in representing individual preferences.

We make a final comment in respect of participating policies. In Sects. 7.3.1 and 7.3.2 we have already commented on the financial options which are embedded. We give here just an example about how to explicit the pay-off of the relevant option. Assume that the accumulation factor $f^{[1]}(s,t)$ is adopted (see (7.3.16)). After some rearrangements, such a quantity can be expressed as follows

$$f^{[1]}(s,t) = (1+i')^{t-s} \prod_{h=s+1}^{t} \left(1+\max\left\{\frac{\eta_t\, g_t - i'}{1+i'}, 0\right\}\right) \tag{7.5.4}$$

The factor $(1+i')^{t-s}$ represents the minimum guaranteed accumulation, while $\prod_{h=s+1}^{t}\left(1+\max\left\{\frac{\eta_t\, g_t - i'}{1+i'}, 0\right\}\right)$ is originated by call options on the yield of the investment fund. The accumulation factor $f^{[1]}(s,t)$ could be rearranged so to explicit the pay-off of put options, but for participating policies the description provided by (7.5.4) is more natural, as first of all a participating policy guarantees a given return, and possibly an extra-yield.

7.5.2 The valuation of financial options in a unit-linked policy

As we have mentioned in Sect. 7.5.1, the valuation of financial options included in life insurance covers is complex. In this Section we aim at providing some remarks on how the different events to which the exercise of such an option is subject should be accounted for.

Refer to a unit-linked endowment insurance, issued with a single premium. The death and maturity benefit are defined as in (7.5.1), with $G_t = G$, constant. We assume that no guarantee is provided on the surrender value (which is then simply the policy fund, possibly net of a surrender fee). The single expense-loaded premium $\Pi^{[T]}$ is split into three components:

- the management fees and the acquisition costs, Θ;
- the invested amount, $\Pi^{[S]}$;
- the fee for ancillary benefits (namely, for the sum at risk), $\Pi^{[R]}$.

The above notation is similar to what adopted for traditional policies (namely, for the expense loading, the savings and the risk premium); however, the meaning of the several quantities is not the same as for traditional policies, and must be meant as specified above.

The quantity $\Pi^{[S]}$ is invested into the selected fund. We assume

$$\Pi^{[S]} = N w_0 \tag{7.5.5}$$

where N is the number of units which are credited to the policy. The quantity N is determined so that the policy fund always consists of N units, i.e.

$$F_t = N w_t \tag{7.5.6}$$

Replacing (7.5.6) into (7.5.1), we can express the benefit at time t as follows

$$B_t = N w_t + \max\{G - N w_t, 0\} \tag{7.5.7}$$

or, setting $G = N \times E$, as

$$B_t = N w_t + N \max\{E - w_t, 0\} \tag{7.5.8}$$

According to (7.5.8), the benefit consists of N units of the reference fund and N put options, each with underlying a unit of the reference fund and strike E. It is easy to rewrite (7.5.8) so to explicit the pay-off of call options.

The present value at time 0 of the benefit payable at time t, which we denote as $V_0(B_t)$, can be assessed as follows

$$V_0(B_t) = N w_0 + N P_{0(t)} \tag{7.5.9}$$

where $P_{0(t)}$ is the value (or price) at time 0 of a put option with maturity at time t, strike E and underlying one unit of the reference fund. The price $P_{0(t)}$ must be assessed through an appropriate financial model; for example, if we accept standard assumptions (namely: the risk-free rate is deterministic and constant, the current value of the underlying follows a geometric standard Brownian motion, and so on), the Black and Scholes formula applies. Quite often standard assumptions are not appropriate, and numerical techniques must be used instead of a closed formula.

The benefit B_t is paid at time t depending on the lifetime of the insured. Given an appropriate life table, through which the mortality rates q_{x+t} are assessed, we expect that

- a proportion ${}_{t-1|1}q_x$ of the policies issued at time 0 will receive the benefit B_t at time t, $t = 1, 2, \ldots, m-1$ (namely, because death occurs in year $(t-1, t)$);
- a proportion ${}_{m-1}p_x = {}_{m-1|1}\,q_x + {}_m p_x$ of the policies issued at time 0 will receive the benefit B_m at time m (namely, in face of the insureds either dying in the last year, or alive at maturity).

To realize the actuarial balance between the premium and the benefit, the following condition must be fulfilled

$$\Pi^{[\mathrm{T}]} - \Theta = \sum_{t=1}^{m-1} {}_{t-1|1}q_x\, V_0(B_t) + {}_{m-1}p_x\, V_0(B_m) \tag{7.5.10}$$

Rearranging, we have

$$\Pi^{[\mathrm{T}]} - \Theta = N w_0 + \left(\sum_{t=1}^{m-1} {}_{t-1|1}q_x N P_{0(t)} + {}_{m-1}p_x N P_{0(m)} \right) \tag{7.5.11}$$

As stated by (7.5.5), the quantity $N w_0$ represents the invested amount; the quantity in brackets represents the amount $\Pi^{[\mathrm{R}]}$ meeting the cost of supplementary benefits, i.e. the cost of mutuality and of the guarantee. According to the fees, the current value of a unit of the reference fund, the price of the financial options and mortality rates, Eq.(7.5.11) allows to determine the number N of units which can be credited to the policy.

Note that in (7.5.10), and then in (7.5.11), independence between the lifetime of the insureds and the return on the reference fund is implicitly assumed. Such an assumption is reasonable; what is not trivial is how the probabilities q_{x+t} should be chosen (actually, we have used a generic notation, not specifying whether they are realistic or prudential). Following the pricing principles of traditional benefits, a conservative choice should be taken; given that we are dealing with an endowment, mortality rates higher than what is realistic should in particular be involved. However, due to the cost of the guarantees, not necessarily this is a choice on the safe-side.

7.6 With-profit policies

With-profit policies represent a traditional UK business. Similarly to participating policies, they guarantee a given return on investment, while distributing (part of) the realized extra-yield to policyholders. They are typically issued with annual constant premiums.

The main difference in respect of participating policies consists in the way profit is assigned; in with-profit policies, a *bonus* is added in each year to the benefit, which is defined according to a given rule. According to the prevailing practice, bonuses are calculated so that the release of profit is smoothed in time. This is obtained adopting parameters which are approximately constant (in particular, they should be constant if the yield on investment is flat). However, as we illustrate below, this may imply that in some years the bonus is too high in respect of the yield realized in that year on the investment of the policyholder. To avoid major costs for the insurer, some rules for the calculation of bonuses are designed so to slow down the distribution of unrealized gains, while maintaining an apparent smoothed release of profit.

Three types of bonus can be identified:

- reversionary bonus;
- terminal bonus;
- guaranteed bonus.

The *reversionary bonus* is funded through financial profit. Once it has been assigned, it is locked-in. Let $B_t^{[\text{rev}]}$ be the reversionary bonus at time t. Similarly to the adjustment of the reserve in participating policies, $B_t^{[\text{rev}]}$ is assigned to in-force policies. Following time t, the benefit amount cannot be lower than

$$G_t = C + \sum_{s=1}^{t} B_s^{[\text{rev}]} \tag{7.6.1}$$

where C is the initial guaranteed amount of the benefit (namely, the amount referred to for premium calculation), given that the current and the previous reversionary bonuses are locked-in.

The benefit paid at time t (in case of death if $t < m$, either in case of survival or death if $t = m$) is defined as follows

$$C_t = G_{t-1} + B_t^{[\text{term}]} \tag{7.6.2}$$

where $B_t^{[\text{term}]}$ is the so-called *terminal bonus*. The goal of the terminal bonus is to pay the profit not yet released; it is required $B_t^{[\text{term}]} \geq 0$, given that reversionary bonuses are locked-in (and the initial benefit amount is guaranteed). Conversely, it may turn out $B_t^{[\text{term}]} \gtreqless B_{t-1}^{[\text{term}]}$, as no specific guarantee is provided on the terminal bonus. It is not unusual that $B_t^{[\text{term}]} > B_t^{[\text{rev}]}$, as we justify below. We finally note that, given the terminal bonus at maturity, no reversionary bonus at maturity is assigned.

Contrarily to the reversionary and terminal bonuses, the *guaranteed bonus* is explicitly funded by premiums, which are determined (at policy issue) accounting for the annual increase of the benefit originated by the guaranteed bonus. Indeed, the guaranteed bonus simply consists in an annual increase of the benefit amount. If a guaranteed bonus has been underwritten, the guaranteed benefit amount since time t is

$$G_t = C + \sum_{s=1}^{t} B_s^{[\text{rev}]} + \sum_{s=1}^{t} B_s^{[\text{guar}]} \tag{7.6.3}$$

where $B_s^{[\text{guar}]}$ is the bonus guaranteed at time s.

As mentioned above, for with-profit policies it is common to obtain (or to show) a smoothed release of profit in time. Rules for the calculation of reversionary bonuses are defined so to slow down the distribution of unrealized gains. We examine some of these rules, just to give an idea on how this target can be reached. We refer to an endowment with-profit policy, with initial benefit amount C and level premium P. Similarly to participating policies, the premium P is calculated as if the policy was with fixed-benefits; namely, P is calculated taking C as a constant benefit (if a guaranteed bonus applies, reference would be made to C for the first year, $C + B_1^{[\text{guar}]}$ for the second year, and so on; for brevity, we disregard this case).

A possible rule for the calculation of the reversionary bonus is the *linear rule*, according to which

$$B_t^{[\text{rev}]} = \alpha_t\, C \tag{7.6.4}$$

where α_t, $\alpha_t \geq 0$, is the bonus proportion at time t. An alternative rule is the *exponential* (or *compound*) *rule*, under which

$$B_t^{[\text{rev}]} = \beta_t\, G_{t-1} = \beta_t \left(C + \sum_{s=0}^{t-1} B_s^{[\text{rev}]} \right) \tag{7.6.5}$$

In principle, α_t or β_t should be assessed referring to the extra-yield on the investment of the policyholder realized in year $(t-1, t)$. However, according to usual practice, they are set more or less constant in time. Basically, the idea is to distribute year by year a share of the total profit which is expected to be realized by maturity. This originates some cross-subsidy effects throughout time and among cohorts, which may produce some costs for the insurer. In order to understand better, it is worth making a comparison with participating policies.

The reversionary bonus $B_t^{[\text{rev}]}$ can be compared with the benefit update $j_t^{[\text{B}]} C_t$ of participating policies. In particular, the proportion β_t in (7.6.5) can be directly compared to $j_t^{[\text{B}]}$. It is useful to refer to the example in Table 7.2.1, where the extra-yield on investment (and then profit, in relative terms) is constant in time. Since $j_t^{[\text{P}]} = 0$ (i.e., premiums are constant, as for with-profit policies), we find that $j_t^{[\text{B}]}$ is increasing in time. Consistently, the proportion β_t in (7.6.5) should be increasing. If, as it occurs in practice, it is set constant in time, then when t is small the proportion $\beta_t = \beta$ is higher than what justified by current profits, whilst when t is close to maturity $\beta_t = \beta$ is smaller than what justified by current profits (we note that β_t is not exactly constant in practice, as to some extent it follows the fluctuations of the experienced investment yield). Overall, at any time t the insurer is assigning a too high bonus to policies recently issued, and a too low bonus to policies close to maturity. If the portfolio composition is appropriate, the insurer can be on balance. Anyhow, a cross-subsidy effect emerges among the different cohorts.

Considering the policyholders' expectation for a constant bonus proportion, rule (7.6.5) is preferable to (7.6.4), from the point of view of the insurer. It is easy to justify why. Assume that (7.6.4) is adopted with $\alpha_t = \alpha$, constant. Then, the guaranteed benefit at time t is

$$G_t = C(1+\alpha t) \tag{7.6.6}$$

Similarly, if we assume $\beta_t = \beta$, constant, in (7.6.5), we find for the guaranteed benefit at time t

$$G_t = C(1+\beta)^t \tag{7.6.7}$$

Assume that the terminal bonus is calculated following the same rule of the reversionary bonus. In this case, the benefit at maturity can be expressed as G_m. Given the total profit realized by maturity, Eqs. (7.6.6) and (7.6.7) should result in the same amount G_m, i.e. we should find

$$(1+\alpha m) = (1+\beta)^m \tag{7.6.8}$$

Condition (7.6.8) requires $\beta < \alpha$. It can be easily checked that if $\beta < \alpha$, then for $t < m$

$$(1+\alpha t) > (1+\beta)^t \tag{7.6.9}$$

Thus, provided that the amount of profit distributed in total is the same, the release of profit in time is slower if an exponential reversionary bonus is adopted (with a constant proportion β).

Example 7.6.1. Refer to a with-profit endowment insurance, with initial sum insured $C = 1000$ and maturity $m = 10$. Assume that the total amount at maturity of the bonuses is 300, i.e. $\sum_{t=1}^{9} B_t^{[\text{rev}]} + B_{10}^{[\text{term}]} = 300$. Expressing the terminal bonus with the same rule of the reversionary bonus and adopting a constant proportion, for the linear rule we find $\alpha = 0.03$, and for the exponential rule $\beta = 0.02658$. It can be easily verified that $1000(1+0.03t) > 1000\,1.02658^t$ at any time $t < 10$. At time 5,

for example, the guaranteed amount with the linear rule is $G_5 = 1\,000\,(1+0.03\times 5) = 1\,150$, while with the exponential rule $G_5 = 1\,000\times 1.02658^5 = 1\,140.16$.
□

A rule further slowing down the release of profit is the so-called *supercompound rule*, defining the reversionary bonus as follows

$$B_t^{[\text{rev}]} = \gamma_t\, C + \delta_t \sum_{s=1}^{t-1} B_s^{[\text{rev}]} \tag{7.6.10}$$

In this case, the reversionary bonus has both a linear and an exponential component. An appropriate choice of the parameters γ_t and δ_t can result in a reduced rate of increase of the guaranteed amount.

Example 7.6.2. Refer to Example 7.6.1. If we set $\gamma_t = \gamma = 0.02$, after some little algebra we find that if $\delta_t = \delta = 0.08732$, then $\sum_{t=1}^{9} B_t^{[\text{rev}]} + B_{10}^{[\text{term}]} = 300$. We can verify that at time $t < 10$, it turns out $G_t < 1\,000\,1.02658^t < 1\,000\,(1+0.03\,t)$. At time 5, for example, the guaranteed amount is $G_5 = 1\,119.06$, which can be compared with the amounts quoted in Example 7.6.1 for the linear and the exponential rule.
□

Of course, given a rule for the calculation of the reversionary bonus, a straightforward way to avoid the release of unrealized profits consists in setting the value of the bonus proportion lower than what would be required by the total profit expected during the policy duration; in other words, the bonus proportion should be chosen with a conservative view. The terminal bonus ensures that in case of immediate payment the beneficiary would receive the extra-yield really gained so far on the investment of the policyholder. Indeed, while the development in time of the minimum guaranteed amount is slowed down, the actual benefit would be in line with the realized gain on investment.

7.7 Index-linked policies

Index-linked policies are endowment-like contracts, funded with a single premium, whose benefit amount is linked to the performance of a stock-market index, the so-called *reference index*. A guarantee is provided for the maturity benefit, as this is defined as the single premium (also called *invested amount*) rolled-up with the highest between an accumulation factor depending on the performance of the reference index and a guaranteed accumulation factor. The reference index is usually based on a wide basket of stocks, so to smooth extreme fluctuations; possibly, a mix of indexes is referred to, with the aim of improving such a smoothing.

Let I_t denote the value at time t of the reference index. A given function Φ, the *participating rule*, defines the accumulation factor based on the performance of the reference index during the policy duration. In principle, the participating rule depends on the whole path of the reference index during the policy duration; the specific form of Φ may address just some aspects of such a path (see below for some examples).

The *maturity benefit* is defined as follows

$$S = \Pi \times \max\{\gamma, \Phi(I_0, I_1, \dots, I_m)\} \tag{7.7.1}$$

where Π is, as usual, the net single premium and γ the guaranteed accumulation factor (as it is reasonable, the expense loading is not accounted for in the rolling-up of the single premium). An alternative expression for the maturity benefit is the following:

$$S = \Pi\gamma + \Pi \times \max\{\Phi(I_0, I_1, \dots, I_m) - \gamma, 0\} \tag{7.7.2}$$

where $\Pi\gamma$ is the guaranteed benefit, while $\Pi \times \max\{\Phi(I_0, I_1, \dots, I_m) - \gamma, 0\}$ is the pay-off of a call option on the reference index, with strike γ and maturity m.

Several choices can be made in respect of the guaranteed accumulation factor γ:

- $\gamma = 0$ (the arrangement is referred to as *index-linked with no explicit guarantee*);
- $0 < \gamma < 1$ (*index-linked with a partial guarantee*);
- $\gamma = 1$ (*index-linked with a guaranteed principal*);
- $\gamma > 1$ (*index-linked with guaranteed interest*).

At a first instance, it could seem difficult to accept $\gamma \leq 1$. However, first it must be noted that through the index-linked policy the policyholder realizes an investment in a stock-market index; as is well-known, stock-market indexes are subject to downwards fluctuations. In this case, a guaranteed principal may be of interest. Further, it must be considered that the premium has to fund both the guaranteed amount and the call option. The lower the guaranteed amount, the higher the amount available for investing in the call option, and then in the participation to the performance of the reference index. We further note that, depending on the option, some guarantees may be embedded into its pay-off, so that a high γ would be unnecessary.

The *integral participating rule* is a very simple example of participation to the performance of the index. Let

$$g_t = \frac{I_t}{I_{t-1}} - 1 \tag{7.7.3}$$

be the rate of change of the reference index in year $(t-1, t)$. Due to the nature of the index I_t, we may experience $g_t \gtreqless 0$. The participating rule is defined as follows

$$\Phi(I_0, I_1, \dots, I_m) = (1 + g_1)(1 + g_2)\dots(1 + g_m) = \frac{I_m}{I_0} \tag{7.7.4}$$

In practice, the option embedded in (7.7.1) is European-style. Clearly, it may turn out $\frac{I_m}{I_0} < 1$. However, if $\frac{I_m}{I_0} < \gamma$ the guaranteed amount would be paid at maturity.

In the *Cliquet participating rule*, the single premium is rolled-up in year $(t-1, t)$ at the rate

$$j_t = \begin{cases} 0 & \text{if } g_t < 0 \\ g_t & \text{if } 0 \leq g_t < g' \\ g' & \text{if } g_t \geq g' \end{cases} \tag{7.7.5}$$

where g' expresses the maximum annual rate of increase of the reference index admitted for the rolling-up of the single premium; for example, $g' = 0.20$. The participating rule is defined as follows

$$\Phi(I_0, I_1, \ldots, I_m) = \alpha\,(1+j_1)\,(1+j_2)\ldots(1+j_m) \tag{7.7.6}$$

where α, $\alpha > 0$, is a participating proportion, amplifying (if $\alpha > 1$) or compressing (if $\alpha < 1$) the change of the reference index. We note that $j_t \geq 0$; thus $\Phi(I_0, I_1, \ldots, I_m) \geq \alpha$, i.e. the Cliquet participating rule embeds a minimum accumulation guarantee. Indeed, past positive jumps of the reference index are locked-in. Depending on the proportion α and on the possible path of the reference index, the option implied by the Cliquet participating rule could be more expensive than that implied by the integral participating rule.

The pay-off described by (7.7.1) is that of a *structured Zero Coupon Bond* or *index-bond*. This is the asset backing the policy. Usually, the insurer purchases index-bonds issued by investment banks; this explains why the index-linked policy is issued at single premium (annual premiums would require that index-bonds with the features specified in (7.7.1) are available also after issue, and the insurer cannot be certain about this). A default risk emerges, which should be borne by the insurer.

In case of early termination of the contract, because of death or surrender, a benefit is paid. The *death benefit* is usually defined as the current value of the index-bond, increased by a given proportion (say, 5 or 10%). The beneficiaries are usually given the possibility of keeping the investment until maturity, if they think that it is not currently convenient to cash the investment. We point out that the amount of the death benefit is not guaranteed, as the factor γ just concerns the maturity benefit. We also note that the insurance component is negligible; basically, the index-linked policy is an investment product. Consistently, the *surrender value* is the current value of the index-bond, possibly reduced by a (small) fee. No guarantee applies to this benefit.

The expense-loaded premium $\Pi^{[T]}$ consists of three components: the expense loading (which is referred to as management fees), the cost of the index-bond and the cost of mutuality. This latter component is usually assessed approximately, due to the small size of the sum at risk. The reserve is simply the value of the index-bond, possibly increased by a small proportion, to account for the death benefit.

7.8 Universal Life policies

Universal Life (UL) policies are typical products of the US market, which can be designed either as participating or unit-linked policies. Their main features consists in a high flexibility available to the policyholder in deciding year by year: the amount of premium, to make a partial withdrawal, the type of investment backing the reserve, and so on. Further, similarly to a bank account, the policyholder receives a periodic statement, showing the costs (acquisition costs, management fees, fees for rider benefits, and so on) that have been charged to her policy account. If the policy is designed on a unit-linked basis, the current value of the policy assets is reported in the statement; if a participating arrangement is designed, the statement reports the annual adjustment which has been credited to the reserve.

The underlying contractual form is a whole life assurance. This way, the contract has no specified maturity; the contract terminates either because of death or full withdrawal. The sum at risk is defined so that it is positive; see examples provided in Sect. 5.4.4.

The UL is a complex product for the insurer. The flexibility granted to the policyholder originates many risks. For example, it is difficult to predict future profits, due to the uncertainty on the premium level; the possibility of partial withdrawal determines a liquidity risk; it is difficult to match the liabilities with appropriate assets, as there is uncertainty on the timing of the former, and so on. Further, a considerable transparency in respect of the information provided to the policyholder is required. On the other hand, the product could be very attractive. The insurer can try to gain the loyalty of the policyholder designing an insurance package, providing capital protection and other insurance benefits during the working life of the insured, and then pension benefits after retirement. This idea is realized by the variable annuities policies, which we describe in the next Section.

Health insurance benefits can also be included in the UL policy: accident insurance, disability benefits, hospitalization benefits, and so on. Thus, the UL policy can be shaped as a package of insurance covers.

7.9 Variable annuities

The term variable annuity is used to refer to a wide range of life insurance products, whose benefits can be protected against investment and mortality / longevity risks by selecting one or more guarantees out of a broad set of possible arrangements. Originally developed for providing a post-retirement income with some degree of flexibility, nowadays accumulation and death benefits constitute important components of the product design. Indeed, the variable annuity can be shaped so as to offer dynamic investment opportunities with some guarantees, protection in case of early death and/or a post-retirement income.

The design of variable annuities matches features of unit-linked life insurance contracts (the investment into a reference fund selected by the policyholder) to those of participating contracts (the guarantees). Basically, the variable annuity is a fund-linked insurance contract, including a package of financial options on the policy fund value. Guarantees are then also looked at as riders to the basic benefit given by the account value. Similarly to unit-linked policies, guarantees are explicit, and then a fee is applied to the policyholder who underwrites them.

As for participating or unit-linked contracts, financial options in variable annuities are non-standard, as their exercise depends not just on economic factors, but also on the lifetime of the insured or on preferences of the policyholder. Thus, their valuation raises several complex issues; difficulties arise also in relation to the time-horizon, which involves many years when post-retirement benefits are dealt with. Some issues in this respect are discussed in Chap. 8. In this Section we only give a description of the most common guarantees; as for participating and unit-linked policies, we do not deal with their valuation.

Guarantees in variable annuities may be first classified into two main broad classes:

- Guaranteed Minimum Death Benefit (GMDB);
- Guaranteed Minimum Living Benefit (GMLB).

The second class can be further arranged into three subclasses:

- Guaranteed Minimum Accumulation Benefit (GMAB);
- Guaranteed Minimum Withdrawal Benefit (GMWB);
- Guaranteed Minimum Income Benefit (GMIB).

The acronym *GMxB* is used to briefly refer to the whole set of guarantees, i.e. *Guaranteed Minimum Benefit of type 'x'*, where 'x' stands for the class of benefits involved: accumulation (A), death (D), withdrawal (W) or income (I).

Variable annuities are generally issued with single premium or single recurrent premiums. The total amount of premiums is also named the principal of the contract or the invested amount. Apart from some upfront costs, premiums are entirely invested into the reference funds chosen by the policyholder. Similarly to unit-linked policies, several investment opportunities are available to the customer, providing different risk/return profiles. The policyholder is allowed to switch from one risk/return solution to another at no cost, if some constraints are fulfilled (for example, the switch is required no more than once a year). Unlike in unit-linked, with profit or participating policies, reference funds backing variable annuities are not required to replicate the guarantees selected by the policyholder, as these are hedged by specific assets. Therefore, reference fund managers have more flexibility in catching investment opportunities.

Guarantees and asset management fees, administrative costs and other expenses are charged year by year to the contract through a reduction of the policy account value. This improves the transparency of the contract, as any deduction to the policy account value must be reported to the policyholder; this follows the tradition of Universal Life policies. Some guarantees can be added or removed, at policyholder's discretion, when the contract is already in-force. Accordingly, the corresponding fees start or stop being charged. The cost of guarantees, as well as other expenses, are typically expressed as a given percentage of the policy account value. In particular when relating to mortality or longevity guarantees, applying a constant percentage may result in some approximations of the real cost. If the sum at risk is positive, such an approximation is usually negligible, as we have commented for unit-linked policies (see Sect. 7.4.2); conversely, when income benefits are involved, and then the sum at risk is negative, major costs may emerge from such approximations, in particular due to the extent of the time-horizon involved. See Sect. 8.6 for some remarks in this regard.

The *Guaranteed Minimum Accumulation Benefit* (GMAB) is usually available prior to retirement. At some specified date, the insured (if alive) is credited the greater between the policy account value and a guaranteed amount. Such guaranteed amount can be stated as follows:

- the amount of premiums paid, net of withdrawals (the so-called *return of premiums*);
- the *roll-up* of premiums, net of withdrawals, at a specified guaranteed interest rate;
- the highest account value recorded at some specified times (prior to the maturity of the GMAB); this is a *ratchet* guarantee, which locks-in the positive performances of the reference fund.

See Fig. 7.9.1 for a graphical representation of the main guarantees; to make clearer the presentation, a single premium has been considered and it has been assumed that no withdrawals occur. A further guarantee which may be attached to the GMAB is the *reset*, which gives the opportunity to renew the GMAB when it reaches maturity.

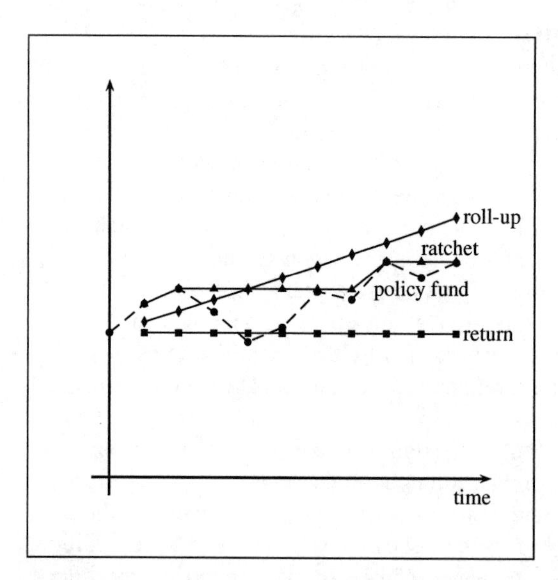

Fig. 7.9.1 Possible choices for the GMAB and the GMDB; single premium (no withdrawal)

Similarly to the GMAB, also the *Guaranteed Minimum Death Benefit* (GMDB) is available during the accumulation period; some insurers are willing to provide a GMDB also after retirement, up to some maximum age (say, 75 years). The structure of the guarantee is similar to the GMAB: in case of death prior to the stated maturity, the insurer will pay the greater between the account value and a stated amount. The guaranteed amount can be either fixed, e.g. equal to

- the amount of premiums paid, net of withdrawals;
- the roll-up of premiums, net of withdrawals, at a specified guaranteed interest rate;

or depending on the account value, such as

- the highest account value recorded at some specified times prior to death (ratchet);
- the account value at some prior specified date (the so-called reset date) plus the total amount of premiums paid following such date, net of withdrawals. This is the *reset* guarantee (whose meaning is different within the GMAB and GMDB).

Fig. 7.9.1 also represents the main GMDB guarantees. The difference between the ratchet and the reset guarantee within the GMDB stands in the behavior of the guaranteed minimum amount: in the ratchet guarantee the minimum amount never decreases, whilst a reduction may occur in the reset, if the account value decreases between two reset dates.

The *Guaranteed Minimum Income Benefit* (GMIB) provides a lifetime annuity from a specified future point in time. The guarantee may be arranged in two different ways:

- the amount to be *annuitized* (namely, the amount to be converted into a life annuity) will be the greater between the account value and a specified amount. Possible ways to specify such an amount are similar to the GMAB. The *annuitization rate* (that is, the ratio between the annual income and the annuitized amount, also called the *conversion coefficient*) will be defined according to market conditions prevailing at the annuitization date;
- the annuitization rate will be the more favorable between a stated rate and what resulting from current conditions. The annuitized amount will be the account value.

The former guarantee is sometimes described as a guarantee on the annual amount, which would suggest an arrangement similar to a deferred life annuity; it is then worthwhile to stress that the guarantee actually concerns the amount to be annuitized, as described above. In principle, it is possible to offer both guarantees, but in practice this is not usual, because of the high cost.

Remark It is worth quoting here a terminology prevailing in the life annuity markets. Although it is not commonly used for variable annuities, it may help in understanding better the features of the GMIB.
An annuitization rate defined according to current market conditions is named *current annuitization rate* (*CAR*). Under a CAR, the annual amount is guaranteed after annuitization (given that the CAR essentially expresses the price of an immediate life annuity), but not prior to this time. Conversely, a *guaranteed annuitization rate* (*GAR*) states the annuitization rate prior to annuitization. A policyholder entitled to a GAR usually has the possibility to choose, at annuitization, the best rate between the CAR and the GAR; this possibility is referred to as a *Guaranteed Annuitization Option* (*GAO*). See also Sect. 8.6.
Referring such terminology to the GMIB, we would say that the GMIB can consists in:

- a guarantee on the amount to be annuitized (while a CAR is adopted for the annuitization);
- a GAO (while the amount to be annuitized is not guaranteed).

As already mentioned, it is possible to underwrite both a guarantee on the amount to be annuitized and a GAO, but this would be very expensive for the policyholder (as the insurer would be exposed to major risks).

If the GMIB is exercised, after annuitization the policyholder loses access to the account value (while prior to annuitization the contract works like an investment product, bearing some guarantees). The guarantee must be selected by the policyholder some years before annuitization; typically, the GMIB may be exercised after a waiting period of 5 to 10 years. The cost of the GMIB is deducted from the account value during the accumulation period. If prior to annuitization the policyholder gives up the guarantee, the insurer stops deducting the relevant fee. Typically, full annuitization is required; however, partial annuitization is admitted in some arrangements. As far as the duration of the annuity is concerned, the following solutions may be available: a traditional life annuity; a last survivor annuity; a life annuity with a minimum number of payments (say, up to 5 or 10 years). *Money-back* (or *capital protection*) arrangements may also be available, providing a death benefit consisting of the residual principal amount, i.e. the annuitized amount net of the annual payments already cashed. The annual amount may be either fixed (either flat or escalating), participating or inflation-linked or linked to stock prices (see also Sect. 4.3.3); a financial risk is borne by the annuitant in the latter case, as the annual amount can fluctuate in time (conversely, in a participating scheme the annual amount never decreases; see Sects. 7.3 and 7.4).

The *Guaranteed Minimum Withdrawal Benefit* (GMWB) guarantees periodical withdrawals from the policy account, also if the account value reduces to zero (either because of bad investment performances or long lifetime of the insured). See Fig. 7.9.2 for a graphical representation.

The guarantee concerns the annual payment and the duration of the income stream. The annual payment is stated as a given percentage of a base amount, which is usually the account value at the date the GMWB is selected. In some arrangements, at specified dates (e.g., every policy anniversary) the base amount may step up to the current value of the policy account, if this is higher; this is a ratchet guarantee, which may be lifetime or limited to some years (e.g., 10 years). Note that, thanks to the ratchet, the guaranteed annual payment may increase in time; in some arrangements, a maximum accepted annual increase is stated in policy conditions. The guaranteed annual payment may be alternatively meant as the exact, the maximum or the minimum amount that the policyholder is allowed to withdraw in each year. In the last case, any withdrawal above the guaranteed level reduces the base amount. The duration of the withdrawals may be fixed (e.g., 20 years) or lifetime. In the former case, if at maturity the account value is positive, it is paid back to the policyholder or, alternatively, the contract stays in-force until exhaustion of the policy account value. The cost of the guarantee is deducted from the account value during the payment period; if the policyholder gives up the guarantee, the relevant fee stops being applied. During the withdrawal period, the policyholder keeps access to the

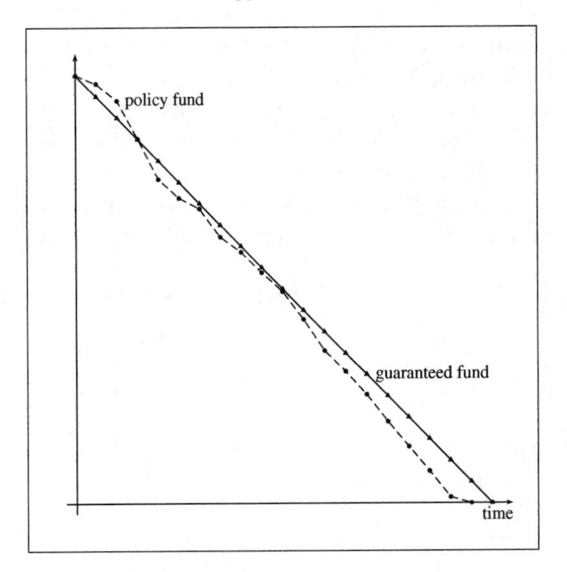

Fig. 7.9.2 Fund available under a GMWB; guaranteed annual withdrawal: 5% of the initial fund, for 20 years

unit-linked fund; if at death the account value is positive, such an amount is paid to the estate of the policyholder.

From the descriptions above, it emerges that GMAB and GMDB are similar to what can be found in participating and unit-linked arrangements, apart from the possible range of guarantees, which is wider in variable annuities than in traditional contracts and is explicit. The GMIB is like a traditional life annuity, possibly participating. The GMWB is the real novelty of variable annuities in respect of traditional life insurance contracts; it provides a benefit which is similar to an income drawdown, but with guarantees (see Sect. 8.5.3 for more details on the income drawdown). When comparing a GMIB to a GMWB, three major differences arise: the duration of the annuity (which is lifetime in the GMIB), the accessibility to the account value (just for the GMWB) and the features of the reference fund (which is unit-linked in the GMWB, but typically participating in the GMIB). Clearly, the presence of death benefits also in the GMIB, a lifetime duration for the withdrawals in the GMWB and other possible features reduce a lot the differences between the GMIB and the GMWB. Apart from the use of one name or the other, policy conditions should suggest the real features of the income provided by the contract.

7.10 References and suggestions for further reading

Literature on insurance benefits linked to the investment performance is very rich. Here, we just mention the textbooks dealing with this topic. A description of the early forms of policies realizing several form of flexibility and linking to the investment performance is given by [18]. Many authors have addressed the valuation of financial options embedded in insurance benefits. Market-valuation methods for the valuation of participating policies are described in [41]. Unit-linked policies, and possible approaches to the valuation of the relevant financial options are dealt with by [29] and [20]. A description of with-profit policies can be found in [9], while for Universal Life reference can be made to [7]. An introduction to variable annuities can be found in [40].

Chapter 8
Pension plans: technical and financial perspectives

8.1 Introduction

In this Chapter we examine some features of private pension programmes, namely those arrangements providing a post-retirement income in addition to the public pension. As we will see, a private pension plan can be designed either on an individual or a group basis. Although in the modern forms the funding of benefits is always realized on individual basis, group pension plans allow for a funding arrangement based on solidarity principles. The post-retirement income is the basic benefit of a pension plan; however, several rider benefits can be underwritten, covering risks to which an individual is exposed either before or after retirement.

It is worth anticipating some of the common terminology adopted when referring to pension plans; further terms will be introduced later on.

- The lifetime of an individual is split into two economic periods: the period before retirement, the so-called *savings period* (or *working period*) and the period after retirement, the so-called *post-retirement period* (or simply *retirement period*); see also Sect. 1.2.5.
- An individual joining a pension plan is referred to as a *member*. The member is *active* during her working life, and *retired* after retirement.
- Similarly to life insurance, benefits must be funded by appropriate payments (the premiums, in life insurance). Such payments are called *contributions*.
- The institution arranging a private pension plan is generically referred to as the *provider*. As we will see, it can be an insurer, another financial institution or a specific institution set up for this purpose.
- The pension income is considered to be the main benefit of a private pension plan. Further benefits can be underwritten, which are looked at as riders. Within pension plans, they are referred to as *ancillary benefits*.

In detail, the main issues dealt with in this Chapter are the following:

- possible technical designs of private pension plans, in particular with reference to the definition of benefits and the relevant funding principles;

A. Olivieri, E. Pitacco, *Introduction to Insurance Mathematics*,
DOI 10.1007/978-3-642-16029-5_8, © Springer-Verlag Berlin Heidelberg 2011

- the accumulation of savings for pension purposes, and the risks to which an individual is exposed before retirement;
- solutions for the post-retirement income available to an individual, and relevant risks;
- risks borne by the provider, depending on the benefits provided before and after retirement.

We point out that the topic discussed in this Chapter is very wide, and a thorough presentation is not possible here. Basically, in this Chapter we aim at carrying forward the discussion started in Sect. 1.2.5 concerning possible solutions for the provision of a post-retirement income.

8.2 Pension programmes

We refer the term *pension programme* (or *pension plan*) to any arrangement aimed at providing a post-retirement income. Pension plans may be classified according to the number of individuals they cover, the rule linking benefits to contributions and the timing of payment of contributions.

8.2.1 Individual and group pension plans

Referring to the number of individuals which are covered by the pension plan, we identify *individual* (or *single-member*) *plans* and *group pension plans*.

An *individual pension plan* is similar to a life insurance contract, although the legal form of the contract may be other than that. The provider can be an insurer or another financial institution with a specific license for dealing with pension benefits. The individual pays contributions during her working life, and receives an income after retirement. Several benefits can be underwritten as riders to the post-retirement income, such as a death benefit during the savings period and in the first years (say, 5–10 years) after retirement, sickness insurance benefits, and so on; see also Sect. 8.5.2. Individual contributions must be on balance with the benefits underwritten by the individual; the way this balance is realized depends on the risks that are transferred to the provider; we comment on this in Sects. 8.2.2 and 8.3. As mentioned in Sect. 1.2.5, the savings period is also called *accumulation phase*, while the post-retirement period is also called *decumulation phase*. These terms follow the idea that during her working life the individual saves money, to be used after retirement.

A *group pension plan* covers a number of individuals who share some common features as regards their occupation. Usually, they either work for the same employer, or in the same economic sector, or are self-employed for the same profession, and so on. Joining or not the pension plan is an individual decision, unless current legislation states otherwise. The employer is referred to as the *sponsor* of the pension plan. The plan may be managed directly by the sponsor; in this case, the sponsor typically underwrites some insurance contract, typically a *group insurance* or some other specific agreement with an insurer, to transfer at least partially its risks. More commonly, a specific institution is set up for managing the liabilities of the pension plan, the so-called *pension fund*. Similarly to the case of individual pension plans, also in the case of group pension plans individuals pay contributions during their working life and receive an income after retirement; some ancillary benefits can be underwritten, typically concerning the event of early death. The balance between contributions and benefits can be realized on an individual basis (similarly to an individual pension plan) or for the whole group. This latter solution implies solidarity effects, as we explain in Sect. 8.2.2. Contributions may be paid also by the sponsor, as an indirect (and deferred) form of salary to its employees.

Social security plans (or *state pension plans*) represent an "extreme" example of group pension plan, as they cover the whole population of a country. Joining a social security plan is not a choice; in particular, it is compulsory to pay contributions to the social security plan. The balance between contributions and benefits is realized on a group basis: the contributions paid currently by active people are used to fund the benefits currently paid to retired people. As opposed to social security plans, individual and group pension plans are considered *private pension plans*. In this Chapter, we only address this type of plans.

Remark According to legislation, it may be compulsory to join some private pension plan, in addition to the social security plan. In particular, this is imposed when the public pension is set at minimum levels, not adequate to ensure to each citizen living standards in line with those during her working life. In a welfare economy, the State Government has to guarantee an adequate income to any retired citizen; if the benefit paid by the social security is kept low, the compulsory membership to some private pension plan ensures that in the future unexpected costs will not be originated by individuals not getting in total an adequate income (apart from a possible default of the provider). This is the idea of what is called a *three-pillar pension system*, namely a pension system arranged on the social security plan (the first pillar), group pension plans (the second pillar) and individual pension plans (the third pillar). The pension legislation contributes to define the importance of each pillar. The second and third pillars are the *private pension solutions*. Usually, the third pillar is not compulsory, while joining the second pillar can be mandatory (for the reasons quoted above). Nevertheless, if it is mandatory to join some pension plan and the individual is not satisfied with the performance of the pension fund supported by her sponsor, she has the possibility to join some other private plan (possibly, an individual one). In the following, we do not take care of the constraints imposed by legislation on the membership to a private pension plan; we just discuss some technical issues of private arrangements.

A fourth pillar is sometimes referred to, the so-called *phased retirement* or *partial retirement*. An individual may decide, at the normal retirement age, to continue to carry on a working activity, but at a slower pace (either taking a part-time position or a lighter job). In this case, she (usually) will receive in total the public pension, but just partially the private pension. The advantage stays in the flexibility gained in respect of the amount accumulated within the private pension plan which has not yet been converted into a post-retirement income. Further remarks in this regard are given in Sect. 8.5.4.

8.2.2 Benefits and contributions

As noted in Sect. 8.2.1, a balance must be realized between contributions and benefits. In particular, each plan must adopt specific rules for the calculation of benefits and contributions. In the following, we refer to the pension benefit only (while we disregard ancillary benefits).

A major distinction exists between defined contribution and defined benefit pension plans.

In a *defined benefit* (*DB*) pension plan, a rule is given for the definition of the benefit, i.e. the post-retirement pension. It can be a fixed annual amount or, more commonly, a proportion of the member's salary prior to retirement. The proportion depends on the number of working years; the salary prior to retirement can be the salary received in the last year prior to retirement or an average of the salary received in a given number of years prior to retirement. The contributions are then calculated so that they are on balance with the specified benefits.

If the balance between contributions and benefits is realized on a individual basis, from a technical point we have to solve an equation similar to (5.2.1) when the member joins the plan; after the initial time, the balance is expressed similarly to (5.3.1). Indeed, the arrangement works like a life insurance contract with fixed-benefits. Let 0 be the time when the individual joins the pension plan, and r the retirement time. In principle, the actuarial balance between contributions and benefits must be assessed at time 0, as follows

$$\mathrm{Prem}(0,r) = \mathrm{Ben}(0,+\infty) \tag{8.2.1}$$

where, similarly to life insurance, the quantity $\mathrm{Prem}(0,r)$ represents the expected present value at time 0 of the contributions of the individual in the time-interval $(0,r)$, while $\mathrm{Ben}(0,+\infty)$ represents the expected present value at time 0 of the benefits which will be paid to the individual (starting from time r and until member's death, given that we are only addressing the pension benefit). Appropriate assumptions are required for the assessment (8.2.1). In particular, an interest rate must be chosen to discount future contributions and future benefits. To understand the other assumptions, we first note that $\mathrm{Ben}(0,+\infty)$ corresponds to the actuarial value at time 0 of a life annuity (with fixed benefits) commencing at time r if the individual is alive and still a member of the plan at that time. During the time-interval $(0,r)$ it may happen that the individual moves to another plan (e.g. because she changes employment), while after retirement she remains a member of the plan, until death. Then we note that contributions are paid in $(0,r)$ if the individual is alive, still belongs to the plan and still receives a salary. Due to a disability, it is possible that the individual is unable to perform the usual work, which inhibits her from receiving the salary, and then paying the contribution.

Summarizing, apart from the choice of the discount rate, to assess (8.2.1) assumptions are required in respect of the lifetime of the individual, the probability that she remains a member of the plan and the probability that she receives a salary without discontinuances. Contributions resulting from all these assumptions should not be changed after time 0, similarly to what happens in life insurance. Several risks emerge for the provider, in particular due to the extent of the time-horizon, which may make difficult to take appropriate assumptions. In practice, the rules of the pension plan may state that assumptions could be updated, if this is required by the evolving economic and demographic scenario; this usually results in an update of contributions (and thus some risks are charged to the member). We finally note that (8.2.1) implies the accumulation of a fund in the interval $(0,r)$, to be used after retirement for paying out the defined pension benefit. So, at time t, $t=0,1,\dots,r-1$, the following balance must be fulfilled

$$\mathrm{Prem}(t,r)+V_t=\mathrm{Ben}(t,+\infty) \tag{8.2.2}$$

while at time t, $t=r,r+1,\dots$

$$V_t=\mathrm{Ben}(t,+\infty) \tag{8.2.3}$$

In both cases, V_t represents the individual fund, whose management is similar to that of a reserve in life insurance (which explains the notation we have adopted). Note that, if the plan's rules allow for this, the cost of an update to the valuation assumptions can be charged to the member just in the time-interval $(0,r)$, as no contribution is paid following time r.

If the balance between contributions and benefits is realized on a group basis, the following condition must be satisfied:

$$\mathrm{Prem}^{[\mathrm{P}]}(t,t+T)+V_t^{[\mathrm{P}]}=\mathrm{Ben}^{[\mathrm{P}]}(t,t+T) \tag{8.2.4}$$

where t is the current time, T is a given time-horizon (namely, the time-horizon in respect of which, according to the plan's rules, the balance between benefits and contributions must be realized), $V_t^{[\mathrm{P}]}$ is the total amount of assets hold by the pension fund at time t, $\mathrm{Prem}^{[\mathrm{P}]}(t,t+T)$ is the present value at time t of the contributions which are expected to be received in the time-interval $(t,t+T)$ by the pension plan, and $\mathrm{Ben}^{[\mathrm{P}]}(t,t+T)$ is the present value at time t of the benefits which are expected to be paid by the pension plan in the time-interval $(t,t+T)$. Besides the assumptions already mentioned for the balance (8.2.1), condition (8.2.4) requires assumptions on the number of members paying contributions in the time-interval $(t,t+T)$, as well as on the number of those who cash benefits in the same time-interval. In particular, as active members reference can be made just to those who are within the plan at time t, or alternatively also to those who will join the plan in the period $(t,t+T)$. It must be noted that $V_t^{[\mathrm{P}]}$ refers to the whole group; in general, it is not possible to split this amount into individual funds. Indeed, when the balance between contributions and benefits is realized on a group basis, it is not clear what contributions are meeting the cost of the benefits of a given individual. To understand better, we can consider that while condition (8.2.1) always implies the accumulation of the individual contributions, the balance (8.2.4) could be realized also with $V_t^{[\mathrm{P}]}=0$ at any time t; in this case, the benefits currently paid to the retired members would be funded by the contributions currently paid by the active members (as it typically happens in social security plans). In other words, the implementation of (8.2.4) involves solidarity effects. We further note that (8.2.4) usually implies less guarantees than (8.2.1). For example, it is natural that the provider updates the valuation assumptions from one year to the other, in particular because the composition of the group is changing in time; the cost of the update is spread over the contributions, so that it is charged to the active members.

We now address *defined contribution* (*DC*) pension plans. In this case, a rule is given for the calculation of the contributions. The simplest choice is to set a fixed annual amount for each member, but more often the annual individual contribution is a proportion of the member's salary. Contributions are accumulated in an individual account, which is used at retirement to obtain a pension income. No guarantee is naturally implied before retirement, unless ancillary benefits have been underwritten; after retirement, a guarantee is provided if the benefit consists of an immediate life annuity. Other choices are possible, as we discuss in more detail in Sect. 8.5. We will come back on the possible guarantees prior and after retirement in Sect. 8.3.

As suggested by the descriptions above, DB pension plans imply several risks for the provider. Conversely, a DC pension plan does not necessarily imply guarantees; the advantage for the member is a greater flexibility, in respect both of investment choices and the type of post-retirement income.

In recent times, DC pension plans have become more popular than DB plans. This is due, in particular, to the fact that the former allow for more flexibility in favor of the member, while reducing risks for the provider. Further, nowadays the member is commonly allowed to move from a plan to another (although some constraints may apply). Thus, plans based on funding arrangements implying solidarity effects, as DB plans do, become unsustainable. In the following we only address DC arrangements.

8.2.3 Timing of the funding

It is clear that an individual pays contributions while she is an active member of the pension plan, and receives an income while she is retired. The payment of contributions may be interrupted in face of specific events (such as a disability that prevents the usual working activity), and some rider benefits may come into payment during the working period. In the following, for simplicity we refer to the pension benefit only and we disregard discontinuances in the payment of contributions.

If we take the point of view of the provider, at any time contributions are being received from the active members and benefits are being paid to the retired members. Depending on the rule linking contributions to benefits, as well as on the principle adopted for the balance between contributions and benefits, there can be an accumulation of assets.

A *funded pension plan* is an arrangement in which contributions are accumulated into a fund. If the balance between contributions and benefits is realized on an individual basis, each member is assigned a specific fund. An important issue concerns how the fund is invested, as well as who is bearing the investment risk.

In an *unfunded pension plan* (or *pay-as-you-go pension plan*) benefits currently paid are met by the contributions currently received by the provider. In this case, the balance between contributions and benefits is realized on a group basis. No fund is accumulated. An intermediate solution, adopted by some social security plans, consists in using the contributions currently paid by the active members to fund the amount backing the liability of the provider in respect of the members who are currently retiring; no fund is accumulated during the working period, while a fund (namely, the reserve of an immediate life annuity) is set up at retirement, and maintained up to death.

As we have already mentioned, unfunded plans are of interest just for social security plans, so that we no further address them. In the following, we just consider funded plans realizing an individual balance between benefits and contributions.

8.3 Transferring risks to the provider

In this Section, we summarize the risks that an individual, who is planning her post-retirement income, can transfer to the provider. We will come back in more detail to some of the issues introduced here in the following Sections.

We refer to a pension plan in which an individual saves money during her working life, in the form of contributions which are credited to her own personal fund. At retirement, the accumulated fund is used to receive a pension income. This can be realized within an individual or a group pension plan. As stated in Sect. 8.2.2, we only refer to DC pension plans. In this case, only an individual balance can be realized between contributions and benefits. From a technical point of view, the specific form of the pension plan (either individual or group) does not matter in this case; for brevity, we then refer only to individual pension plans.

In a DC pension plan, the working and the post-retirement period are addressed separately when defining the benefits. During the working period, the money is accumulated in the individual fund; the investment risk is naturally borne by the member. The advantage consists in the possibility for the member to select the asset composition she prefers, in particular in terms of risk/return profile (see also Sect. 8.4). Financial guarantees may be underwritten, so to transfer part of the financial risk to the provider; a fee is usually required. The availability of guarantees depends on who is the provider; insurers offer financial guarantees on their pension products (similarly to those examined in Chap. 7 for participating, unit-linked and variable annuity policies), while a group pension plan usually does not.

Ancillary benefits available during the working period are death benefits, disability benefits and other health insurance benefits. The death benefit can either be a lump sum benefit (a fixed amount or a multiple of the pensionable salary at death), or a pension in favor of the member's spouse. Death benefits can also take the form of a financial guarantee, similarly to what available within a unit-linked or a variable annuity product. These latter benefits are usually offered by an insurer, while a lump sum or a pension to the spouse are offered by any pension fund. Disability benefits may consist in the possibility to interrupt the payment of contributions in the case of a disability, or may be given by a disability income replacing the salary if the member is unable to work because of sickness or injury. Health insurance benefits are offered by insurers, or by a sponsor getting protection by an insurer. A fee is required for the ancillary benefits, whose cost is assessed counting on the possibility for the provider to realize mutuality effects. Risks originated by mutuality are borne by the provider. Withdrawals prior to retirement are allowed just in face of specific events (such as the purchase of a house, the wedding of a child, a critical illness requiring special medical care, the change of the pension plan in face of a new employment, and so on).

At retirement, the member has to select the form of the pension income. In some cases, it is possible to cash the accumulated amount. The member can simply plan a sequence of withdrawals from her account, as long as money is left. This is the so-called *income drawdown* (see Sect. 8.5.3). The investment risk and the risk connected to her longevity (see Sect. 1.2.5 and 8.5.3) are borne by the individual. The advantage is that she has access to her fund, in particular for the selection of the asset composition; further, in case of early death the residual fund belongs to her estate. Alternatively, the individual fund at retirement can be *annuitized*, i.e. converted into a life annuity. All risks are transferred to the provider, in particular the longevity risk, with the disadvantage of loosing access to the individual fund (for example, in case of early death the residual fund is used by the insurer for mutuality purposes). Intermediate solutions are possible; the fund at retirement can be partially annuitized. The advantage is to get some guarantees from the life annuity, while keeping some flexibility on the fund not annuitized. See Sect. 8.5.3 for more details. Ancillary benefits during retirement are typically death benefits, and can be obtained in respect of the fund which has been annuitized. In particular, the death benefit is implied by the type of life annuity selected by the individual (see Sect. 8.5.2). We mention the benefit provided by a capital protection, under which at death the estate receive the difference between the fund annuitized at retirement and the total income received by the annuitant up to death, and the pension in favor of the member's spouse, the so-called last-survivor annuity. Death benefits similar to those packaged in a variable annuity product can also be available, typically for some years after retirement (see Sect. 7.9). The funding of a death benefit is based on mutuality, similarly to what examined for life insurance products; the relevant risk is charged to the provider. However, we note that the death benefit mitigates the longevity risk taken by the provider; see also Sect. 8.6. The disadvantage for the individual of a death benefit taken as a rider to a life annuity is the cost: given the fund to be annuitized, the annual amount available if a rider benefit is underwritten is lower than in the case of a standard life annuity.

8.4 Pension savings before retirement

As we have mentioned in Sect. 8.3, during the accumulation period the investment risk is naturally borne by the individual. It is then desirable that the individual has some control over the investment of her fund. In principle, the member can select the asset composition more suitable to her preferences in terms of risk/return profile. It often happens that members do not have the required expertise for selecting appropriately the investment, so that the provider gives advice; in particular, it prearranges some lines of investment, which are characterized by different risk/return profiles. What is usually recommended is a *lifestyle investment strategy*. While young, the member should try to maximize the investment return by including in the assets an appropriate proportion of stocks. When approaching retirement time, a defensive strategy is preferable, and thus the investment should consists mainly of bonds. The shift from the former to the latter asset composition should be clearly progressive in time.

Several guarantees may protect the investment, but this typically requires the payment of a fee. Underwriting a guarantee corresponds to underwriting a financial option, as mentioned in Sects. 7.5 and 7.9. Since the guarantees imply a risk for the provider, some constraints may then be imposed on the asset composition. More often, the guarantees are hedged with appropriate assets, as we have commented for variable annuities.

8.5 Arranging the post-retirement income

As mentioned in Sect. 8.3, at retirement time the individual can usually choose among several alternatives to obtain the post-retirement income. Immediate life annuities and income drawdown constitute typical solutions. "Mixtures" of life annuities and income drawdown also provide practicable solutions.

Life annuities have been described in Sect. 4.3.3. In this Section we first turn again on this insurance product, looking at the life annuity as a (possible) element in post-retirement income arrangements. Then, alternatives to the life annuity are examined. In what follows, we just refer to the net cost of benefits, i.e. we disregard expenses.

8.5.1 Some basic features of life annuities

When planning the post-retirement income, some basic features of the life annuity product should be carefully accounted for. In particular, we note the following aspects.

1. The life annuity product relies on the mutuality mechanism, like the pure endowment insurance (see Sect. 1.7.4, and Fig. 1.7.7 in particular). This means that:
   a. the amounts released by the deceased annuitants are shared among the annuitants who are still alive;
   b. on the annuitant's death, her estate is not credited with any amount, and hence no bequest is available.
2. A life annuity provides the annuitants with an "inflexible" post-retirement income, in the sense that the annual amounts must be in line with the payment profile, as stated by the policy conditions.

Both features 1b and 2 can be perceived as disadvantages, and hence weaken the propensity to immediately annuitize the whole amount available at retirement. We now illustrate how these disadvantages can be mitigated, at least to some extent, either by purchasing life insurance products in which other benefits are packaged, or adopting a specific annuitization strategy.

In the following, we denote by y the age at retirement; the retirement time is denoted as time 0. The amount available at retirement, resulting from an accumulation process, is denoted by S.

8.5.2 Packaging benefits into the life annuity product

If the annuitant dies soon after the (ordinary) life annuity commencement, neither the annuitant nor the annuitant's estate receive much benefit from the purchase of the life annuity. In order to mitigate this risk, it is possible to buy a *life annuity with a guarantee period* (5 or 10 years, say), in which case the benefit is paid for the guarantee period regardless of whether the annuitant is alive or not. For a guarantee period of s years, and an amount S to be converted into an annuity (so that S represents the single premium), the resulting annual benefit fulfills the following relation:

$$S = b\, a'_{\overline{s}|} + b\, {}_{s|}a'_x \tag{8.5.1}$$

where $a'_{\overline{s}|}$ denotes the present value of a temporary annuity-certain, according to interest rate i'. Thus, the insurance product results in a deferred life annuity combined with a temporary annuity-certain.

Capital protection represents an interesting feature of some life annuity products, usually called *value-protected life annuities* or *money-back life annuities*. Consider, for example, a single-premium, level annuity. In the case of early death of the annuitant, a value-protected annuity will pay to the annuitant's estate the difference (if positive) between the single premium and the cumulated benefits paid to the annuitant. Usually, capital protection expires at some given age (75, say), after which nothing is paid even though the difference above mentioned is positive.

A *last-survivor annuity* is an annuity payable as long as at least one of two individuals (the annuitants), say (1) and (2), is alive. It can be stated that the annuity continues with the same annual benefit, say b, until the death of the last survivor. A modified form provides that the amount, initially set to b, will be reduced following the first death: to b' if individual (2) dies first, and to b'' if individual (1) dies first, clearly with $b' < b$, $b'' < b$. Conversely, in many pension plans the last-survivor annuity provides that the annual benefit is reduced only if the retiree, say individual (1), dies first. Formally, $b' = b$ (instead of $b' < b$). Whatever the arrangement, the expected duration of a last-survivor annuity is longer than that of an ordinary life annuity (that is, with just one annuitant).

8.5.3 Life annuities versus income drawdown

A *temporary withdrawal* (or *drawdown*) *process* can mitigate both disadvantages 1b and 2, mentioned in Sect. 8.5.1. Assume that the retiree can choose between the two following alternatives:

1. to purchase an immediate life annuity, with annual benefit b, such that $b\,a'_y = S$, namely to *annuitize* the available amount;
2. to leave the amount S in a fund, and then
    a. withdraw the amount $b^{(1)}$ at times $h = 1, 2, \dots, k$ (say, with $k = 5$ or $k = 10$);
    b. (provided she is alive) convert at time k the remaining amount R into an immediate life annuity with annual benefit $b^{(2)}$.

(see Fig. 8.5.1).

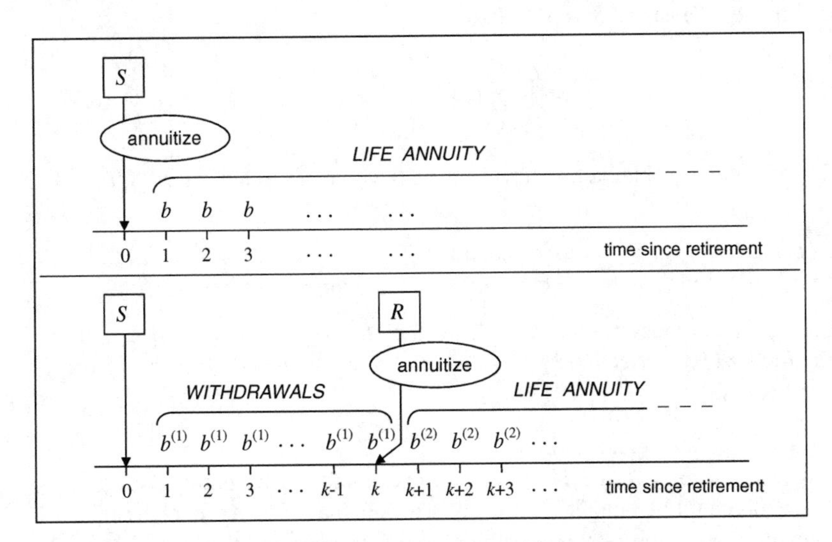

Fig. 8.5.1 Immediate annuitization versus delayed annuitization

If the retiree chooses the second alternative, the amount R available at time k to buy the life annuity depends on the annual withdrawal $b^{(1)}$ and the interest rate, g, credited to the non-annuitized fund. If $g = i'$, namely the interest rate assumed in the pricing basis of the life annuity, and $b^{(1)} = b$ then, the amount R is not sufficient to purchase a life annuity with annual benefit $b^{(2)} = b$, because of the absence of mutuality during the withdrawal period.

However, the absence of mutuality can be compensated (at least in principle) by a higher investment yield, namely if $g > i'$. We note the analogy between this problem and the one we have addressed while dealing with the pure endowment (see Sect. 4.3.2).

In formal terms, we can find relations among the quantities g, i', b, $b^{(1)}$, $b^{(2)}$, and k. In the case a life annuity (in arrears) is purchased at retirement time, we obviously have:

$$S = b\, a'_y \tag{8.5.2}$$

In the case of k-year delay, the amount R available at time k is given by:

$$R = S(1+g)^k - b^{(1)} \sum_{h=1}^{k} (1+g)^{k-h} \tag{8.5.3}$$

and the resulting annuity benefit $b^{(2)}$ fulfills the following equation:

$$R = b^{(2)}\, a'_{y+k} \tag{8.5.4}$$

in which it is assumed that the underlying technical basis coincides with the one adopted in Eq. (8.5.2) (see below for comments on this aspect).

From Eqs. (8.5.3) and (8.5.4), we obtain:

$$S(1+g)^k - b^{(1)} \sum_{h=1}^{k} (1+g)^{k-h} = b^{(2)}\, a'_{y+k} \tag{8.5.5}$$

Several results can be obtained by using Eq. (8.5.5). For example, given S, i', b, k, and

- given g and $b^{(1)}$ (e.g. $b^{(1)} = b$), calculate $b^{(2)}$;
- given $b^{(1)}$ and $b^{(2)}$ (e.g. $b^{(1)} = b^{(2)} = b$), calculate the interest rate g.

The spread $g - i'$ compensates the mutuality effect (for a given delay k), and is often called the *Implied Longevity Yield (ILY)*[1]; we note that g corresponds to the rate $g_{x,m}$ defined in Sect. 4.3.2.

Example 8.5.1. Assume that the amount $S = 1706.88$ is available at age $y = 65$. Use the technical basis $\text{TB1} = (0.02, \text{LT4})$. Hence, an immediate life annuity with annual benefit $b = 100$ could be bought, as it results from Table 4.3.7. As an alternative to the immediate conversion of S into a life annuity, assume that the annual amount $b^{(1)} = b$ is withdrawn from a fund (whose initial value is S). Table 8.5.1 displays

[1] Registered trademarks and property of CANNEX Financial Exchanges.

the annuity benefit $b^{(2)}$ as a function of the delay k, and the interest rate g credited to the fund throughout the delay period. We note that the technical basis TB1 = (0.02, LT4) is adopted, whatever the delay k. If $g = i' = 0.02$, then we have, of course, $b^{(2)} < b$; further, $b^{(2)}$ decreases as the delay k increases. If $g > i'$, we can have situations in which the higher yield during the delay period implies $b^{(2)} > b$, that is, a higher annuity benefit.

Table 8.5.1 Life annuity benefit $b^{(2)}$ after the delay period; $b^{(1)} = b$; TB1 = (0.02, LT4)

k	$g = 0.02$	$g = 0.025$	$g = 0.03$	$g = 0.035$
5	95.63	98.54	101.50	104.53
10	85.79	92.65	99.87	107.45
15	64.09	76.61	90.21	104.96
20	16.40	37.29	60.88	87.42

Table 8.5.2 shows, for various delays k (and still assuming $b^{(1)} = b$), the "equivalent rate", namely the investment yield g required to have $b^{(2)} = b$, hence compensating exactly the absence of mutuality during the withdrawal period.

Table 8.5.2 Equivalent rates; $b^{(1)} = b$; TB1 = (0.02, LT4)

k	g
5	0.02748
10	0.03009
15	0.03336
20	0.03718

□

The delay in the purchase of the life annuity has some advantages. In particular:

- in the case of death before time k, the fund available constitutes a bequest (which is not provided by a life annuity purchased at time 0, because of the mutuality effect);
- more flexibility is gained, as the annuitant may change the income profile modifying the withdrawal sequence (however, with a possible change in the fund available at time k).

Conversely, a disadvantage is due to the risk of a shift to a different mortality assumption in the pricing basis of life annuities, leading to a conversion rate at time k which is less favorable to the life annuity purchaser than that in-force at time 0. Further, if k is high, it may be difficult to gain the required investment yield (in particular, avoiding too risky investments) to cover the absence of mutuality.

The ideas underlying the delayed annuitization can be generalized, leading to the so-called *staggered annuitization*. As shown in Fig. 8.5.2, the staggered annuitization can be defined as a process according to which

- no life annuity is purchased at retirement time (time 0), so that an income drawdown process starts at that time;
- a first life annuity is purchased at time k', by using part of the remaining amount R';
- a second life annuity is purchased at time k'', by using part of the remaining amount R'';
- ... ...

The staggered annuitization implies that (after time k') a share of the post-retirement income consists of withdrawals whereas the remaining share is provided by a (set of) life annuities. Advantages and disadvantages of this arrangement can be easily understood looking at what noted above in relation to the delayed annuitization.

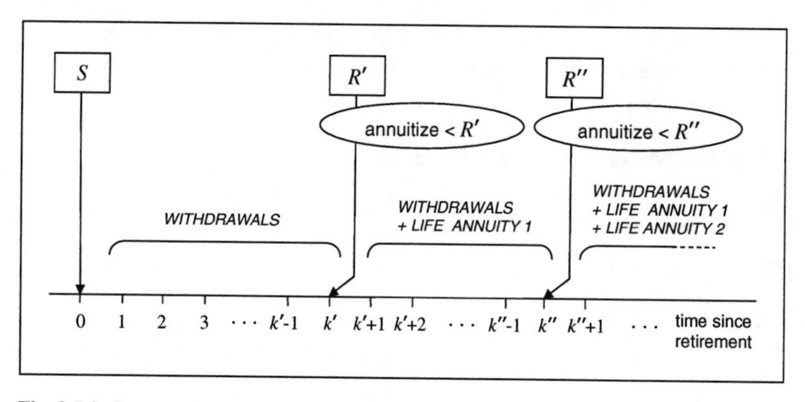

Fig. 8.5.2 Staggered annuitization

8.5.4 Phased retirement

Several employment arrangements allow an employee to gradually move from the working period to the retirement period. Such a progressive shift from full-time work to full-time retirement is usually denoted as *phased retirement* (see also the Remark in Sect. 8.2.1).

The phased retirement can be implemented in several ways (according to possible constraints imposed by current legislation). For example:

1. an employee who is approaching retirement age continues working with a reduced working load, until the transition to full-time retirement;

2. an employee who reaches retirement age y asks for partially continuing her working activity, or starting a similar activity, anyway with a limited working load.

We focus on solution 2, which in particular allows to maintain a higher income than that received, as post-retirement income, if the employee quits work entirely.

We assume that the employee chooses to obtain her income via an immediate life annuity. However, thanks to partial retirement, an annual benefit is chosen, lower than that needed in the case of total retirement. Hence, only a part of the available amount S is annuitized at age y, namely at the beginning of the partial retirement phase. Let $b^{(A)}$ denote the annual benefit which is paid from the beginning of this phase onwards. Clearly $b^{(A)} < b$, where b denotes the annual benefit provided by the full annuitization of S (see Eq. (8.5.2)). The amount required to purchase a whole life annuity with benefit $b^{(A)}$ is given by $b^{(A)}\, a'_y$. Assume that the total duration of the partial retirement phase is m years. At time m the following amount, R, will be available

$$R = (S - b^{(A)}\, a'_y)\,(1+g)^m \tag{8.5.6}$$

where g denotes the interest rate credited on the non-annuitized fund throughout the partial retirement phase. The amount R can be annuitized to obtain a further life annuity with annual benefit $b^{(B)}$, determined by the following relation:

$$R = b^{(B)}\, a'_{y+m} \tag{8.5.7}$$

Hence, during the total retirement phase, the retiree will cash the annual benefit $b^{(A)} + b^{(B)}$, which clearly depends on the interest rate g. Figure 8.5.3 shows the annuitization process related to phase retirement.

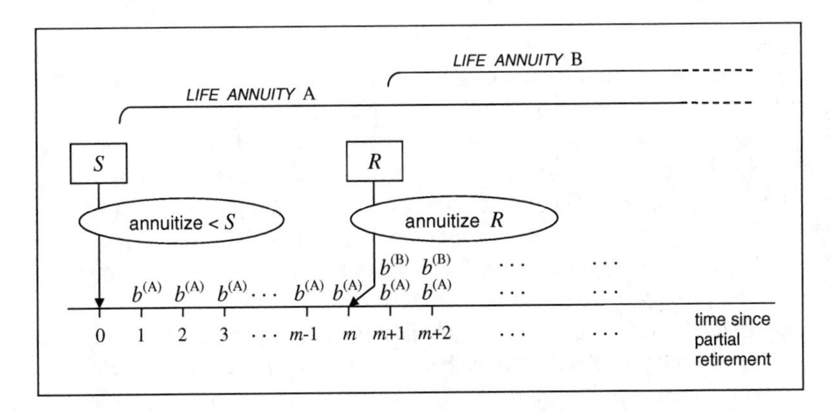

Fig. 8.5.3 Annuitization in phased retirement

Note that, as in the staggered annuitization process, the individual bears the risk of an unfavorable change in the technical basis adopted at time m to determine the benefit $b^{(B)}$ (while keeping access to the non-annuitized fund over the whole partial retirement period).

The phased retirement process and the related annuitization process can be generalized in several ways. For example:

- more than just one phase of partial retirement can be envisaged, to implement a more gradual shift from full-time work to full-time retirement;
- annuitization and income drawdown can coexist during the various phases (according to arrangements like those described in Sect. 8.5.3).

8.6 Risks for the provider

As we have mentioned in Sect. 8.3, several risks can be transferred to the provider prior and after retirement, which require an appropriate management. Basically, in this Section we summarize what are the risks, and when they are located. Most of the comments quoted below have already been developed previously in this book, with reference to life insurance.

In the following we address both the working and the retirement period of an individual; thus, time 0 denotes the time when the individual joins the plan (during her working period), while r ($r > 0$) is the retirement time. The individual age at time 0 is x, while at retirement it is $y = x + r$.

Let us first address the working period. For an individual joining the plan at time 0, and retiring at time r, the following fund is accumulated at time t, $t = 1, 2, \dots, r$, if no guarantee applies and no rider benefit is underwritten

$$F_t = (F_{t-1} - EX_{t-1})\,(1 + g_t) + c_t \tag{8.6.1}$$

where $F_0 \geq 0$, c_t is the contribution paid at time t, g_t is the investment return in year $(t-1, t)$, and EX_{t-1} are the expenses and other fees charged to the individual account at time $t-1$. Following the notation adopted in Sects. 1.2.5 and 8.5, the value $F_r = S$ of the fund at time r is converted into a sequence of periodic amounts. Note that in (8.6.1) we have assumed, similarly to Sect. 1.2.5, that the contribution is paid at the end of the year, once the annual salary has been gained. In practice, contributions may be paid at the end of each month, given that the salary is received monthly; to shorten the notation, we prefer to make reference to annual contributions. Due to the fees charged to the individual account at the beginning of each year, it is required $F_0 > 0$ (namely, an entry fee is applied to new members). Further, at the beginning of each year the fund must be large enough to cover the current fee. For management fees this is always realized, as they are expressed as a proportion of the value of the fund.

A financial guarantee affects the investment return. If the financial guarantee concerns the annual return, then instead of (8.6.1) we should consider

$$F_t = (F_{t-1} - EX_{t-1})\,(1 + \max\{g_t, i'\}) + c_t \tag{8.6.2}$$

where i' is the guaranteed annual return. The financial option embedded in (8.6.2) is a cliquet option, and the financial risk borne by the provider is similar to what emerges in participating policies. Note that no participation proportion is applied, as here the option is explicit, and then a specific fee is applied. We assume that the fee for the financial guarantee is included in EX_{t-1}; if the current fund value is not large enough to meet the cost of the guarantee, then the guarantee is not provided (or an additional contribution is required to the member).

The financial guarantee, instead of the annual return as in (8.6.2), could concern the average return in a given period, such as the guarantee described by the accumulation factor defined by (7.3.28). Similarly to what we have commented in Sect. 7.5, other types of guarantees can be arranged, following the pay-off of the financial options traded on the market. A fee is applied, which reflects the cost of the financial option. Of course, an appropriate hedging of the financial guarantee must be realized by the provider, through an adequate investment strategy.

Assume now that a lump sum death benefit C_t is underwritten, in case of death in year $(t-1,t)$ before retirement. The amount C_t can be chosen in one of the forms examined for life insurances; see in particular Sect. 5.4.4. An actuarial balance must be realized by the provider, as follows:

$$F_t = (F_{t-1} - EX_{t-1})(1+g_t) + c_t - (C_t - F_t)\, q'_{x+t-1} \tag{8.6.3}$$

Equation (8.6.3) can be easily interpreted if compared to the recursive equation of the reserve (5.4.8). Equation (8.6.3) shows us that the individual fund at time t for a member still alive is the result of the annual contribution, of the investment of the individual fund at the beginning of the year net of expenses (quantity $(F_{t-1} - EX_{t-1})(1+g_t)$) and net of the cost of mutuality originated by the death benefit (quantity $(C_t - F_t)\, q'_{x+t-1}$). Similarly to life insurance, the cost of mutuality is assessed on the basis of a life table (from which the mortality rate q'_{x+t} is derived), which is guaranteed during the coverage period. A mortality risk then emerges for the provider. If the observed frequency of death is higher than q'_{x+t}, then an unexpected cost emerges for the provider. Given that we are addressing the working period, which involves young adult ages, the risk is usually originated by random fluctuations (see Sect. 2.3.1), and can be diversified by increasing the size of the pool or by taking an appropriate reinsurance arrangement (see Sects. 2.4 and 2.5).

The death benefit could consist, instead of a lump sum, of a life annuity in favor of the member's spouse. The amount C_t in (8.6.3) would correspond to the actuarial value of a life annuity depending on the lifetime of the spouse. A financial risk and a mortality risk would be involved, similarly to any life annuity (see below). Overall, two lives would be involved; in particular, a second life table would be required, for the estimate of the spouse's lifetime.

Disability benefits or other health insurance benefits can be underwritten as riders during the working period. A disability benefit, in particular, could provide an annual income to the member if, because of a sickness or an injury, the member is unable to work; several policy conditions state the nature and the severity of the disability which is covered. Further benefits could consist in a lump sum paid in case of an accident causing a permanent injury or the death of the member, a refund of medical expenses, an so on. All these benefits are managed by the provider on the basis of the mutuality principle; a risk of random fluctuations emerges. If the provider is not an insurer, usually protection is obtained from an insurer by underwriting an appropriate insurance contract (a group insurance contract).

Let us now address the post-retirement period. As described in Sect. 8.5, the member can select among a life annuity, an income drawdown, a combination of the two or a phased retirement. As long as the fund is not annuitized, i.e. a life annuity has not been underwritten, risks are borne by the member. Thus, the development of the fund can be described as

$$F_t = (F_{t-1} - EX_{t-1})\,(1 + g_t) - b_t^{(1)} \tag{8.6.4}$$

where $b_t^{(1)}$ is the withdrawal at time t (note that (8.6.4) generalizes (1.2.17) in Sect. 1.2.5). A financial guarantee can be underwritten, for example

$$F_t = (F_{t-1} - EX_{t-1})\,(1 + \max\{g_t, i'\}) - b_t^{(1)} \tag{8.6.5}$$

The annual fee EX_{t-1} includes also the cost of the guarantee.

Assume now that a fixed-life annuity is underwritten at retirement time, i.e. that the fund available at maturity is fully annuitized, with the guarantee of receiving the annual amount b at the end of each year, until death. The amount F_r is transferred to the provider (typically, an insurer), which has to set up an individual reserve in face of its liabilities. The development in time of the individual reserve is described as follows:

$$V_t + b = V_{t-1}\,(1 + i') + (V_t + b)\,q'_{x+t-1} \tag{8.6.6}$$

where V_t, as usual in life insurance, is the individual reserve. As noted in Sect. 8.5.3, contrarily to the amount $F_t + b_t^{(1)}$ in (8.6.4) or (8.6.5), which in case of death of the member in year $(t-1, t)$ is available to her estate (clearly if $F_t + b_t > 0$), the quantity $V_t + b$ is available to the insurer in case of death of the member in year $(t-1, t)$, for the funding of mutuality. Equation (8.6.6) is the recursive equation of the reserve (see Sect. 5.4.2). The following interpretation is useful, to understand the risks taken by the insurer (see also Example 5.4.4). The quantity $V_t + b$ represents the amount the insurer must hold at time t if the member is alive: V_t is used to carrying on the contract, while b must be paid to the member. This amount is funded by the assets available for the policy at the beginning of the year, V_{t-1}, joint to the interest guaranteed on their investment, $V_{t-1}\,i'$, and by the mutuality contribution $(V_t + b)\,q'_{x+t-1}$. We note that q'_{x+t-1} expresses the expected frequency of death, which is estimated according to a given (projected and conservative) life table. If the observed frequency of death is lower than q'_{x+t-1}, then the insurer experiences a longevity risk. The risk may be originated by random fluctuations, as well as by systematic deviations (see Sect. 2.3.1). Systematic deviations, in particular, can be originated by an unanticipated mortality dynamics. The term *aggregate longevity risk* is used to refer to the systematic component of the longevity risk.

We point out that in Eq. (8.6.6) we have disregarded expenses; just the net reserve has been addressed. As described for a life insurance contract, a provision is set up for meeting the annual expenses charged to the contract (see Sect. 5.6). At time r, the individual fund $F_r = S$ is used to meet the cost of the annuity, namely V_0, and the loading for expenses, $\Theta^{[A]} + \Theta^{[G]}$.

The rate i' in Eq. (8.6.6) is a technical interest rate, so it is guaranteed. The provider has to assign an annual return which is exactly i', and this originates a financial risk. Given that usually i' is set at a low level, the risk is not severe. However, a participating life annuity is more usual than a fixed-benefit life annuity. In this case, the development in time of the individual reserve is described as follows:

$$V_t + b_t = V_{t-1}\,(1 + \max\{\eta_t\, g_t, i'\}) + (V_t + b_t)\, q'_{x+t-1} \tag{8.6.7}$$

where we have adopted the notation introduced for participating policies (see Sect. 7.3); note, in particular, that we have considered the standard revaluation rate $r_t^{[1]}$ (defined by (7.3.6)). The quantity b_t is the annual amount to be paid at time t, which includes the adjustments at previous years (see Sect. 7.2.2). As noted in Sect. 7.3, the interest rate i' in (8.6.7) is a minimum guaranteed annual return; the financial risk to which the insurer is exposed requires an appropriate hedging.

The longevity risk implied by (8.6.6) or (8.6.7), which is originated by the longevity of the annuitants, can be mitigated by a death benefit. Assume that a lump sum C_t is paid at time t in case of death in year $(t-1,t)$. We refer to the case of a fixed annual amount (i.e., to (8.6.6)). First, we note that given the fund available at time r, $F_r = S$, if a death benefit is underwritten, then the annual amount is lower than the amount b in (8.6.6); we denote the new amount by b'. The development in time of the individual reserve is now described as follows:

$$V_t + b' = V_{t-1}\,(1 + i') + (V_t + b' - C_t)\, q'_{x+t-1} \tag{8.6.8}$$

In face of reasonable choices for C_t, the quantity $(V_t + b' - C_t)$ is positive, so that the provider is still exposed to the longevity risk, but lower then $V_t + b$ in (8.6.6). This reduces the need for mutuality, and then the importance of longevity risk. If the death benefit consists of a life annuity in favor of the annuitant's spouse, than C_t would correspond to the actuarial value of a life annuity, which originates further longevity risk for the provider (given that two lives are involved).

From the discussion above, it emerges that the main risks for a pension provider are the financial and the mortality/longevity risks. The mortality risk, in particular, arises during the working period, while the longevity risk in the post-retirement period. While during the working period the mortality risk is not too important (due to the range of ages involved), after retirement the longevity risk, and in particular the systematic component, may become considerable. After retirement, it is worth noting that when t is small (i.e., not too far away from the retirement time), the expected frequency of death is low, so that the contribution expected from mutuality in (8.6.6) or (8.6.7) is small (and it is even smaller in (8.6.8)); conversely, the individual reserve is high, the size of the gain on investments is expected to be large (given that a large amount of money is invested), and then the financial risk may be important. When t is high (namely, far away from the retirement time), the rate q'_{x+t-1} is high, so that a major contribution is expected from mutuality, and this increases the importance of the longevity risk; at the same time, the financial risk is moderate, as the size of the assets is small. In order to understand how the importance of the financial risk versus the longevity risk evolves in time in a life annuity, we suggest to look at Example 5.4.4.

As a final source of risk, we mention the *GAO* (*Guaranteed Annuity Option*; see the Remark in Sect. 7.9). With reference to the possible choice at retirement of a life annuity, a *guaranteed annuitization rate* (*GAR*) $\frac{1}{a'_{x+r}}$ may be underwritten before retirement time. Since the rate $\frac{1}{a'_{x+r}}$ requires the choice of an interest rate and a life table, the provider is exposed to financial and longevity risk. The financial risk is originated by the possibility that the interest rate included in the GAR is too high in relation to the market rates at retirement time; the longevity risk is originated by the possibility that at retirement time a new (projected) life table is available, according to which the life table adopted in the GAR is considered to be no longer conservative. The exercise of the GAO is affected by the comparison between the *current annuitization rate* (*CAR*) and the GAR, but also by the preferences of the member in respect of receiving a life annuity (instead of entering into an income drawdown process). In any case, the GAO implies a financial option, whose underlying is given by the current annuitization rate (CAR). A fee must be applied by the provider, but calculating this fee is hard work, as the financial option is very particular (for example, the underlying is an annuitization rate) and its value depends on interest rates, life tables, as well as on the member's preferences in respect of the life annuity.

The management of the risks taken by the provider should follow the guidelines described in Sect. 1.3. Risks must be identified, their importance must be assessed, and appropriate actions must be taken either for controlling or financing the loss. Monitoring is also an important step of the risk management, as the importance of the several risks may change in time, as we have mentioned above.

8.7 References and suggestions for further reading

The book [40] is specifically devoted to post-retirement income planning, and life annuities and pensions in particular. Aging and post-retirement solutions are discussed by [6].

In this Chapter we have not dealt with methods for funding benefits in group pension plans, under an actuarial perspective. The reader interested in these issues can refer to [9] (Part IV), [1], and [58]. Actuarial aspects of pension plans are dealt with also by [10] (Chap. 20).

Financial risks in pension plans and related risk management solutions are focussed by [24].

Finally, we recall that [46] also addresses the impact of future mortality trends on the costs of pensions and life annuities.

Chapter 9
Non-life insurance: pricing and reserving

9.1 Introduction

The purpose of this Chapter is to introduce the fundamentals of the actuarial valuation of non-life insurance covers. First we give an overview of the contents of non-life insurance products, then we focus on premium calculation and reserving issues. While numerical examples are provided, specific covers are not dealt with in detail. To develop premiums and reserves for specific lines of business, further reading is required (some suggestions are provided in Sect. 9.13).

In detail, the main issues dealt with in this Chapter are the following:

- general aspects of non-life insurance products;
- main policy conditions limiting the liability of the insurer;
- premium calculation and related statistical bases;
- general aspects of the stochastic modeling of the payment of the insurer;
- technical reserves:
- profit assessment.

Problems other than those focused in life insurance technique will emerge. In particular, while investment perspectives can be disregarded, modeling the uncertainty of the payout of the insurer is a major issue. For a contract, such uncertainty concerns the number of events originating a payment by the insurer, the amount of each payment and the time of each payment. Clearly, in face of such uncertainty, a stochastic modeling of the insurer's payout could be considered more coherent than a deterministic representation. Indeed, for some lines of business (e.g., those subject to extreme events), a stochastic modeling is necessary to avoid biased valuations. However, a deterministic modeling is satisfactory in many cases. Given the introductory character of this Chapter, we mainly discuss deterministic models.

A. Olivieri, E. Pitacco, *Introduction to Insurance Mathematics*,
DOI 10.1007/978-3-642-16029-5_9, © Springer-Verlag Berlin Heidelberg 2011

9.2 Non-life insurance products

A short description of the main features of non-life insurance products is provided in this Section, mainly aiming at introducing the basic items involved in premium and reserve calculation.

9.2.1 General aspects

The contents of non-life insurance (also named general insurance or property/casualty insurance) is compensating a person or an organization for a loss or a damage to her property or for the liability to indemnify a third party for a loss or a damage arising from specified contingencies such as fire, theft, injury, negligence, and so on. Health insurance is the term used when the purpose is to compensate a person or her family for the economic consequences of an alteration of the health status originated by a sickness or an accident.

In a non-life insurance contract, the benefit amount is not stated in advance. Except for covers with forfeiture benefits (see Sect. 9.2.2), the amount paid by the insurer depends on the severity of the loss or damage suffered by the insured or by a third party, in respect of which the insured is liable. Further, the total payout of the insurer for one policy depends not just on the size of each loss, but also on the number of events determining a loss or a damage to the insured or to a third party. Both are unknown at the time of issue.

The insurance coverage period is usually short, typically one year; a single premium is the common arrangement. The contract may be subject to automatic renewal, so that the contractual relationship between the insured and the insurer extends over more than one year. However, premiums keep on being paid yearly, and each of them is the single premium for the relevant year. The times of occurrence of the adverse events are unknown, as well as the time of the relevant settlement. In respect of the latter, the time-span of an annual policy may extend well beyond one year, for example in case of litigation in setting the eligibility to or the size of the benefit. This aspect must be allowed for, in particular, in the calculation of technical reserves. The term *policy year* is used to refer to the coverage period, i.e. the period in which claims are covered by the insurer.

9.2.2 Main categories of non-life insurance products

Non-life insurance includes a wide range of products, offering protection in respect of many risks. We do not aim to provide a comprehensive and detailed presentation of the possible contents of non-life insurance covers; we just give some information, which are useful to understand the fundamentals of pricing and reserving.

The non-life business may be segmented according to different perspectives. Considering the possible contractor, we may distinguish between *personal insurance*, addressed to individuals or families (e.g., motor insurance, health insurance, homeowners insurance, and so on), and *commercial insurance*, addressed to business entities (e.g., transportation insurance, workers compensation, and so on). In relation to the possible beneficiary, we may classify property insurance, liability (or casualty) insurance and health insurance. *Property insurance* provides financial protection against a possible loss of or damage to the property of the insured, including loss of profits or emergence of costs; *liability* (or *casualty*) *insurance* offers financial protection against various liability claims; *health insurance*, as stated above (see Sect. 9.2.1), offers financial protection for expenses or loss of income originated by a sickness or an accident (we note that some forms of health insurance, typically those with forfeiture benefits and a duration of more than one year, are classified within life insurance).

Going into greater detail in respect of the insured contingency, we can identify the following main classes of non-life insurance products: in the framework of health insurance, personal accident insurance (providing forfeiture benefits in case of bodily injury or dismemberment), and sickness insurance (providing hospitalization benefits and reimbursement of medical expenses); motor insurance (merging liability and property insurance benefits in favor of car owners); marine and transportation insurance (usually a separate line in respect of personal motor insurance); insurance against fire and other damages to property; liability insurance; credit insurance.

With regard to the timing of claim settlement, the business may be short-tail or long-tail; usually, liability business is long-tail, given possible litigations concerning the existence and the size of the claim, while property insurance is short-tail, as it is relatively easy to verify the existence and the size of a loss. Personal insurance lines tend to be less volatile than commercial lines; some lines of business may be severely exposed to catastrophe risk, e.g. homeowners business located in geographical areas subject to earthquakes or hurricanes.

From the presentation above, the large variety of products that fall within non-life insurance emerges. While the general principles for pricing and reserving are common to all the business lines, the specific methods applied in practice may differ significantly, consistent with the features of the particular line of business dealt with. As mentioned previously, we aim just at describing the general principles.

As mentioned in Sect. 9.1, an essential component of pricing and reserving models for non-life insurance is the representation of uncertainty, i.e. of the random occurrence and amount of claims. However, a stochastic approach is not always strictly required; for many purposes, deterministic models provide a satisfactory representation.

9.3 Loss and claim amount

As a first step in the valuation of a non-life insurance contract, in particular for premium calculation, the possible amount of claims must be assessed. Not necessarily a loss suffered by the insured (emerging either from a damage to her property, a liability, or medical expenses) is covered in full by the insurer. Limitations to the insurer's payment may be introduced through appropriate policy conditions.

Let us refer to a policy covering a given risk (for example, motor insurance), with term one year. During the year, the policy will record a random number N of claims. The possible outcomes of N are $0, 1, \dots$ (similarly to Case 3d in Sect. 1.2.4). Each claim will cause a random loss to the insured. We denote by X_k the loss to the insured caused by claim k, $k = 1, 2, \dots$. According to policy conditions, the insurer will assess the claim amount Y_k for claim k. Reasonably, $Y_k \leq X_k$, to prevent moral hazard. In general terms, the claim amount Y_k is a given function of the loss amount X_k; such a function is called the *claim function*. Under the same contract, a different claim function could be selected for each claim, so that (for example) the higher is the number of claims reported so far, the more restrictive is the policy condition applied to the current claim. For brevity, we will disregard this possibility, so that the same claim function f will apply to any claim, i.e. $Y_k = f(X_k)$.

Remark The settlement of a claim originates some expenses, the so-called *claim settlement* or *claim processing expenses* (see also Sect. 9.6). The total cost of a claim to the insurer, i.e. Y_k and claim settlement expenses, is sometimes called the *loss amount*. Further, sometimes Y_k is meant to already include claim settlement expenses, and hence it is Y_k to be called loss amount. To avoid any misunderstanding, we prefer to use the term "loss" just to refer to X_k; in the following (unless it is necessary, due to the prevailing terminology in practice) we will refer to Y_k as to the *claim amount*, not inclusive of claim settlement expenses.

Under the *full compensation* arrangement, the insurer pays in full the loss suffered by the insured or by a third party; thus, the claim function is defined as follows

$$Y_k = X_k \tag{9.3.1}$$

In property insurance, arrangement (9.3.1) is known as *full value*, while in liability insurance as *unlimited liability*. Figure 9.3.1 provides a graphical representation. In the case of property insurance, the maximum loss amount and then the maximum payment by the insurer are given by the value V of the property (so the graph in Fig. 9.3.1 should be read for $X_k \leq V$); conversely, no natural cap is provided for the payment by the insurer in the case of liability insurance (in this case, the graph in Fig. 9.3.1 must be read for $X_k > 0$).

Remark Experience could suggest that some extreme values for the loss amount are unrealistic. The *maximum probable loss* (or *MPL*), in particular, is defined as

$$\text{MPL} = \inf\{x : \mathbb{P}[X_k \leq x] = 1\} \tag{9.3.2}$$

In words: the MPL is the highest value for the loss originated by a (single) claim for which the probability to occur is positive. In the case of property insurance, it may turn out MPL $< V$; in the case of liability insurance, we exclude to observe loss amounts higher than the MPL.

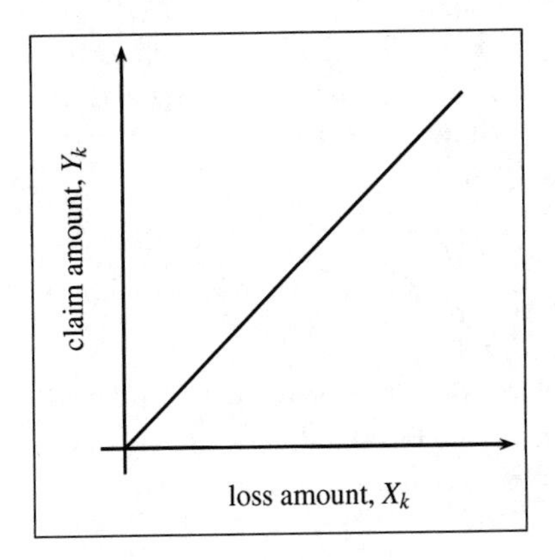

Fig. 9.3.1 Claim amount according to full compensation

Arrangement (9.3.1) is clearly unsatisfactory for the insurer. Not only it is exposed to the risk of large claims, but he is also facing small claims, which are usually high in numbers and carry processing costs which may exceed the benefit amount. Further, the insured could be careless in preventing accidents, given that the cost of a claim is fully charged to the insurer.

Small claims can be avoided through *deductibles*. In particular, according to a *franchise* (or *minimum*) *deductible* the insurer only intervenes if the loss amount is above a given threshold, the deductible d. The claim amount is then defined as follows (see also Fig. 9.3.2):

$$Y_k = \begin{cases} 0 & \text{if } X_k \leq d \\ X_k & \text{if } X_k > d \end{cases} \tag{9.3.3}$$

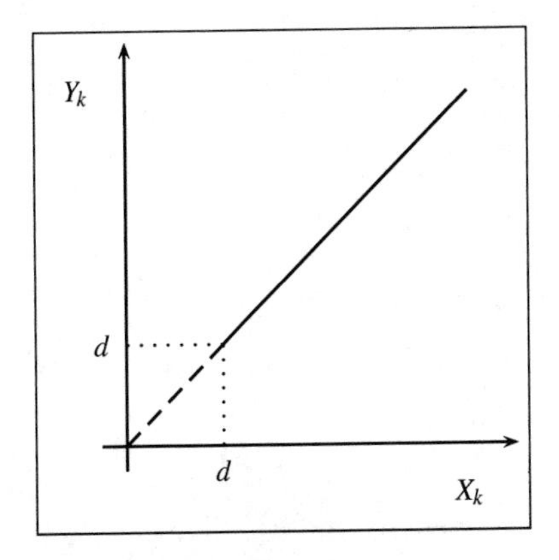

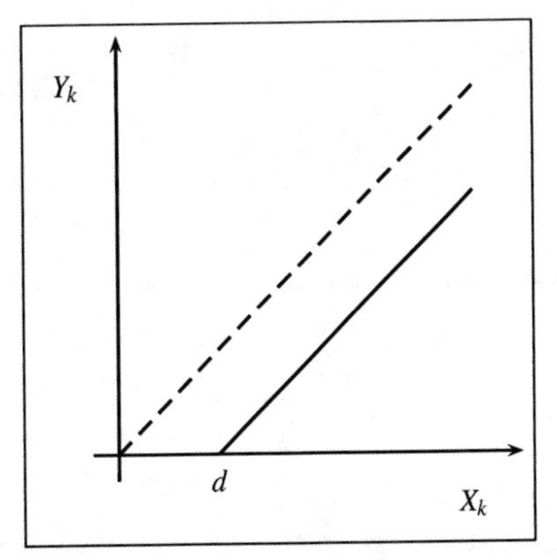

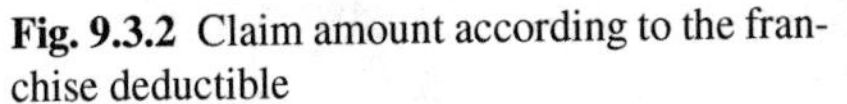

Fig. 9.3.2 Claim amount according to the franchise deductible

Fig. 9.3.3 Claim amount according to the fixed-amount deductible

According to a *fixed-amount deductible*, an amount d is always charged to the policyholder; clearly, if the loss amount is lower than d, there is no payment by the insurer. The claim amount is then defined as follows (see also Fig. 9.3.3 and, in Sect. 1.3.4, Fig. 1.3.4 and Eqs. (1.3.4a) and (1.3.4b)):

$$Y_k = \begin{cases} 0 & \text{if } X_k \le d \\ X_k - d & \text{if } X_k > d \end{cases} \tag{9.3.4}$$

A proportion α of the loss $(0 \le \alpha < 1)$ is charged to the insured under the *proportional* (or *fixed-percentage*) *deductible*; in this case, the claim amount is defined as follows (see also Fig. 9.3.4 and, in Sect. 1.3.4, Fig. 1.3.3 and Eqs. (1.3.3a) and (1.3.3b)):

$$Y_k = (1-\alpha) X_k\,; \qquad 0 \le \alpha < 1 \tag{9.3.5}$$

Note that the higher is the loss amount, the higher is the cost charged to the insured. The arrangement is usual in property insurance, in case the insured value, V', is lower than the current value of the property, V. In this case, $\alpha = \max\{1 - \frac{V'}{V}, 0\}$. Note that V is usually ascertained at the time of claim occurrence, while V' is set at policy issue (or renewal time); due to a depreciation or a revaluation of the property, it may well turn out $V \gtreqless V'$. In case $V' < V$, the insurer reduces accordingly the claim amount, to avoid that at issue the insured reports an underestimated value of the property, so to pay a lower premium. Of course, underinsurance (i.e., $V' < V$) can be a specific choice of the insured. The proportional deductible is applied also in covers where the behavior of the insured can affect the claim cost, such as sickness insurance, theft insurance, all risks motor insurance, and so on.

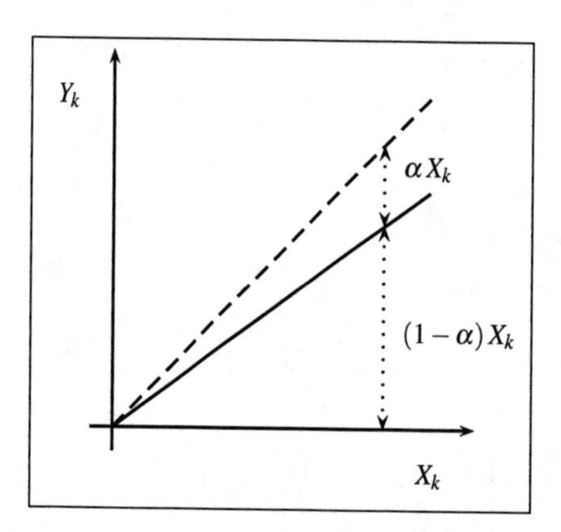

Fig. 9.3.4 Claim amount according to the proportional deductible

In order to avoid large claims, the insurer may apply upper limits. If a *limit value* M is adopted, the claim amount is defined as follows (see also Fig. 9.3.5):

$$Y_k = \min\{X_k, M\} \tag{9.3.6}$$

In liability insurance, the limit value is also called the *capacity* of the policy; in property insurance (where $M < V$), the arrangement is also called *first loss*.

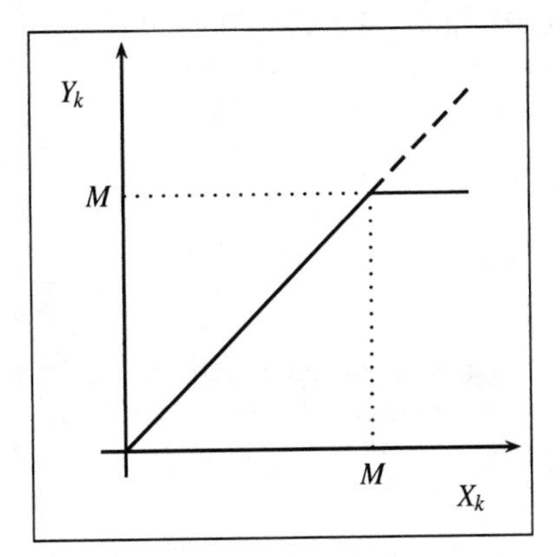

Fig. 9.3.5 Claim amount according to the limit value

Policy conditions are usually arranged in various combinations: for example, in property insurance the proportional deductible is usually joint to a franchise deductible; in liability insurance, a deductible (either franchise or fixed-amount) and a limit value are usual (in property insurance, the possible claim of the insurer is naturally capped by the value of the property, if no limit is explicitly set; conversely, in liability insurance, the maximum claim amount is completely unknown if a limit value is not adopted). Table 9.3.1 provides an example of claim amount determined according to alternative policy conditions, for three possible loss values.

Table 9.3.1 Claim amount under alternative policy conditions

Policy conditions	Loss		
	100	500	1 000
Full compensation	100	500	1 000
Franchise deductible, $d = 150$	0	500	1 000
Fixed-amount deductible, $d = 150$	0	350	850
Proportional deductible, $\alpha = 5\%$	95	475	950
Limit value, $M = 900$	100	500	900
Proportional deductible, $\alpha = 5\%$, and limit value, $M = 900$	95	475	900
Franchise deductible, $d = 150$, and limit value, $M = 900$	0	500	900
Franchise deductible, $d = 150$, proportional deductible, $\alpha = 5\%$, and limit value, $M = 900$	0	475	900

The claim functions described above represent the most common forms of limitation to the insurer's liability. Insurance practice provides further examples of policy conditions; some of them are in particular suitable for a specific line of business (and not for others). Overall, deductibles and limit values result in a premium reduction, as the cost of the benefit turns out to be reduced. As already noted, deductibles allow the insurer to avoid the settlement costs of small claims; thus, they also originate a reduction of the expense loading. Conversely, limit values, avoiding too large claims, reduce the risk profile of the insurer and thus may lead to a lower safety loading.

9.4 The equivalence premium

By definition, the equivalence premium is the expected present value of the insurer's payout (see Sect. 1.7.4). To calculate this value, specific items must be allowed for, according to the type of insurance cover dealt with.

9.4.1 The items of the equivalence premium

In non-life insurance, the following items must be considered for the calculation of the equivalence premium:

- the number of claims which may be reported by a policy during the coverage period;
- the claim amount of each possible claim;
- the time of payment of each claim;
- the value of time.

As discussed in Sect. 9.3, the amount of a claim is a function of the loss suffered by the insured, as it is defined by the policy conditions applied. To some extent, also the number of claims is affected by policy conditions, as deductibles exclude small claims. The description of the possible time-pattern of a claim, as well as assumptions for its representation are discussed in the next Section.

9.4.2 The time-pattern of a claim

Figure 9.4.1 provides an example of the possible development of a single claim. The claim occurs at time t_1; such a time must fall before the policy term (which we assume to be one year); if we denote by 0 the time of policy issue, then we must have: $0 < t_1 \leq 1$. Conversely, times t_2, t_3 and t_4 may fall after the maturity of the policy. The time-lag between occurrence and settlement may be due to the administrative processing of the claim, to possible litigations, and so on. The time-lag between occurrence and notification may run from some days (e.g. for property or motor insurance) to years (e.g. for liability in health insurance, due to the nature of the damage, which may be perceived just after some time since when it was incurred). Between times t_1 and t_2, the claim is *incurred but not reported* (*IBNR*). Between times t_2 and t_4 it is *outstanding*. At time t_4, the liability of the insurer (for this specific claim) is written off. In normal situations, the closure of the claim occurs right after settlement; in case of litigation, the closure may take place after some time, due to the possibility of a further payment required to the insurer, or also of a partial recovery of the settled amount.

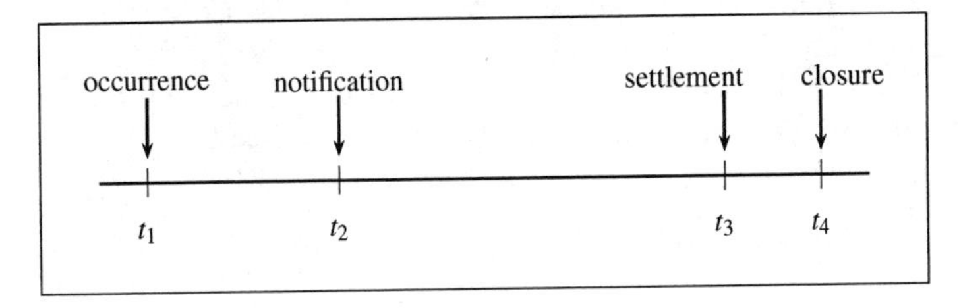

Fig. 9.4.1 The possible time-pattern of a claim

Fig. 9.4.1 suggests that the settlement of a claim may occur far away since occurrence. However, to the purpose of premium calculation, usually the following two assumptions are adopted:

- the claim is reported as soon as it occurs, and it is immediately settled (in the example of Fig. 9.4.1 we would assume: $t_1 = t_2 = t_3$);
- the times of occurrence of claims are uniformly distributed over the year, so that on average claims occur in the middle of the policy year (in the example of Fig. 9.4.1 we would then assume: $t_1 = \frac{1}{2}$).

9.4.3 The expected aggregate claim amount

Let us refer to a policy with term one year. During the policy year, N claims may be reported, $N = 0, 1, \ldots$. Realistically, the number of claims is limited, so that we can refer to a maximum (reasonable) outcome $n_{\max}$ for N, as suggested by the physical features of the insured risk (e.g., the risk cannot report more than one claim per day), by policy conditions (e.g., no compensation is acknowledged if the number of claims exceeds a given threshold), or by empirical evidence. Within a deterministic setting, we do not need to take care of this aspect, so that we will let N take any possible integer value. Then, X_k is the loss amount (to the insured) for claim k, $k = 1, 2, \ldots$; as discussed in Sect. 1.2.4 (see Case 3b), X_k is a random variable that can take only positive values, given that the case of a null loss is expressed by $N = 0$.

Remark To be more precise, X_k is defined conditional on having observed $N \geq k$. From a practical point of view, this means that X_k is the loss amount of a claim which has occurred. However, the random variable X_k is defined at issue, when the number of claims is not known. So, we should better assume that X_k is defined whatever the number of claims will be, but necessarily $X_k = 0$ if $N < k$, whilst $X_k > 0$ if $N \geq k$. From a probabilistic point of view, this latter approach is more accurate. However, to the purpose of simplifying the presentation, we can accept the idea that X_k is the claim amount of a claim that has occurred; the main conclusions of the discussion are not affected by this simplification.

According to policy conditions, the claim amount Y_k (i.e., the payment made by the insurer) is defined as a function of X_k, $k = 1,2,\dots,N$; we note that, because of deductibles, it is possible that $Y_k = 0$, while $X_k > 0$. In practice, it often happens that the insured does not report the claim if the loss amount is below the deductible; in this case, also Y_k would only take positive values.

Assuming that claims are immediately reported and settled, and that they occur on average at the same time, i.e. in the middle of the policy year (see Sect. 9.4.2), the *aggregate claim amount* (or *total payout* of the insurer) in a year for a policy is defined as follows:

$$S = \begin{cases} 0 & \text{if } N = 0 \\ Y_1 + Y_2 + \cdots + Y_N & \text{if } N > 0 \end{cases} \tag{9.4.1}$$

The *equivalence premium*, P, by definition the expected present value of the insurer's payout, is then assessed as

$$P = \mathbb{E}[S]\,(1+i')^{-1/2} \tag{9.4.2}$$

where the expected value $\mathbb{E}[S]$ is calculated according to realistic assumptions for the number of claims N and for the claim amounts $Y_1, Y_2, \dots, Y_N$, whilst i' is the annual interest rate expressing the time-value of money. Usually, a conservative assumption is adopted for i'; possibly, $i' = 0$, due to the short duration of the policy. For brevity, in the following we set $i' = 0$ (a quite usual choice also in practice).

As commented in Sect. 1.4.4, the calculation of $\mathbb{E}[S]$ is usually performed accepting the following assumptions:

1. the random variables X_k are independent of the random number N;
2. whatever the outcome n of N, the random variables $X_1, X_2, \dots, X_n$ are
   a. mutually independent;
   b. identically distributed (and hence with a common expected value, say $\mathbb{E}[X_1]$).

We further assume that

3. the same policy conditions are applied to any claim, so that also the random variables $Y_1, Y_2, \dots, Y_n$ are identically distributed (it follows, in particular, that if we assign the probability distribution of Y_1, we also hold the probability distribution of any claim amount Y_k, $k = 1,2,\dots,n$). The random variables $Y_1, Y_2, \dots, Y_n$, then, have a common expected value (say, $\mathbb{E}[Y_1]$).

Thanks to such assumptions, S has a *compound distribution*, with components Y_1, the so-called *claim severity*, and N, the so-called *claim frequency*. Then, following steps similar to those described in Sect. 1.4.4, the expected aggregate claim amount $\mathbb{E}[S]$ can be factorized as follows (see (1.4.37)):

$$\mathbb{E}[S] = \mathbb{E}[N]\,E[Y_1] \tag{9.4.3}$$

We note that, for the assessment of the expected total payout, and then of the equivalence premium, all what we need is an estimate of the expected claim frequency, $\mathbb{E}[N]$, and the expected claim severity, $\mathbb{E}[Y_1]$; see also Sect. 9.7 and Case 3d in Sect. 1.2.4. No specific distributional assumption is required for N and Y_1 (however, we stress that result (9.4.3) is underpinned by assumptions (1)–(3), which in fact relate to the probability distribution of S).

9.5 The net premium

In order to determine the net premium, a safety loading must be added to the equivalence premium. As is well-known (see Sect. 1.7.4 and 2.3.5), the safety loading is a reward for the risks borne by the insurer; meanwhile, the safety loading represents the expected profit to the insurer. In non-life insurance, an explicit assessment of the safety loading is usually performed, as the data on which the claim frequency and the claim severity are estimated are based on insurance experience (see Sect. 9.7), so that they originate a realistic valuation of the insurer's liability.

The rule adopted to determine the safety loading is called *premium principle*. To be precise, a premium principle (as is suggested by the name) is a formula for the calculation of the net premium; the safety loading can then be assessed subtracting the equivalence to the net premium. A description of the more practical premium principles follows.

According to the *expected value principle*, the net premium is calculated as follows:

$$\Pi = (1+\alpha)\,\mathbb{E}[S] \tag{9.5.1}$$

where α is a given proportion ($\alpha > 0$). The safety loading, $\alpha\,\mathbb{E}[S]$, is proportional to the expected total payout of the insurer. The advantage of this rule is that data required for the calculation of the net premium coincide with those used for the equivalence premium; the disadvantage is that the safety loading is not based on a risk measure. An expression alternative to (9.5.1) is as follows:

$$\Pi = \mathbb{E}[S] + \kappa\Pi \tag{9.5.2}$$

where the safety loading, $\kappa\Pi$, is expressed as a proportion of the net premium. It turns out:

$$\Pi = \mathbb{E}[S] \times \frac{1}{1-\kappa} \tag{9.5.3}$$

Clearly, (9.5.1) and (9.5.2) are equivalent, provided that $\alpha = \frac{\kappa}{1-\kappa}$.

A safety loading proportional to a risk measure is originated by the *variance principle*. In this case, the net premium is assessed as follows:

$$\Pi = \mathbb{E}[S] + \lambda\,\mathbb{V}\mathrm{ar}[S] \tag{9.5.4}$$

where λ ($\lambda > 0$) is a given intensity.

Remark We note that $\lambda\,\mathbb{V}\mathrm{ar}[S]$ must be an amount; since $\mathbb{V}\mathrm{ar}[S]$ is an amount to the square, the dimension of λ must be that of $\frac{1}{\text{amount}}$. Otherwise said, λ is an intensity.

The safety loading, $\lambda\,\mathbb{V}\mathrm{ar}[S]$, is proportional to the variance; its quality as a risk reward depends on the appropriateness of the variance in quantifying the risks originated by S. To understand if this is the case, the probability distribution of S should be analysed. If it is "regular enough", i.e. it is symmetric and short tailed, then the variance is a good risk measure; see Sect. 9.8 for comments in this regard. We point out that (9.5.4) requires further data in respect of those used for the calculation of the equivalence premium; this justifies the large preference, in practice, for the equivalence principle.

Remark It is interesting to obtain the expression of $\mathbb{V}\mathrm{ar}[S]$, to understand in detail what information are required for implementing rule (9.5.4). To shorten the notation, as in Sect. 1.4.4 we let $\mathbb{P}[N=h]=\pi_h$. In general terms

$$\mathbb{V}\mathrm{ar}[S]=\mathbb{E}[S^2]-(\mathbb{E}[S])^2 \tag{9.5.5}$$

We already know the expression of $\mathbb{E}[S]$ (see (9.4.3)), so now we need to work out $\mathbb{E}[S^2]$. We have

$$\begin{aligned}\mathbb{E}[S^2] &= \textstyle\sum_{h=1}^{\infty}\pi_h\,\mathbb{E}[S^2|N=h]=\sum_{h=1}^{\infty}\pi_h\,\mathbb{E}\left[\left(\sum_{i=1}^{h}Y_i\right)^2|N=h\right]\\ &= \textstyle\sum_{h=1}^{\infty}\pi_h\left(\sum_{i=1}^{h}\mathbb{E}[Y_i^2|N=h]+\sum_{i=1}^{h}\sum_{j:j\neq i}\mathbb{E}[Y_iY_j|N=h]\right)\end{aligned} \tag{9.5.6}$$

Thanks to assumption (1) (and (3)), $\mathbb{E}[Y_i^2|N=h]=\mathbb{E}[Y_i^2]$ for all i and $\mathbb{E}[Y_iY_j|N=h]=\mathbb{E}[Y_iY_j]$ for all i,j. Thanks to assumption (2a) (and (3)), $\mathbb{E}[Y_iY_j]=\mathbb{E}[Y_i]\,\mathbb{E}[Y_j]$ for all i,j. Finally, thanks to assumption (2b) (and (3)), $\mathbb{E}[Y_j^2]=\mathbb{E}[Y_1^2]$ for all i and $\mathbb{E}[Y_i]=\mathbb{E}[Y_1]$ for all i. Replacing into (9.5.6), we obtain

$$\begin{aligned}\mathbb{E}[S^2] &= \textstyle\sum_{h=1}^{\infty}\pi_h\,h\,\mathbb{E}[Y_1^2]+\sum_{h=1}^{\infty}\pi_h\,h\,(h-1)\,(\mathbb{E}[Y_1])^2\\ &= \mathbb{E}[N]\,\mathbb{E}[Y_1^2]+\mathbb{E}[N\,(N-1)]\,(\mathbb{E}[Y_1])^2=\mathbb{E}[N]\,\mathbb{V}\mathrm{ar}[Y_1]+\mathbb{E}[N^2]\,(\mathbb{E}[Y_1])^2\end{aligned} \tag{9.5.7}$$

When we plug (9.5.7) into (9.5.5), we finally find

$$\mathbb{V}\mathrm{ar}[S]=\mathbb{E}[N]\,\mathbb{V}\mathrm{ar}[Y_1]+\mathbb{V}\mathrm{ar}[N]\,(\mathbb{E}[Y_1])^2 \tag{9.5.8}$$

from which we learn that to implement rule (9.5.4) we first need an estimate of the expected claim frequency, $\mathbb{E}[N]$, and the expected claim severity, $\mathbb{E}[Y_1]$, i.e. the same data required for the equivalence premium; we further need an estimate of the variance of the claim frequency, $\mathbb{V}\mathrm{ar}[N]$, and the variance of the claim severity, $\mathbb{V}\mathrm{ar}[Y_1]$.

Quite similar to the variance principle, the *standard deviation principle* assesses the net premium as follows:

$$\Pi=\mathbb{E}[S]+\beta\,\sqrt{\mathbb{V}\mathrm{ar}[S]} \tag{9.5.9}$$

where β ($\beta > 0$) is a given proportion. The main advantage of (9.5.9) in respect to (9.5.4) consists in the fact that the parameter β is unit-free (while, as recalled above, λ is an intensity). Apart from this, the rationale of the two rules is similar; in particular, the same amount for the net premium could be determined under the two rules, provided that $\beta=\lambda\,\sqrt{\mathbb{V}\mathrm{ar}[S]}$.

Example 9.5.1. Assume $\mathbb{E}[S] = 1.30$ and $\mathbb{V}\mathrm{ar}[S] = 13$. The equivalence premium is: $P = \mathbb{E}[S] = 1.30$. Then assume that the net premium is: $\Pi = 1.40$; trivially, the safety loading is: $\Pi - P = 0.10$. Such a value could have been obtained (alternatively) as follows:

- through the expected value principle, taking: $\alpha = \frac{\Pi}{\mathbb{E}[S]} - 1 = 7.692\%$ or $\kappa = 1 - \frac{\mathbb{E}[S]}{\Pi} = 7.143\%$;
- through the variance principle, taking: $\lambda = \frac{\Pi - \mathbb{E}[S]}{\mathbb{V}\mathrm{ar}[S]} = 0.00769$;
- through the standard deviation principle, taking: $\beta = \frac{\Pi - \mathbb{E}[S]}{\sqrt{\mathbb{V}\mathrm{ar}[S]}} = 2.774\%$.

□

We have commented above on some practical implications of the various premium principles. It is worthwhile to note that a premium principle defines a functional $\mathbb{H}$ which assigns a positive real number (namely, the net premium Π) to the distribution function of the aggregate claim amount S; thus, $\Pi = \mathbb{H}[S]$. Some mathematical properties should be satisfied by $\mathbb{H}$, which are relevant from a practical point of view. We recall the main properties.

(P1) For any S, it must turn out:

$$\mathbb{H}[S] > \mathbb{E}[S] \tag{9.5.10}$$

This is an obvious requirement: the safety loading must be positive.

(P2) If S_1 and S_2 are two independent risks (i.e., the aggregate claim amounts of two independent risks), we require:

$$\mathbb{H}[S_1 + S_2] \leq \mathbb{H}[S_1] + \mathbb{H}[S_2] \tag{9.5.11}$$

This prevents the insured to find convenience in fragmenting the risk.

(P3) Given two independent risks S_1 and S_2, we require:

$$\mathbb{H}[S_1] \leq \mathbb{H}[S_1 + S_2] \tag{9.5.12}$$

If the cover protects against a wider range of risks, the premium should be higher.

(P4) Given two positive real numbers a and b, we require:

$$\mathbb{H}[aS + b] \geq a\mathbb{H}[S] + b \tag{9.5.13}$$

The constant b represents an increase of the claim amount, common to all the possible claims; similarly, a represents a proportional increase of any possible claim. If the possible amount of any claim increases, we expect a similar increase in the premium. We note that the property is not satisfied by the variance and the standard deviation principles (the variance principle fulfils (9.5.13) only if $b = 0$).

(P5) If the claim amount cannot exceed an amount K, or if there exists a positive amount K such that $\mathbb{P}[S \leq K] = 1$, then it must turn out:

$$\mathbb{H}[S] \leq K \tag{9.5.14}$$

Also this property is quite obvious: no insured would rationally be willing to pay a premium higher than the maximum compensation she can realistically obtain from the insurer.

We mention a last premium principle. The event $S > \Pi$ represents a situation of (economic) loss to the insurer. According to the *percentile principle*, the net premium Π must be such that

$$\mathbb{P}[S > \Pi] = \varepsilon \tag{9.5.15}$$

where ε ($\varepsilon > 0$) is the accepted loss probability. To apply (9.5.15), the probability distribution of S must be assigned; the technical implementation of rule (9.5.15) may be time-consuming, and clearly data for the estimate of the whole probability distribution of S are required. In practice, simpler rules are preferred, unless extreme risks are transferred to the insurer.

9.6 The expense-loaded premium

Expenses charged to a non-life insurance policy include: the initial commission; administrative and other expenses; claim settlement expenses. The latter are sometimes reported joint to claim amounts, so that an explicit loading is not applied (to avoid a double charge). Expenses may be fixed or floating; in this latter case, their amount may depend either on the number of claims, the amount of premiums or the amount of claims. In most practice, a forfeiture loading rule is adopted, which however may be justified only considering in detail the several types of expenses which may be charged to the policy. In the following, we examine such a rule, assuming that the net premium has been calculated according to the expected value principle.

We consider the following classes of *expenses*:

- initial commission: $\Theta^{[A]}$ (stated as a fixed amount);
- administrative and other expenses: $\Theta^{[G]}$ (stated as a fixed amount);
- claim settlement expenses: $\Theta^{[S]}$ (stated as an amount per claim).

The *expense-loaded* (or *gross*) *premium*, $\Pi^{[T]}$, is defined as follows:

$$\Pi^{[T]} = \mathbb{E}[S] + \kappa \Pi^{[T]} + \Theta^{[A]} + \Theta^{[G]} + \Theta^{[S]} \mathbb{E}[N] \tag{9.6.1}$$

where κ is the safety-loading proportion, applied to the expense-loaded premium (instead of the net premium). We note that if claim settlement expenses are included in the cost of claims, then $\Theta^{[S]} = 0$.

Replacing (9.4.3) into (9.6.1) and rearranging, we obtain

$$\Pi^{[T]} = \mathbb{E}[S] \frac{1 + \frac{\Theta^{[S]}}{\mathbb{E}[Y_1]}}{1 - \kappa} + \frac{\Theta^{[A]} + \Theta^{[G]}}{1 - \kappa} \tag{9.6.2}$$

Setting: $\delta = \frac{1 + \frac{\Theta^{[S]}}{\mathbb{E}[Y_1]}}{1 - \kappa}$ and $e = \frac{\Theta^{[A]} + \Theta^{[G]}}{1 - \kappa}$, we finally get to the forfeiture formula

$$\Pi^{[T]} = \delta\,\mathbb{E}[S] + e \tag{9.6.3}$$

quite common in practice. We note that, in principle, the parameters δ and e should reflect the various expenses loaded, as well as the safety-loading proportion. In practice, some approximated choices could be adopted.

Example 9.6.1. Assume $\mathbb{E}[S] = 1.30$ and $\mathbb{E}[N] = 0.13$. Let the expense-loaded premium be: $\Pi^{[T]} = 1.50$. Such a value could have been obtained assuming the following loading parameters:

- safety loading: $\kappa = 7\%$ (per unit of expense-loaded premium);
- initial and administrative expenses: $\Theta^{[A]} + \Theta^{[G]} = 0.0924$ (fixed amount);
- claim settlement expenses: $\Theta^{[S]} = 0.02$ (amount per claim).

We find: $\delta = 1.07742$ (proportion of the expected aggregate claim amount) and $e = 0.10$ (fixed amount).
□

9.7 Statistical data for the equivalence premium

In this Section we illustrate some quantities which can be used to estimate the expected claim frequency, $\mathbb{E}[N]$, the expected claim severity, $\mathbb{E}[Y_1]$, and then the expected total payout for a policy, $\mathbb{E}[S]$. Data are collected from a set of policies with specified features.

Remark Assume that all policies are termed one year. As already mentioned in Sect. 9.2.1, the time between the issue (or renewal) time of a policy and its maturity (or next renewal time) is called *policy year*. Reasonably, such a period does not coincide with the *calendar year* (unless the policy is issued on January 1). Data on claims may be collected either on a calendar or a policy year basis. For pricing, policy year data are appropriate, as the premium has to match the cost of claims arising during the life of the policy. Conversely, when reporting the result of the management of the portfolio, the natural reference is to the calendar year. In the following, we will specify which is the form of data we are referring to.

9.7.1 Risk premium, claim frequency, loss severity

In this Section, we refer to a homogeneous portfolio, consisting of r policies (or insured risks), all issued at the same time and all with duration one year. Homogeneity of the policies means, in particular, that they are similar in respect of: the type of risk covered (e.g., fire insurance, motor insurance, or others), policy conditions (deductibles, limit values or insured valued), the propensity to incur into a claim, the possible severity of a claim, and so on. The policy year is the same for all the policies, so that we can easily collect data on this basis. We stress that all the policies are exposed for one year (the common policy period) to the risk of incurring into one or more claims.

Assume that, during the (policy) year, policies report z claims in total, $z \gtreqless r$, with claim amounts $y_1, y_2, \dots, y_z$. Note that the information is aggregate, as we just know that z claims have been reported in the portfolio, while we do not know which policies have reported such claims.

The ratio between the total payout for the portfolio and the number of policies, i.e. the *claim amount per policy*

$$Q = \frac{y_1 + y_2 + \cdots + y_z}{r} \tag{9.7.1}$$

is called *risk premium* or *average claim cost*. Should each policy have paid a (net) premium $\Pi = Q$, then the insurer would be on balance, as the total inflow amount would be rQ, the same as the outflow amount, $y_1 + y_2 + \cdots + y_z$; for this reason, the quantity Q is looked at as an "observed premium".

The quantity Q provides an estimate of $\mathbb{E}[S]$ (it is reasonable to add a safety loading to Q, in face of random fluctuations). It is interesting to split Q as follows. The ratio

$$\bar{n} = \frac{z}{r} \tag{9.7.2}$$

represents the *average number of claims per policy*, or the *average claim frequency*. Conversely, the ratio

$$\bar{y} = \frac{y_1 + y_2 + \cdots + y_z}{z} \tag{9.7.3}$$

represents the *average claim amount per claim*, or the *average claim severity*. Note, in particular, that $\bar{n}$ expresses an estimate of $\mathbb{E}[N]$, while $\bar{y}$ provides an estimate of $\mathbb{E}[Y_1]$. Then we have

$$Q = \bar{n} \times \bar{y} \tag{9.7.4}$$

which is the statistical estimate of (9.4.3).

With regard to the average claim frequency, the following splitting is of interest. Let $z_{\max}$ be the maximum number of claims reported by one policy (clearly, $z_{\max} \leq z$) and r_h the number of policies realizing h claims ($h = 0, 1, \dots, z_{\max}$). The number of policies can be split as follows:

$$r = r_0 + r_1 + \cdots + r_{z_{\max}} \tag{9.7.5}$$

while the number of claims can be written as:

$$z = r_1 + 2 r_2 + \cdots + z_{\max} r_{z_{\max}} \tag{9.7.6}$$

The average claim frequency can then be factorized as follows:

$$\bar{n} = \frac{r_1 + 2 r_2 + \cdots + z_{\max} r_{z_{\max}}}{r_1 + r_2 + \cdots + r_{z_{\max}}} \times \left(1 - \frac{r_0}{r}\right) \tag{9.7.7}$$

The first ratio (which could also be written as $\frac{z}{r-r_0}$), represents the average number of claims per policy reporting claims; the quantity $\frac{r_0}{r}$ expresses the no-claim frequency. Thus, the quantity in brackets in (9.7.7) represents the average frequency of at least one claim. It is interesting to read (9.7.7) as the statistical estimate of $\mathbb{E}[N]$, appropriately expressed. As recalled in (1.4.31) (see Sect. 1.4.3), the expected number of claims per policy is defined as follows:

$$\mathbb{E}[N] = \sum_{n=0}^{+\infty} n \times \mathbb{P}[N = n] = \sum_{n=1}^{+\infty} n \times \mathbb{P}[N = n] \tag{9.7.8}$$

Such a quantity can be decomposed as

$$\mathbb{E}[N] = \mathbb{E}[N|N = 0] \times \mathbb{P}[N = 0] + \mathbb{E}[N|N \geq 1] \times \mathbb{P}[N \geq 1] \tag{9.7.9}$$

which reduces to

$$\mathbb{E}[N] = \mathbb{E}[N|N \geq 1] \times \mathbb{P}[N \geq 1] \tag{9.7.10}$$

given that $\mathbb{E}[N|N = 0] = 0$. It is easy to see that (9.7.7) provides a statistical estimate of the factors in the right-hand side of (9.7.10).

Equation (9.7.7) is useful to get some information (at least at an aggregate level) about the concentration of claims on few policies, and then on the acceptability of the independence assumptions underlying (9.4.3). Indeed, we note that for a given value of the average claim frequency $\bar{n}$, the higher is the ratio $\frac{r_1+2r_2+\cdots+z_{\max} r_{z_{\max}}}{r_1+r_2+\cdots+r_{z_{\max}}}$, the stronger is the concentration of claims on few policies. Clearly, if a high concentration emerges, the independence assumptions should be checked through further investigations, as correlation effects could be present when several claims are reported by a policy.

Example 9.7.1. In Table 9.7.1 the average claim frequency experienced in two portfolios is reported. Both portfolios consist of 100 000 policies and have reported the same number of claims. However, portfolio B experiences a higher concentration of claims on few policies, as witnessed by the average number of claims per policy with claims. Whilst for portfolio A the low value of the average number of claims per policy with claims suggests that the independence assumptions could be considered reasonable, for portfolio B some further investigation could be necessary in this respect.

□

9.7.2 Units of exposure: the case of heterogeneous portfolios

An *exposure unit* is a measure of some feature of the insured risk which has proved to bear a close correspondence to the claim experience. Examples of exposure units are as follows: insured value (suitable for property insurance), time spent in the

Table 9.7.1 Claim experience in two portfolios

	both portfolios	
number of policies, r	100 000	
number of claims, z	13 000	
total claim amount, $y_1 + y_2 + \cdots + y_z$	13 000 000	
	portfolio A	portfolio B
risk premium, Q	130	130
average claim severity, $\bar{y}$	1 000	1 000
average claim frequency, $\bar{n}$	0.13	0.13
average number of claims per policy with claims, $\frac{z}{r-r_0}$	1.08	1.80
average frequency of at least one claim, $1 - \frac{r_0}{r}$	0.118	0.072

portfolio in a given calendar year (used in motor insurance), payroll (for workers compensation insurance). Exposure units are used to summarize appropriately the cost of claims incurred or the amount of premiums earned. One example in this respect is provided by the risk premium, introduced in Sect. 9.7.1.

Refer to a property insurance coverage. The risk premium, as defined by (9.7.1), requires that policies are homogeneous in respect of the insured value, the time of entry and the duration; in these circumstances, to get an average information about the claim cost, we simply divide the total portfolio payout by the number of policies. We now address how we should measure the average claim cost if the insured values are different.

Let $V'^{(1)}, V'^{(2)}, \ldots, V'^{(r)}$ be the insured values of the r policies (for which we still assume the same type of cover, the same time of entry and the same duration). Reasonably, the higher is the insured value $V'^{(j)}$, the higher should be the possible claim amount that we expect from a policy. The average claim cost should then be measured as follows:

$$\theta = \frac{y_1 + y_2 + \cdots + y_z}{V'^{(1)} + V'^{(2)} + \cdots + V'^{(r)}} \tag{9.7.11}$$

i.e. as an average claim amount per unit of exposure (clearly, θ is unit-free). We note that it is reasonable that those policies with a higher insured value pay a higher premium. In particular, the same premium rate (i.e., the same premium per unit of insured value) could be applied to all the policies (given that, apart from the insured value, they are similar); the premium amount would then be proportional to the insured value. If θ is the premium rate applied, then the total inflow amount of the insurer would coincide with the total outflow amount:

$$\theta\,(V'^{(1)} + V'^{(2)} + \cdots + V'^{(r)}) = y_1 + y_2 + \cdots + y_z \tag{9.7.12}$$

In this perspective, θ can be looked at as an observed premium rate.

Let define the *average insured value* as follows:

$$\bar{V}' = \frac{V'^{(1)} + V'^{(2)} + \cdots + V'^{(r)}}{r} \tag{9.7.13}$$

which, clearly, represents the average exposure per policy. The observed premium rate θ can be split as follows:

$$\theta = \bar{n} \times \frac{\bar{y}}{\bar{V}'} = \frac{Q}{\bar{V}'} \tag{9.7.14}$$

The quantity $\bar{y}/\bar{V}'$ is named *average claim degree*. We note that, similarly to the case of the homogeneous portfolio (see Sect. 9.7.1), the quantity $Q = \bar{n} \times \bar{y}$ still expresses the average claim amount per policy; however, due to the different insured values, such a piece of information is not appropriate neither for pricing, nor for summarizing the cost of claims incurred.

9.7.3 Units of exposure: the number of policy years

So far, we have assumed that policies are issued (or renewed) at the same time; more realistically, issue or renewal times are different. As a first consequence, policy years are different. It may then become easier, or more natural, to collect claim on a calendar year basis (which, trivially, is a term of reference common to all policies), and this is what we will assume from now on.

A second consequence of the different times of issue (or renewal) is the following. When the policy year is the same for all the policies, the number of policies which are in-force at a given time (say, at issue) also represents the number of policies which are overall in-force during the year we are referring to. Conversely, this correspondence does not hold when policies have different issue or renewal times. This should be considered when calculating summaries of the cost of claims. For example, in the risk premium (see (9.7.1)), the total amount of claims incurred in one year is compared to the number of policies which, during the year, have been exposed to the risk of generating those claims. In this Section, we discuss how we should assess the denominator of Q when policies have different policy years. As stated above, we assume that data are collected on a calendar year basis; in particular, then, $y_1 + y_2 + \cdots + y_z$ is the total payout for a portfolio in a given calendar year.

We call *number of the exposed to risk* (or *number of policy years* or, in the specific case of motor insurance, *number of car years*) the time totally spent in the portfolio during the calendar year by the policies which are in-force for a part (at least) of such a year. For example, if a policy is issued on July 1 and a second policy is issued on February 1 of year t, the number of the exposed to risk during year t is $\frac{6}{12} + \frac{11}{12} = \frac{17}{12}$; actually, the policy issued on July 1 stays in the portfolio for half a year during year t (and for half a year during year $t+1$), while the policy issued on February 1 spends 11 months in the portfolio in year t (and one month in year $t+1$).

The calculation of the number of policy years may be performed exactly, considering for each policy the exact time spent in the portfolio during the year, or approximately. Clearly, exact calculation techniques do not require any further comment.

As far as approximate methods are concerned, there are some alternative solutions. The method to be preferred depends on the type of data available. We illustrate two common approaches; for brevity, we do not give formal details (which would be cumbersome), but we introduce such approaches through two examples. This is enough to understand how the approximate methods work.

Example 9.7.2. Assume that we are provided with the information regarding the number of policies entering a given portfolio, on a monthly basis; see Table 9.7.2 for an example. Policies may be newly issued or renewed. All are assumed to have term one year, and to stay in the portfolio for one year.

Table 9.7.2 Number of policies according to the period of issue or renewal

	Number of policies	
Month	Year $t-1$	Year t
1/1–31/1	74	75
1/2–28/2	89	82
1/3–31/3	82	87
1/4–30/4	69	75
1/5–31/5	81	75
1/6–30/6	95	90
1/7–31/7	98	95
1/8–31/8	79	83
1/9–30/9	85	90
1/10–31/10	93	90
1/11–30/11	90	98
1/12–31/12	70	80

We can assume that, within each month, policy anniversaries are uniformly spread. Thus, on average each policy enters in the middle of the relevant month. Split each year in 24 periods, and let 0 be January 1; then, in each year, the times of possible issue or renewal of a policy are: 1, 3, ..., 23. The 74 policies issued or renewed on January (i.e., at time 1) of year $t-1$ spend in the portfolio 23 periods (over 24) in year $t-1$, and 1 (over 24) period in year t; the 89 policies issued (or renewed) on February (i.e., at time 3) of year $t-1$ spend in the portfolio 21 periods (over 24) in year $t-1$, and 3 (over 24) periods in year t; ...; the 80 policies issued (or renewed) on December (i.e., at time 23) of year t spend in the portfolio 1 period (over 24) in year t, and 23 periods (over 24) in year $t+1$. See Fig. 9.7.1 for a graphical representation.

The total time spent by policies in the portfolio in year t (namely, the number of policy years in year t) can then be calculated as follows:

$$\begin{aligned} &74 \times \tfrac{1}{24} + 89 \times \tfrac{3}{24} + \cdots + 70 \times \tfrac{23}{24} \\ &+75 \times \tfrac{23}{24} + 82 \times \tfrac{21}{24} + \cdots + 80 \times \tfrac{1}{24} = 1003.96 \end{aligned} \tag{9.7.15}$$

□

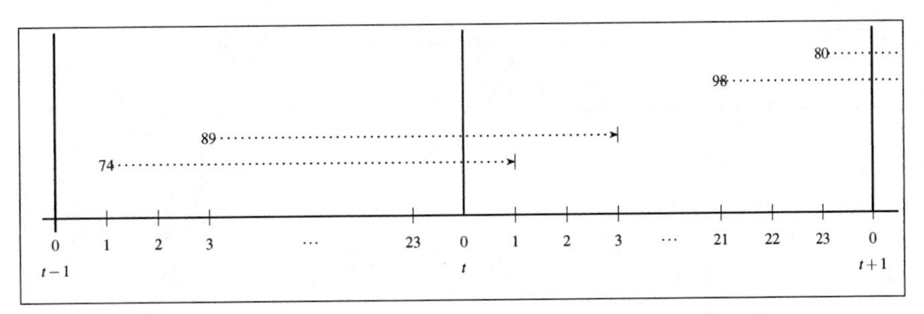

Fig. 9.7.1 Graphical representation of the time spent by each group of policies in the portfolio

The method described in Example 9.7.2 is called *method of the 24-ths*. Possible adjustments can be easily introduced if data on the policy inception times are more rare (e.g., if they are available bimonthly or quarterly), or if policies have a duration shorter than one year. In case data are available on a quarterly basis, the method is called of the 8-ths (as the year would be split into 8 periods, in this case); if they are available on a bimonthly basis, the year would be split into 12 periods, and then the method would be called of the 12-ths. In general, if data are available on a $\frac{k}{2}$ basis, the method is called of the k-ths (as the year is then split into k periods).

Example 9.7.3. We now give an example of the *census method*. Assume that data provide us with the information about the number of policies in-force at some specific dates; see Table 9.7.3 for an example. Again, we assume that all the policies are termed one year, and that they remain in the portfolio for one whole year.

Table 9.7.3 Number of policies in-force in year t, on a monthly basis

Time	Number of policies in-force
1/1	1 200
1/2	1 320
1/3	1 405
1/4	1 380
1/5	1 300
1/6	1 400
1/7	1 450
1/8	1 350
1/9	1 500
1/10	1 650
1/11	1 700
1/12	1 800
31/12	1 750

We can first calculate the average number of policies in-force in each month, as depicted in Table 9.7.4.

Table 9.7.4 Average number of policies in-force in year t, on a monthly basis

Month	Average number of policies in-force
1/1–31/1	$\frac{1200+1320}{2} = 1260.00$
1/2–28/2	$\frac{1320+1405}{2} = 1362.50$
1/3–31/3	$\frac{1405+1380}{2} = 1392.50$
1/4–30/4	$\frac{1380+1300}{2} = 1340.00$
1/5–31/5	$\frac{1300+1400}{2} = 1350.00$
1/6–30/6	$\frac{1400+1450}{2} = 1425.00$
1/7–31/7	$\frac{1450+1350}{2} = 1400.00$
1/8–31/8	$\frac{1350+1500}{2} = 1425.00$
1/9–30/9	$\frac{1500+1650}{2} = 1575.00$
1/10–31/10	$\frac{1650+1700}{2} = 1675.00$
1/11–30/11	$\frac{1700+1800}{2} = 1675.00$
1/12–31/12	$\frac{1800+1750}{2} = 1775.00$

Each of the groups of policies quoted in Table 9.7.4 spend on average one month in the portfolio. Thus, the number of policy years can be calculated as follows:

$$1260 \times \frac{1}{12} + 1362.50 \times \frac{1}{12} + \cdots + 1775 \times \frac{1}{12} = 1477.50 \tag{9.7.16}$$

Note that the number of policy years assessed through (9.7.16) corresponds to the simple arithmetic mean of the average number of policies in-force in each month, as each group of policies is assumed to spend the same time (i.e., one month) in the portfolio.
□

Also the census method may be easily adjusted if data are more rare, or policies do not spend one whole year in the portfolio.

9.7.4 Updating the risk premium to portfolio experience

The data set expressing the claim experience of the insurer in a given portfolio could be inadequate for pricing, either because:

a. the portfolio has been recently issued, and thus has not yet gained an adequate experience;
b. the behavior of claims is not stable in time, but evolves according to some trend (possibly unknown);
c. data are sparse and the sample of the observed claims is considered to be too small.

In case (a) and (c), the problem relates to the size of the sample, which is considered to be too small. Data for premium calculation are then usually obtained from other portfolios (possibly belonging to other insurers), taking care that they have features similar to the portfolio dealt with. In case (b), the problem has a different nature. The inadequacy of the data base can be traced to the underlying (unknown) dynamics; experience could be rich enough but, because of the trend, the observed data reflect old information. Appropriate adjustments are required before such data can be used for estimating the cost of future claims. The two situations ((a) and (c) on the one hand, (b) on the other) require a different treatment; a similar methodological structure can be designed, but with different implementing profiles. The dynamic problem can be considered an advanced topic, which for non-life insurance is of particular interest just for some lines of business; given the introductory character of this Chapter, we do not give details in this regard. In the following, we refer to a static framework and illustrate the idea of updating in time the pricing basis to new experience; we make explicit reference to case (a) above.

We refer to an insurer issuing a coverage for which it has no direct experience, and thus no data. To set the premium, a *reference population* must be selected, which has already (or almost) reached a steady state in respect of claim experience, and can thus provide reliable data. Typically, the reference population is the portfolio of another insurer, who deals with the same or a similar coverage; the relevant experience is assumed to be consistent with what will emerge from the new portfolio. Henceforth, we assume for brevity that policies are homogeneous in respect of the insured value and that the appropriate exposure unit is the number of policy years.

Let 0 be the time at which the new portfolio is issued and Q_0 the risk premium observed in the reference population. Given that Q_0 is the only available information, the (equivalence) premium for the new portfolio is set simply as

$$P_0 = Q_0 \tag{9.7.17}$$

At time 1, the new portfolio has gained some experience; let Q_1 be the average claim cost observed in the time-interval $(0,1)$. At time 1, the insurer has to decide how to set the premium for the next year, say P_1. Three choices are available:

1. the premium is not revised, and thus $P_1 = P_0$;
2. the premium is revised, accounting for the new information only, i.e. $P_1 = Q_1$;
3. the premium is revised according to the new experience, but continuing to account also for the initial information.

Choice 1 has the advantage of providing stability to the premium, which is good from a commercial point of view; however, comparing Q_0 with Q_1 one can perceive some differences between the claim experience of the new portfolio and the reference population, which would be better not to disregard. On the other hand, choice 2 has the disadvantage that Q_1 may turn out to be exceptionally high or low, for accidental reasons (e.g.: the portfolio is not yet large enough; the policies issued in the first year are self-selected, and hence they do not yet express appropriately the average claim experience of the portfolio; and so on). Choice 3 clearly represents an intermediate solution. In particular, a sound way to set the premium for the second year, i.e. at time 1, is:

$$P_1 = \alpha_1 Q_1 + (1-\alpha_1) Q_0 \tag{9.7.18}$$

where α_1, $0 < \alpha_1 < 1$, is a given proportion expressing the weight assigned to the new information (the cases $\alpha_1 = 0$ and $\alpha = 1$ are excluded, as they correspond, respectively, to choice 1 and choice 2 above). Reasonably α_1 is closer to 0 than 1, given that the experience gained on the new business is not yet stable.

At time 2, the average claim cost can be assessed with reference to the experience gained in $(0,2)$. We let Q_2 denote the ratio (9.7.1) based on the data collected in $(0,2)$. Such a quantity embeds a wider experience than Q_1, but can still be considered subject to more fluctuations than Q_0, as the experience of the insurer is less rich than that relating to the reference population. Similarly to time 1, the equivalence premium at time 2 is set as follows

$$P_2 = \alpha_2 Q_2 + (1-\alpha_2) Q_0 \tag{9.7.19}$$

where α_2 is the new weight assigned to the portfolio experience. Reasonably, $\alpha_2 > \alpha_1$, but still $0 < \alpha_2 < 1$. In general, a reasonable rule for setting the premium at time t is the following:

$$P_t = \alpha_t Q_t + (1-\alpha_t) Q_0 \tag{9.7.20}$$

where Q_t is the average claim cost experienced within the new portfolio in the period $(0,t)$ and α_t is the weight assigned at time t to such information. Reasonably, $0 < \alpha_1 < \alpha_2 < \cdots < \alpha_t \leq 1$.

Remark In Sect. 9.7.1, we have described the risk premium as a quantity based on observations collected in one year. Clearly, the ratio Q could be referred to a wider time span. The advantage of increasing the time-interval of observation consists in enlarging the data-set. Conversely, some disadvantages follow: claim amounts could be subject to inflation; given that policies are termed one year, the homogeneity of the portfolio may be weakened by new entries (possibly joint to a reduction of renewals); the claim frequency may be subject to changes in time (due to the development of new technologies, the introduction of a new regulation, and so on). Thus, the average claim amount per policy may be exposed to systematic deviations, which are not detected if the risk premium is assessed with reference to the average experience over more than one year. When adopting approach (9.7.20) for premium calculation, clearly one assumes that systematic deviations are either not present or negligible.

A premium calculated through (9.7.20) is called *experience premium*, and the approach described by (9.7.20) an *experience-rating system*; more precisely, since the premium turns out to be updated on the experience gained on a portfolio, the system is referred to as a *collective experience-rating*. Fig. 9.7.2 illustrates the process of gradually updating the equivalence premium to portfolio experience, which is realized through (9.7.20).

Formula (9.7.20) is an example of a *credibility model*, in which information collected from some external source are gradually merged with those collected on a specific population. The coefficient α_t is called a *credibility factor* and it expresses the relative reliability (or "credibility") of the specific information. The wider is the volume of the specific data relative to the volume of those obtained from the external

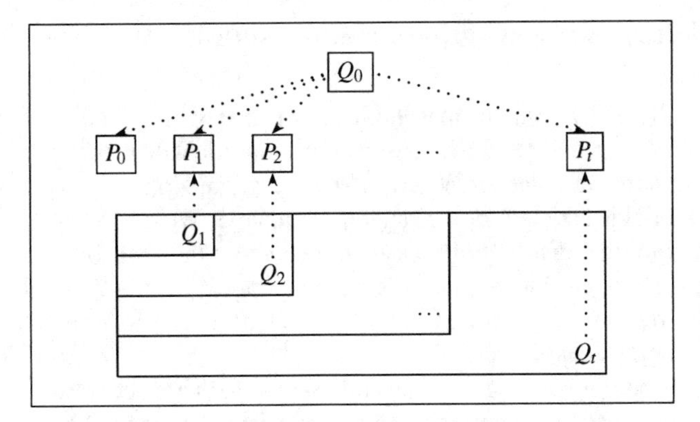

Fig. 9.7.2 Updating the premium according to collective experience-rating

source, the higher is the credibility acknowledged to the former. The relation

$$\alpha_1 < \alpha_2 < \cdots < \alpha_t < \ldots \tag{9.7.21}$$

expresses a (reasonable) increase in time of the credibility acknowledged to the specific experience, in respect of that witnessed by the reference population. When $\alpha_t = 1$, *full credibility* has been reached, and the information obtained from the external source can be disregarded.

Several theoretical models can support the choice of the credibility factors α_t. A distributional assumption must be adopted for the aggregate claim amount S (see, for some remarks in this regard, Sect. 9.8). It goes beyond the scope of this presentation to deal with a detailed stochastic modeling of non-life insurance; thus, we do not develop such a discussion. We just give some comments. We recall that the average claim cost Q (Q_t, in the discussion above) represents an estimate of the expected aggregate claim amount, $\mathbb{E}[S]$. One idea is to assign full credibility to Q_t when the probability that the estimate Q_t is close enough to the true underlying value of the aggregate claim amount is at a given (high) level (say, 0.95). The notion of "close enough" clearly requires to be formalized, as follows

$$\mathbb{P}[(1-a)\,Q_t < S \leq (1+a)\,Q_t] = \varepsilon \tag{9.7.22}$$

where a states the width of the band around S (reasonably, a should be low, say 0.01), and ε is the required probability level (say, 0.95). For an example, see Sect. 9.8.3. An alternative way, more widely known, to assess the credibility factors involves Bayesian statistic techniques; for some references, see Sect. 9.13.

9.8 Stochastic modeling of the aggregate claim amount

Most of the discussion in Sect. 9.7 assumes that the insurer can rely on an appropriate data set, in which case assessments can be based on the empirical distribution of the aggregate claim amount. In some situations, however, calculation are only possible if a theoretical model is adopted for the random variable S. We stress that, even when dealing with empirical distributions, assumptions about the probability distribution of S cannot be avoided. Indeed, in the simple setting of Sect. 9.7, not invoking a specific choice for the distribution of S, it is anyhow necessary to accept a compound distribution for S; see also Sect. 9.4.3.

When a theoretical distribution is introduced, several assumptions are accepted; from a practical point of view, each assumption implies some simplification, which leads to a representation more or less far away from (or more or less close to) real situations. Further, dealing with theoretical distributions may require some analytical expertise (and, because of this, someone may consider that working with empirical distributions is preferable to the adoption of theoretical models). However, the properties of theoretical distributions facilitate the analysis of many problems, or even make such an analysis possible. We also note that a theoretical distribution is summarized by a small number of parameters, while an empirical distribution requires to work always with a large amount of data.

As mentioned earlier, it goes beyond the aim of this book to deal in details with the theoretical distribution of the aggregate claim amount. However, we think that some information, and some examples, may be useful to understand which kind of analyses may be developed through this approach.

9.8.1 Modeling the claim frequency

From definition (9.4.1) for the aggregate claim amount, it emerges that to model S we first need to model the number of claims, N. Several choices are possible, some of which are more interesting (or useful) for practical applications.

We start from an elementary case. If a policy may experience at most one claim during the coverage period, then N follows a Bernoulli distribution, i.e.

$$N = \begin{cases} 0 & \text{with probability } 1-p \\ 1 & \text{with probability } p \end{cases} \tag{9.8.1}$$

(see Sect. 1.2.3 and 1.4.2), where p is the claim probability. The law has one parameter, namely p, which should be estimated through the average claim frequency $\bar{n}$.

In the non-life insurance business, just for few lines (and, possibly, just under some restrictive policy conditions) the assumption that each policy may experience at most one claim is consistent with evidence. In the more realistic case in which a policy may experience more than one claim, we must first wonder whether claims are independent or not. The usual assumption is that they are independent; intuitively, this is a reasonable assumption when we refer to claims reported by different policies. But now we are referring to one policy, and in this case some form of correlation among claims may be present; henceforth, we will keep the assumption of independence among claims also when referring to a policy. If we know that the maximum number of claims per policy is $n_{\max}$, and that each claim has the same probability p to occur, then N follows a Binomial distribution, i.e.

$$\mathbb{P}[N=n] = \binom{n_{\max}}{n} p^n (1-p)^{n_{\max}-n}\,; \qquad n=0,1,\dots,n_{\max} \tag{9.8.2}$$

where $n_{\max}$ and p are the parameters of the law. The choice of $n_{\max}$ should be suggested by the features of the contract (in particular, by its policy conditions); conversely, p can be estimated, once again, through the average claim frequency $\bar{n}$. Noting that under (9.8.2) we have $\mathbb{E}[N] = p\,n_{\max}$, the ratio $\bar{n}/n_{\max}$ provides us with an estimate for p.

A better fitting to empirical data is usually provided by the Poisson distribution, according to which

$$\mathbb{P}[N=n] = \mathrm{e}^{-\lambda}\frac{\lambda^n}{n!}\,; \qquad n=0,1,\dots \tag{9.8.3}$$

where λ, $\lambda > 0$, is the parameter of the law. We recall that the Binomial distribution with parameters $n_{\max}, p$ is well-approximated by a Poisson distribution with parameter $\lambda = p\,n_{\max}$ when $n_{\max}$ is large enough and p is small enough. The Poisson law has been originally developed for rare events; considering that in non-life insurance most of the insured risks bear a low claim probability, it is not surprising that the Poisson law turns out to be more appropriate than the Binomial one. The Poisson law is, for example, more realistic in respect of the maximum number of claims, which does not need to be stated in advance. We recall that under (9.8.3) we have $\mathbb{E}[N] = \mathbb{V}\mathrm{ar}[N] = \lambda$.

The Poisson law offers several analytical advantages. Let N_t be the number of claims for a policy in a period of t years ($t > 0$); consistently with the previous notation, we let $N_1 = N$ whenever $t = 1$. If claims occur independently one from the other, whatever is the time of their occurrence, from $N \sim \mathrm{Poi}(\lambda)$ it follows $N_t \sim \mathrm{Poi}(\lambda\,t)$; indeed, the sum of a given number of independent Poisson random variables is a Poisson random variable, whose parameter is the sum of the parameters of the original random variables. This result is more fruitful when referred to a portfolio. If $N \sim \mathrm{Poi}(\lambda)$ for any policy in the portfolio and if claims reported by different policies are independent, then $N^{[\mathrm{P}]} \sim \mathrm{Poi}(\lambda\,r)$ (where $N^{[\mathrm{P}]}$ is the total number of claims in the portfolio in one year and r is the number of policy years for that year). More generally, if $N^{(j)} \sim \mathrm{Poi}(\lambda^{(j)})$ is the number of claims reported by policy j in one year, then $N^{[\mathrm{P}]} \sim \mathrm{Poi}(\sum_{j=1}^{r}\lambda^{(j)})$ is the number of claims reported within the portfolio. Extensions to time-intervals shorter or longer than one year are straightforward (we would denote by $N_t^{[\mathrm{P}]}$ the total number of claims in the portfolio in a period of t years, $t > 0$).

Example 9.8.1. Let $N \sim \text{Poi}(0.13)$ be the number of claims for a policy in one year. Then $N^{[P]} \sim \text{Poi}(13\,000)$ is the number of claims in one year for 100 000 homogeneous and independent policies. Let us split the year into terms and assume that the number of claims in each term is independent and identically distributed in respect of the previous terms; then, for example, the number of claims for the portfolio in the first term of the year is $N^{[P]}_{0.25} \sim \text{Poi}(3\,250)$. Further examples can be easily derived.

Once we know the probability distribution of the number of claims, and the relevant parameter as well, the probability of several events of interest can be easily assessed. For example, if $N \sim \text{Poi}(0.13)$, the probability that a policy reports no claim in one year is: $\mathbb{P}[N=0] = \mathrm{e}^{-0.13} = 0.878$; the probability that no claim is reported in one year by a portfolio consisting of 100 000 independent and homogeneous policies is: $\mathbb{P}[N^{[P]}] = \mathrm{e}^{-13000} \approx 0$. The probability that $13\,000 = \mathbb{E}[N^{[P]}]$ claims are reported by the portfolio in one year is: $\mathbb{P}[N^{[P]} = 13\,000] = 0.00364$.
□

The parameter λ in (9.8.3) represents the expected number of claims per policy: $\mathbb{E}[N] = \lambda$. Thus, it can be estimated through the average claim frequency $\bar{n}$. We point out that, when calculating this quantity, the underlying (implicit) assumption is that all risks in the portfolio have the same attitude to report claims, i.e. they are homogenous in respect of the claim frequency. More realistically, policies may be (more or less) heterogeneous in this respect: for some policies, we should expect a claim frequency higher than $\bar{n}$, while for others the opposite is true. Thus, we should think that $N^{[P]} \sim \text{Poi}(\sum_{j=1}^{r} \lambda^{(j)})$ and set an appropriate $\lambda^{(j)}$ for each policy. However, the piece of information commonly available is the average claim frequency $\bar{n}$, which expresses an estimate for the whole population, i.e. for $\frac{\sum_{j=1}^{r} \lambda^{(j)}}{r}$. In such a situation, when we model the number of claims per policy, we should consider the parameter of the Poisson distribution (9.8.3) as a random one. Usually, it is assumed that λ follows a Gamma distribution, with parameters $(\rho, \frac{p}{1-p})$; then it can be shown that N follows a Negative Binomial distribution, i.e.

$$\mathbb{P}[N=n] = \frac{\Gamma(\rho+n)}{n!\,\Gamma(\rho)}\, p^{\rho}\,(1-p)^{n} \tag{9.8.4}$$

where $\Gamma(s) = \int_0^{\infty} t^{s-1}\,\mathrm{e}^{-t}\,dt$ is the Gamma function, and ρ and p are the parameters of the Negative Binomial distribution ($0 < p < 1$ and $\rho > 0$). The analytical advantages of the Poisson assumption are missed when adopting (9.8.4); however, a better fitting to data may emerge, in particular in respect of dispersion (we note that for the Poisson distribution we have to accept necessarily $\mathbb{E}[N] = \mathbb{V}\text{ar}[N]$; conversely, the Negative Binomial distribution admits $\mathbb{V}\text{ar}[N] > \mathbb{E}[N]$, as it emerges in many empirical distributions).

Example 9.8.2. Tables 9.8.1 and 9.8.2 quote two empirical distributions and the corresponding Poisson and Negative Binomial fitted distributions. Both the Poisson and the Negative Binomial distribution represent appropriately the magnitude of the number of claims per policy. However, the Negative Binomial distribution better captures the dispersion, both in terms of variance and right tail. In portfolio A (see Table 9.8.1), the heterogeneity of policies is not very strong, so that a Poisson approximation may be satisfactory. For portfolio B (see Table 9.8.2), adoption of the Poisson distribution could lead to an underestimate of the extreme cases, i.e. of the right tail.

Table 9.8.1 Empirical distribution of the number of claims per policy, and two fitted distributions; portfolio A

	Empirical distribution		Poisson	Negative Binomial
# claims in a year	# of policies	frequency	probability	probability
0	87 897	0.87897	0.87810	0.87906
1	11 263	0.11263	0.11415	0.11236
2	785	0.00785	0.00742	0.00812
3	53	0.00053	0.00032	0.00044
4	2	0.00002	0.00001	0.00002
5	0	0	0	0
6	0	0	0	0
7 or more	0	0	0	0
all	100 000	1	1	1
mean		0.13	0.13	0.13
variance		0.13222	0.13	0.13222

Table 9.8.2 Empirical distribution of the number of claims per policy, and two fitted distributions; portfolio B

	Empirical distribution		Poisson	Negative Binomial
# claims in a year	# of policies	frequency	probability	probability
0	88 146	0.88146	0.87810	0.88152
1	10 799	0.10799	0.11415	0.10788
2	973	0.00973	0.00742	0.00976
3	76	0.00076	0.00032	0.00078
4	4	0.00004	0.00001	0.00006
5	1	0.00001	0	0
6	1	0.00001	0	0
7 or more	0	0	0	0
all	100 000	1	1	1
mean		0.13	0.13	0.13
variance		0.1381	0.13	0.1381

□

9.8.2 Modeling the claim severity

If we accept the assumptions originating a compound probability distribution for S (see Sect. 9.4.3), in order to define the probability distribution of the aggregate claim amount, we can separately define the probability distribution of the claim frequency N (see Sect. 9.8.1) and of the claim severity Y_1.

Realistically, the set of possible values of Y_1 is limited; however, usually probability distributions taking value in $[0, +\infty)$ are selected. Clearly, continuous distributions are considered. Common choices include the Gamma, Lognormal, Pareto and Loggamma distributions. The specific choice is suggested by the particular features of the line of business dealt with. The Normal distribution can be assumed as a limit case, if the Central Limit Theorem applies.

The actuarial application of continuous positive probability distributions representing the claim severity does not raise special issues; some probabilistic and statistical expertise is clearly required. Of course, the actuarial analyses that can be performed through the modeling of the claim severity are important for many purposes. Given the introductory character of this Chapter, we are not going into details in this respect. We just provide some examples.

Example 9.8.3. Table 9.8.3 provides the empirical distribution of the claim severity for a given portfolio. The distribution is clearly asymmetric. Some investigations could be performed just through the empirical distribution; for example, we could assess the probability that the claim size is above a given class among those displayed in the table. However, several information are missed (for example, we do not know neither what is the average size of claims whose amount is higher than 50, nor the average claim size inside each class).

Table 9.8.3 Empirical distribution of the claim severity

claim size	# of claims
0–5	3 116
5–10	6 446
10–20	2 084
20–30	731
30–40	450
40–50	120
50 and over	53
all	13 000
mean	10
variance	76.38

□

Example 9.8.4. Fig. 9.8.1 plots two theoretical distributions, namely a Lognormal and a Gamma, keeping the same expected value and variance of the empirical distribution quoted in Table 9.8.3. While both distributions are asymmetric, differences in the shape are apparent. Clearly, any theoretical distribution implies some approximations in respect of the observed data. However, with a theoretical distribution we gain in generality.

As mentioned above, the choice of the theoretical distribution depends on the features of the line of business dealt with. The Lognormal and the Gamma distributions are appropriate in many cases; alternative distributions, already mentioned, are the Pareto (useful in particular for representing very large claims) and the Loggamma.

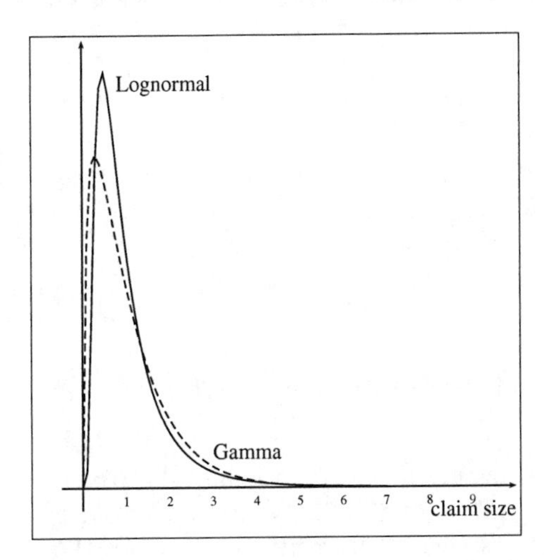

Fig. 9.8.1 Theoretical distributions (density functions) of the claim severity

□

9.8.3 *Modeling the aggregate claim amount*

The aggregate claim amount S, as defined through (9.4.1), is a function of the stochastic process $\{N, Y_1, Y_2, \dots\}$. As stated in Sect. 9.8.1, we assume for S a compound distribution. In particular, if N has a Poisson distribution, then S has a Poisson compound distribution; if N has a Negative Binomial distribution, then S has a Negative Binomial compound distribution; and so on.

Investigations which require to deal with the probability distribution of S typically involve the assessment of the probability that S falls in a given range of values. See, for example, (9.7.22), concerning the update of the pricing basis to the claim experience. Another example is given by solvency investigations, in which we are interested in the probability that the aggregate claim amount of a portfolio (and not just for one policy) is above a given threshold.

Analytical results are difficult to obtain. For example, assume that we want to assess $\mathbb{P}[S > s]$, where s is a given value. Let $F_S(s)$ be the probability distribution function of S, namely $F_S(s) = \mathbb{P}[S \le s]$, and $F_{Y_1}(y)$ be the probability distribution function of Y_1, namely $F_{Y_1}(y) = \mathbb{P}[Y_1 \le y]$. Clearly, $\mathbb{P}[S > s] = 1 - F_S(s)$. We have

$$F_S(s) = \sum_{n=0}^{+\infty} \mathbb{P}[N = n] \times \mathbb{P}[(Y_1 + Y_2 + \cdots + Y_n) \le s | N = n] \tag{9.8.5}$$

Given that $Y_1, Y_2 \ldots, Y_n$ are assumed to be independent of N, identically distributed and reciprocally independent, we have

$$\mathbb{P}[(Y_1 + Y_2 + \cdots + Y_n) \le s | N = n] = \mathbb{P}[(Y_1 + Y_2 + \cdots + Y_n) \le s] = F_{Y_1}^{*(n)}(s) \tag{9.8.6}$$

where $F_{Y_1}^{*(n)}(s)$ is the n-th convolution of $F_{Y_1}(y)$ (for $n = 0$, it is conventionally assumed that $F_{Y_1}^{*(0)}(s) = 0$ if $s < 0$ and $F_{Y_1}^{*(0)}(s) = 1$ if $s \ge 0$). Replacing in (9.8.5), we find

$$F_S(s) = \sum_{n=0}^{+\infty} \mathbb{P}[N = n] \times F_{Y_1}^{*(n)}(s) \tag{9.8.7}$$

Computing (9.8.7) analytically is hard work; numerical or simulation techniques are usually adopted.

Example 9.8.5. To provide an example of investigation performed through the probability distribution of S, we go back to probability (9.7.22), which allows one to assess whether full credibility can be acknowledged to the portfolio experience. We recall that in (9.7.22) the quantity Q_t expresses an estimate of $\mathbb{E}[S]$. So, we rewrite as follows

$$\mathbb{P}[(1-a)\mathbb{E}[S] < S \le (1+a)\mathbb{E}[S]] = \varepsilon \tag{9.8.8}$$

or also as follows

$$\mathbb{P}\left[\frac{-a\mathbb{E}[S]}{\sqrt{\mathbb{V}\mathrm{ar}[S]}} < \frac{S - \mathbb{E}[S]}{\sqrt{\mathbb{V}\mathrm{ar}[S]}} \le \frac{a\mathbb{E}[S]}{\sqrt{\mathbb{V}\mathrm{ar}[S]}}\right] = \varepsilon \tag{9.8.9}$$

where $\frac{S-\mathbb{E}[S]}{\sqrt{\mathbb{V}\mathrm{ar}[S]}}$ is the standardized random variable. When the experience is reasonably large, a standard Normal distribution can be assumed for $\frac{S-\mathbb{E}[S]}{\sqrt{\mathbb{V}\mathrm{ar}[S]}}$. Let $a = 0.1$ and $\varepsilon = 0.95$. Then

$$\frac{0.1\,\mathbb{E}[S]}{\sqrt{\mathbb{V}\mathrm{ar}[S]}} = 1.96 \tag{9.8.10}$$

from which we get

$$\mathbb{E}[S] = 19.6\sqrt{\mathbb{V}\mathrm{ar}[S]} \tag{9.8.11}$$

When Q_t, which estimates $\mathbb{E}[S]$, fulfils (9.8.11), then we can assign full credibility to the portfolio experience.

A more detailed conclusion can be reached if we assume a specific probability distribution for the number of claims. Assume that N follows a Poisson distribution. Then, both the expected number of claims, $\mathbb{E}[N]$, and their variance, $\mathbb{V}\text{ar}[N]$, can be estimated by the average claim frequency $\bar{n}$, given that for the Poisson law $\mathbb{E}[N] = \mathbb{V}\text{ar}[N]$. When we plug this into (9.5.8), we obtain the following expression for $\mathbb{V}\text{ar}[S]$

$$\mathbb{V}\text{ar}[S] = \mathbb{E}[N]\,(\mathbb{V}\text{ar}[Y_1] + (\mathbb{E}[Y_1])^2) = \bar{n}\,(\sigma^2 + (\bar{y})^2) \tag{9.8.12}$$

where σ^2 denotes the estimate for $\mathbb{V}\text{ar}[Y_1]$. Recalling that we can estimate $\mathbb{E}[S]$ as follows

$$\mathbb{E}[S] = \bar{n} \times \bar{y} \tag{9.8.13}$$

we can rewrite (9.8.11) as

$$\sqrt{\bar{n}} = 19.6\sqrt{\frac{\sigma^2}{(\bar{y})^2} + 1} \tag{9.8.14}$$

expressing the minimum number of expected claims required for full credibility. The coefficient 19.6 would of course be different if we make choices other than $a = 0.1$ and $\varepsilon = 0.95$.

□

We note that (9.8.7) refers to one policy only; if we are dealing with a solvency investigation, we should rather refer to the aggregate claim amount for the portfolio. Let $S^{(j)}$ be the aggregate claim amount for policy j, and $S^{[\mathrm{P}]}$ the aggregate claim amount for the portfolio; clearly, $S^{[\mathrm{P}]} = S^{(1)} + S^{(2)} + \cdots + S^{(r)}$, where r is the number of policies. If policies represent independent risk, and if $S^{(j)}$ has a Poisson compound distribution with Poisson parameter $\lambda^{(j)}$ and claim probability distribution $F_{Y_1^{(j)}}(y)$, then it can be shown that also $S^{[\mathrm{P}]}$ has a compound Poisson distribution, with Poisson parameter $\lambda = \lambda^{(1)} + \lambda^{(2)} + \cdots + \lambda^{(r)}$ and claim probability distribution $F_{Y_1}(y) = \sum_{j=1}^{r} \frac{\lambda^{(j)}}{\lambda} F_{Y_1^{(j)}}(y)$. This result contributes to understand the large preference, in practice, for the adoption of a Poisson distribution for the modeling of the claim frequency.

Example 9.8.6. Refer to a homogeneous portfolio, consisting of r policies, which represent independent risks. Each policy may report N claims, in a year, and we assume $N \sim \text{Poi}(0.13)$ for each policy. The claim amount is fixed to 10; Y_1 then has a degenerate probability distribution. The number of claims in the portfolio, $N^{[\mathrm{P}]}$, has a Poisson distribution with parameter $0.13\,r$. The aggregate claim amount for the portfolio, $S^{[\mathrm{P}]}$, is simply defined as $S^{[\mathrm{P}]} = 10N^{[\mathrm{P}]}$. Table 9.8.4 quotes the probability that the aggregate claim amount is higher than the net premium, i.e. the probability of loss, for several portfolio sizes. As in Example 9.5.1, we set $\Pi = 1.4$. Due to the assumptions, we have: $\mathbb{P}[S^{[\mathrm{P}]} > r\Pi] = \mathbb{P}[N^{[\mathrm{P}]} > \frac{r\Pi}{10}]$. As the number of policies r increases, such a probability decreases, as a result of the pooling effect.

□

Table 9.8.4 Probability of loss in a portfolio: $\mathbb{P}[S^{[P]} > r\Pi]$

# of policies, r	$\mathbb{P}[S^{[P]} > r\Pi]$
1	0.12190
10	0.37318
100	0.32487
1 000	0.17791
10 000	0.00292

9.9 Risk classification and experience-rating

In Sect. 9.4–9.7 we dealt with premium calculation assuming that the same premium rate is applied to each policy; this is justified when policies are similar, i.e. homogeneous (except possibly for the sum insured and the time of issue or renewal). However, as emerged in Sect. 9.8.1, policies always differ for some features; in some cases, such differences suggest the adoption, within the same line of business, of specific premium rates.

9.9.1 Risk classes and rating classes

Policies for which the insurer can assume the same attitude to record claims are usually grouped into a *risk class*. For example, in fire insurance buildings are classified according to use (e.g.: domestic, commercial, industrial building), location (e.g.: urban, industrial, rural area), building materials (e.g.: cement, bricks, wood), number of floors (e.g.: one, two, three, four, five, six or more). The basics of risk classification have already been described in Sect. 2.2.6. Policies to which the same premium rate is applied are grouped into a *rating* (or *premium*) *class*. Usually, premium classes are fewer than risk classes, for the reasons discussed in Sect. 2.2.6. The consequences in terms of mutuality and solidarity of a rating system for heterogeneous risks were discussed in Sect. 2.2.7. In this Section we focus on some implementing aspects of risk classification, with specific reference to the non-life business.

The definition of a risk class is based on:

- *risk factors*, i.e. the features of a risk which prove to explain the claim experience (in the example above, the risk factors are: use, location, building materials and number of floors);
- the *outcomes* (or *modes*) *of each risk factor*, which can be either qualitative or quantitative (in the example, the possible outcomes of the risk factor use are: domestic, commercial, industrial).

The selection of the risk factors and their outcomes is based on a statistical investigation, which we do not discuss. We just describe how the selected risk factors can be accounted for in order to define *differentiated* (or *specific*) *premium rates*.

At issue, some risk factors are observable, while others are unobservable. Some information in respect of the latter emerge from the specific claim story of the policy. Observable risk factors originate a specific premium rate at issue; in respect of unobservable risk factors, an *individual experience-rating* system can be adopted, through which the premium rate is updated in time according to the individual claim experience of the policy.

9.9.2 Risk classification at issue

We first focus on the possibility to differentiate premium rates at issue, consistently with the risk factors observable at that time. We refer to the example of fire insurance, and consider the four risk factors mentioned above:

- occupation, with $c_1 = 3$ possible modes, i.e. domestic, commercial and industrial building;
- location, with $c_2 = 3$ possible modes, i.e. urban, industrial and rural area;
- building materials, with $c_3 = 3$ possible modes, i.e. cement, bricks and wood;
- number of floors, with $c_4 = 6$ possible modes, i.e. one, two, three, four, five and six or more.

Combining the possible outcomes of the four risk factors, we can define $c = c_1 \times c_2 \times c_3 \times c_4 = 162$ risk classes. Possibly due to some inconsistencies among some modes of the risk factors (e.g.: industrial building in wood), the actual number of risk classes could be $c' < c$. In what follows, we assume that rating classes coincide with risk classes.

At issue, as a part of the underwriting process, the policy (or, better, the risk) is selected and assigned to an appropriate risk class; thus, an *a-priori risk classification* is determined. The risk class is identified by the outcome of each risk factor (e.g., domestic building, located in an urban area, built in bricks, with one floor). Shortly, we denote the risk class by (i, j, h, k) (each index referring to an outcome of the relevant risk factor). Experience gained in risk class (i, j, h, k) allows the insurer to estimate a risk premium $Q_{i,j,h,k}$ or a risk premium rate $\theta_{i,j,h,k}$, specific of that risk class. The insurer can further summarize the average experience in the portfolio through the risk premium Q or the risk premium rate θ, calculated accounting for the experience of the whole portfolio. For some risk classes, it will turn out $Q_{i,j,h,k} < Q$ (or $\theta_{i,j,h,k} < \theta$), while for others $Q_{i,j,h,k} > Q$ (or $\theta_{i,j,h,k} > \theta$). The problem we want to focus on concerns the calculation of the premium to be applied to risks assigned to class (i, j, h, k), considering the information provided by the specific and the average risk premium (rate). We note that while the premium rate is the same for all the policies belonging to the same risk class, the premium amount may be different because of a different insured value. To shorten the notation, we refer to the calculation of premium rates only; with $p_{i,j,h,k}$ we denote the equivalence premium rate applied to policies in risk class (i,j,h,k).

Retracing what discussed in Sect. 9.7, the equivalence premium rate $p_{i,j,h,k}$ for class (i,j,h,k) should be estimated through the risk premium rate $\theta_{i,j,h,k}$. However, due to the low number of policies in some classes, some risk premium rates $\theta_{i,j,h,k}$ could be unreliable, because too heavily subject to random fluctuations. Conversely, the information provided by θ should be stable enough, given that it is collected over the whole portfolio. So it is wiser to assess $p_{i,j,h,k}$ as a function of θ. Common choices are as follows:

$$p_{i,j,h,k} = \theta + a_i + b_j + d_h + g_k \tag{9.9.1}$$

known as the *additive* (or *linear*) *rule*, and

$$p_{i,j,h,k} = \theta \, \alpha_i \, \beta_j \, \delta_h \, \gamma_k \tag{9.9.2}$$

known as the *multiplicative* (or *exponential*) *rule*. The parameters a_i, b_j, d_h, g_k in (9.9.1), α_i, β_j, δ_h, γ_k in (9.9.2) are the so-called *relativities*: they relate the premium rate of a class to the features of that class. Apart from the advantage provided by θ (in respect of $\theta_{i,j,h,k}$), rules (9.9.1) and (9.9.2) require a lower number of parameters than what would be required by estimating $p_{i,j,h,k}$ just through $\theta_{i,j,h,k}$. In this latter case, the number of parameters would be c (i.e., one risk premium rate $\theta_{i,j,h,k}$ for each risk class; we recall that $c = 162$ in our example); when adopting (9.9.1) or (9.9.2) the number of parameters reduces to $c_1 + c_2 + c_3 + c_4 + 1$ (i.e., 16 in our example): one for each mode of the four risk factors, and one represented by the average risk premium rate θ. Each of the relativities in (9.9.1) and (9.9.2) is estimated on a wider data set than $\theta_{i,j,h,k}$; for example, a_i must be estimated on all the risk classes in which the first risk factor takes outcome i. Addressing one risk factor at a time, however, could result in disregarding some possible correlations among the risk factors.

9.9.3 *Risk classification at renewal times: individual experience rating*

When an individual experience-rating system is adopted, the insurer is willing to reduce the premium for a policy if its claim experience is below the average; conversely, the insured must be willing to accept a premium increase if her claim experience is above the average. Such an arrangement is very common for motor insurance.

Premium rates for new policies are the same for all policies, unless observable risk factors suggest the application of some relativities (see Sect. 9.9.2). According to the individual experience, year by year the premium rate is updated, either increased or decreased, so that at renewal the policy is applied a specific premium rate. Thus, an *a-posteriori risk classification* is determined.

Let p_t be the premium rate applied to a policy after t years since issue. Further, let p denote a reference premium rate, typically representing the premium rate applied at issue. Individual experience could be reported in terms either of the number of claims or the claim amounts. Usually, reference is to the number of claims. More specifically, if p_{t-1} is the premium rate applied at time $t-1$, then the premium rate at time t is defined as

$$p_t = f(p_{t-1}, n_t) \tag{9.9.3}$$

where n_t is the number of claims reported in year $(t-1,t)$ and f is an increasing function of n_t. This is how a *Bonus-Malus* (*BM*) system, possibly the most well-known individual experience rating arrangement, works. We point out that, instead of changing the premium rate, the individual experience could result in a revision of policy conditions. For example, the deductible could be decreased if no claim occurs, or increased in the opposite case, thus rewarding the insureds who report less claims.

It is worthwhile to give some information on a Bonus-Malus system, due to its wide application, in the motor insurance business in particular. The risk class to which a risk is assigned, in relation to the number of claims occurred previous to the current year, is called *merit class*. The premium rate is revised each year as a function of the number of claims reported in the latest year and the current merit class. In more detail, the items of a BM system are the following.

- The set $\{1,2,\ldots,m\}$ of merit classes.
- The *reference premium* rate p (possibly, a net premium rate, π, namely including a safety loading).
- The *premium coefficient* $\gamma(j)$ for merit class j, $j = 1,2,\ldots,m$. The premium applied to policies in class j is defined as $p\,\gamma(j)$. For some classes, the so-called bonus classes, $\gamma(j) < 1$; for others, the so-called malus classes, $\gamma(j) > 1$. Typically, bonus and malus classes are defined so that $\gamma(1) < \gamma(2) < \cdots < \gamma(m)$. Thus, classes with a low ranking are bonus classes (a premium discount is applied), while those with a high ranking are malus classes (a premium increase is applied).
- The *entry class* i, $1 < i \leq m$, to which new policies (for which no previous experience is available) are assigned. The premium coefficient is set so that $\gamma(i) \geq 1$.
- The matrix of the *transition rules*, stating the new merit class c_{j,n_t} for a risk previously in merit class j, which has reported n_t claims in the latest year (see also Table 9.9.1).

The premium rate to be applied at time t to a policy coming from class j is defined as follows: $p_t = p\,\gamma(c_{j,n_t})$.

In some systems, $\gamma(j) < 1$ for $j = 1,2,\ldots,m-1$, while $\gamma(m) = 1$. In this case, there is only one malus class, where the full premium is required. The arrangement is called *No-Claim Discount* (*NCD*) system: policies which receive a discount are those that did not report any claim in the latest year. The longer is the period free of claims, the higher is the discount applied to the premium.

Table 9.9.1 Matrix of the transition rules

previous merit class	# claims in current year: 0	1	2	...
1	$c_{1,0}$	$c_{1,1}$	$c_{1,2}$	...
2	$c_{2,0}$	$c_{2,1}$	$c_{2,2}$	...
...	...	...	...	...
i	$c_{i,0}$	$c_{i,1}$	$c_{i,2}$	...
...	...	...	...	...
m	$c_{m,0}$	$c_{m,1}$	$c_{m,2}$	...

Example 9.9.1. Table 9.9.2 describes the matrix of the transition rules of a BM system, and the relevant premium coefficients (the example is not taken from a real BM system, but anyhow it reflects a realistic arrangement). There are 9 merit classes; briefly, the transition rule is defined as follows:

$$c_{j,n_t} = \begin{cases} \max\{j-1,1\} & \text{if } n_t = 0 \\ \min\{j+2n_t,9\} & \text{if } n_t > 0 \end{cases} \tag{9.9.4}$$

A policy in the highest class is applied a premium which is more than 4 times that required to a policy in the lowest class.

Table 9.9.2 Matrix of the transition rules for a BM system

previous merit class	# claims in current year: 0	1	2	3	4	...
1	1	3	5	7	9	...
2	1	4	6	8	9	...
3	2	5	7	9	9	...
4	3	6	8	9	9	...
5	4	7	9	9	9	...
6	5	8	9	9	9	...
7	6	9	9	9	9	...
8	7	9	9	9	9	...
9	8	9	9	9	9	...

merit class j	premium coefficient $\gamma(j)$
1	35%
2	50%
3	55%
4	70%
5	85%
6	100%
7	110%
8	130%
9	150%

Table 9.9.3 describes the matrix of the transition rules of a NCD system, and the relevant premium coefficients (neither in this case the example is taken from real data). There are 6 merit classes; briefly, the transition rule is defined as follows:

$$c_{j,n_t} = \begin{cases} \max\{j-1,1\} & \text{if } n_t = 0 \\ 6 & \text{if } n_t > 0 \end{cases} \tag{9.9.5}$$

A policy in the highest class is applied a premium which is 2.5 times that required to a policy in the lowest class.

□

Table 9.9.3 Matrix of the transition rules for a NCD system

previous merit class	# claims in current year: 0	1 or more
1	1	6
2	1	6
3	2	6
4	3	6
5	4	6
6	5	6

merit class j	premium coefficient $\gamma(j)$
1	40%
2	75%
3	80%
4	85%
5	90%
6	100%

The ultimate goal of a BM or a NCD system is to define specific premium rates; however, one can guess that some heterogeneity remains among the policies assigned to the same merit class. For example, referring to the NCD system in Table 9.9.3, policies in class 6 may have reported just one claim in the latest year, or two claims, or may have reported one claim in each of the latest two years. Thus, some form of solidarity is anyhow present. Further, solidarity effects may occur among different classes; indeed, the BM or NCD premium system follow the idea described by (2.2.15) (or (2.2.16); see Sect. 2.2.4). Choices in respect of the number of merit classes, the transition rules, the premium coefficient do affect such solidarity effects. Intuitively the solidarity effects are stronger in face of a lower number of merit classes, a narrower range of variation of the premium coefficients, a faster transition backward to the lowest classes. As noted in Sect. 2.2.7, solidarity effects may originate adverse selection; on the other hand, a strong personalization of premiums may reduce the mutuality effect inside each class (given that we should expect a lower number of policies in each class), or also lead to unsustainable premium rates for the worst risks. When designing a BM or a NCD discount system, such aspects require careful consideration.

Remark A further aspect which is investigated when designing a BM or a NCD system is the so-called *stationary distribution*, i.e. the composition of the portfolio (in terms of the number of policies in each class, as a percentage of the total number of policies in the portfolio) when the portfolio itself reaches a steady state. The premium coefficients should be defined considering that under the stationary distribution the insurer should be on balance (see (2.2.15) in Sect. 2.2.4).
The stationary distribution depends on the claim frequency and on the transition rules. Refer, for example, to the NCD system in Example 9.9.1, and let r_j be the percentage of policies in class j when the stationary distribution is reached. The following conditions must be fulfilled in the steady state:

$$\begin{aligned} & r_1 + r_2 + \cdots + r_6 = 1 \\ & r_1 = r_1\,\mathbb{P}[N=0] + r_2\mathbb{P}[N=0] \\ & r_2 = r_3\,\mathbb{P}[N=0] \\ & \dots \\ & r_6 = (r_1 + r_2 + \cdots + r_6)\,\mathbb{P}[N>0] \end{aligned} \tag{9.9.6}$$

If we assume, for example, that the number of claims for a policy in one year follows a Poisson distribution with parameter λ, then the probabilities $\mathbb{P}[N=0]$ and $\mathbb{P}[N>0]$ in (9.9.6) are easy to assess (namely, $\mathbb{P}[N=0]=\mathrm{e}^{-\lambda}$, $\mathbb{P}[N>0]=1-\mathrm{e}^{-\lambda}$). Solving the linear system (9.9.6) is then little algebra. We stress, however, that the steady state is a notional scenario, in particular due to the possibility of adverse selection.

9.10 Technical reserves: an introduction

In this Section we discuss a simple dynamic model which allows us to introduce the fundamentals of the technical annual management of a non-life portfolio. We refer to a homogeneous portfolio (say, fire insurance or motor insurance), consisting of policies holding the same policy year. The investigation is developed with reference to the policy year $(0,1)$, where $t=0$ is the (first) time of issue of the policies.

Let $\Pi^{[\mathrm{T}][\mathrm{P}]}$ be the total amount of the expense-loaded premiums cashed from the policies at time 0. With such an amount of money, the insurer faces:

- the initial commission, to be paid at time 0;
- annual expenses (overhead and other administrative expenses), to be paid during the year;
- the cost of claims occurring during the year.

We assume that $\alpha\,\Pi^{[\mathrm{T}][\mathrm{P}]}$ is the amount of the initial commission at time 0, while $\beta\,\Pi^{[\mathrm{T}][\mathrm{P}]}$ is the total amount of the annual expenses. Such expenses are paid gradually in time; it is usually acceptable to assume that their payment is uniformly spread over the year, so that $t\,\beta\,\Pi^{[\mathrm{T}][\mathrm{P}]}$ is the total amount paid in $(0,t)$. Finally, by $S_t^{[\mathrm{P}]}$ we denote the aggregate claim amount reported within the portfolio in $(0,t)$.

The quantity

$$\Pi^{[\mathrm{T}][\mathrm{P}]}\,(1-\alpha-\beta\,t) \tag{9.10.1}$$

represents the residual amount of premiums, once expenses up to time t, $0<t\leq 1$, have been paid. Fig. 9.10.1 plots the typical behavior of such a quantity. Note that we are disregarding the time-value of money (so that no accrual due to interest gained on investments is accounted for); this is justified by the short-term nature of the non-life business (and by the fact that we are referring to one year only).

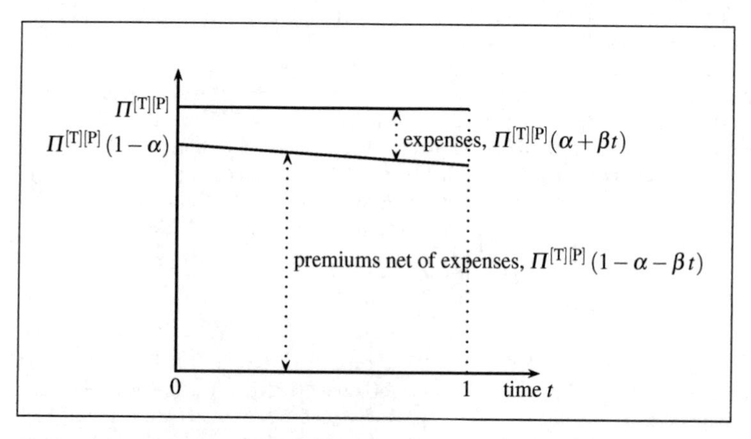

Fig. 9.10.1 Premium amount net of expenses

We define the portfolio assets at time t as the amount

$$A_t = \Pi^{[T][P]}(1-\alpha-\beta t) - S_t^{[P]} \tag{9.10.2}$$

Once more, we point out that we are disregarding investment perspectives; we are further disregarding specific capital allocation to the portfolio; thus, A_t simply represents the residual amount of the premiums cashed by the insurer, once claims and expenses have been paid. Fig. 9.10.2 provides an example (left panel), in which it is assumed that claims (in terms both of frequency and amount) occur continuously and uniformly in time (right panel). It is further assumed that claims are immediately settled. Reasonably, claims do not occur uniformly in time; Fig. 9.10.3 suggests a more realistic path for the aggregate claim amount, assuming that claims occur just at some (random) times during the year, and in an amount which is not always the same.

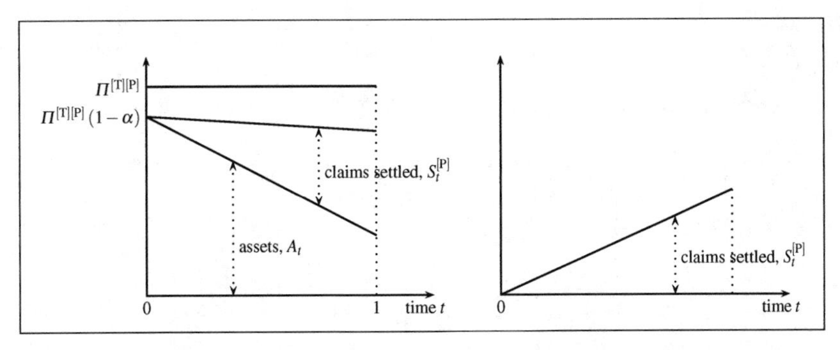

Fig. 9.10.2 Portfolio assets (left panel) and portfolio aggregate claim amount (right panel); uniform distribution of the aggregate claim amount

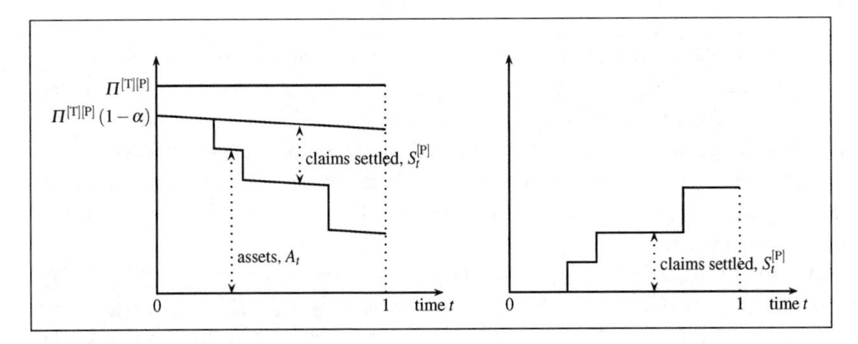

Fig. 9.10.3 Portfolio assets (left panel) and portfolio aggregate claim amount (right panel); discrete distribution of the aggregate claim amount

At any time t, $0 \le t < 1$, a share of the assets A_t must be reserved to face future claims and expenses, i.e. those possibly emerging in $(t,1)$. Future expenses consist of the annual expenses (not yet paid), which as stated above are assumed to emerge uniformly in time. For future claims, we can make an assumption similar to that accepted when calculating the premium (see Sect. 9.4), i.e. that they also occur uniformly in time. The reserve set up to meet future claims and expenses, which is called *(unearned) premium reserve*, is a proportion $1-t$ (i.e., the time to maturity of the policies) of the premiums, net of the initial expenses (which were fully paid at time 0). Thus, the (unearned) premium reserve, $R_t^{[\Pi]}$, is defined as follows:

$$R_t^{[\Pi]} = (1-t)\,\Pi^{[\mathrm{T}][\mathrm{P}]}\,(1-\alpha) \tag{9.10.3}$$

The behavior in time is clearly linear, as sketched in Fig. 9.10.4. For some lines of business claim occurrence may be affected by cyclical or seasonal effects; in this case, the proportion of the initial premium set aside would be other than $1-t$.

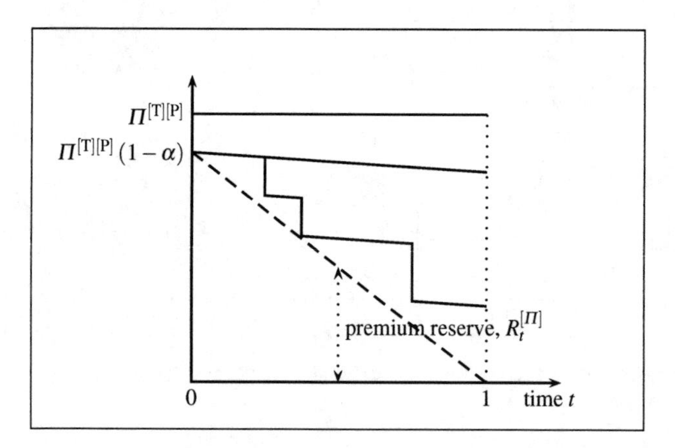

Fig. 9.10.4 Portfolio assets and (unearned) premium reserve

We point out that (9.10.3) defines a portfolio reserve. When policies hold different policy years, the calculation of the premium reserve may be performed exactly for each policy; then, the portfolio reserve can be obtained by summing up the relevant individual values. An alternative consists in grouping policies whose policy anniversary falls in a given period of the year (e.g., in January) and then, following the method of the k-ths described in Sect. 9.7.3, estimating for each group the average time to maturity.

As mentioned in Sect. 9.4.2, usually claims are not immediately settled (see the example provided in Fig. 9.4.1). In Fig. 9.10.5 it is assumed that claims occurring at the second claim occurrence are not immediately settled. A reserve must be set up, given that the insurer's obligation has become due; uncertainty may remain in respect of the amount to be settled and the time of payment. The relevant reserve is called the *claim reserve*, which we denote by $R_t^{[S]}$, and its amount is given by the estimated amount of the claims which have already occurred, but have not yet been settled.

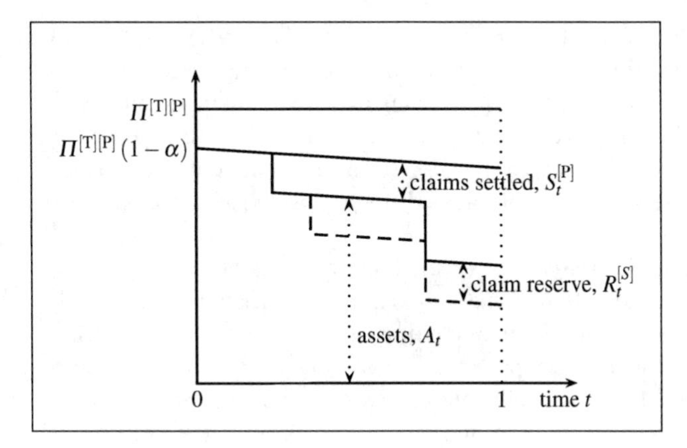

Fig. 9.10.5 Portfolio assets and claim reserve

When calculating the claim reserve, the time between claim occurrence and claim settlement must be accounted for. The approximate assumption adopted for premium calculation (namely, claims occur on average in the middle of the year, and are immediately settled) is no longer acceptable. Statistical procedures, either deterministic or stochastic, are available. Deterministic models, in particular (which we briefly describe in Sect. 9.12) are typically satisfactory for lines of business with frequent claims. Ad hoc estimates may be required for very specific claims, in particular when extreme events occur.

The term claim reserve is somewhat generic, and can be better specified. The *outstanding claim reserve* refers to claims which have been reported to the insurer, but have not yet been settled (referring to Fig. 9.4.1, the outstanding claim reserve is calculated at some time between t_2 and t_3); the calculation can be based either on experience or specific estimates. The *IBNR (Incurred But Not Reported) claim reserve* refers, instead, to claims which are expected to have already occurred, but have not yet been reported to the insurer (referring to Fig. 9.4.1, this reserve is assessed at some time between t_1 and t_2, time t_1 being unknown to the insurer). It is anyhow appropriate to set up a reserve, whose calculation can just be based on experience. The outstanding and the IBNR reserves are the most important items of claim reserves. Further items are: the *IBNER (Incurred But Not Enough Reported) claim reserve*, which concerns claims which have already been notified, but whose damage has just been partially reported to the insurer; the *reopened claim reserve*, which concerns claims which need to be reopened (after a first settlement), possibly because of litigation or further information gained after settlement; the *notified (open) claim reserve*, which refers to claims which have already been reported, but have not yet received an accurate assessment by the insurer; other items are possible, depending on market practice.

Turning to expenses, there is clearly a time-lag between the income of the expense loading and the payment of expenses. The premium reserve already accounts for future expenses, namely overhead and other administrative expenses and processing expenses for claims which have not yet occurred. Depending on the way the claim cost is assessed (either inclusive or not of settlement expenses), the claim reserve may (or not) already account for processing expenses relating to claims already occurred. If the claim reserve does not include claim settlement expenses, or if it is felt that the amount of expenses is underestimated within current reserves, a specific reserve may be set up, usually named the *provision for claim handling costs*.

Finally, we mention the *contingent reserves*. Such reserves are usually set up to provide additional funds should the emerging claim experience differ adversely from the assumptions underlying the main claim reserve. The idea is to set aside money in years in which the claim experience is favorable, to face adverse fluctuations in some years. Examples of contingent reserves are the *catastrophe reserve* and the *claim equalization reserve*. The underlying idea is to spread the cost of large claims not just on the year of occurrence, but on more than one year. This way, contingent reserves provide a smoothing of the annual economic results obtained through the management of a non-life portfolio. Indeed, in many countries they are treated as capital reserve (for example, for tax purposes).

9.11 Earned premiums, incurred claim amounts and profit assessment

In this Section we examine the role of technical reserves on the emergence of annual profits, i.e on *profit timing*. We make reference to the premium reserve and, generically, to the claim reserve.

Consider Fig. 9.11.1, which summarizes the example discussed in Sect. 9.10, introducing the premium and the claim reserve.

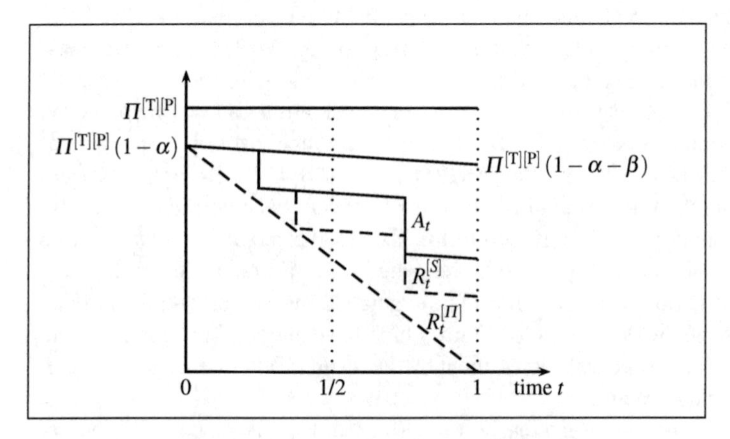

Fig. 9.11.1 Portfolio assets, premium reserve and claim reserve

Still considering the same policy year for policies in the portfolio, we can reasonably assume that time 0 falls in the middle of the calendar year, i.e. that policies enter on average in the middle of the year. Let τ be the calendar year. At the end of the calendar year τ, i.e. at time $\frac{1}{2}$ in the picture, the annual profit must be reported in the balance sheet. Intuitively

$$A_{1/2} - R_{1/2}^{[\Pi]} - R_{1/2}^{[S]} \tag{9.11.1}$$

represents the annual profit for this cohort in calendar year τ, i.e. in the calendar year of issue. Replacing (9.10.2) into (9.11.1), we find

$$\Pi^{[T][P]}\,(1 - \alpha - \beta\,\frac{1}{2}) - S_{1/2}^{[P]} - R_{1/2}^{[\Pi]} - R_{1/2}^{[S]} \tag{9.11.2}$$

where

- the quantity $\Pi^{[T][P]} - R_{1/2}^{[\Pi]}$, the so-called earned premiums, represents the amount of premiums contributing to profit in year τ;
- the quantity $S_{1/2}^{[P]} + R_{1/2}^{[S]}$, the so-called incurred claim amount, represents the cost of claims occurred in year τ, either settled or not;
- the quantity $\Pi^{[T][P]}\,(\alpha + \beta\,\frac{1}{2})$ represents the expenses incurred in year τ.

We stress that the premium reserve $R_{1/2}^{[\Pi]}$ represents the amount of premiums cashed in year τ that will contribute to profit in year $\tau+1$; similarly, the claim reserve $R_{1/2}^{[S]}$ represents the amount of claims which will be settled in the future, but whose cost contributes to profit in year τ.

We now generalize the definitions of earned premiums, incurred claim amount and annual profit, introduced above. We refer to a calendar year, which we denote as year $(t-1,t)$, in which policies enter at different times (thus, we now address the more realistic case in which policies do not hold the same policy year). Let $\Pi_t^{[T][P]}$ be the total amount of premiums cashed in year $(t-1,t)$, also called *written premiums*. Such premiums do not contribute entirely to profit in year t, and thus they must be reduced by the premium reserve at time t. Further, some of the policies in-force during year t paid the relevant premium in the previous year; the part of such premiums contributing to profit in current year is measured by the premium reserve at time $t-1$. Thus, we define *earned premiums* in year t the quantity

$$\Pi_t^{[P][\text{earned}]} = \Pi_t^{[P]} + R_{t-1}^{[\Pi]} - R_t^{[\Pi]} \tag{9.11.3}$$

As far as claims are concerned, we first note that a part of the claim amount settled in year t, $S_t^{[P]}$, refers to claims incurred in previous years; the estimate of their cost was included in the claim reserve at time $t-1$. Further, claims incurred in year t but not settled are accounted for in the claim reserve at time t. Thus, we define *incurred claim amount* in year t the quantity

$$S_t^{[P][\text{incurred}]} = S_t^{[P]} + R_t^{[S]} - R_{t-1}^{[S]} \tag{9.11.4}$$

Remark We can now give a general definition of the risk premium Q (see Sect. 9.7.1). In general terms, with reference to a given calendar year, the risk premium is defined as follows:

$$Q = \frac{\text{incurred claim amount}}{\text{units of earned exposure}} \tag{9.11.5}$$

The units of earned exposure can consist of the number of policy years, the total amount of the insured value in the year, and so on.

In (9.11.4), it is interesting to note that if a claim which occurred in the past is settled in the current year, the possible difference between the settled and the reserved amount contributes to the incurred claim amount in the year of settlement. Similarly, if a claim has occurred previous to time $t-1$ and has not yet been settled at time t, it is accounted for in the claim reserve both at time $t-1$ and at time t. In principle, such a claim must not contribute to the incurred claim amount of the current year (as the amount set up in the reserve at the end of the year, i.e. the final reserve, should be offset by the relevant allocation in the reserve at the beginning of the year, i.e. the initial reserve); however, if the estimate of the cost of the claim is updated in the final reserve, either because of new information or simply because of a different methodology for the calculation of the claim reserve, the update to the cost contributes to the incurred claim amount of the current year. The calculation of the claim reserve should then be maintained stable in time, so to avoid unnecessary updates to the cost of claims, and then biased assessments of the annual profit.

The *annual profit* (or *loss*) can be expressed as follows

$$PL_t^{[\mathrm{P}]} = \Pi_t^{[\mathrm{P}][\mathrm{earned}]} - S_t^{[\mathrm{P}][\mathrm{incurred}]} - EX_t \tag{9.11.6}$$

or more explicitly as

$$PL_t^{[\mathrm{P}]} = \Pi_t^{[\mathrm{P}]} - S_t^{[\mathrm{P}]} + R_{t-1}^{[\Pi]} + R_{t-1}^{[S]} - R_t^{[\Pi]} - R_t^{[S]} - EX_t \tag{9.11.7}$$

In both cases, EX_t represents the expenses paid during year t.

Remark The annual profit defined above is also named *industrial profit*, as it originates from the "industrial activity" of the insurer, consisting in the creation of a pool of individual risks (see Sect. 1.6.3). It is interesting to compare expression (9.11.7) to (6.4.9) for life insurance. The two quantities consist of similar terms: premiums received, benefits paid, change in the reserve value. In the case of life insurance, also incomes on the investment of the reserve are considered. Indeed, while for a life insurer the investment activity is considered part of the obligation taken in respect of the policyholder, for a non-life insurer only the transfer of individual risks is addressed as the main task of the insurer.

Several indexes are usually calculated, in order to give a summary of the performance of the portfolio. The *claim ratio* (or *loss ratio*) is defined as follows

$$LR_t = \frac{S_t^{[\mathrm{P}][\mathrm{incurred}]}}{\Pi_t^{[\mathrm{P}][\mathrm{earned}]}} \tag{9.11.8}$$

A ratio lower than 1 informs us that the claim costs incurred in the year have been covered by the earned premiums.

Remark We note that the term loss ratio is more common than claim ratio, as suggested also by the notation adopted. Contrarily to the choice made so far to refer to the payout of the insurer as to the claim amount (see also Sect. 9.3), in the following we will then refer to ratio (9.11.8) as to the loss ratio.

The *expense ratio* is defined as follows

$$ER_t = \frac{EX_t}{\Pi_t^{[P][earned]}} \tag{9.11.9}$$

and represents the part of the earned premiums which must be used to cover expenses. The *combined ratio*, defined as follows

$$CR_t = \frac{S_t^{[P][incurred]} + EX_t}{\Pi_t^{[P][earned]}} \tag{9.11.10}$$

summarizes the industrial profitability of the portfolio; for example, a combined ratio lower than 1 would detect a situation of positive industrial profit.

9.12 Deterministic models for claim reserves

As mentioned in Sect. 9.10, the claim reserve is originated by the delay between claim occurrence and claim settlement. Depending on the line of business, such a delay may run from some weeks (e.g., in property insurance and for small claims) to several years (e.g. in liability insurance, and in general if the claim is large). Reasons of the delay are to be found in the time required for processing the claim, the need for ascertaining the responsibility and the size of the damage, the delayed reporting of the claim, litigation, and so on. For the largest claims, a custom estimate is usually worked out; for the other claims, statistical assessments are performed. In this Section we address statistical methods.

Deterministic methods for claim reserves are based on an average assessment of the time-pattern of a claim; conversely, stochastic methods make explicit reference to its randomness. Deterministic methods offer the advantage of simplicity, and thus their implementation is straightforward; at the same time, they are simplified in many respects, and hence they may lead to a biased assessment. Overall, a considerable degree of judgment is required for claim reserves; in many situations it is appropriate to compare several methods to get to a reasonable estimate of the claim reserve. In this Section, a description of the main deterministic methods is provided, so to introduce the main issues involved in the calculation of the claim reserve; for stochastic methods, some references are quoted in Sect. 9.13.

9.12.1 Run-off triangles

A *run-off triangle* collects data on outstanding claims, classifying the available information in respect of both the year of claim occurrence and the year of claim settlement. Table 9.12.1 provides an example, where $S_{i,j}^{[P]}$ is the aggregate claim amount paid up to j years since occurrence for claims originating in year i. Alternatively, the run-off triangle could quote the incremental claim amounts (namely, $S_{i,j}^{[P]} - S_{i-1,j}^{[P]}$), the number $N_{i,j}^{[P]}$ of claims reported up to year j for claims incurred in year i, or other information. In the following, we just refer to run-off triangles with information as those provided in Table 9.12.1.

Table 9.12.1 Run-off triangle

year of origin (or accident year)	time to settlement (or development year)								
	0	1	...	j	...	$\tau-i$	...	$\tau-1$	τ
0	$S_{0,0}^{[P]}$	$S_{0,1}^{[P]}$	...	$S_{0,j}^{[P]}$	...	$S_{0,\tau-i}^{[P]}$	...	$S_{0,\tau-1}^{[P]}$	$S_{0,\tau}^{[P]}$
1	$S_{1,0}^{[P]}$	$S_{1,1}^{[P]}$	...	$S_{1,j}^{[P]}$	...	$S_{1,\tau-i}^{[P]}$	...	$S_{1,\tau-1}^{[P]}$	
...	...	...							
i	$S_{i,0}^{[P]}$	$S_{i,1}^{[P]}$	...	$S_{i,j}^{[P]}$	...	$S_{i,\tau-i}^{[P]}$			
...	...								
τ	$S_{\tau,0}^{[P]}$								

Assume that within τ years since occurrence all claims are fully settled. In Table 9.12.1, which is filled in at time τ, the quantity $S_{0,\tau}^{[P]}$ represents the ultimate aggregate claim amount for claims originating in year 0; for such claims, we do not expect to have further settlements in the future. For year i, $i = 1,2,\dots,\tau$, the amount $S_{i,\tau-i}^{[P]}$ is provisional, as further settlements are expected in the next i years. Let U_i be the ultimate aggregate claim amount estimated according to an appropriate method. The claim reserve set up at time τ is

$$R_{\tau}^{[S]} = \sum_{i=1}^{\tau} (U_i - S_{i,\tau-i}^{[P]}) \tag{9.12.1}$$

In the following, we describe some approaches to the estimate of the ultimate aggregate claim amount U_i.

Example 9.12.1. In the next Sections, numerical examples are based on the run-off triangle quoted in Table 9.12.2, where we have also reported the premium amount earned in the various years of origin of claims. We assume that all claims are settled within 4 years since occurrence; thus, $S_{0,4}^{[P]} = 2627$ is the ultimate aggregate claim amount for claims occurred in year 0. A reserve must be set up for claims occurred in year i, $i = 1,2,3,4$, as further settlements are expected in the next i years. Thus, the reserve at time 4 is: $R_4^{[S]} = \Sigma_{i=1}^{4}(U_i - S_{i,4-i}^{[P]})$.
□

Table 9.12.2 Run-off triangle and earned premiums

year of origin	time to settlement 0	1	2	3	4
0	790	1 422	2 275	2 502	2 627
1	910	1 729	2 680	2 921	
2	995	1 841	3 038		
3	1 200	2 100			
4	1 100				

year i	earned premiums $\Pi_i^{[P][earned]}$
0	3 400
1	4 000
2	4 300
3	5 200
4	5 000

9.12.2 The Expected Loss Ratio method

Assume that a block of business was initiated setting a given target for the loss (or claim) ratio, i.e. for the ratio between the incurred claim amount and the earned premium (see (9.11.8)). Such a target is referred to as the *expected loss ratio*; we will denote it as *ELR*. If $\Pi_i^{[P][earned]}$ represents the amount of premiums earned in year i, then the ultimate aggregate claim amount for claims originating in year i can be estimated as follows

$$U_i^{[ELR]} = ELR \times \Pi_i^{[P][earned]} \tag{9.12.2}$$

Example 9.12.2. Refer to data in Table 9.12.2, and assume that the expected loss ratio is 75%. For each year i of origin, $i = 0, 1, \ldots, 4$, Table 9.12.3 quotes the ultimate aggregate claim amount estimated according to the expected loss ratio, $U_i^{[ELR]} = 0.75\,\Pi_i^{[P][earned]}$, the amount to be reserved, $U_i^{[ELR]} - S_{i,4-i}^{[P]}$, and the claim reserve at time 4 assessed through the ELR approach, $R_4^{[S][ELR]}$. Note that for year 0 the difference $U_i^{[ELR]} - S_{0,4}^{[P]}$ has been set to 0, also in face of $U_i^{[ELR]} \neq S_{0,4}^{[P]}$; having assumed that claims are fully settled within 4 years since occurrence, no reserve needs to be set up for claims occurred in year 0. More in general, should $U_i^{[ELR]} < S_{i,4-i}^{[P]}$, the difference $U_i^{[ELR]} - S_{i,4-i}^{[P]}$ must be set to 0, as it is not possible to contribute to the reserve with a negative term.

Table 9.12.3 Expected ultimate aggregate claim amount and claim reserve through the ELR method

year i	$U_i^{[ELR]}$	$U_i^{[ELR]} - S_{i,4-i}^{[P]}$
0	2550.00	0.00
1	3000.00	79.00
2	3225.00	187.00
3	3900.00	1800.00
4	3750.00	2650.00
	$R_4^{[S][ELR]}$	4716.00

□

As it emerges also from Example 9.12.2, the ELR method is very simple and requires few data. Disadvantages are given by the subjectivity of the assessment of *ELR* and the static nature of the model; the development of claim amounts may suggest that the quantity *ELR* is more and more unlikely, but no update to such a quantity is implied by the methodology. In particular, the estimated ultimate aggregate claim amount, $U_i^{[\mathrm{ELR}]}$, depends only on the year of origin, and not on the time j passed since then. The approach may be useful when dealing with a new business, for which no previous experience is available on the likely time-pattern of claims.

9.12.3 The Chain-Ladder method

The *chain-ladder* (*CL*) method assumes that the time-pattern of claims is stable in time, apart from possible random fluctuations. In particular, the following assumption is accepted

$$S_{i,j+1}^{[\mathrm{P}]} = S_{i,j}^{[\mathrm{P}]}\, d_j \qquad i = 0, 1, \ldots \tau;\; j = 0, 1, \ldots, \tau - 1 \tag{9.12.3}$$

where d_j ($d_j \geq 1$) is the *development factor* of the cumulative aggregate claim amount from year j to year $j+1$ since claim occurrence. The development factors d_j are also known as *link ratios*. Note that they do not depend on the year of origin i, but just on the time to settlement j. Assuming that claims are fully settled within τ years since occurrence, it turns out $d_t = 1$ for $t = \tau, \tau+1, \ldots$.

Example 9.12.3. With reference to data in Table 9.12.2, Table 9.12.4 quotes the observed development factors, i.e. the experienced ratios $\frac{S_{i,j+1}^{[\mathrm{P}]}}{S_{i,j}^{[\mathrm{P}]}}$. For any j, the observed development factors seem to be subject to random fluctuations only; taking their average, such random fluctuations should then be smoothed away.

Table 9.12.4 Observed development factors

year of origin	time to settlement 0	1	2	3	4
0	$\frac{1422}{790} = 1.800$	$\frac{2275}{1422} = 1.600$	$\frac{2502}{2275} = 1.100$	$\frac{2627}{2502} = 1.050$	
1	$\frac{1729}{910} = 1.900$	$\frac{2680}{1729} = 1.550$	$\frac{2921}{2680} = 1.090$		
2	$\frac{1841}{995} = 1.850$	$\frac{3038}{1841} = 1.650$			
3	$\frac{2100}{1200} = 1.750$				
4					
average	1.821	1.601	1.094	1.050	

The last row of Table 9.12.4 quotes the average observed development ratio, for each year j since occurrence. More precisely, it is a weighted average of the annual development ratios, with weights given by the current cumulative aggregate claim amounts; for example:

$$1.821 = \frac{1.800 \times 790 + 1.900 \times 910 + 1.850 \times 995 + 1.750 \times 1\,200}{790 + 910 + 995 + 1\,200} \tag{9.12.4}$$

Replacing into (9.12.4) the expression of the observed development factors, we find quite easily

$$1.821 = \frac{1422 + 1729 + 1841 + 2\,100}{790 + 910 + 995 + 1\,200} \tag{9.12.5}$$

which is the usual way to estimate the link ratios.

□

According to data in the run-off triangle, the development factor for year j is estimated as follows (see Example 9.12.3):

$$\bar{d}_j = \frac{\sum_{i=0}^{\tau-1-j} S_{i,j+1}^{[\mathrm{P}]}}{\sum_{i=0}^{\tau-1-j} S_{i,j}^{[\mathrm{P}]}} \tag{9.12.6}$$

The development factor d_j describes, for any origin year i, the increase of the cumulative aggregate claim amount from time j to time $j+1$ since occurrence. We further define f_j as the development factor to the full settlement of a claim, i.e.

$$f_j = d_j \times d_{j+1} \times \cdots \times d_{\tau-1} \tag{9.12.7}$$

with $f_t = 1$ for $t = \tau, \tau+1, \ldots$. Clearly, the factor f_j is estimated through the estimated development factor $\bar{d}_{j+h}$; we will denote by $\bar{f}_j$ the estimated value of f_j. The ultimate aggregate claim amount for claims originated in year i once j year have passed since occurrence is estimated as follows through the chain-ladder approach:

$$U_{i,j}^{[\mathrm{CL}]} = S_{i,j}^{[\mathrm{P}]} \bar{f}_j \tag{9.12.8}$$

Note that $U_{i,j}^{[\mathrm{CL}]}$ is proportional to the accumulated aggregate claim amount observed to date, and then depends both on the year of origin and the time passed since then.

Example 9.12.4. Refer to data in Table 9.12.2. For each year i of origin, $i = 0, 1, \ldots, 4$, Table 9.12.5 quotes the estimated development factor to full development, $\bar{f}_{4-i}$, the ultimate aggregate claim amount estimated according to the chain-ladder approach, $U_{i,4-i}^{[\mathrm{CL}]} = \bar{f}_{4-1} S_{i,4-i}^{[\mathrm{P}]}$, the amount to be reserved, $U_{i,4-i}^{[\mathrm{CL}]} - S_{i,4-i}^{[\mathrm{P}]}$, and the claim reserve at time 4 assessed through the chain-ladder approach, $R_4^{[\mathrm{S}][\mathrm{CL}]}$. Note that, contrarily to the ELR approach, $U_{i,4-i}^{[\mathrm{CL}]} - S_{i,4-i}^{[\mathrm{P}]} \geq 0$ and in particular $U_{0,4}^{[\mathrm{CL}]} - S_{0,4}^{[\mathrm{P}]} = 0$.

□

Table 9.12.5 Expected ultimate aggregate claim amount and claim reserve through the CL method

year i	factor $\bar{f}_{4-i}$	$U_{i,4-i}^{[\mathrm{CL}]}$	$U_{i,4-i}^{[\mathrm{CL}]} - S_{i,4-i}^{[\mathrm{P}]}$
0	1.000	2627.00	0.00
1	1.050	3066.93	145.93
2	1.149	3491.05	453.05
3	1.840	3863.88	1763.88
4	3.350	3685.17	2585.17
	$R_4^{[S][\mathrm{CL}]}$		4948.04

When comparing the findings in Examples 9.12.2 and 9.12.4, we might conclude that the chain-ladder approach leads to a more accurate assessment of the ultimate aggregate claim amount than what is the case for the expected loss ratio method. In some respects, this is the appropriate conclusion; certainly, for each year of origin, through the chain-ladder method the estimate of the ultimate aggregate claim amount is updated to the information collected so far, as the estimate is expressed as a proportion of the current accumulated amount (see (9.12.8)). However, the assumptions underlying the chain-ladder approach may be unsatisfactory in some cases. The basic assumption concerns the time-pattern of each accident year, assumed to be stable in time; on the contrary, the development in time of claims may change. The estimate of the ultimate aggregate claim amount may be distorted by a different dynamics of the claim payment patterns; for example, if the administrative processing of claims is speeded up, the total claim amount is overestimated. In the extreme case that no claim has been settled to date, the method predicts a total claim amount which is 0. Further, the claim amounts could be affected by inflation, which would imply a trend in the behavior of the cumulative aggregate claim amount; this problem, however, can be easily dealt with, by adjusting appropriately the observed aggregate claim amounts.

Overall, the chain-ladder method is simple and practicable, and has been largely used in practice. In this Section, we have described the basic version; many extensions have been proposed, so to overcome some of the main limitations of the approach. See Sect.9.13 for some references.

9.12.4 The Bornhuetter-Ferguson method

The *Bornhuetter-Ferguson* (*BF*) method merges the findings of the expected loss ratio with those of a projected method, such as the chain-ladder method.

First refer to (9.12.8), from which we obtain

$$S_{i,j}^{[\mathrm{P}]} = U_{i,j}^{[\mathrm{CL}]} \frac{1}{\bar{f}_j} \tag{9.12.9}$$

If we accept that $\bar{f}_j$ is a good indicator of how the aggregate claim amount should evolve in time, then Eq. (9.12.9) shows us that the coefficient $1/\bar{f}_j$ represents the share of the ultimate aggregate claim amount that have already been settled to date. We note that $S_{i,j}^{[P]}$ represents the liability reported to date.

Now take $U_i^{[ELR]}$, which represents the ultimate liability that we expect at the beginning of the year of origin of claims, i.e. at issue. After j years since issue (or, since the year of origin of claims), the quantity

$$U_i^{[ELR]} \times \left(1 - \frac{1}{\bar{f}_j}\right) \tag{9.12.10}$$

represents the liability still to emerge for claims originated in year i.

The Bornhuetter-Ferguson method estimates the ultimate aggregate claim amount as the sum of the reported and the emerging liability, namely

$$U_{i,j}^{[BF]} = S_{i,j}^{[P]} + U_i^{[ELR]} \times \left(1 - \frac{1}{\bar{f}_j}\right) \tag{9.12.11}$$

Replacing (9.12.9) into (9.12.11), we obtain the alternative expression

$$U_{i,j}^{[BF]} = U_{i,j}^{[CL]} \times \frac{1}{\bar{f}_j} + U_i^{[ELR]} \times \left(1 - \frac{1}{\bar{f}_j}\right) \tag{9.12.12}$$

which shows us that $U_{i,j}^{[BF]}$ is the weighted average of the ultimate aggregate claim amount estimated through the chain-ladder and the expected loss ratio approach. Since $\bar{f}_j$ decreases in time, the higher is j, the higher is the weight assigned to claim information data (i.e., to the estimate obtained through the chain-ladder method). The Bornhuetter-Ferguson method then uses the initial ELR estimate as long as claims are not paid or reported. Further, it assumes that past experience is not fully representative of the future.

Example 9.12.5. Still referring to data in Table 9.12.2, for each year i of origin, $i = 0, 1, \dots, 4$, Table 9.12.6 quotes the ultimate aggregate claim amount estimated according to the expected loss ratio, the chain-ladder and the Bornhuetter-Ferguson approach, and the amount to be reserved according to the three methods. Note that when few years have passed since the year of origin, the quantity $U_{i,j}^{[BF]}$ is closer to $U_i^{[ELR]}$ than to $U_{i,j}^{[CL]}$; vice versa when many years have already passed. As it emerges from Table 9.12.6, alternative methods result in different estimates of the claim reserve. After having investigated the reasons of the differences and considered the specific features of the line of business dealt with, a final value should be assessed by the reserving actuary. In particular, we point out that the actuary could find it is necessary to set up a reserve for claims still to be settled after τ years since occurrence; personal judgement is required in this case, as data are not available (or are not statistically reliable).

Table 9.12.6 Expected ultimate aggregate claim amount and claim reserve through the ELR, CL and BF methods

year i	factor $\bar{f}_{4-i}$	$U_i^{[\mathrm{ELR}]}$	$U_i^{[\mathrm{CL}]} - S_{i,4-i}^{[\mathrm{P}]}$	$U_{i,4-i}^{[\mathrm{CL}]}$	$U_{i,4-i}^{[\mathrm{CL}]} - S_{i,4-i}^{[\mathrm{P}]}$	$U_{i,4-i}^{[\mathrm{BF}]}$	$U_{i,4-i}^{[\mathrm{BF}]} - S_{i,4-i}^{[\mathrm{P}]}$
0	1.000	2550.00	0.00	2627.00	0.00	2627.00	0.00
1	1.050	3000.00	79.00	3066.93	145.93	3063.75	142.75
2	1.149	3225.00	187.00	3491.05	453.05	3456.53	418.53
3	1.840	3900.00	1800.00	3863.88	1763.88	3880.37	1780.37
4	3.350	3750.00	2650.00	3685.17	2585.17	3730.65	2630.65
		$R_4^{[S][\mathrm{ELR}]}$	4716.00	$R_4^{[S][\mathrm{CL}]}$	4948.04	$R_4^{[S][\mathrm{BF}]}$	4972.29

□

9.12.5 Further aspects

As mentioned in Sect. 9.12, the methods examined above for the calculation of claim reserves are deterministic; indeed, no explicit assumption about the stochastic path of the aggregate claim amount is introduced. Formal statistical models could be adopted, whose presentation goes beyond the purpose of this book.

In the previous discussion, the data referred to concern the aggregate claim amount. However, the patterns of the claim settlement may behave differently than the pattern of the number of claims. The *average cost per claim method*, which is deterministic, considers two run-off triangles: one for the number of the incurred claims and one for the average payment per claim. The chain-ladder method, possibly with extensions, is applied separately to the two run-off triangles; then, the ultimate aggregate claim amount is estimated by multiplying the ultimate number of claims and the ultimate average cost per claim (thus following the splitting in (9.4.3)).

In the example referred to above, four years are required to reach the full settlement of claims. In practice, the time-pattern of claims could require a longer time. From an economic point of view, it would make sense to allow for the value of time, i.e. to discount future liabilities consistently with the timing of their emergence. We note that, for example, the chain-ladder method allows us to estimate such a timing; indeed, the quantities $S_{i,j}^{[\mathrm{P}]}\,(d_j-1)$, $S_{i,j}^{[\mathrm{P}]}\,(d_{j+1}-d_j)$, ... represent (respectively) the amounts estimated to be settled in 1,2,... years from now for claims originated in year i. However, in many legislations it is not admitted to discount future liabilities when estimating the claim reserve, which must be rather assessed according to the ultimate cost; clearly, an implicit safety-loading is thus embedded in the valuation. In general, the investment activity is not considered to fall within the traditional business of a non-life insurer. For example, contrarily to the life insurance business, the annual industrial profit (see (9.11.7) and the remark following such Equation) does not include investment earnings. Clearly, this does not mean that non-life insurers do not make investments. Assets backing the claim reserve must be appropriately invested, and they originate financial earnings which are reported within the annual general profit (or loss) of the insurance company.

9.13 References and suggestions for further reading

A number of actuarial textbooks deal with the technical aspects of non-life insurance. Textbook [12] provides a general introduction to ratemaking and reserving, while [31] describes several aspects of actuarial practice for non-life insurance. Several practical aspects of health insurance are provided by [7], dealing also with life insurance.

Risk classification and experience-based ratemaking, in particular bonus-malus systems, are dealt with by [19] and [36]. Basic ratemaking concepts and techniques are described in [56]. The calculation of reserves is addressed by [53], [59], [23].

A simple introduction to stochastic models for non-life insurance is provided by [8] and [33]. Loss distributions are described in [32], while [39] makes use of stochastic processes.

Finally, we mention [27] for historical remarks on the development of actuarial science, including contributions to the actuarial technique for non-life insurance.

References

The following reference list only includes textbooks and monographs, thus disregarding papers in scientific journals, congress proceedings, research and technical reports, and so on. Our choice aims at limiting the number of citations, in line with the teaching orientation of this work. For more detailed presentations of the various topics, the reader can also refer to the lists of references provided by the several monographs, all quoted below, we have cited in the final section of each chapter.

1. Anderson, A.W.: Pension Mathematics for Actuaries. ACTEX Publications (2006)
2. Aspinwall, J., Chaplin, G., Venn, M.: Life Settlements and Longevity Structures. John Wiley & Sons (2009)
3. Barrieu, P., Albertini, L. (eds.): The Handbook of Insurance-linked Securities. John Wiley & Sons (2009)
4. Bellis, C., Shepherd, J., Lyon, R. (eds.): Understanding Actuarial Management: the Actuarial Control Cycle. The Institute of Actuaries of Australia (2003)
5. Benjamin, B., Pollard, J.H.: The Analysis of Mortality and Other Actuarial Statistics. The Institute of Actuaries, Oxford (1993)
6. Bertocchi, M., Schwartz, S.L., Ziemba, W.T.: Optimizing the Aging, Retirement and Pensions Dilemma. John Wiley & Sons (2010)
7. Black, K., Skipper, H.D.: Life & Health Insurance. Prentice Hall, New Jersey (2000)
8. Boland, P.J.: Statistical and Probabilistic Methods in Actuarial Science. Chapman & Hall / CRC (2007)
9. Booth, P., Chadburn, R., Haberman, S., James, D., Khorasanee, Z., Plumb, R.H., Rickayzen, B.: Modern Actuarial Theory and Practice. Chapman & Hall / CRC (2005)
10. Bowers, N.L., Gerber, H.U., Hickman, J.C., Jones, D.A., Nesbitt, C.J.: Actuarial Mathematics. The Society of Actuaries, Schaumburg, Illinois (1997)
11. Broverman, S.A.: Mathematics of Investment and Credit. ACTEX Publications (2008)
12. Brown, R.L., Gottlieb, L.R.: Introduction to Ratemaking and Loss Reserving for Property and Casualty Insurance. ACTEX Publications (2007)
13. Carter, R.L.: Reinsurance Essentials. Euromoney Institutional Investor PLC (2004)
14. Crouhy, M., Galai, D., Mark, R.: Risk Management. McGraw - Hill (2001)
15. Cummins, J.D., Smith, B.D., Vance, R.N., VanDerhei, J.L.: Risk Classification in Life Insurance. Kluwer - Nijhoff Publishing (1983)
16. Cunningham, R., Herzog, T., London, R.: Models for Quantifying Risks. ACTEX Publications (2008)
17. Daykin, C.D., Pentikäinen, T., Pesonen, M.: Practical Risk Theory for Actuaries. Chapman & Hall (1994)

A. Olivieri, E. Pitacco, *Introduction to Insurance Mathematics*,
DOI 10.1007/978-3-642-16029-5, © Springer-Verlag Berlin Heidelberg 2011

18. Delvaux, T., Magnée, M.: Les nouveaux produits d'assurance-vie. Editions de l'Université de Bruxelles (1991)
19. Denuit, M., Maréchal, X., Pitrebois, S., Walhin, J.F.: Actuarial Modelling of Claim Counts: Risk Classification, Credibility and Bonus-malus Scales. John Wiley & Sons (2006)
20. Dickson, D.C.M., Hardy, M.R., Waters, H.R.: Actuarial Mathematics for Life Contingent Risks. Cambridge University Press (2009)
21. Doff, R.: Risk Management for Insurers. Risk Control, Economic Capital and Solvency II. Risk Books (2007)
22. Dowd, K.: Beyond Value at Risk. The New Science of Risk Management. John Wiley & Sons (1998)
23. Friedland, J.: Estimating unpaid claims using basic techniques. Casualty Actuarial Society (2009)
24. Gajek, L., Ostaszewski, K.M.: Financial Risk Management for Pension Plans. Elsevier (2004)
25. Gerber, H.U.: Life Insurance Mathematics. Springer-Verlag (1995)
26. Gupta, A.K., Varga, T.: An Introduction to Actuarial Mathematics. Kluwer Academic Publishers (2002)
27. Haberman, S.: Landmarks in the history of actuarial science (up to 1919). Actuarial Research Paper No. 84, Dept. of Actuarial Science and Statistics, City University, London, http://www.cass.city.ac.uk/arc/reports/84ARC.pdf (1996)
28. Haberman, S., Pitacco, E.: Actuarial Models for Disability Insurance. Chapman & Hall / CRC (1999)
29. Hardy, M.: Investment guarantees. Modeling and Risk Management for Equity-Linked life insurance. John Wiley & Sons (2003)
30. Harrington, S.E., Niehaus, G.R.: Risk Management and Insurance. Irwin / McGraw-Hill (1999)
31. Hart, D.G., Buchanan, R.A., Howe, B.A.: The Actuarial Practice of General Insurance. The Institute of Actuaries of Australia (1996)
32. Hogg, R.V., Klugman, S.A.: Loss Distributions. John Wiley & Sons (1984)
33. Hossack, I.B., Pollard, J.H., Zehnwirth, B.: Introductory Statistics with Applications in General Insurance. Cambridge University Press (1983)
34. Jorion, P.: Value at Risk. McGraw-Hill (2007)
35. Koller, G.: Risk Assessment and Decision Making in Business and Industry. A Practical Guide. CRC Press (1999)
36. Lemaire, J.: Bonus-malus Systems in Automobile Insurance. Kluwer Academic Publishers (1995)
37. Luenberger, D.G.: Investment Science. Oxford University Press (1998)
38. Mangiero, S.M.: Risk Management for Pensions, Endowment and Foundations. John Wiley & Sons (2005)
39. Mikosch, T.: Non-Life Insurance Mathematics. Springer-Verlag (2006)
40. Milevsky, M.A.: The Calculus of Retirement Income. Cambridge University Press (2006)
41. Möller, T., Steffensen, M.: Market-valuation methods in life and pension insurance. Cambridge University Press (2007)
42. Olivieri, A., Pitacco, E.: L'Assurance-vie. Évaluer les Contrats et les Portefeuilles. Pearson Education (2008)
43. Outreville, J.F.: Theory and Practice of Insurance. Kluwer Academic Publishers (1998)
44. Parmenter, M.M.: Theory of Interest and Life Contingencies, with Pension Applications: a Problem-solving Approach. ACTEX Publications (1999)
45. Pearson, N.D.: Risk Budgeting. John Wiley & Sons (2002)
46. Pitacco, E., Denuit, M., Haberman, S., Olivieri, A.: Modelling Longevity Dynamics for Pensions and Annuity Business. Oxford University Press (2009)
47. Promislow, S.D.: Fundamental of Actuarial Mathematics. John Wiley & Sons (2006)
48. Rejda, G.E.: Principles of Risk Management and Insurance. Pearson (2010)
49. Rotar, V.I.: Actuarial Models. The Mathematics of Insurance. Chapman & Hall / CRC (2007)
50. Sandström, A.: Solvency. Models, Assessment and Regulation. Chapman & Hall / CRC (2006)

51. Seog, S.H.: The Economics of Risk and Insurance. Wiley-Blackwell (2010)
52. Tabeau, E., van den Berg Jeths, A., Heathcote, C. (eds.): Forecasting Mortality in Developed Countries. Kluwer Academic Publishers (2001)
53. Taylor, G.: Loss reserving. Kluwer Academic Press (2000)
54. Vaaler, L.J.F., Daniel, J.W.: Mathematical Interest Theory. Mathematical Association of America (2007)
55. Vaughan, E.J., Vaughan, T.: Fundamentals of Risk and Insurance. John Wiley & Sons (2008)
56. Werner, G., Modlin, C.: Basic ratemaking. Casualty Actuarial Society (2010)
57. Williams, C.A., Smith, M.L., Young, P.C.: Risk Management and Insurance. Irwin / McGraw-Hill (1998)
58. Winklevoss, H.E.: Pension Mathematics with Numerical Illustrations. Irwin (1993)
59. Wüthrich, M.V., Merz, M.: Stochastic claims reserving methods in insurance. John Wiley & Sons (2008)

Index

[1] Registered trademarks and property of CANNEX Financial Exchanges.

P